SPECTROSCOPY

FUNDAMENTALS AND DATA INTERPRETATION

Neeraj Kumar Fuloria
Shivkanya Fuloria

2013

Studium Press (India) Pvt. Ltd.

SPECTROSCOPY

Fundamentals and Data Interpretation

ISBN: 978-93-80012-58-2

Published by:

Studium Press (India) Pvt. Ltd.
4735/22, 2nd Floor, Prakash Deep Building
(Near Delhi Medical Association),
Ansari Road, Darya Ganj, New Delhi-110 002
Tel.: 91-11-43240200-15 (15 lines), Fax: 91-11-43240215;
E-mail: studiumpress@gmail.com

Printed at: Akash Press

ASMOH
Laboratories Ltd.
(A Division of **ASMOH** Group)

Foreword

The scientific needs, developments and advancements have currently posed spectroscopy as an essential element of science. It is such that spectroscopy subject is now incorporated as a separate subject in many of the national and overseas universities undergraduate and postgraduates courses of pharmacy, chemistry, physics, engineering, medical wing, and also many other courses. However, it is a common need of students and professionals to have a single book of spectroscopy with easy language and more detailed information about theory-principle, more of pictorial representation of instruments/ components, interpretation, and advance techniques of IR, NMR, MASS, U.V-visible, X-Ray spectroscopy.

So, with a view to have a book that would fulfil the need of readers or students, this separate book has been designed for spectroscopy.

After going through the content and subjects matter of whole book of Spectroscopy: fundamentals and data interpretation, a few qualities and features of this book are identified such as simple language, detailed information of theory/principle, classified representation of all spectroscopic information, maximum possible images for instrumentation and working principles, use of specialized data charts/data tables to make interpretation more easier, detailed

Regd. Office : 10/148, Geeta Colony, Delhi- 110031 (INDIA)
Corporate Office / Mailing Address : A-165, Sector- 63, Industrial Area, Noida - 201 301 (INDIA)
Ph. : + 91-120-4323861, 4348302, Fax : + 91-120-4336931 | Email : info@asmohlab.com | Web : www.asmohlab.com

interpretation data, and current/advance techniques of IR, NMR, MASS spectroscopy.

This book possesses four key strengths firstly the pictorial representation of all spectroscopic instruments and their components; secondly the complete and in-depth interpretation of all types of spectra; thirdly explanation of maximum necessary and currently used techniques of spectroscopy; fourthly, the use of simpler language for easy understanding of spectroscopic fundamentals and concepts.

Looking on the use of simplified language and strong pictorial representation in this book, I consider this book to be user-friendly. So, I wish all the success for this book.

Dr. Rajul Gupta
Manager Regulatory Affairs
Asmoh Laboratories Ltd.

Regd. Office : 10/148, Geeta Colony, Delhi- 110031 (INDIA)
Corporate Office / Mailing Address : A-165, Sector- 63, Industrial Area, Noida - 201 301 (INDIA)
Ph. : + 91-120-4323861, 4348302, Fax : + 91-120-4336931 | Email : info@asmohlab.com | Web : www.asmohlab.com

About the Author(s)

Neeraj Kumar Fuloria

Neeraj Kumar Fuloria, M. Pharm (Pharmaceutical Chemistry), Ph.D., M.B.A., P.D.C.T.M., is working as Professor & Head of Department of M. Pharm (Q.A.), Anuradha College of Pharmacy, Chikhli, Dist. Buldana, Maharashtra, India. He is having an experience of 13 years in academics and industry as well. He has published several research papers of National and International repute. He has authored two books of international repute.

Shivkanya Fuloria

Shivkanya Fuloria, M. Pharm (Phytochemistry), Ph.D., P.D.C.T.M., is working as an Associate Professor, at Anuradha College of Pharmacy, Chikhli, Dist. Buldana, Maharashtra, India. She is having an experience of 10 years in academics. She has published several research papers of National and International repute. She has authored two books of international repute.

Acknowledgements

Authors are solely thankful to Almighty eyes for blessing this idea. Solely thanks to Dr. Lalit Golani, Pittsburg, Pennsylvania; Dr. Asha Budakoti, Department of Chemistry HNB Garhwal University; Dr. G.K. Sinha, for creating the urge to write this book. Authors are highly grateful to Dr. Mohammad Ali, Professor & Ex-Dean, Faculty of Pharmacy, Jamia Hamdard; Dr. Mohammad Shaharyar, Assoc. Professor, Faculty of Pharmacy, Jamia Hamdard; and Dr. Vijender Singh, Head, BBS College of Pharmacy, Noida, for their technical support and valuable suggestions in developing this book.

Thanks are to Shri Shankersinh Vaghela, Ex-Chief Minister of Gujarat, from whose institute the journey of this book writing began. Thanks are also due to Dept. of NMR-IIT Delhi for technical assistance, R.V. Northland Institute and Rameesh Institute, Greater Noida, for providing instrumental assistance.

Sincere thanks to Shri Siddhi Vinayak Ji Bondre, Chairman, Anuradha College of Pharmacy, and Prof. Kamal Singh, Ex-Vice Chancellor, Amravati University, for their support and encouragement while writing this book.

At last but not the least thanks to all our colleagues, friends, our entire family, our parents, and especially Shivansh to stand with us all the time for successful completion of this book.

Preface

As per the advancements in the research field of Chemistry, Physics, Pharmacy and Medical science, today it is the need to get acquaintance with spectrometric analysis especially in the techniques like IR , MASS, NMR, UV-Visible and X-Ray spectroscopy.

In this book we have tried to create the readers interest by using simple language and definition/meaning for almost every possible technical word that is used in our book. This book emphasizes especially on three parts that is theory/principle very simplified, instrumentation with every possible image and interpretation for a maximum number of simple and complex compounds.

The speciality of this book is that students will receive knowledge of various theories, principles, instrumentations, applications and data interpretation of various analytical techniques like IR, MASS, NMR, UV-Visible, and X-Ray spectroscopy. Also students/readers will come to know about the current advances in spectroscopy 2D, 3D, and MDNMR etc. Apart from it, in this book we have tried to highlight and interpretate all individual peaks of IR, NMR and MASS spectra of simple as well as complex molecules.

This book is strongly supported with pictorial representation which will act as an effective tools for easy understanding of the fundamentals of spectroscopy. It is our belief and expectation that this book will lay down its own standards and will aid in easy learning for Under Graduate and Post Graduate students of (Chemistry, Physics, Analysis, Pharmaceutical Chemistry, Pharmacy, Chemical Engineering, Medical); Research professionals; Quality Control Department of Pharmaceutical/Chemical industries and also teachers as well.

The Authors are highly grateful, to Dr. J.N. Govil, Chief editor, Publications Director & Managing Editor, Studium Press, for shaping and publishing this book.

Neeraj Kumar Fuloria
Shivkanya Fuloria

Table of Contents

Abbreviations and Acronyms

A	:	Absorbance
ADC	:	Analog-to-Digital Converter
ATR	:	Attenuated Total Reflectance
CaF_2	:	Calcium Difluoride
$CDCl_3$	:	Deuterated Chloroforms
CIMS	:	Chemical Ionization Mass Spectrometry
COSY	:	Correlation Spectroscopy
cps	:	Cycles Per Second
CW-NMR	:	Continuous Wave Nuclear Magnetic Resonance
D_2O	:	Deuterated Water
DEPT	:	Distortion-less Enhancement by Polarization Transfer
DMSO	:	Dimethyl Sulfoxide
DOF	:	Degree of Freedom
dp	:	Depth of Penetration
2DNMR	:	Two Dimensional Nuclear Magnetic Resonance
3DNMR	:	Three Dimensional Nuclear Magnetic Resonance
4DNMR	:	Four Dimensional Nuclear Magnetic Resonance
5DNMR	:	Five Dimensional Nuclear Magnetic Resonance
6DNMR	:	Six Dimensional Nuclear Magnetic Resonance
DQFCOSY	:	Double Quantum Filtered Correlation Spectroscopy
DRT	:	Diffused Reflectance Technique
1D TOCSY	:	One dimensional Total Correlation Spectroscopy
2D TOCSY	:	Two dimensional Total Correlation Spectroscopy
EIMS	:	Electron Impact Method
EMR	:	Electro Magnetic Radiation
ESIMS	:	Electron Spray Ionization Mass Spectroscopy
EXSY	:	Exchange Spectroscopy
FABMS	:	Fast Atomic Bombardment Mass Spectrometry
FDMS	:	Field Desorption Mass Spectrometry
FID	:	Free Induction Decay
FIMS	:	Field Ionization Mass Spectrometry
FTICRMS	:	Fourier Transformed Ion Cyclotron Resonance Mass

FTIR	:	Fourier Transformed Infrared
FT-NMR	:	Fourier Transformed Nuclear Magnetic Resonance
GCMS	:	Gas Chromatographic Mass Spectrometry
GFTNMR	:	G-Matrix Fourier Transform
HETCOR	:	Heteronuclear Correlation Spectroscopy
HMBC	:	Heteronuclear Multiple Bond Coherence
HMQC	:	Heteronuclear Multiple Quantum Coherence
HOESY	:	Heteronuclear Overhauser Effect Spectroscopy
HOHAHA	:	Homonuclear Hartmann Hann
HSQC	:	Heteronuclear Single Quantum Coherence
Hz	:	Hertz
IA	:	Indium Antimonide
ICRMS	:	Ion Cyclotron Resonance Mass Spectrometry
INADEQUATE	:	Incredible Natural Abundance Double Quantum Transfer
IR	:	Infrared
^3J, ^4J & ^5J	:	Ortho, Meta and Para coupling
KBr	:	Potassium Bromide
LCMS	:	Liquid Chromatographic Mass Spectrometry
LDMS	:	Laser Desorption Mass Spectroscopy
LIMA	:	Laser Ionization Mass Analysis
LTT	:	Lead Tin Telluride
M^+	:	Molecular Ion
M+1, M+2 & M+4	:	Isotope Peaks
M-1, M-2, & M-3	:	Fragment Ion Peaks
MALDI	:	Matrix Assisted Laser Desorption Ionization
MCT	:	Mercury Cadmium Telluride
MDNMR	:	Multiple Dimensional Nuclear Magnetic Resonance
MHz	:	Mega Hertz
MIR	:	Multiple Internal Reflectances
MQFCOSY	:	Multiple Quantum Filtered Correlation Spectroscopy
NaCl	:	Sodium Chloride
NMR	:	Nuclear Magnetic Resonance
NOE	:	Nuclear Overhauser Effect
NOE Effect	:	Nuclear Over Hauser Effect
NOESY	:	Nuclear Overhauser Effect Spectroscopy
OOP	:	Out of Plane
PDMS	:	Plasma Desorption Mass Spectroscopy
PFG NMR	:	Pulse Field Gradient Nuclear Magnetic Resonance
pH	:	Negative Logarithm of Hydrogen Ion Concentration
pKa	:	Ionization Constant
QIS	:	Quadrupole Ion Storage

QMF	:	Quadrupole Mass Filter
RDNMR	:	Relaxation Dispersion NMR
RF	:	Radio Frequency
RI	:	Refractive Index
ROESY	:	Rotating Frame Overhauser Effect Spectroscopy
SIMS	:	Secondary Ion Mass spectrometry
SO-FAST NMR	:	Selective Optimized Flip-Angle Short-Transient NMR
SQFCOSY	:	Single Quantum Filtered Correlation Spectroscopy
SRT	:	Specular Reflectance Technique
SSNMR	:	Solid State Nuclear Magnetic Resonance
T	:	Transmittance
TES	:	Tri Ethyl Silane
ThBr	:	Thorium Bromide
TMS	:	Tetra Methyl Silane
TOCSY	:	Total Correlation Spectroscopy
TOFMS	:	Time of Flight Mass Spectroscopy
TQFCOSY	:	Triple Quantum Filtered Correlation Spectroscopy
TRNOE	:	Transferred Nuclear Overhauser Effect
UV	:	Ultra Violet
VIBGYOR	:	Violet Indigo Blue Green Yellow Orange Red
XRA	:	X-Ray Absorbance
XRD	:	X-Ray Diffraction
XRF	:	X-Ray Fluorescence
XRT	:	X-Ray Tube
ZnS	:	Zinc Sulphide

SYMBOLS AND UNITS

A°	:	Angstrom
B	:	Magnetic Field
^{10}B, ^{11}B	:	Boron Isotopes
^{79}Br	:	Bromine Isotope
c	:	Speed of Light
^{13}C, ^{12}C	:	Carbon Isotopes
^{113}Cd	:	Cadmium Isotope
^{35}Cl	:	Chlorine Isotope
cm^{-1}	:	Unit of Wavenumber
δ value	:	Chemical shift in Parts per million
ΔE	:	Transition Energy
ΔE_{vib}	:	Difference between vibrational energy levels
E	:	Potential Energy of Precessing Nucleus

E_{max}	:	Maximum Extinction Coefficient
^{19}F	:	Fluorine Isotope
γ	:	Magnetogyric Ratio
G_s	:	Space of Grating
gm	:	Gram
h	:	Planks constant
^1H, ^2H	:	Hydrogen Isotope
J value	:	Coupling Constant Value
$^1J_{H\text{-}H}$, $^2J_{H\text{-}H}$, $^3J_{H\text{-}H}$	:	Coupling between two hydrogen separated by one, two or three bonds
$^1J_{C\text{-}H}$, $^2J_{C\text{-}H}$, $^3J_{C\text{-}H}$	:	Coupling between hydrogen and carbon separated by one, two or three bonds
k	:	Bond force constant (bond strength)
K	:	Kelvin Unit of Temperature
^6Li	:	Lithium Isotope
λ_{max}	:	Maximum Wavelength
m	:	Unit Measurement in Metres
M_o	:	Net Magnetization
mm	:	Millimetre
m_1 & m_2	:	Mass of Atoms in gram of Particular Bond
mμ	:	Milli microns
mg	:	Milligram
μ	:	Reduced mass
μm	:	Micrometre
N	:	Avogadro's number
$N+1$ & 2^N	:	Rules for equivalent or Non-equivalent Peak Splitting
^{14}N, ^{15}N	:	Nitrogen Isotopes
^{23}Na	:	Sodium Isotope
α	:	Wavenumber
υ	:	Frequency
^{17}O, ^{16}O	:	Oxygen Isotopes
ω	:	Angle of incident radiation
^{31}P	:	Phosphorus Isotope
^{195}Pt	:	Platinum Isotope
r	:	Reflection Angle
^{129}Xe	:	Xenon Isotope
σ^* & π^*	:	Antibonding Sigma and Pi Orbital's
σ, π, & n	:	Sigma, Pi and Non-Bonding Electrons
ppm	:	Parts Per Million
p.s.i.	:	Per Square Inch
ρ	:	Density

S	:	Sulphur
Sec.	:	Seconds
^{29}Si	:	Silicon Isotope
τ value	:	Chemical shift in Parts per million opposite to ä value
Tesla	:	Magnetic Field Unit
ν	:	Radiation Velocity
W	:	Tungsten

Chapter 1

Overview of Spectroscopy

Objective

To make readers learn

- Overview of Spectroscopy

Spectroscopy is a qualitative or quantitative tool of analysis for determination of matter or physical processes, that involves study of interaction between electromagnetic radiations and matter in the form of absorption, emission, or scattering. Electromagnetic radiation (EMR) is a form of energy emitted and absorbed by charged particles, that exhibits wave-like property when travels through space.

Few characteristics of EMR are as follows:

- Electromagnetic waves are also called as electromagnetic radiations because these are produced by motion of electrically charged particles.
- Electromagnetic radiations can travel through empty space, air and other substances.
- All the electromagnetic radiations contains some discrete packets of energy, called photons.
- Measured in wavelength, frequency or wave numbers.
- All EMR's travels with same velocity and does not requires any medium for propagation.
- Passage of a visible EMR through a prism shows dispersion into seven colour *i.e.*, VIBGYOR.

EMR is comprised of photons that further comprise an oscillating electric component (*E*), and oscillating magnetic component (*M*). The two electric and magnetic fields are perpendicular to each other, and also perpendicular to the direction of travelling of photon. The electric and magnetic fields of a photon flips or oscillates in travelling direction of photon. The number of oscillations (*flips*) which occurs in one second is called as frequency (n). The unit of frequency is described as oscillations per second, or s^{-1} or Hertz.

All photons in a non-absorbing medium travels at the same velocity (v).

The distance in the direction of propagation of photon over which electric and magnetic fields of a photon undergoes one complete oscillation is called the wavelength (λ), of the electromagnetic radiation.

The relationship between the light velocity, wavelength, and frequency is given as

$$\nu = \lambda \,.\, \nu \qquad \text{...(1.1)}$$

The electromagnetic properties of all photons are same, but their frequencies can be different.

The energy, *E*, of a photon depends upon its frequency of oscillation, which is expressed as

$$E = h\,.\,\nu = \frac{h\,.\,c}{\lambda} = \overline{\nu} \qquad \text{...(1.2)}$$

Where, h is Planck's constant & $\bar{\nu}$ is wavenumber

As radiation/light acts as a wave so it also possess particle-like properties. Low-energy waves possess longer wavelengths (λ), whereas high-energy waves possess shorter wavelengths. As per the energies or wavelength, the light waves can be easily classified with the help of an electromagnetic spectrum. An electromagnetic spectrum represents arrangement of all types of radiations/light waves in the increasing or decreasing order of energy or wavelength. Figure 1.1 and Table 1.1, shows an electromagnetic spectrum for all types of radiations and their respective transitions.

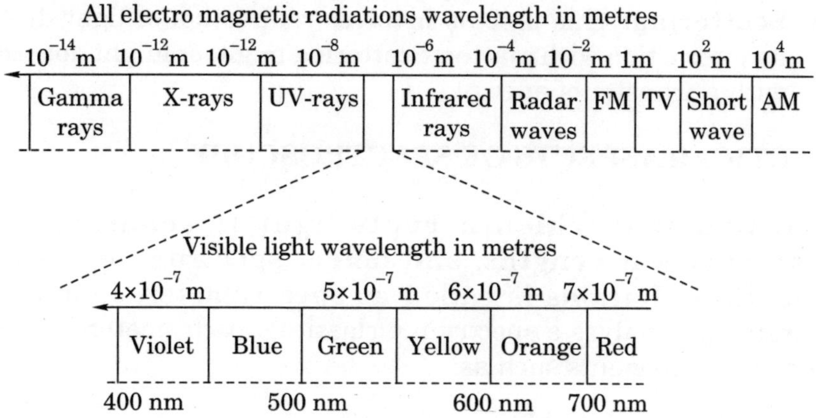

Figure 1.1: Electromagnetic spectrum.

Table 1.1: Electromagnetic radiations and type of transitions.

Radiation	***Type of Transition***
Gamma-rays	Nuclear
X-Rays	Inner electron
Ultraviolet	Outer electron
Visible	Outer electron
Near-Infrared	Outer electron molecular vibrations
Infrared	Molecular vibrations
Microwaves	Molecular rotations, electron spin flipping
Radio waves	Nuclear spin flipping

An electromagnetic radiation can interact with matter by three ways such as

- **Absorption:** This process involves transfer of energy from radiations to matter (atom, molecule, or absorber), which undergoes transitions from a lower level to a higher level.
- **Emission:** This process involves transitions from a higher level to a lower level along with transfer of energy/radiations from emitter to radiation field. *Process in which no radiation is emitted during transition from higher to lower energy levels is termed as non radiative decay.*

- **Scattering:** This process involves redirection of light due to its interaction with matter. Scattering might or might not occur with a transfer of energy.

PRACTICAL ASPECTS OF SPECTROSCOPY

An instrument which accepts light to separate into component wavelengths, and detect spectrum of electro-magnetic radiations is called as spectrometer. In order to generate and analyze a spectrum a classical spectrometer requires three key components such as:

- A light source for electromagnetic radiation.
- A dispersing device to separate the light into its component wavelengths.
- A detector to sense the presence of light after dispersion.

Apart from the above three important components a spectrometer also make use of several other additional components such as slit, mirrors and computer interface.

In spectroscopy the spectra can be obtained either in the form of emission spectra (possess one or more bright lines or bands on dark background) or absorption spectra (possess continuously bright background except for one or more dark lines).

In spectroscopy three types of principle measurements are used such as:

- **Absorption Spectroscopy:** This technique involves illumination of sample with EMR and further measurement of loss of electromagnetic energy after passing through the sample.

  For example, a light (*containing broad band of wavelengths*) is passed through a sample (*vapor of atoms*, *ions*, or *molecules*). Thereby sample will absorb some wavelengths that could excite the sample from one quantum state to another. Due to which absorbed wavelengths will be absent from the original light spectrum. Thereby, the measurement of the missing absorption lines assists in identification of the sample.

- **Emission Spectroscopy:** This technique involves excitement of sample atoms, ions, or molecules under investigation. After excitement, the excited sample atoms, ions, or molecules are allowed to relax to lower energy state. While such relaxation phase the sample atoms, ions, or molecules emits radiation corresponding to the energy difference, such as $\Delta E = h\nu = hc/\lambda$, between the various energy levels of the quantum system.

  This fluorescence (emission) radiation acts as a analytical tool, as this is the component of missing wavelengths in absorption spectroscopy. Therefore, the emission lines are the characteristic fingerprint that is associated with a unique atom, ion, or molecule. The earlier excitation methods involved placement of sample in a flame or an electric-arc discharge. The atoms or molecules were excited by collisions with electrons, the broadband light in the excitation source, or collisions with energetic atoms. The analysis of the emission lines is done with the same types of spectrometer as used in absorption spectroscopy.
- **Diffraction Spectroscopy:** This technique involves direction of radiations on the plane of solid (atomic plane) sample which causes scattering or redirection of light on a photographic screen. Therefore diffraction spectroscopy acts as analytical tool for determination of crystal shape or physical structure of any solid compound. Scattering might or might not occur with a transfer of energy.

SPECTROSCOPIC APPLICATIONS

Spectroscopy possess a very wide scope and application to science such as:

- **To study the structures of atoms and molecules:** A large number of wavelengths range makes it useful to investigate the structures in detail, which includes electronic configurations in ground and excited states.
- **To determine the constituents of a given substance with unknown chemical composition:** It is an accurate and precise method to determine concentration of trace elements

in a substance in few parts per million detected through its emission spectrum.

- **In Astronomy:** The study of the spectral emission lines of distant galaxies determined that universe is expanding rapidly and isotropically (independent of direction).

This book covers and highlights various theories, principles, instrumentation, applications, advancements and data interpretation of different analytical techniques like Infrared (IR), MASS, Nuclear Magnetic Resonance (NMR), UV-visible and X-Ray spectroscopy.

Chapter 2

Infrared Spectroscopy

Objective

To make readers learn

- Basic Concepts, Theory and Underlying Principles of IR Spectroscopy.
- Instrumentation of IR Spectroscopy.
- IR Spectra/Data Interpretation for Structural Determination of Compounds.

2.1 INTRODUCTION

Infrared (IR) spectroscopy is a form of absorption or vibrational spectroscopy, which involves treatment and absorption of infrared radiations onto the organic compounds. This causes molecular excitation of bonds which tends to undergo stretching or bending along the chemical bonds of a compound called vibrations. A plot of absorbed or transmitted vibrational frequency is drawn, which generates an IR spectrum that contains numerous IR absorption bands. Every molecule vibrates on a specific frequency. So, different molecules with different structures vibrate at different frequencies. That is why IR spectrometry is considered as a powerful tool for distinguishing and identification of different molecules. As per the literature it has been revealed that none of the two compounds (except enantiomers) can have the same IR spectrum. For identification of organic compounds the best region is between 4000 and 400 cm^{-1}. To use IR as an analytical tool, it is necessary to understand the theory and basic principles of IR spectroscopy.

2.2 THEORY/BASIC PRINCIPLES

In IR spectroscopy for molecular identification there are three IR regions as mentioned below.

1. 10 to 400 cm^{-1} (Far IR region).
2. 400 to 4000 cm^{-1} (Practical organic region).
3. 4000 to 14290 cm^{-1} (Near IR region).

Among these three regions, the practical organic region is of very high importance to investigators for molecular structure elucidation. **Initially, when a molecule is irradiated with IR radiations of wavenumber of around 100 cm^{-1}, it undergoes only molecular rotation which causes appearance of some discrete lines in the IR spectrum. On the other hand when the same molecule is irradiated with IR radiations of wavenumber range between 10,000 cm^{-1} and 100 cm^{-1}, thereby rather than only rotations, the molecule undergoes molecular vibrations also. As a consequence there appear some bands (vibrational-rotational) in the IR spectrum of the same molecule.** Hence, IR spectrum of a molecule comprises various vibrational-rotational bands. The position of these vibration bands in the IR spectrum of a molecule depends upon various characteristics of molecule such as:

- Bond Strength
- Mass of Atoms
- And Atomic Arrangement

2.2.1 Vibrational Frequency

Position of vibrational-rotational bands is expressed in wavenumber ($\overline{\nu}$) that determines the vibrational energy. This ($\overline{\nu}$) is reciprocal of wavelength (λ) and is further expressed in cm^{-1}.

Such that, $(\overline{\nu}) = cm^{-1} = 10000/\mu m$

Application of Hooke's Law

The fact that vibrational frequency for any particular bonds depends upon bond strength and reduced mass is best supported by Hooke's law. According to this law stretching vibrational frequency of a bond of a molecule can be calculated as:

$$\bar{\nu} = \frac{\nu}{c} \qquad \text{...(2.1)}$$

$$\bar{\nu} = \frac{1}{2.\pi.c}.\sqrt{\frac{k}{\frac{m_1.m_2}{m_1 + m_2}}} \qquad \text{...(2.2)}$$

$$\bar{\nu} = \frac{1}{2.\pi.c}.\sqrt{\frac{k}{\mu}} \qquad \text{...(2.3)}$$

Where, μ = Reduced mass

m_1 & m_2 = Mass of atoms in gram of particular bond

k = Bond force constant (bond strength)

c = Radiation velocity

Therefore, this Hooke's law supports that the value of vibrational frequency depends upon bond strength and reduced mass. Hence, an increase in bond length and decrease in reduced mass makes the vibrational frequency to increase.

For example in case of C–H stretching of alkenes

μ = Reduced mass

Mass of carbon atom (m_1) = 20×10^{-24} gm

Mass of hydrogen atom (m_2) = 1.6×10^{-24} gm

$k = 5 \times 10^5$ gm . sec^{-2}

$c = 3.0 \times 10^8$ m . sec^{-1}

Applying all the values in the Hooke's law equation offers,

$$\bar{\nu} = \frac{\nu}{c} = \frac{1}{2.\pi.c}.\sqrt{\frac{k}{\frac{m_1.m_2}{m_1 + m_2}}} = \frac{1}{2.\pi.c}.\sqrt{\frac{k}{\mu}}$$

$$= 3.1 \times 10^5 \text{m}^{-1} = 3100 \text{ cm}^{-1}$$

A few more examples that justify the dependence of vibrational frequency on bond strength and mass are as follows:

Example: *Alkenes*—In this case the C = C is always absorbed at higher frequency than C – C because of high k value (*bond strength*).

Merits

- Resolution is better than mull technique.
- Can be used for quantitative analysis.
- KBr pellets can be stored for longer time.

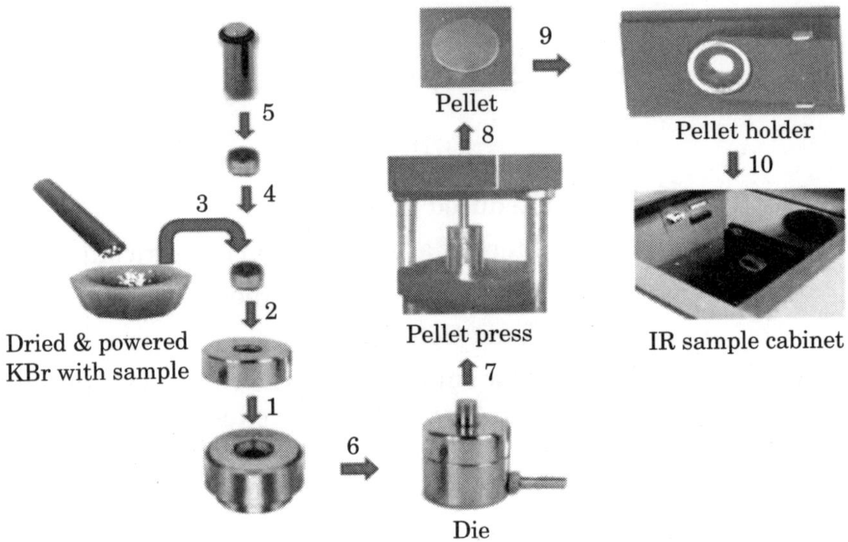

Figure 2.21: Sequential steps for solids sampling as pressed pellet for analysis.

Demerits

- High pressure in pellet formation may cause polymorphic change in crystallinity of sample, which results in astonishing complex spectrum.
- Usually shows a band near 3450 cm^{-1} due to moisture (O–H stretching).
- Not a suitable method for polymers.

B. Liquid Sample Handling

As samples that are in liquid state at room temperature do not require any extra preparation, hence they are directly subjected into the rectangular cells that are made up of metal halides such as NaCl, KBr, and ThBr etc. In liquid samples, commonly a thin-film cell is used. This thin film cell consists of two salt plates with a very thin

(*a*) Solids Sampling As Solution: While performing the method of solids sample as solution the investigation of the solid sample is done for its solubility in an IR transparent, non-interacting and non-aqueous solvent. If such suitable solvent is found, than a solution of solid sample and suitable solvent is prepared. After this there are two procedures available that are usually followed while performing the solid sampling as solution such as:

- To directly place the solid sample solution in the liquid sample cell of IR spectrometer for investigations.
- To Place the drop of solid solution on the alkali metal disk. Nextly, evaporate the solution to get a thin film of solid solute and subjected to the IR spectrometer.

The Figure 2.18, describes the method for solids sampling in solution form for IR analysis.

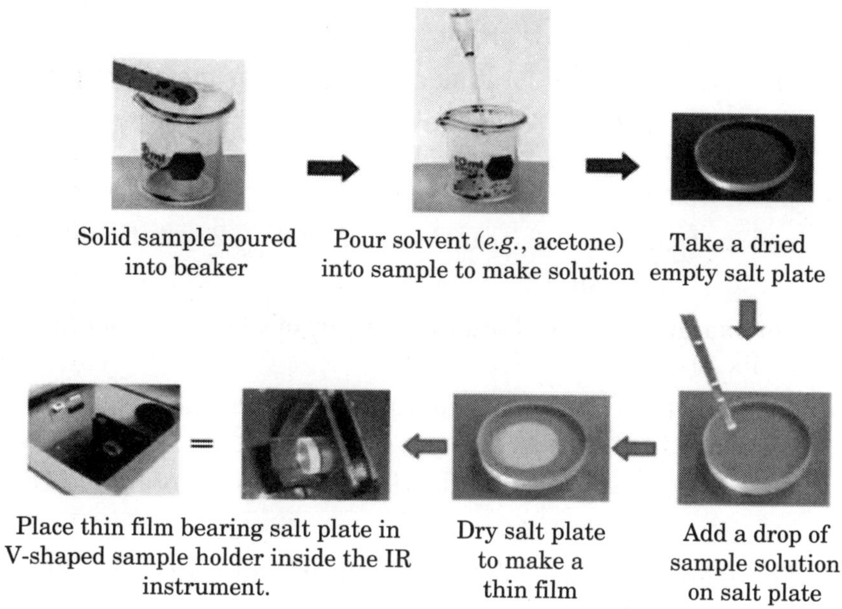

Figure 2.18: Method for solids sampling in solution form for IR analysis.

Demerits

- In this method of solids investigation in solution form, a blank solvent analysis in a cell of same thickness has to be carried out using double beam IR spectrometer.

difference of these bands (E_{vib}) is $h.\nu$. Polyatomic molecules exhibits more than one fundamental band. The number of fundamental bands for polyatomic molecule can be calculated as per following guidelines.

- Number of fundamental bands depends upon DOF (Degree of freedom *i.e.*, sum of coordinates necessary to locate atoms of a molecule in space) of a molecule.
- An isolated atom possesses only one DOF *i.e.*, translational DOF. But when the isolated atom combines with other atom to form a molecule it possess three DOF, *i.e.*, translational, rotational and vibrational DOF.
- Hence, a molecule possesses $3n$ DOF (where n = number of atoms).
- As a molecule possess three coordinates in space so it exhibits three translational DOF.
- The rotational DOF for linear and nonlinear molecules are two and three respectively.
- Hence, for a linear molecule fundamental vibrations can be calculated as

  $$\text{Vib. DOF} = 3n - 3\ (\text{translational DOF}) - 2\ (\text{rotational DOF})$$
  $$= 3n - 5$$

  For example in case of linear molecule of CO_2, Vibrational DOF $= (3 \times 3) - 5 = 04$.
- Whereas for a nonlinear molecule fundamental vibrations can be calculated as

  $$\text{Vib. DOF} = 3n - 3\ (\text{translational DOF}) - 3\ (\text{rotational DOF})$$
  $$= 3n - 6$$

  For example in case of C_6H_6,Vib. DOF $= (3 \times 12) - 6 = 30$.

  Practically there should be 30 fundamental bands in the IR spectrum of benzene. But it does not happen so because of some instances such as:

  - Overlapping of bands as they may be very close.
  - Some fundamental bands may appear out of main IR region.

- Fundamental bands may be too weak to observe.
- Presence of some additional bands such as combination bands, overtone and difference bands.

B. Overtone Bands

Bands that arises due to electronic transitions from ground state ($\nu = 0$) to the second excited state ($\nu = 2$) or above are termed as overtone bands. These bands exhibit weak absorption of IR radiations. Electronic transitions from ν_0 to ν_2 produces first overtone bands. Similarly, depending upon excited state number there exists other overtone bands also. Intensity of these bands is below 100 cm^{-1}. The energy difference of these bands (E_{vib}) is $2\ h.\nu$. for first overtone bands. In IR spectrum if there are two fundamental bands such as '*a*' and '*b*', then overtone bands are expressed as '2*a*' and '2*b*'.

C. Combination Bands

In an IR spectrum if there are two fundamental bands such as '*a*' and '*b*', then combination bands are expressed as '*a* + *b*', '*a* + 2*b*', and '2*b* + *a*'.

D. Difference Bands

If an IR spectrum is comprised of two fundamental bands '*a*' and '*b*', then difference bands are expressed as '*a* – *b*', '*a* – 2*b*', and '2*b* – *a*'.

The intensity and position of IR absorption bands depends upon the difference between the dipole moment of the molecule in the ground state and excited state. Greater will be the difference between two states the higher would be the intensity of absorption band.

In the IR spectrum of a molecule intensity of bands is expressed in transmittance (T) or absorbance (A).

Absorbance is expressed as $A = \log_{10} \cdot \frac{1}{T}$ ...(2.7)

Where, T = Radiation transmitted by sample

A simple non dispersive IR instrument is comprised of at least three circular and variable interference filters (selects different wavelength and covers a range of 2.5-15 μm) that are mounted on the shaft of a potentiometer of high resolution power.

These instruments are used for selective compounds online process control, such that these can be connected to liquid chromatograph to monitor effluents. The Figure 2.25, shows the optical schematic diagram of non dispersive IR instrument.

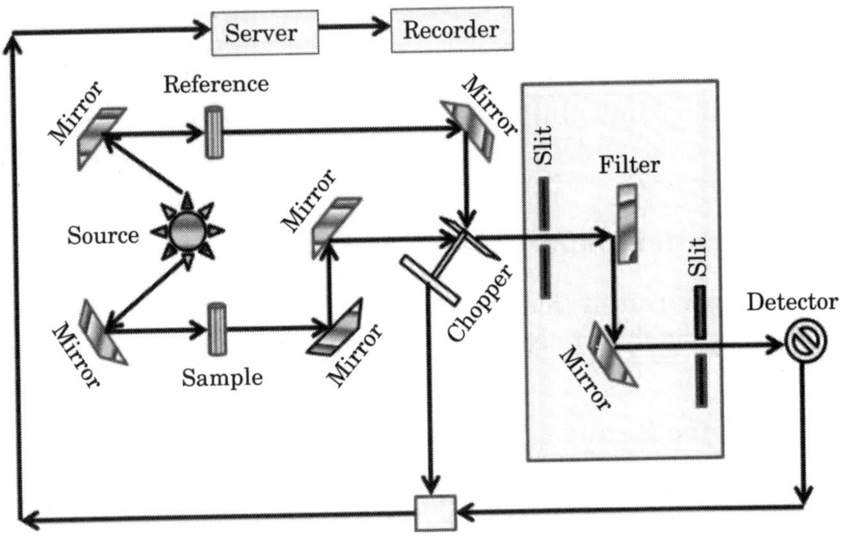

Figure 2.25: Optical schematic diagram of nondispersive IR spectrometer.

2.3.3.3 *Fourier Transformed Infra Red (FT-IR) Instruments*

Commercially, Fourier transform infrared (FT-IR) spectrometers were introduced in 1960's. But because of the high cost during that time, the FTIR's were limited to area of advance research only. With the advancements in technology and instruments, the cost of these instruments was reduced which enhanced the FT-IR capabilities. Today, FT-IR spectrometer is the standard for identification of organic compounds.

(*a*) **Sampling in FTIR:** As FTIR instrument is very sensitive, most convenient and satisfactory method, so it involves formulation

space in between them. Two channels allow liquid to be injected and then subsequently removed. The windows on these cells are made by a wide variety of IR optical materials. For example in case of water-based solutions CaF_2 is most commonly used (as insoluble in water). Cleaning of cells includes removal of solutions by firstly flushing the cell with fresh solvent and nextly removing the solvent by syringe. A precaution is required to be taken is not to blow air or nitrogen through the ports as this can cause mechanical deformation of salt window on high pressure. It is important to have a constant path length for the radiation to travel through and also the sample to be homogeneous. For large volume of liquid sample specialized couvettes can be used. The Figure 2.22, shows various components of liquid sample holder.

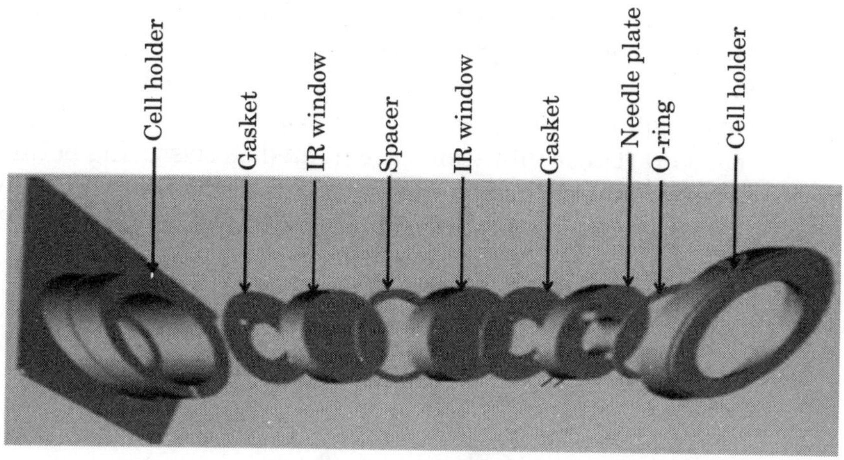

Figure 2.22: Various components of liquid sample holder.

C. Gas Sample Handling

The materials for construction for gas sample cells are similar to liquid sample cells such as NaCl, KBr, and ThBr etc. The cells for gas samples are relatively very large in size *i.e.*, from 5 – 10 cm. The requirements for sample gas IR analysis firstly includes that they should be nonreactive to cell windows or reflecting surfaces. Secondly moisture should be avoided. The Figure 2.23, shows the image of gas sample holder.

occurs below region of 1500 cm^{-1}. Bending vibrations are further classified into two types such as:

a. Inphase Bending Vibrations

b. Out of Phase Bending Vibrations

(*a*) Inphase Bending Vibrations: *Bending vibrations that causes bending of bonds of molecule, so that two atoms remains in same molecular plane but undergoes bending* (*i.e.*, changes the bond angle) *are termed as inphase bending vibrations.* These are further categorized into types:

(i) Scissoring Vibrations

(ii) Rocking Vibrations

(*i*) Scissoring: *Scissoring are the type of bending vibrations in a molecule* (example CH_2 molecule) *which causes bending of bonds of a molecule, so that the two atoms* (hydrogens) *moves towards each other, but distance between two atoms* (hydrogens) *remain same with respect to a central atom* (carbon). These vibrations are named as scissoring because here the two atoms approach towards each other which give the resemblance of a scissor. For example in case of CH_2, scissoring band occurs near 1465 cm^{-1}.

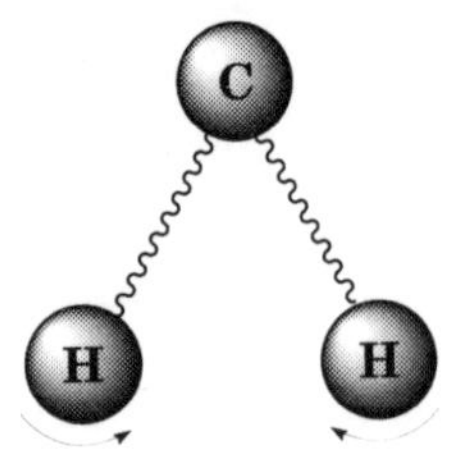

Two hydrogen atoms approach towards each other like a scissor

(*ii*) Rocking: *Rocking are the type of bending vibrations in a molecule* (example CH_2 molecule) *which causes bending of bonds of molecule, so that the distance between two atoms* (hydrogens) *remain same with respect to a central atom* (carbon) *but the two atoms* (hydrogens) *moves in the same direction.* These vibrations are named as rocking because here the two atoms rocks together. For example in case of CH_2, rocking band occurs near 720 cm^{-1}.

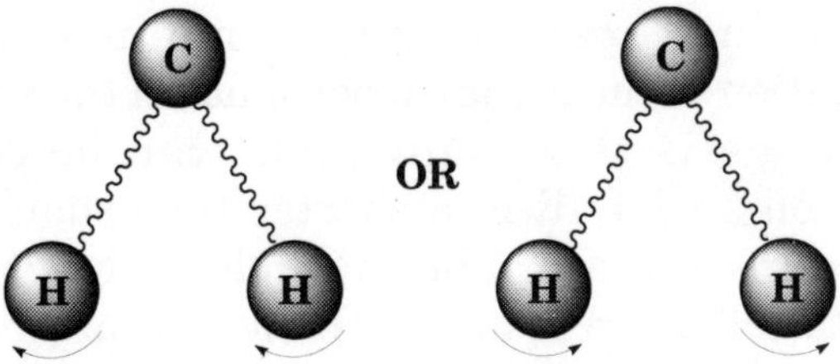

Two hydrogen atoms moves in a pendulum fashion with same bond angle

(b) Out of Phase (OOP) Bending Vibrations: *OOP are the type of bending vibrations which causes bending of bonds of molecule, so that the two atoms changes their plane with respect to the central atom.* These are of two types viz.

(i) Wagging Vibrations

(ii) Twisting Vibrations

(i) Wagging Vibrations: *Wagging are the bending vibrations in a molecule* (example CH_2 molecule) *which causes bending of bonds of molecule, so that the two atoms moves either above or below with respect to the plane of central atom.* For example in case of CH_2, wagging band appears between 1350–1150 cm^{-1}.

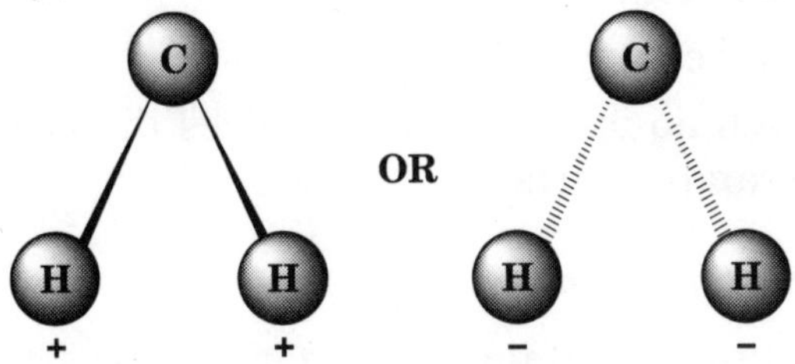

Both hydrogen atoms are in same plane either above or below to plane of central atom

(ii) Twisting Vibrations: *Twisting are the bending vibrations in a molecule* (example CH_2 molecule) *which causes bending of bonds of molecule, so that among two atoms one moves above and the other moves below the molecular plane with respect to the plane of central atom.* For example in case of CH_2, twisting band appears between 1350–1150 cm^{-1}.

and individual scans can be combined to give better representation of the actual absorbance of the sample. The FT-IR advantage is that many scans can be completed and combined on an FT-IR in a shorter time than one scan on a dispersive instrument. The multiplex advantage results in faster data collection of an FT-IR spectrum.

- *Constant Spectral Resolution:* In an FTIR spectrometer, the resolution of the measured spectrum is the same for all frequencies, not varied throughout the spectrum as is often true with dispersive instruments.

2.3.4 IR Reflectance Techniques

There exists a variety of samples where IR transmission is difficult or impossible, and to overcome such problems reflectance techniques are used. The reflectance techniques replaces the constant path transmission cells and salt plates. These are used for the analysis of liquid, semi-liquid, films, pastes, pliable solids and fine powders.

Commonly in IR experiment two types of reflectance techniques are used such as:

- Internal Reflectance and
- External Reflectance

These two techniques generally involves direct reflection of IR beam from the sample surface.

2.3.4.1 *Internal Reflectance*

Internal reflectance technique further involves two techniques namely:

a. ATR (Attenuated total reflectance) and
b. MIR (Multiple internal reflectance).

(*a*) Attenuated Total Reflectance (ATR): ATR uses the property of total internal reflection known as the evanescent wave. This concept of total internal reflection using ATR, can be best understood from the diagram of an attenuated total reflectance cell given in Figure 2.27.

of solution of the sample (in chloroform, ether, dichloromethane, or $CDCl_3$ etc.) and evaporation on a KBr salt plate to form a thin film. Thereby on the remaining film IR spectra is generated. This method provides excellent spectra with flat baseline. The sample may alternatively be prepared as a Nujol mull.

(*b*) **Interferometer:** *FT-IR instrument works on a system called interferometer to collect a spectrum through an interferogram (pattern of optical pathway). Interferogram is a complex signal. As interferogram generates time domain curve, and requirement is for frequency domain curve, hence a mathematical operation called Fourier transformation is applied to convert a time domain curve into a frequency domain curve.* The components of an interferometer are

- A source
- A beam splitter
- Two mirrors
- A laser, and
- A detector

The series of events in an interferometer can be best understood with the optical schematic diagram of Michelson interferometer given in Figure 2.26.

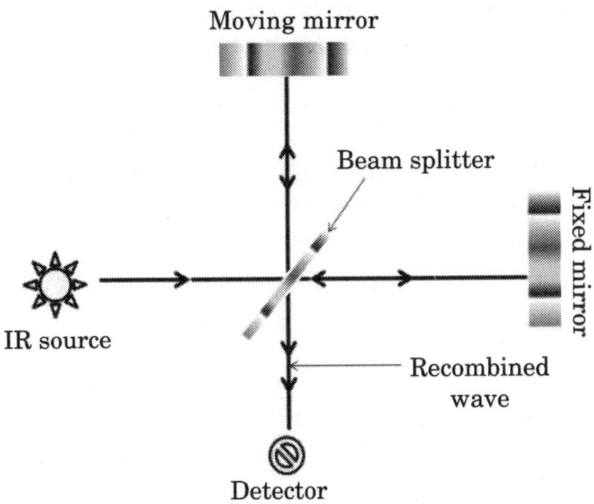

Figure 2.26: Optical schematic diagram of Michelson interferometer.

For example in case of CO_2 molecule, the two bending vibrations appears at 667 cm^{-1} with an overtone at 1334 cm^{-1} (*i.e.*, 2 × 667 cm^{-1}), and a symmetrical stretching fundamental vibrations at 1340 cm^{-1}. As these two vibrations of overtone and fundamental levels are very close, causes phenomenon of Fermi resonance to occur *i.e.*, interaction of fundamental and overtone vibrations. This leads to generation of two bands at 1286 cm^{-1} and 1388 cm^{-1}.

A few more examples that exhibits Fermi resonance phenomenon are described as:

- Butyl vinyl ether shows two bands at 1640 cm^{-1} and 1630 cm^{-1} (*due to mixing of one fundamental band at 810 cm^{-1} and one overtone band at 1640 cm^{-1}*).
- Aldehydes shows doublet for C–H stretch at 2820 cm^{-1} and 2720 cm^{-1}.
- Cyclopentanone shows two bands 1746 cm^{-1} and 1750 cm^{-1} due to Fermi resonance.

Phenomenon of Fermi resonance occurs very frequently in IR and Raman spectra. General conditions for occurrence of Fermi resonance are as follows:

- The two vibrational bands should be of the same symmetry.
- Interacting group should be so located in the molecule such that it can favour resonance to occur.

C. Inductive Effect

When an electron withdrawing or an electron donating group is introduced near to the group under IR investigation, it causes increase or decrease in force constant. Such an electronic effect is called as inductive effect. The increase or decrease in force constant further affects vibrational frequency.

As per the increase or decrease in force constant, inductive effect is categorized into two types namely,

(i) Positive Inductive Effect

(ii) Negative Inductive Effect

(*i*) **Positive Inductive Effect:** *Case where an electron donating group* (for example: alkyl group) *is introduced near to group under IR investigation, it causes lengthening of bond and thereby reduces the force constant. Such type of electronic effect is known as positive inductive effect.* This reduction in force constant decreases the wavenumber of vibrational frequency. Below mentioned are some of the examples of positive inductive effect to cause decrease in wavenumber of its C=O stretching.

- HCHO – 1750 cm^{-1}
- CH_3CHO – 1745 cm^{-1}
- CH_3COCH_3 – 1715 cm^{-1}

(*ii*) **Negative Inductive Effect:** *When an electron withdrawing group* (for example: chlorine group) *is introduced near to the group under IR investigation it causes shortening of the bond and thereby increases the force constant. Such type of electronic effect is called as negative inductive effect.* This increase in force constant increases the wavenumber of vibrational frequency. Below mentioned are some of the examples of negative inductive effect to cause increase in wavenumber of C=O stretching.

- CH_3COCH_3 – 1715 cm^{-1}
- CH_3COCH_2Cl – 1725 cm^{-1}
- $CH_3COCHCl_2$ – 1740 cm^{-1}
- $Cl_2CHCOCHCl_2$ – 1750 cm^{-1}

D. Mesomeric Effect

In case a compound contains a lone pair of electron on one of its atom that contributes in extending conjugation with the double bond of a group and thereby lowers the vibrational frequency of its group. This effect is named as mesomeric effect. Mesomeric effect through conjugation causes lengthening and weakening of bond of a particular group. This thereby lowers the absorption frequency. Below mentioned are some of the cases where mesomeric effect dominates over inductive effect to cause decrease in wavenumber of its C = O stretching.

- If an ethene group, adjacent to carbonyl group is replaced with a phenylgroup, it leads to lowering of the vibrational frequency.

in the study of surface coatings, such as surface treated metals, paints and polymers etc. The accessories for specular range of reflectance include both fixed and variable incidence angle accessories. To increase the sensitivity of specular reflectance experiment the path length can be increased through thin coatings using grazing angles of incidence (85°). Coatings of higher thickness can be studied using angles of typically 30°. Also monolayer films at the liquid—air interface can also be studied using a grazing angle of incidence. Band intensities of specular reflectance depends upon the type and degree of polarization. Figure 2.29, shows the optical diagram of specular reflectance.

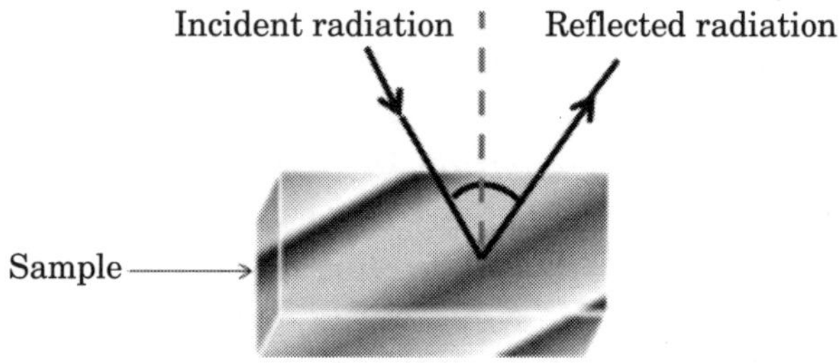

Figure 2.29: Optical diagram of specular reflectance.

(b) **Diffuse Reflectance Technique (DRT):** *The energy which penetrates one or more particles is reflected in all directions. This component is called diffuse reflectance.* On a rough surface, such as a powder, specularly reflected light is a minor contributor to the overall signal. Specific accessories are optimized to increase collection of the diffuse reflectance and decrease the specular components. Figure 2.30, shows the optical diagram of diffused reflectance.

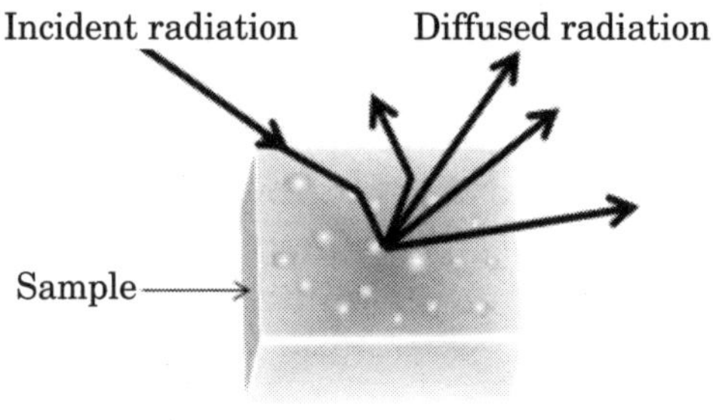

Figure 2.30: Optical diagram of diffused reflectance.

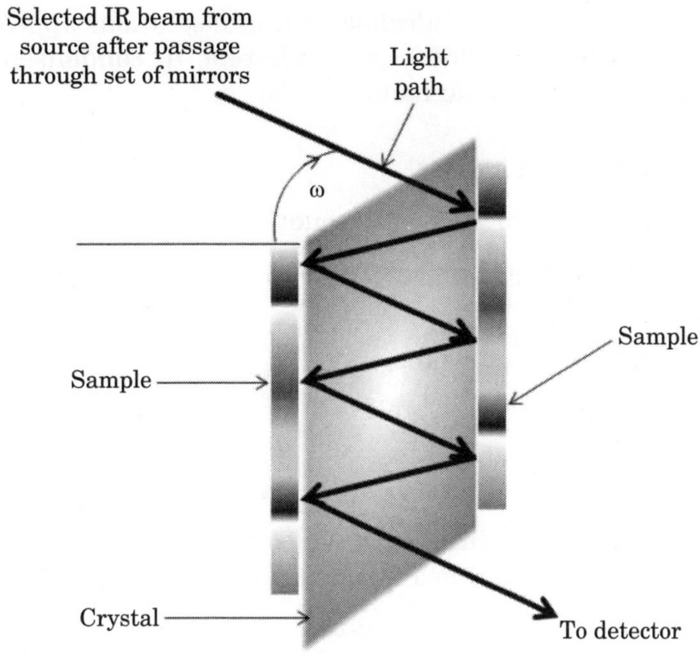

Figure 2.27: An Attenuated Total Reflectance Cell.

Keys Concepts

- In ATR technique, an IR beam after passage through a set of mirrors, is allowed to pass through one end of a crystal.
- The crystal causes complete internal reflection of incident IR light (several times) till the IR light is exit through its other end. This occurs only when the angle of incidence is greater than the critical angle, which is a function of refractive indices of two surfaces.
- The internal reflections inside the crystal create an evanescent wave, which extends further into the sample which is placed on the crystal surface.
- As evanescent wave decays rapidly with distance from the surface, therefore the sample should always be in direct contact of crystal surface.
- When the beam strikes the crystal surface it can penetrate the surface slightly and, if the substance absorbs light at

- O–H stretching of alcohols in dilute solutions appears as a sharp band at 3650 cm^{-1}, whereas in condensed phase appears as a broad band at 3350 cm^{-1}.

2.3 INSTRUMENTATION

2.3.1 Components of IR Spectrometer

Commonly, the list of components used in IR spectrometer are as follows:

- Source (UV and visible)
- Monochromator
- Detector

2.3.1.1 *Sources*

In all the IR sources an inert solid is electrically heated in the range of 1500 – 2000 K. Thereby heated material emits infra-red radiations.

Ideal characteristics of an IR Source

- Should emit radiations on entire IR range.
- Should emit stable and continuous IR radiations over the entire wavelength region.
- Must be of sufficient intensity for the transmitted energy to be detected.
- Temperature should be as high as possible to maximize the results.
- Should be in direct contact with the atmosphere.

Types of IR Sources

a. Nernst Glower
b. Silicon Carbide Rod (Globar Source)
c. Incandescent Light Bulb
d. Incandescent Wire Source
e. Mercury Arc
f. Carbon Arcs (An Unsuitable IR Source)
g. Kanthal Wire Coils
h. Carbon Dioxide Laser

(*a*) **Nernst Glower:** A Nernst glower is composed of a cylinder that is fabricated from a mixture of earth oxides. Diameter of cylinder is 1 – 2 mm diameter, with an approximate length of 20 – 50 mm. The ends of cylinder are sealed with platinum lead, which permits electrical connectivity in conformance to resistive heating element. A Nernst glower can reach to a temperature of 2200 K. Nernst glower possess good capacity for hotter temperature. Figure 2.1, shows the image of Nernst glower.

Figure 2.1: Nernst glower.

Working: Current is passed through the Nernst glower, which raises the temperature to about 2200 K. Nernst glower possesses a negative temperature coefficient of electrical resistance. Hence, before current is sufficient to produce desired temperature, the instrument is heated externally till it is dull red hot.

Merits: Frequency of IR Nernst glower is exceptional below 2000 cm^{-1} but on the entire mid IR range it results in low signal to noise ratios.

Demerits: Despite being capable of higher temperature than Globar source the Nernst glower is not capable of producing IR radiation above 2000 cm^{-1}.

(*b*) **Silicon Carbide Rod (Globar Source):** The most commonly used IR source in FTIR is a resistively heated silicon carbide rod/bar termed as Globar source. Globar possess a 5 mm diameter, 50 mm long that is heated electrically. Globar source possess positive coefficient of resistance. The earlier models of Globar source required water cooling to safeguard the damage of electrical components. Now advance Globar sources makes use of ceramic metal alloys. The Figure 2.2, shows the image of Globar source.

Figure 2.2: Globar source.

Working: When an electric current is passed through the bar to make it extremely hot *i.e.*, about 1700°C, thereby bar emits large amounts of IR radiation. A Globar source can reach to a temperature of 1300 K. The maximum IR radiation emission of this instrument is about 5200 cm^{-1}.

Merits: These are self-starting and convenient to control.

Demerits

- These newer Globars cannot be operated on temperature higher than 1300 K.
- These are less intense than Nernst glower.

(*c*) Incandescent Light Bulb: Incandescent bulb is composed of a tungsten filament which is sealed into vacuum through a glass bulb. This source is structurally similar to a electric bulb. This bulb comprises vacuum which is necessary to keep the tungsten free from oxidation on high temperature. This IR source is electrically heated (*generates white heat*) to make it glow at a temperature of 3000 k, and emits massive amounts of IR. The intensity of radiation is usually low or of shorter wavelength (below 350 nm). Moreover intensity of lamp varies with wavelength. The Figure 2.3, shows the image of an incandescent lamp.

Figure 2.3: Incandescent lamp.

Demerits

- As glass serves as an IR absorber, so its use in incandescent lamp causes blockage of IR path of sample.
- As source is in direct contact with the atmosphere, so causes drastic limits on operative temperature.

(*d*) **Incandescent Wire Source:** Incandescent wire source is composed of a tightly wound coil of nichrome wire on the ceramic cylinder, that is electrically heated to 1100 K. It produces a lower intensity of radiation than the Nernst or Globar sources, but possess a longer working life. Instead of nichrome wire a rhodium wire can be used to give similar properties but it becomes more expensive. The Figure 2.4, shows the image of an incandescent wire source.

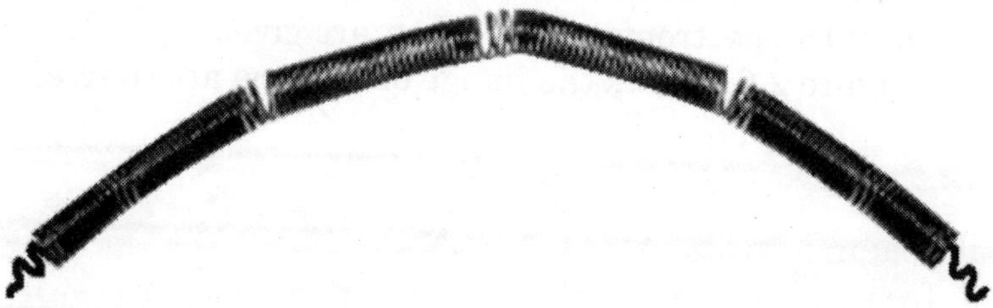

Figure 2.4: Incandescent wire source.

Merits

- Does not require cooling.
- Minimum maintenance is required.

Demerits: These are less intense than many other IR sources.

(*e*) **Mercury Arc:** Mercury source is composed of a quartz jacketed tube that contains mercury vapours with a pressure of more than one atmospheric. When the electric current is passed through the tube, the vapors form an internal plasma source to generate IR radiations. Figure 2.5, shows the image of a mercury tube.

Merits: None of IR sources can provide sufficient/continuous radiation power comparable to mercury arc in far IR region.

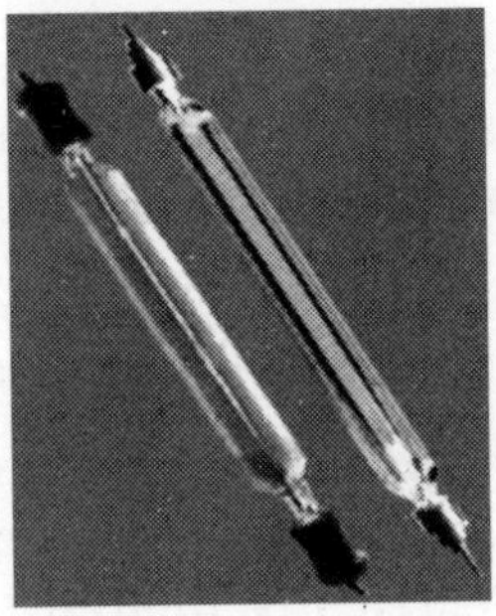

Figure 2.5: Mercury tube.

(*f*) **Carbon Arcs (An Unsuitable IR Source):** A carbon arc source involves introduction of electrical discharge between two carbon electrodes. These sparks are incredibly bright reaching temperature as hot as 6000 K. The carbon IR sources are used in many IR spectrometers. Carbon arc gives high IR outputs. The Figure 2.6, shows the image of carbons arc source.

Figure 2.6: Carbon arc source.

Demerits

- Requires more maintenance because carbon electrodes are consumed in the arcing process, so it is necessary to continuously feed new rod forward to maintain the arc.
- The rod also requires an inert atmosphere to avoid combustion of the carbon.

(*g*) **Kanthal Wire Coils:** Kanthal wire coils were also once popular IR sources. These doesn't require any water cooling, and runs at lower temperatures than a Globar source, and possesses lower emissivity. The Figure 2.7, shows the diagram of a kanthal wire coils.

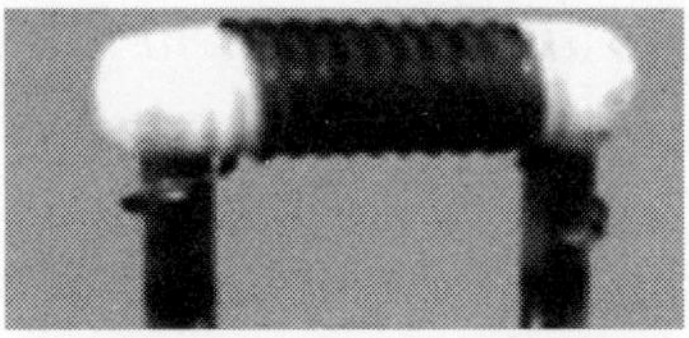

Figure 2.7: Kanthal Wire Coils.

Demerits: As these runs at a lower temperature than Globar source so less recognized than Globar sources.

(h) **Carbon Dioxide Laser:** Carbon dioxide laser source involves introduction of a tunable carbon dioxide laser, especially to determine concentration of atmospheric pollutants and absorbing species in aqueous solutions. This laser source produces IR radiations in the range of 900 – 1100 cm^{-1}. The Figure 2.8, shows the diagram of a carbon dioxide laser.

Figure 2.8: Image of carbon dioxide laser source.

Merits: It is useful for quantitative determination of number of important species such as ethanol, benzene, NO_2, ammonia, and butadiene etc.

2.3.1.2 *Monochromators*

A device which converts polychromatic light into a monochromatic light is called as monochromator. In IR spectrometers in order to allow only the desired IR frequency to reach to sample, in practice monochromators are commonly used.

The monochromator assembly is constituted by following components:

- An Entrance slit (*makes a narrow optical image of radiation source*).
- A Collimator (*renders radiation from entrance slit parallel*).

- A grating or prism (*disperse incident radiation*).
- Another collimator (*reforms the images of entrance slit on exit slit*).
- An exit slit (*isolates the desired spectral band by blocking the other dispersed radiations*).

In IR spectrometer two types of monochromator devices are commonly used such as:

a. Prism Monochromators
b. Grating Monochromators

A. Prism Monochromators

In an IR spectrometer to disperse (*conversion of poly chromatic light into a monochromatic light*) IR radiations, the prisms made up of salts of metal halides are commonly used. Prisms made up of metal halides are of great importance due to their property of high dispersion and almost transparency in the region of 4 to 15 μm (*region that is important for functional group studies*). The major drawbacks of these metal halides are their mechanical or thermal instability and water solubility. So, utmost care is usually undertaken while operating the IR spectrometer.

For construction of IR monochromator prisms a wide variety of metal halides are available such as potassium bromide, cesium iodide, sodium chloride and thallium bromide etc. The Figure 2.9, shows the image of prism monochromator.

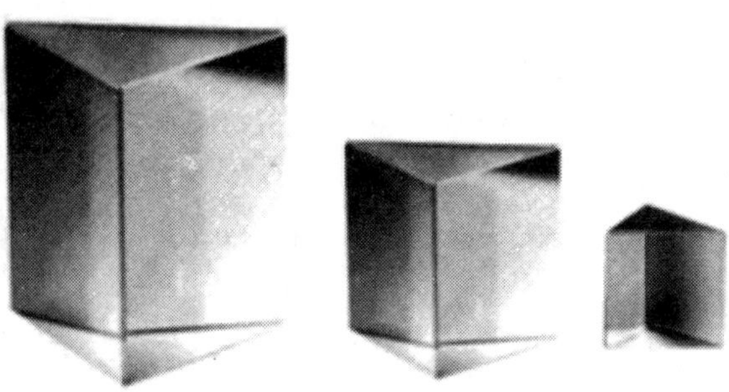

Figure 2.9: Prisms monochromators.

Based on the arrangement of monochromator components, prism monochromators can be of two types such as:

a. Single Pass Monochromator
b. Double Pass Monochromator

(a) Single Pass Prism Monochromator: *Monochromators which allows one time to and fro passage of radiations through their prism component are termed as single pass monochromators.* The series of events in these single pass prism monochromators are shown in the Figure 2.10.

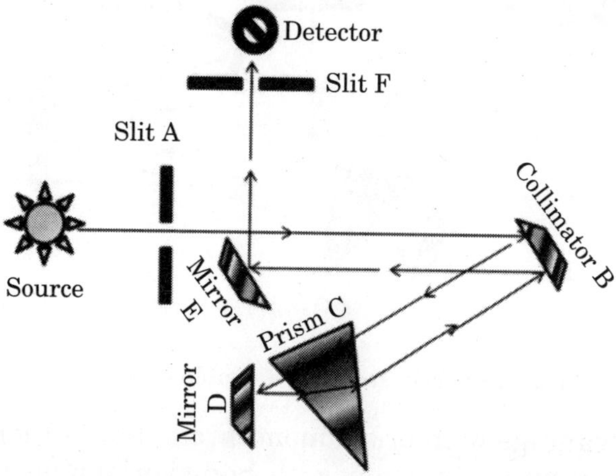

Figure 2.10: Optical schematic diagram of single pass prism monochromator.

The sequence of events in single pass monochromator is as follows:

- Radiations pass from the entrance slit A.
- After entrance slit exposure, radiations strikes onto the collimating mirror B to become parallel.
- After collimator exposure, radiation travels through prism C for the first time and gets dispersed.
- After prism exposure, radiation strikes on to the Littrow mirror D which again returns the radiations to prism for the second time.
- Again these radiations pass through the collimator B.

- Finally, radiations are focused onto the exit slit so as to pass on to the detector.

(b) Double Pass Prism Monochromator: *Monochromators that allows two times to and fro passage of radiations through their prism component are called as double pass monochromators.* This is made possible by incorporating a mirror after the collimator. Optical schematic diagram of double pass prism monochromator is shown in the Figure 2.11.

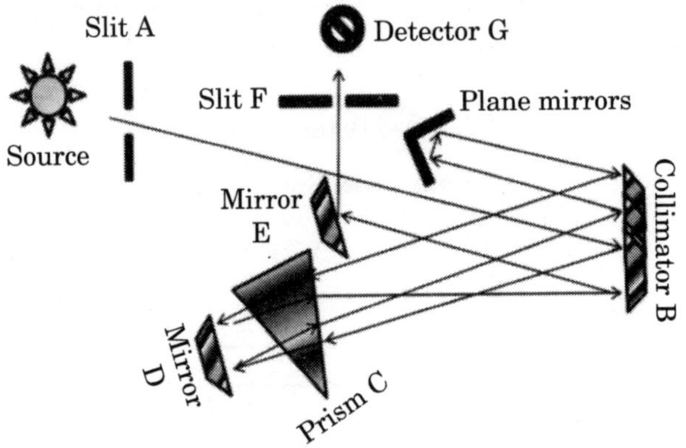

Figure 2.11: Optical schematic diagram of double pass prism monochromator.

The advantage with prism monochromators is their capability of undergoing dispersion with non-overlapping wavelength. Whereas the major drawback of prism monochromators is that these produces a nonlinear dispersion.

B. Grating Monochromators

Gratings are the most efficient devices which contains an array of very large number of equidistant/parallel slits (seen as rulings or linings made on a surface) **on a highly polished surface that reflects or disperse the incident radiations** (converts a polychromatic light into a monochromatic light). Usually the number of rulings in a grating ranges from 15000 to 30000 grooves/lines per square inch or more. Now a days in the newer instruments gratings are most commonly used. The slowly rotating gratings diffracts the IR radiations before reaching the detector. Figure 2.12, shows microscopic and macroscopic images of grating.

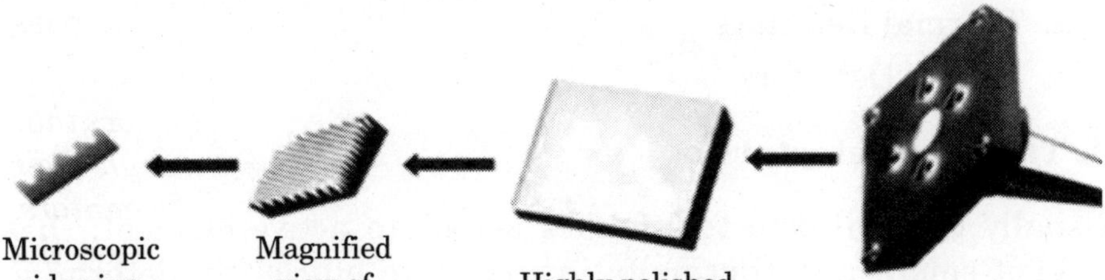

Figure 2.12: Microscopic and macroscopic images of grating.

As alkyl halides are transparent in the important IR region, so for construction of IR spectrometer gratings also alkyl halides such as potassium bromide, sodium chloride etc. are used. Generally, gratings are very difficult to be prepared, so replica gratings are constructed from master gratings. In order to produce replicated grating, a master grating is coated with epoxy resin which is removed after setting to give a replica. Nextly, the produced replica is made reflective by polishing the replica surface with aluminium. The mechanism of diffraction that produces reinforcement can be best understood with the below mentioned equation that is

$$m\lambda = G_s \,.\, (\sin i \pm \sin r) \qquad \text{...(2.8)}$$

Where,

λ = wavelength of radiation

G_s = Space of grating

i = incident angle

r = reflection angle

m = order (*i.e.*, 0, 1, 2, 3 etc.)

The biggest benefit of grating monochromators is that these produce a linear dispersion. Whereas the major drawback associated with these monochromators is that they undergo overlapping of spectral orders.

2.3.1.3 *Detectors*

In IR spectrometers to detect the signals of IR radiations exhibited from the sample, in practice two types of detector devices are commonly used viz.

a. Thermal Detectors
b. Photon Detectors

A. Thermal Detectors

Usually thermal detectors makes use of an active element, that is kept small, and thermally insulated to efficiently determine the change in temperature for any power of incident IR radiation. These operates at room temperature and possess slow response and sensitivity. These exhibits detection through expansion of state (*i.e.*, *solid*, *liquid* and *gases*), electrical resistance, voltage and polarization. Following are the few types of thermal detectors:

a. Golay Cell
b. Thermocouple
c. Bolometer
d. Thermistor
e. Pyroelectric Detectors

(*a*) Golay Cell: This device is composed of a small metal cylinder whose one end is closed by blackened metal plate and the other end is closed with a mirror (*i.e.*, flexible metallic diaphragm). The cylinder is filled with xenon gas and sealed. When IR radiation is passed through a small IR transmission window of the cylinder, it is absorbed by the blackened metal plate, which leads to heating and expansion of gas. The expansion of gas leads to increase in pressure of gas and thereby deforms the mirror. The distortions of mirror are amplified by arranging a lamp inside the detector. The lamp focuses its light on the mirror so as to reflect the light beam onto the phototube/photocell. So, with the motion of mirror, the output of photocell is modified. Hence the detection of signals by phototube depends upon the power of IR radiations falling on the Golay cell. In mid IR region, the sensitivity of this cell is nearly same as of thermocouple detector. Response time is 10^{-2} seconds. The Figure 2.13, shows the diagram of a Golay cell.

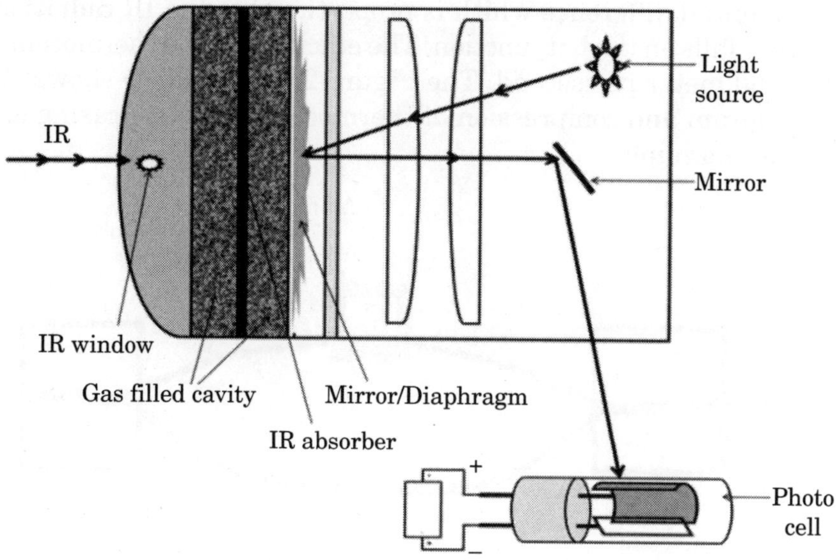

Figure 2.13: Diagram of Golay Cell.

Merits

- Useful IR range is very wide
- Linear response

Demerits

- Less convenient and bulky
- More expensive

(b) Thermocouple: This device consists of an assembly of two junctions that are formed by end to end welding of two different semiconductor wires made up of different materials (*i.e.*, bismuth and antimony). The two junctions are kept on different temperature such that an electric potential is developed between the joints. This junction assembly is enclosed in an evacuated steel casing containing a potassium bromide window, which is meant to avoid loss of energy through convection. In this device one of the junction is called as cold junction (*because kept at constant temperature and is unexposed to IR radiation*) and the other is hot junction (*because kept at variable temperature and is exposed to IR radiation*). This difference in the temperature of the two junctions results in development of potential difference which is proportional to the IR radiations that falls on the hot junction. The efficiency of the thermocouple is 60 meter per second. The Figure 2.14(*a*) and (*b*) shows the diagram and compression of thermocouple in steel casing of a thermocouple.

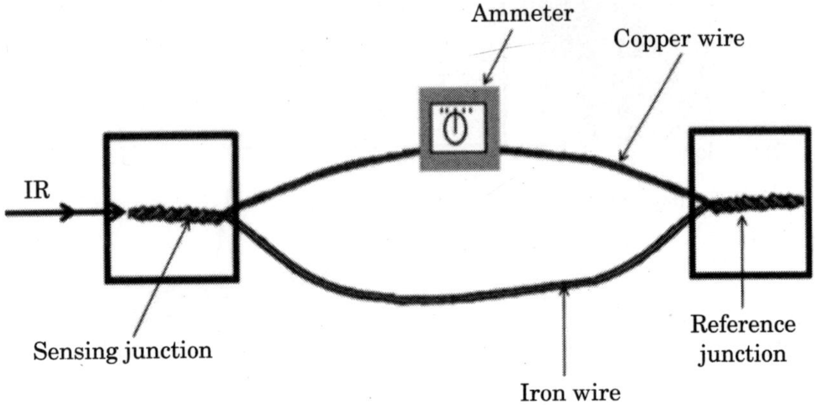

(*a*)

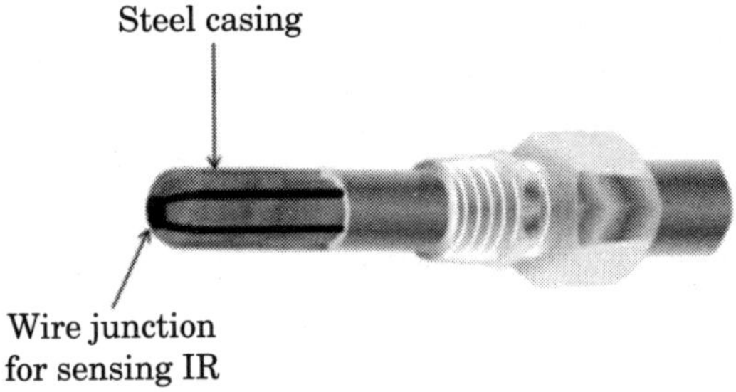

(*b*)

Figure 2.14: (*a*) Diagram of thermocouple and (*b*) Compression of thermocouple in a steel casing.

(*c*) Bolometer: This is a resistance thermometer that is usually constructed from a conductor (such as thin metal strip of platinum or nickel) or semiconductor. Bolometer contains an absorptive element (*i.e.*, thin layer of metal), which is connected to a thermal reservoir. When any radiation falls on the absorptive element, it raises the temperature more than reservoir (higher the absorptive power, the greater the temperature). The temperature change can be measured directly with an attached resistive thermometer or resistance of the absorptive element itself can be used as a thermometer. As the temperature of metallic conductor increases its resistance also gets increased. This degree of change in resistance is proportional to the amount of IR radiation that falls on the bolometer. Therefore a bolometer incorporates wheat stone bridge. Commonly, semiconductor exhibit's a large change in resistance as a function of conductance. The efficiency of the bolometer is 4 meter per second. The Figure 2.15, shows the schematic diagram of a bolometer.

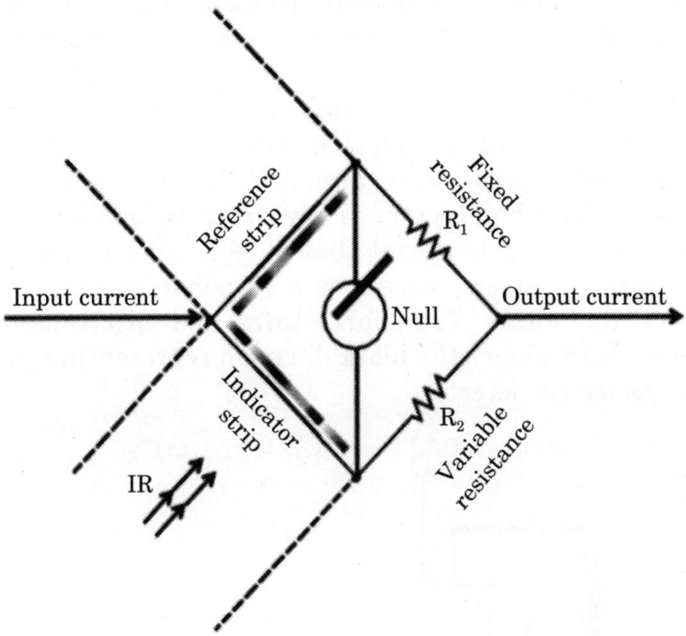

Figure 2.15: Schematic diagram of bolometer.

(*d*) Thermistor: This is also a resistance thermometer (same as Bolometer) that is usually made up of fused mixture of metal oxides (such as sintered oxides of manganese, cobalt and nickel). This device also works by changing the resistance on heating. When conductor is exposed to the IR radiations it undergoes change in its temperature. As the temperature of metallic conductor increases its resistance gets decreased. This degree of change in resistance is proportional to the amount of IR radiation that falls on the thermistor. Thermistors possess slow response time.

(*e*) **Pyroelectric Detectors:** *Those substances which undergo polarization on exposure to the electrostatic field and still retain strong residual polarization on removal of electrostatic field are called as ferromagnetic substances. Whereas, those substances whose residual polarization is some time temperature sensitive are called as pyroelectric substance.* Based on the facts about pyroelectric substances, pyroelectric detectors are designed. Pyroelectric detector contains a thin dielectric flake (*i.e.*, cut from a single crystal), which is connected to an electrode. Hence when a detector is exposed to IR radiation, it leads to change in temperature that changes the electrostatic charge on the face of dielectric flake. The electrode connected to the flake collects the charge and thereby creates a voltage. In pyroelectric detectors triglycerine sulphate is the most commonly used material that is generally cooled in liquid nitrogen. As these possess fast response so commonly used in FTIR (Fourier Transform Infra-Red spectrometer). The Figure 2.16, shows the block diagram representing working of a pyroelectric detector.

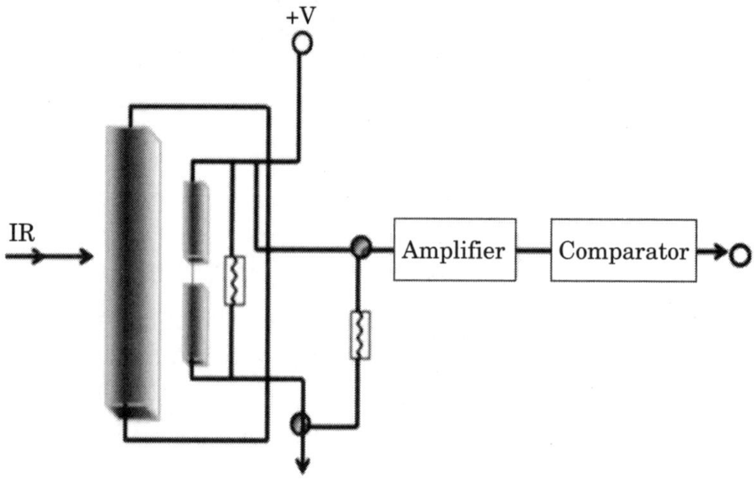

Figure 2.16: Block diagram representing working of a pyroelectric detector.

B. Photon Detectors

Usually, photon detectors comprise quantum interaction of semiconductor with an incident photon, to cause production of electrons and mediate a photoelectric effect.

Photon detectors are of two types such as:

a. Photo Conductivity Detector
b. Photovoltaic Detector

(*a*) **Photo Conductivity Detector:** This is a non-thermal detector that contains a semiconductor which exhibits lowering of resistance due to presence of electrons in the conduction band. The excitation can be done intrinsically (*i.e.*, raising of electrons from valence to conduction band) or extrinsically (*i.e.*, from or to impurity doping levels within the forbidden bands of semiconductor).

(*b*) **Photovoltaic Detector:** This is also a non-thermal detector that makes use of semiconductor material like Indium antimonide (IA) or lead tin telluride (LTT) or mercury cadmium telluride (MCT). Based on the type of semiconductor used, the photovoltaic detectors are categorized into three types *i.e.*, IA, LTT and MCT. In all these detectors the thin layer of p-type of semiconductor (*i.e.*, of IA or LTT or MCT) is kept over the n-type surface to make a diffused p-n junction. The p-type surface is exposed to IR radiation which generates hole and electron pairs. These hole and electron pairs are separated by internal field existing on p-n junction. As a result voltage is generated. This assembly of semiconductor is arranged inside a Dewar type vacuum bottle cooling unit. The cooling is done through liquid nitrogen or Joule Thomson coolers of compressed nitrogen gas or liquid helium. IA, LTT and MCT detector are cooled with liquid nitrogen, or liquid helium. The optimum sensitivity of IA and LTT in IR region are between 5 – 13 μm and 6.6 – 18 μm respectively. The sensitivity of MCT for IR is much more than IA and LTT. The photovoltaic detectors exhibit a maximum response time of 20 Nano seconds. The Figure 2.17, shows the diagram of a photovoltaic detector.

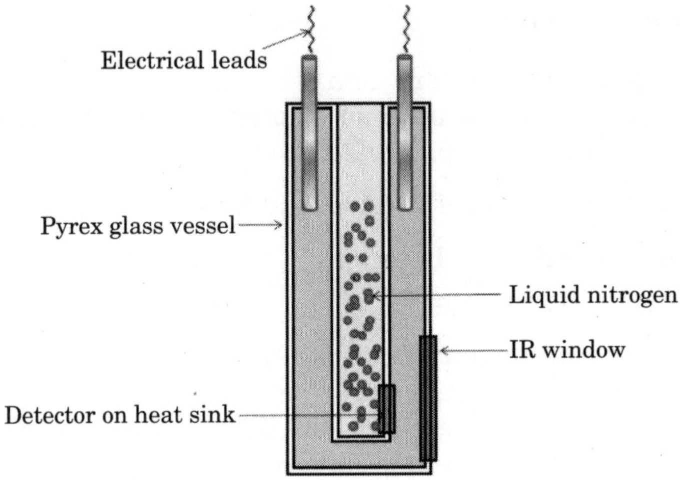

Figure 2.17: Diagram of a photovoltaic detector.

2.3.2 Sample Handling

The basic requirement while sample handling for all the three phases of sample (*i.e.* solid, liquid and gases) is that material which holds sample should be transparent to IR radiations. Actually none of the material is transparent in IR region but metal halides are the one which are almost transparent in between important IR region from 4 to 15 μm. According to nature of sample, handling of sample is categorized into three types namely,

a. Solid Sample Handling
b. Liquid Sample Handling
c. Gas Sample Handling

A. Solid Sample Handling

For solids sampling can be done by four ways such as:

a. Solids as Solution
b. Solids as Films
c. Solids as Mull
d. Solids as Pressed Pellets

(a) Solids Sampling As Solution: While performing the method of solids sample as solution the investigation of the solid sample is done for its solubility in an IR transparent, non-interacting and non-aqueous solvent. If such suitable solvent is found, than a solution of solid sample and suitable solvent is prepared. After this there are two procedures available that are usually followed while performing the solid sampling as solution such as:

- To directly place the solid sample solution in the liquid sample cell of IR spectrometer for investigations.
- To Place the drop of solid solution on the alkali metal disk. Nextly, evaporate the solution to get a thin film of solid solute and subjected to the IR spectrometer.

The Figure 2.18, describes the method for solids sampling in solution form for IR analysis.

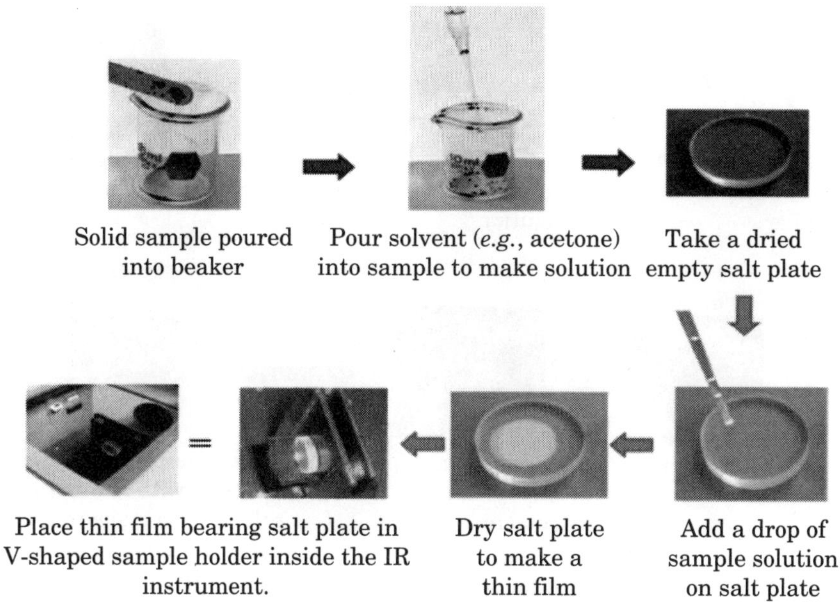

Figure 2.18: Method for solids sampling in solution form for IR analysis.

Demerits

- In this method of solids investigation in solution form, a blank solvent analysis in a cell of same thickness has to be carried out using double beam IR spectrometer.

- Selection of IR transparent, non-interacting and non-aqueous solvent for every solid is tedious and not possible always.

(*b*) Solids Sampling As Film: Solids sampling as film method is done on the amorphous solids. In this method firstly a solution of solid sample and solvent is formed, which is subjected to the evaporation to deposit a thin film on the surface of KBr or NaCl cell. This method is used in case of rapid qualitative analysis but it is of lesser use in case of quantitative analysis. The Figure 2.19, shows the method for solids sampling in film form for IR analysis.

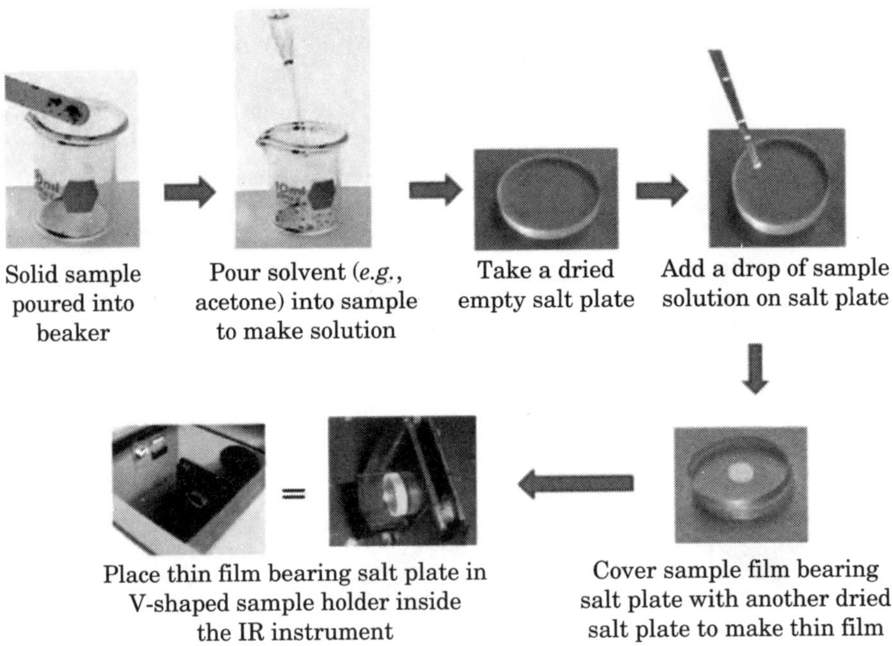

Figure 2.19: Method for solids sampling as film for IR analysis.

Demerits: The search of IR transparent, non-interacting and non-aqueous solvent for every solid is tedious.

(*c*) Solids Sampling As Mull: In this method of solids sampling as a mull, firstly a solid sample not more than 20 mg is finely grounded and then one or two drops of Nujol (mineral oil) are added to formulate into a thick paste. Secondly this paste is spread onto the IR transmitting windows. This method also holds good for qualitative analysis but is not appreciable in case of quantitative analysis. The Figure 2.20, shows the method for solids sampling in mull form for IR analysis.

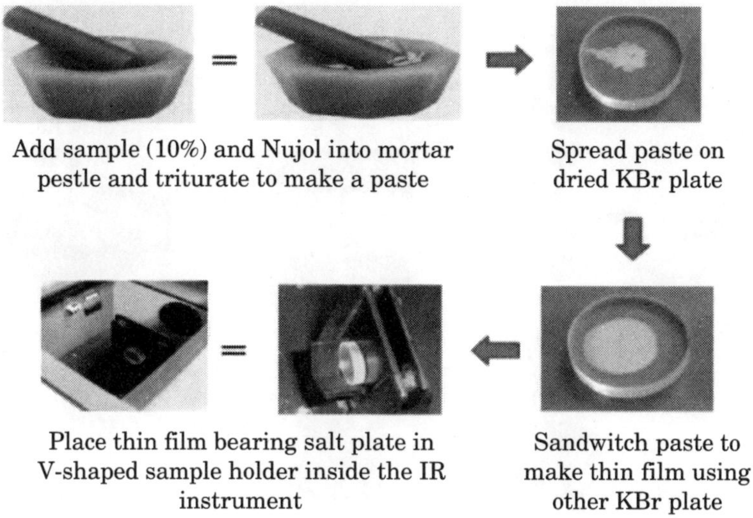

Figure 2.20: Method for solids sampling as Nujol mull for IR analysis.

Merits: This method holds good for crystalline compounds which are difficult to be ground with KBr.

Demerits: Nujol is transparent throughout the IR region, but shows absorption maxima on 2915 cm^{-1}, 1462 cm^{-1}, 1377 cm^{-1} and 720 cm^{-1}.

(*d*) **Solids Sampling As Pressed Pellet:** In this method of solids sampling as a pressed pellet, firstly a solid sample of about 1 – 3 mg is grounded and mixed thoroughly with about 350 mg of ground KBr. Nextly, the mixture is transferred to a die that possess a barrel diameter of around 13 mm. This is then placed in a suitable press and pressed (evacuation is optional) at around 25,000 psi for 1 – 2 minutes. Re-crystallization of the KBr results in a clear glassy disk which can be analyzed by transmission. The Figure 2.21, shows sequential steps for solids sampling as pressed pellet for analysis.

Merits

- Resolution is better than mull technique.
- Can be used for quantitative analysis.
- KBr pellets can be stored for longer time.

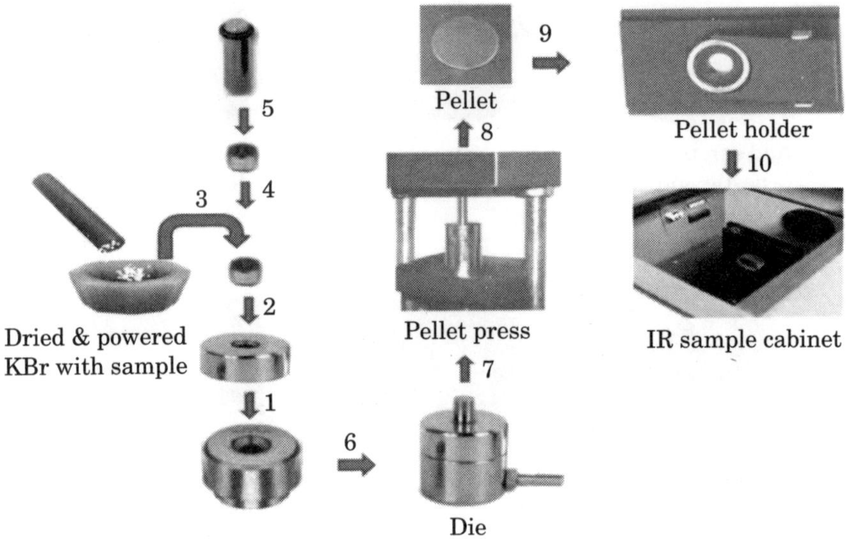

Figure 2.21: Sequential steps for solids sampling as pressed pellet for analysis.

Demerits

- High pressure in pellet formation may cause polymorphic change in crystallinity of sample, which results in astonishing complex spectrum.
- Usually shows a band near 3450 cm^{-1} due to moisture (O–H stretching).
- Not a suitable method for polymers.

B. Liquid Sample Handling

As samples that are in liquid state at room temperature do not require any extra preparation, hence they are directly subjected into the rectangular cells that are made up of metal halides such as NaCl, KBr, and ThBr etc. In liquid samples, commonly a thin-film cell is used. This thin film cell consists of two salt plates with a very thin space in between them. Two channels allow liquid to be injected and then subsequently removed. The windows on these cells are made by a wide variety of IR optical materials. For example in case of water-based solutions CaF_2 is most commonly used (as insoluble in water). Cleaning of cells includes removal of solutions by firstly flushing the cell with fresh solvent and nextly removing the solvent by syringe. A precaution is required to be taken is not to blow air or nitrogen through the ports as this can cause mechanical deformation of salt window on high pressure. It is important to have a constant path length for the radiation to travel through and also the sample to be homogeneous. For large volume of liquid sample specialized couvettes can be used. The Figure 2.22, shows various components of liquid sample holder.

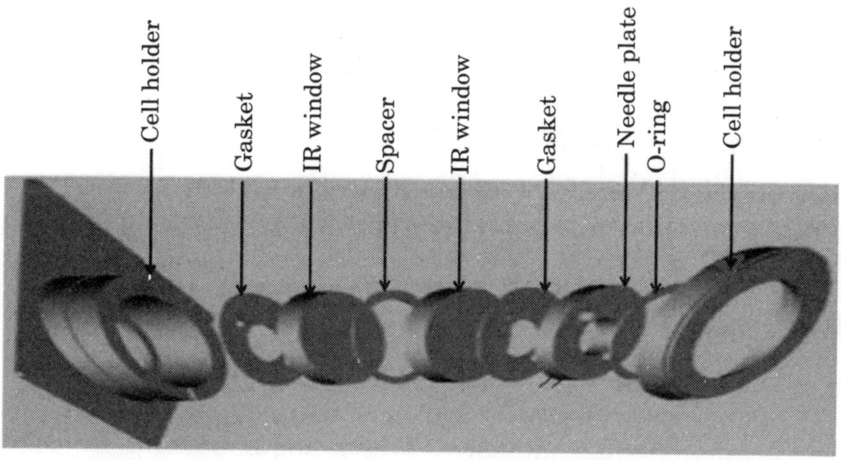

Figure 2.22: Various components of liquid sample holder.

C. Gas Sample Handling

The materials for construction for gas sample cells are similar to liquid sample cells such as NaCl, KBr, and ThBr etc. The cells for gas samples are relatively very large in size *i.e.*, from 5 – 10 cm. The requirements for sample gas IR analysis firstly includes that they should be nonreactive to cell windows or reflecting surfaces. Secondly moisture should be avoided. The Figure 2.23, shows the image of gas sample holder.

Figure 2.23: Image of a gas sample holder.

2.3.3 Types of IR Instruments

Commonly three types of spectrometers are used in the laboratories viz.

- Dispersive IR Instruments
- Non-dispersive IR Instruments
- FT-IR Instruments

2.3.3.1 *Dispersive IR Instruments*

Dispersive infrared instruments are sometimes called as the grating or scanning spectrometers that generates a frequency domain IR plots.

Key Concepts

- In dispersive spectrometers (*most commonly double beam*) IR radiations produced by source are divided into two parallel IR beams of equal intensity. This is made possible by using a rotating mirror, so as to pass radiation through one sample and one reference.
- The two beams after sample and reference are allowed to pass and disperse through the monochromators such as metal halide prisms or plane reflectance gratings (newer instruments) to pass into the detector.
- The detector senses the ratios of intensities of radiations that are emerged through the sample and reference.
- After amplification of IR signals through detector, the IR signals are recorded as an IR spectrum of sample with the help of a recorder pen.

- The use of microprocessor makes the instruments of high resolution and instantaneous.
- While recording the IR characteristics of any given sample, it is important to plot the frequency of IR incident (*i.e.*, in cm^{-1}) against the transmitted IR (as percentage transmittance) through the given sample.
- Although single beam dispersive instrument can also produce high resolution, accurate and instantaneous results when used with a microprocessor.
- As in dispersive instrument IR spectrum for any given sample is recorded in frequency domain, hence dispersive instruments are called frequency domain IR recorders. Figure 2.24, shows optical schematic diagram of dispersive IR instrument.

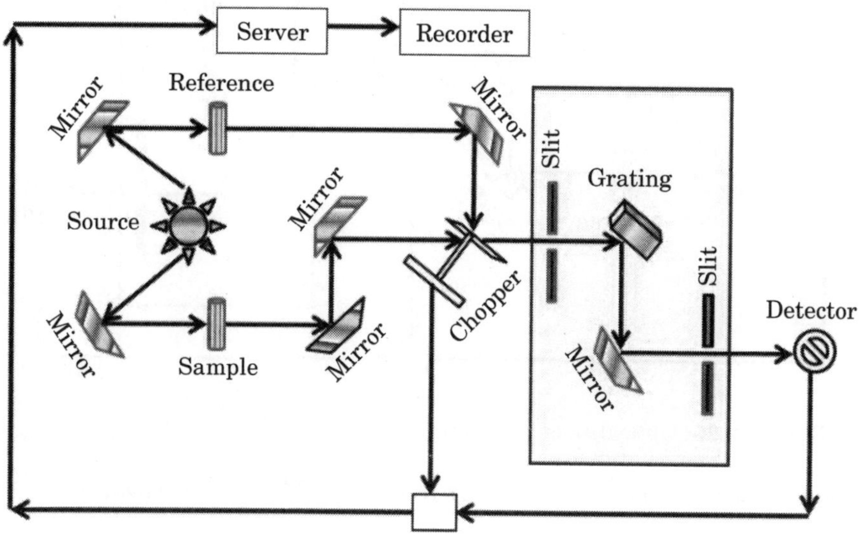

Figure 2.24: Optical schematic diagram of dispersive IR spectrometer.

2.3.3.2 *Non-dispersive IR Instruments*

Non-dispersive IR spectrometers are used in conditions where same determinations are to be done in a repetitive manner.

A non dispersive IR instrument commonly makes use of filters to isolate the desired wavelength of IR radiations.

A simple non dispersive IR instrument is comprised of at least three circular and variable interference filters (selects different wavelength and covers a range of 2.5-15 μm) that are mounted on the shaft of a potentiometer of high resolution power.

These instruments are used for selective compounds online process control, such that these can be connected to liquid chromatograph to monitor effluents. The Figure 2.25, shows the optical schematic diagram of non dispersive IR instrument.

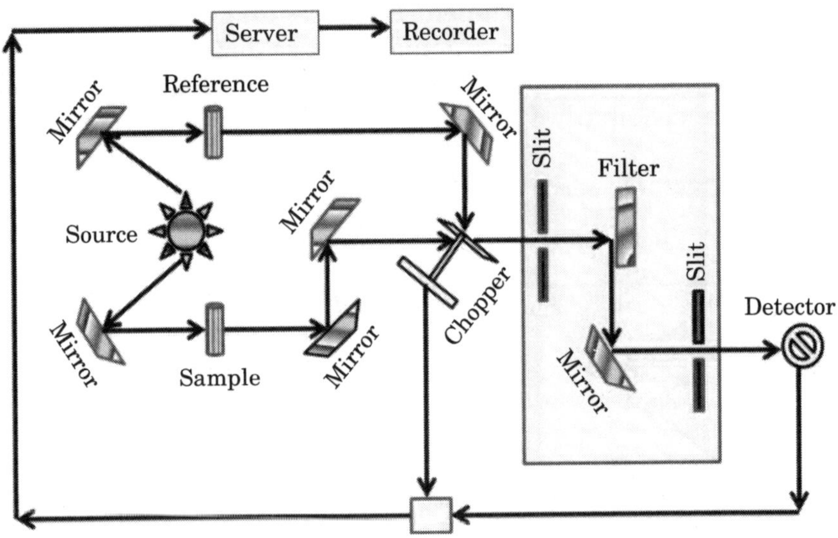

Figure 2.25: Optical schematic diagram of nondispersive IR spectrometer.

2.3.3.3 *Fourier Transformed Infra Red (FT-IR) Instruments*

Commercially, Fourier transform infrared (FT-IR) spectrometers were introduced in 1960's. But because of the high cost during that time, the FTIR's were limited to area of advance research only. With the advancements in technology and instruments, the cost of these instruments was reduced which enhanced the FT-IR capabilities. Today, FT-IR spectrometer is the standard for identification of organic compounds.

(*a*) **Sampling in FTIR:** As FTIR instrument is very sensitive, most convenient and satisfactory method, so it involves formulation of solution of the sample (in chloroform, ether, dichloromethane, or $CDCl_3$ etc.) and evaporation on a KBr salt plate to form a thin film. Thereby on the remaining film IR spectra is generated. This method provides excellent spectra with flat baseline. The sample may alternatively be prepared as a Nujol mull.

(b) Interferometer: *FT-IR instrument works on a system called interferometer to collect a spectrum through an interferogram (pattern of optical pathway). Interferogram is a complex signal. As interferogram generates time domain curve, and requirement is for frequency domain curve, hence a mathematical operation called Fourier transformation is applied to convert a time domain curve into a frequency domain curve.* The components of an interferometer are

- A source
- A beam splitter
- Two mirrors
- A laser, and
- A detector

The series of events in an interferometer can be best understood with the optical schematic diagram of Michelson interferometer given in Figure 2.26.

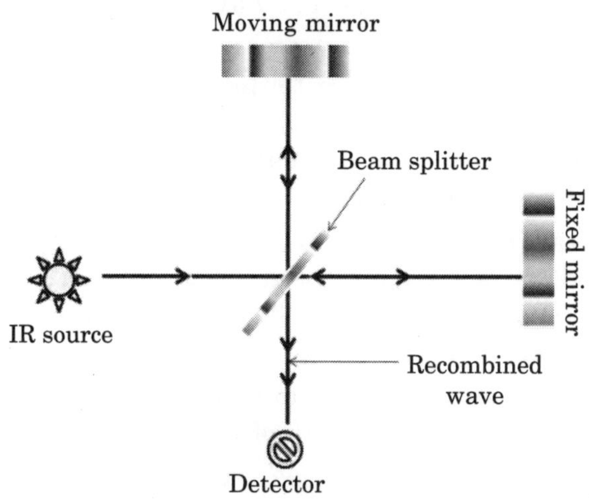

Figure 2.26: Optical schematic diagram of Michelson interferometer.

Sequential Events Followed in FTIR

- Inside the interferometer, firstly, energy travels from a source to the beam splitter to split into two beams. By this one beam is transmitted to a moving mirror and the other beam is reflected to a fixed mirror.
- The moving mirror moves back and forth with a constant velocity. This velocity is timed with very precise laser wavelength in the system that also acts as an internal wavelength calibration.
- Two beams are reflected from respective mirrors and recombined at beam splitter.
- The beam from the moving mirror traverse a different distance than the beam from the fixed mirror. Hence, when the beams are combined together interference occurs, which is attributed to recombination of some of the constructive and destructive wavelengths. This interference pattern is called an interferogram.
- This interferogram then travels from the beam splitter to the sample, where some energy is absorbed and some is transmitted.
- The transmitted energy reaches to the detector. The detector reads information about every wavelength in the infrared range simultaneously.
- The detector signal is sent to the computer, and an algorithm called a Fourier transform is performed on the interferogram to convert it into a single beam spectrum which provides the infrared spectrum.
- Apart from it a reference or "background" single beam is also collected without a sample and the sample single beam is ratio-ed to the background single beam to produce a transmittance or '%T' spectrum.
- This transmittance spectrum can be converted to absorbance by taking the negative log10 of the data points.
- The X-axis of the FT-IR spectrum is commonly recorded in "wavenumbers", or cm^{-1}.

The computer interfaced FTIR spectrometers works in single beam mode. Before generation of an IR spectrum of any given compound, FTIR firstly scans the background (*such as IR active gases, CO_2, moisture etc.*) thereby generates a background Fourier transformed spectrum. Nextly, the sample is subjected for FTIR scanning. Now the spectrum contains both background and pure sample spectrum. The computer automatically removes the background IR bands from sample spectrum.

Merits of FTIR

- *Speed:* The IR spectrum can be generated in less than a second.
- *Accuracy and Precision:* In FT-IR spectrometer laser is used for three functions. Firstly to control the velocity of the moving mirror; Secondly, to time the collection of data points throughout the mirror stroke length for each scan; Thirdly, as a source for wavelength calibration within the instrument. The laser wavelength is a constant value, and the X-axis data points of the FT-IR spectrum are automatically referenced to this known value to maintain internal precision and accuracy of the wavelength positions. Spectra collected with an FT-IR spectrometer can be compared after five minutes or five years also. Accuracy and precision in infrared spectra are much higher in FT-IR instrument.
- *Throughput:* As FT-IR instrument does not use a slit and there occurs fewer mirror surfaces, so there are less reflection losses. Therefore more energy reaches to sample and detector. Due to this signal-to-noise ratio of FT-IR is more than the dispersive instrument, as a result sensitivity of small peaks in FTIR is greater. Hence, FT-IR spectrum is much detailed, highly resolved and clearer than the dispersive spectrum of the same sample.
- *Multiplicity:* FT-IR interferometer does not separate energy into individual frequencies. Each point in the interferogram contains information from each wavelength of light being measured. Every stroke of the moving mirror in the interferometer equals one scan of the entire infrared spectrum,

and individual scans can be combined to give better representation of the actual absorbance of the sample. The FT-IR advantage is that many scans can be completed and combined on an FT-IR in a shorter time than one scan on a dispersive instrument. The multiplex advantage results in faster data collection of an FT-IR spectrum.

- *Constant Spectral Resolution:* In an FTIR spectrometer, the resolution of the measured spectrum is the same for all frequencies, not varied throughout the spectrum as is often true with dispersive instruments.

2.3.4 IR Reflectance Techniques

There exists a variety of samples where IR transmission is difficult or impossible, and to overcome such problems reflectance techniques are used. The reflectance techniques replaces the constant path transmission cells and salt plates. These are used for the analysis of liquid, semi-liquid, films, pastes, pliable solids and fine powders.

Commonly in IR experiment two types of reflectance techniques are used such as:

- Internal Reflectance and
- External Reflectance

These two techniques generally involves direct reflection of IR beam from the sample surface.

2.3.4.1 *Internal Reflectance*

Internal reflectance technique further involves two techniques namely:

a. ATR (Attenuated total reflectance) and
b. MIR (Multiple internal reflectance).

(*a*) **Attenuated Total Reflectance (ATR):** ATR uses the property of total internal reflection known as the evanescent wave. This concept of total internal reflection using ATR, can be best understood from the diagram of an attenuated total reflectance cell given in Figure 2.27.

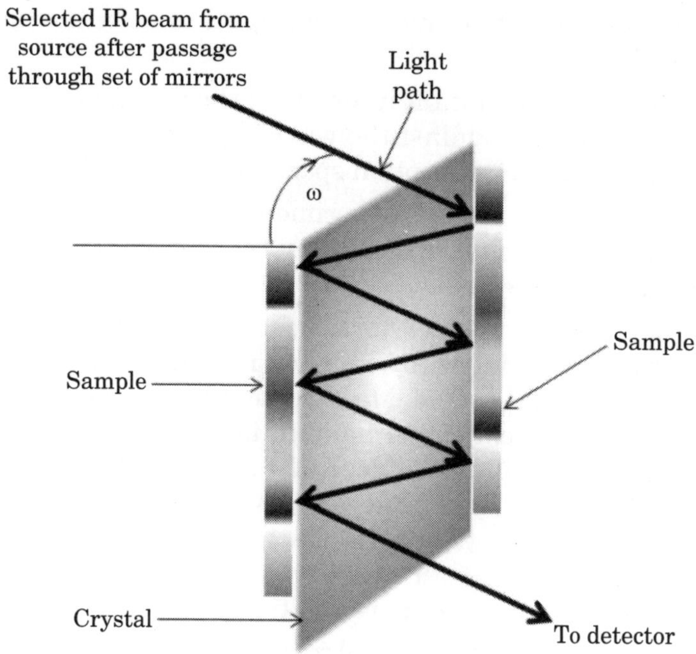

Figure 2.27: An Attenuated Total Reflectance Cell.

Keys Concepts

- In ATR technique, an IR beam after passage through a set of mirrors, is allowed to pass through one end of a crystal.
- The crystal causes complete internal reflection of incident IR light (several times) till the IR light is exit through its other end. This occurs only when the angle of incidence is greater than the critical angle, which is a function of refractive indices of two surfaces.
- The internal reflections inside the crystal create an evanescent wave, which extends further into the sample which is placed on the crystal surface.
- As evanescent wave decays rapidly with distance from the surface, therefore the sample should always be in direct contact of crystal surface.
- When the beam strikes the crystal surface it can penetrate the surface slightly and, if the substance absorbs light at

the wavelength of the incident light, some of the light will be absorbed.

- Thereby the resultant intensity of the attenuated radiation can be plotted against the wavelength of the incident light to generate an absorption spectrum.
- Hence, in a typical IR experiment, a part of the evanescent radiation which is absorbed by the sample produces an IR absorption spectrum.
- The evanescent effect works best if the crystal is made of an optical material with a higher refractive index than the sample being studied. Most commonly zinc selenide (RI 2.4, 20000 – 500 cm^{-1}), germanium iodide (RI 4.0, 5000 – 550 cm^{-1}) and thallium iodide (RI 2.4, 17000 – 250 cm^{-1}) are the typical materials that are used as crystals for total reflectance spectroscopy. The speciality of all these three materials is that all these typical materials possess high refractive indices and are relatively insoluble in water.
- ATR most commonly employs prisms for reflectance.
- With a liquid sample, pouring a shallow amount over the surface of the crystal is sufficient. If it is a solid sample, it is pressed into direct contact with the crystal. Because the evanescent wave into the solid sample improves with intimate contact, solid samples are usually clamped against the ATR crystal so that trapped air does not distort the results.

The depth of penetration (dp) can be represented by following equation

$$d_p = \frac{\lambda}{2.\pi.n_1\left[\sin\omega - \left(\frac{n_1}{n_2}\right)^2\right]^{0.5}} \qquad ...(2.9)$$

Where, λ = Wavelength

n_1 & n_2 = Refractive Indices of the sample & crystal respectively

ω = Angle of incident radiation

Hence, depth of penetration of IR radiation depends upon wavelength, refractive indices and angle of incident radiation

(*b*) Multiple Internal Reflectance (MIR): Multiple internal reflectance techniques produce more intense spectra as a result of multiple reflections. In comparison to ATR that usually employs a prism, MIR techniques employ specially shaped crystals that allow multiple reflections. Such a crystal is shown in Figure 2.28. Such crystals may produce up to 25 multiple reflections.

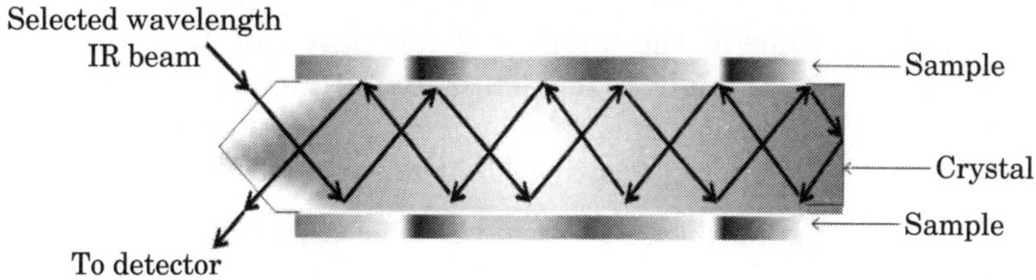

Figure 2.28: A Multiple Internal Reflectance crystal.

2.3.4.2 *External Reflectance Techniques*

When an IR radiation is focused on the external surface of a sample it may cause two types of reflectance to occur such as specular reflectance and diffuse reflectance. Commonly both reflectances can be used to produce IR spectra from a sample. To have effective and useful reflectance, the surface should be made highly reflective or should be attached to a reflective supporting surface. This external reflectance is used for the examination of surface coatings such as paints polymer and metal surface coatings. Commonly external reflectance techniques involve two methods such as:

a. Specular Reflectance Technique (SRT)
b. Diffuse Reflectance Technique (DRT)

(*a*) Specular Reflectance Technique (SRT): *When the angle of reflection of reflected light from the surface of sample becomes equals to angle of incidence then it mediates specular reflectance.* The amount of IR reflection in specular reflectance depends upon the angle of incidence, the refractive index of sample, the surface roughness and the adsorption properties of the sample.

Specular reflectance is a non-destructive method of surface measurement that uses a mirror-like reflection from the shiny surface of the sample. Specular reflectance is particularly used in the study of surface coatings, such as surface treated metals, paints and polymers etc. The accessories for specular range of reflectance include both fixed and variable incidence angle accessories. To increase the sensitivity of specular reflectance experiment the path length can be increased through thin coatings using grazing angles of incidence (85°). Coatings of higher thickness can be studied using angles of typically 30°. Also monolayer films at the liquid—air interface can also be studied using a grazing angle of incidence. Band intensities of specular reflectance depends upon the type and degree of polarization. Figure 2.29, shows the optical diagram of specular reflectance.

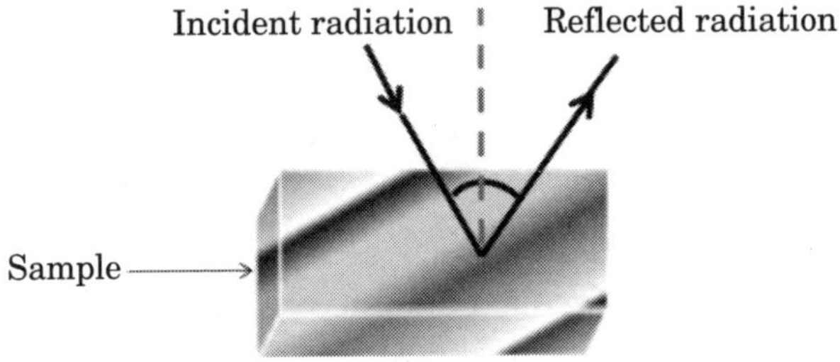

Figure 2.29: Optical diagram of specular reflectance.

(*b*) Diffuse Reflectance Technique (DRT): *The energy which penetrates one or more particles is reflected in all directions. This component is called diffuse reflectance.* On a rough surface, such as a powder, specularly reflected light is a minor contributor to the overall signal. Specific accessories are optimized to increase collection of the diffuse reflectance and decrease the specular components. Figure 2.30, shows the optical diagram of diffused reflectance.

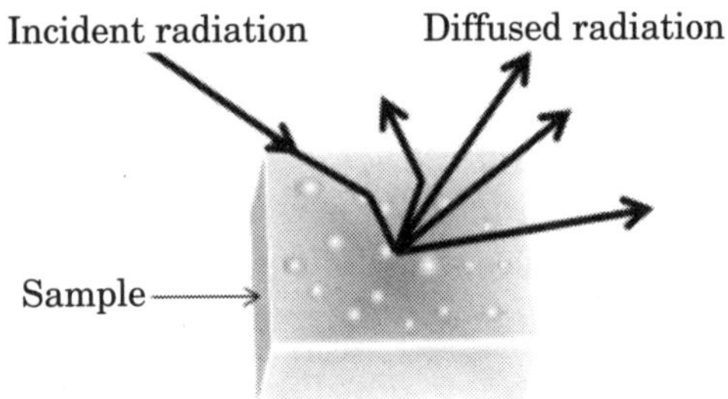

Figure 2.30: Optical diagram of diffused reflectance.

2.4 APPLICATIONS OF IR SPECTROSCOPY

IR spectroscopy has created its own place among all spectroscopic techniques in the field of analysis and science. Below mentioned are some of the important applications of IR spectroscopy.

- ***In Establishment of Structures***: Analysis of peaks corresponding to stretching and bending vibrations of different functional groups in practical IR region *i.e.*, from 4000 cm^{-1} to 400 cm^{-1} determines and confirms the structure of compounds.
- ***In Quantitative Analysis***: By making use of baseline technique, that involves comparison of area of peaks of standard and sample as well, the quantity of any given analyte can be easily determined. Quantitative determination of any given sample can be done either in pure form or in a mixture form. Concentration can be determined by using Beer-lamberts law also.
- ***In Hydrogen Bonding Studies***: By commencing IR variable spectroscopy *i.e.*, by undergoing the destruction of the hydrogen bonding, the nature of hydrogen bond present in a molecule can be easily determined.
- ***In Detection of Impurities***: By comparing or matching the IR spectrum of a pharmaceutical analyte with standard (pure compound), the additional peaks if any can be identified. Thereby presence of impurities can be determined.
- ***In Polymer Science***: IR spectroscopy is very well suited for qualitative and quantitative analysis of polymers as individual and also in mixture form as well.
- ***In Identification of Pharmaceutical Substances***: By comparing or matching the IR spectrum of a pharmaceutical analyte with standard, it becomes easy to identify any pharmaceutical compound. Identical IR spectra supports to identification and confirmation of analyte. Spectral matching is a common practice in industry for confirmation of pharmaceutical substances.
- ***Cis-Trans Isomers Identification***: When the compound under investigation is di-substituted around the C = C. A trans C = C gives a strong C–H bending band near 970 cm^{-1} while a cis C = C gives a wedge shaped C–H bending band near 700

cm^{-1}. Apart from it is also possible to determine the ratio of Cis-trans isomers in a mixture of compound.

- ***Recent Applications:*** Recently IR spectroscopy has been found to have best use in imaging of human hairs; atmosphere and environment; combustion & fire detection; measuring electric and magnetic Fields; spectroscopy and technique development; process monitoring and industrial hygiene; and also forensic investigations.

2.5 INTERPRETATION OF IR SPECTRA

This section describes the general IR absorptions of various classes of compounds, along with their IR spectra and significant group absorption values of some relevant examples.

2.5.1 Alkanes (Paraffins)—Straight Chain Alkanes

IR Spectral Characteristics of Straight Chain Alkanes

In the IR spectrum of straight chain alkanes, vibrations for stretching and bending of C–H and C–C bonds are identified. Some characteristic vibrational bands that are observed for straight chain alkanes are as follows.

- ***C–C bond stretching:*** Exhibits least important weak band between 400 – 800 cm^{-1}.
- ***C–C bond bending vibrations:*** Exhibits least important bands below 500 cm^{-1}.
- ***C–H bond stretching for methyl groups:*** Exhibits two distinct bands at 2962 cm^{-1} (asymmetrical stretch) and 2872 cm^{-1} (symmetrical stretch). Presence of several methyl groups makes these bands to appear strong in IR spectrum.
- ***C–H bond stretching for methylene groups:*** Exhibits two distinct bands at 2926 cm^{-1} (asymmetrical stretch) and 2853 cm^{-1} (symmetrical stretch).
- ***C–H bond bending for methyl groups:*** Exhibits two distinct bands near 1450 cm^{-1} (asymmetrical bending) and 1375 cm^{-1} (symmetrical stretch). Presence of several methyl groups makes these bands to appear strong in IR spectrum.

- ***C–H bond bending for methylene groups:*** Exhibits four distinct bands near 1465 cm^{-1} (scissoring), near 720 cm^{-1} (rocking), and between 1350 – 1150 cm^{-1} (wagging and twisting).

Example: ***Octane*** (IR spectrum given in Figure 2.31)

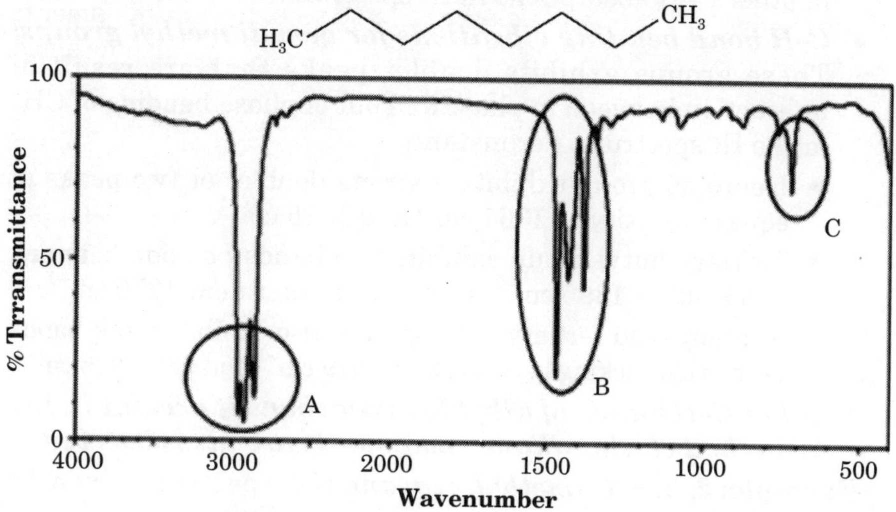

Figure 2.31: IR spectrum of octane.

Spectral Interpretation

Region A : 2971 cm^{-1} (C–H asymmetrical stretch of CH_3), 2863 cm^{-1} (C–H symmetrical stretch of CH_3), 2924 cm^{-1} (C–H asymmetrical stretch of CH_2), 2853 cm^{-1} (C–H symmetrical stretch of CH_2).

Region B : 1470 cm^{-1} (C–H scissoring of CH_2), 1452 cm^{-1} (C–H asymmetrical bending of CH_3), 1383 cm^{-1} (C–H scissoring of CH_3).

Region C : 726 cm^{-1} (C–H rocking of CH_2) a long chain band.

2.5.2 Branched Chain Alkanes

IR Spectral Characteristics of Branched Chain Alkanes

In the IR spectrum of branched chain alkanes, generally vibrations for stretching and bending of C–H bonds are identified. Some characteristic vibrational bands that are observed for branched chain alkanes are as follows:

- ***C–H bond stretching vibrations for tertiary groups:*** Exhibits a distinct weak band at 2890 cm^{-1}. This band is absent in other C–H absorptions in IR spectrum.
- ***C–H bond bending vibrations for gem di methyl groups:*** These groups exhibits doublet peaks that are result of interaction between in phase and out of phase bending of CH_3 in the IR spectrum. For instance
    - Isopropyl group exhibits a strong doublet of two peaks of equal intensity at 1365 cm^{-1} and 1385 cm^{-1}.
    - Tertiary butyl group exhibits two bands *i.e.*, one between 1385 cm^{-1} – 1395 cm^{-1}, whereas another near 1370 cm^{-1}.
    - Isopropyl and tertiary butyl groups also exhibits weak bands for methyl rocking between 922 – 919 cm^{-1} and 932 – 936 cm^{-1}.
- ***Other C–H bands of alkyl hydrocarbons if present in the branched chain alkane compounds are also observed.***

Example: *2, 2, 4-Trimethyl pentane* (IR spectrum given in Figure 2.32).

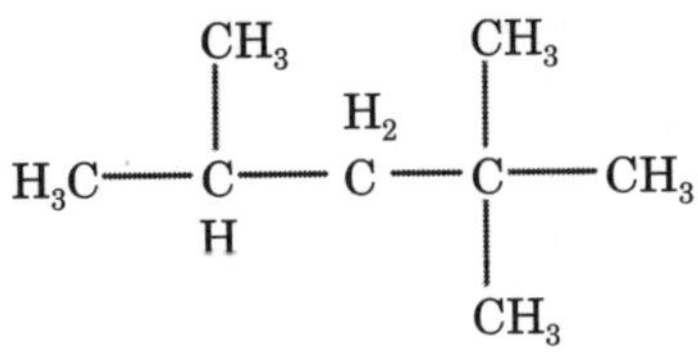

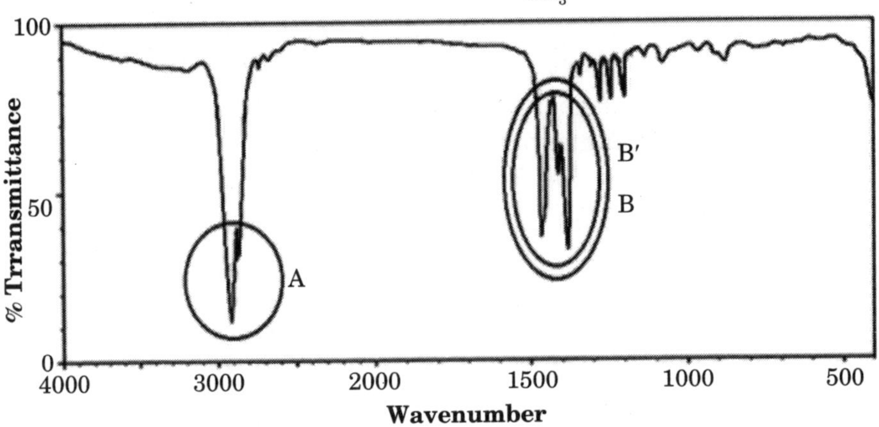

Figure 2.32: IR spectrum of 2, 2, 4-trimethyl pentane.

Spectral Interpretation

Region A : 2971 cm^{-1} (C–H asymmetrical stretch of CH_3), 2863 cm^{-1} (C–H symmetrical stretch of CH_3), 2924 cm^{-1} (C–H asymmetrical stretch of CH_2), and 2853 cm^{-1} (C–H symmetrical CH_2 stretch).

Region B : 1470 cm^{-1} (C–H scissoring of CH_2), 1452 cm^{-1} (C–H asymmetrical bending of CH_3), 1383 cm^{-1} (C–H rocking of CH_3).

Region B′ : 1340 – 1400 cm^{-1} (overlapping doublets for isopropyl and tertiary butyl groups).

2.5.3 Cyclic Alkanes

IR Spectral Characteristics of Cyclic Alkanes

In the IR spectrum of cyclic alkanes, generally vibrations for stretching and bending of C–H are identified. Some characteristic vibrational bands that are observed for cyclic alkanes are as follows:

- ***C–H bond stretching vibrations for methylene group in unstrained ring:*** Exhibits two distinct bands at 2926 cm^{-1} (asymmetrical stretch) and 2853 cm^{-1} (symmetrical stretch).
- ***C–H bond stretching vibrations for methylene group in strained ring:*** As ring strain increases the stretching absorption frequency so exhibits absorption band between 3100 – 2990 cm^{-1}.
- ***C–H bond bending vibrations for methylene groups:*** Ring strain decreases the bending absorption frequency so in case of cyclohexane exhibits absorption bands near 1452 cm^{-1} due to scissoring (which is at 1468 cm^{-1} in normal hexane).
- ***Other C–H bands of alkyl hydrocarbons if present in the cyclo alkane compounds are also observed.***

Example: ***Cyclohexane*** (IR spectrum given in Figure 2.33)

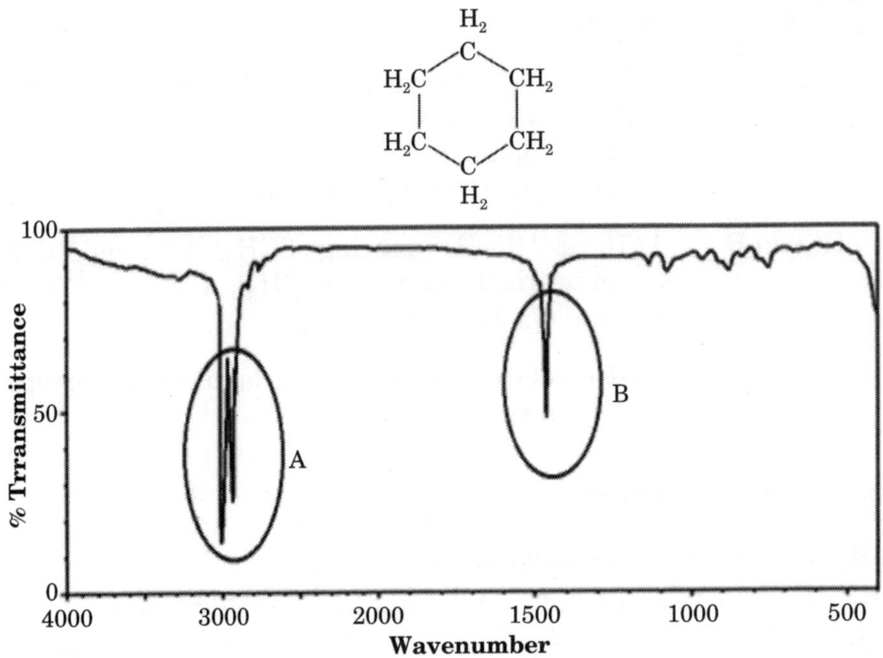

Figure 2.33: IR spectrum of cyclohexane.

Spectral Interpretation

Region A : 2926 cm^{-1} (C–H asymmetrical stretch of CH_2), and 2876 cm^{-1} (C–H symmetrical CH_2 stretch).

Region B : 1452 cm^{-1} (C–H scissoring of CH_2).

2.5.4 Alkenes (Olefins)

IR Spectral Characteristics of Alkenes

In the IR spectrum of alkenes, generally stretching vibrations (for C=C, alkenyl C–H), bending vibrations (for alkenyl C–H) and other alkyl groups vibrations are identified. Some characteristic bands that are observed for alkenes are as follows.

- ***C=C bond stretching vibrations in unconjugated molecule***: Exhibits medium to weak band between 1667 – 1640 cm^{-1}. Some instances of C=C band shifts are

- *Mono alkyl substitution in C=C:* Exhibits a medium band at 1640 cm^{-1}.
- *Di tri & tetra alkyl substitution in trans C=C:* Exhibits a band near 1670 cm^{-1}.
- *Di alkyl substitution in cis C=C, and vinyl C=C:* Exhibits a band near 1650 cm^{-1}.
- *Halogenation in C=C:* For example $CH = CF_2$ and $CF = CF_2$ exhibits a band near 1754 cm^{-1} and 1786 cm^{-1} respectively.
- *Reduction in angle of cyclized C=C:* Causes increase in frequency band value. For example cyclobutene and cyclopropene absorbs at 1566 cm^{-1} and 1641 cm^{-1}.
- *Alkyl substitution in cycloalkene:* Causes increase in absorption frequency. For example methyl cyclobutene absorbs at 1641 cm^{-1}.
- *Exo cyclic alkene:* Absorption increases with decrease in ring size. For example methylene cyclohexane and methylene cyclopropane absorbs at 1650 cm^{-1} and 1781 cm^{-1}.
- *Unsymmetrical conjugated dienes:* (For example 1, 3-pentadiene) exhibits two bands at 1650 cm^{-1} and 1600 cm^{-1}. Whereas symmetrical conjugated diene (for example 1, 3-butadiene) shows one band at 1600 cm^{-1}.
- *Cumulated alkene:* Exhibits band between 200 – 1900 cm^{-1}.

- ***Alkenyl = C–H stretching vibrations:*** Exhibits bands above 3000 cm^{-1}.
- ***Alkenyl = C–H bending vibrations:*** Exhibits a band near 1416 cm^{-1} (due to scissoring of CH_2) and between 1000 – 650 cm^{-1} (due to out of plane bending).
- ***Other C–H bands of alkyl hydrocarbons present in the alkene compounds are also observed.***

Example: ***1–Octene*** (IR spectrum given in Figure 2.34).

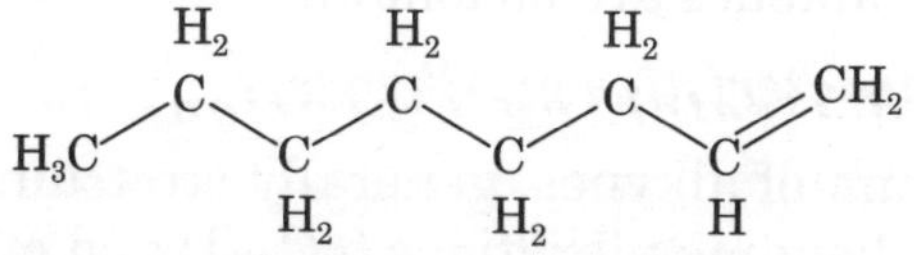

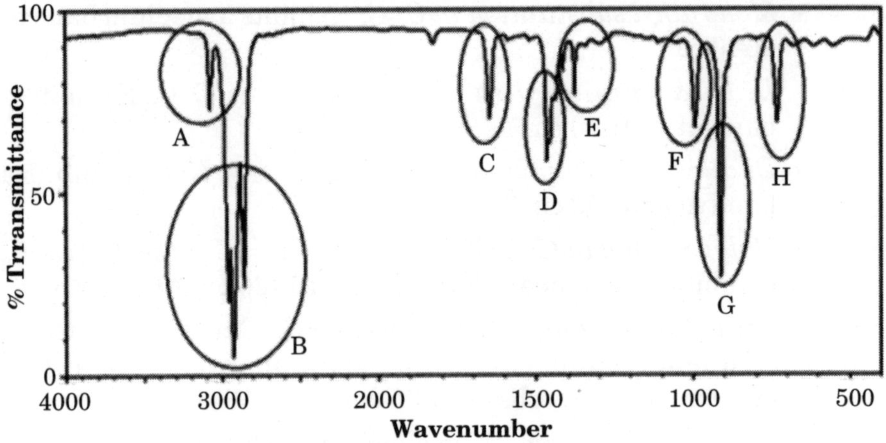

Figure 2.34: IR spectrum of 1-octene.

Spectral Interpretation

Region A : 3086 cm^{-1} (=C–H Stretch).

Region B : 2966 cm^{-1} (C–H asymmetrical stretch of CH_3), 2863 cm^{-1} (C–H symmetrical stretch of CH_3), 2924 cm^{-1} (C–H asymmetrical stretch of CH_2), 2853 cm^{-1} (C–H symmetrical stretch of CH_2).

Region C : 1642 cm^{-1} (C=C stretch).

Region D : 1465 cm^{-1} (C–H methylene scissoring).

Region E : 1378 cm^{-1} (C–H methyl bending).

Region F : 991 cm^{-1} (=C–H scissoring).

Region G : 917 cm^{-1} (=C–H bending).

Region H : 726 cm^{-1} (Methylene rocking a long chain band).

2.5.5 Alkynes (Acetylenes)

IR Spectral Characteristics of Alkynes

In the IR spectrum of alkynes, generally stretching vibrations (for C≡C and ≡C–H), bending vibrations (≡C–H) and other alkyl groups vibrations are identified. Some characteristic vibrational bands that are observed for alkynes are as follows:

- ***C≡C bond stretching vibrations:*** Exhibits a weak band between 2260 – 2100 cm^{-1}. Some instances of C≡C band shifts are
  - *Mono substitution in C≡C:* Exhibits weak band between 2140 – 2100 cm^{-1}.
  - *Substitution with two different groups in C≡C:* Exhibits a band near 2260 – 2190 cm^{-1}.
- ***≡C–H bond stretching vibrations:*** Exhibits a strong band (narrower than O–H and N–H band) between 3333 – 3267 cm^{-1}.
- ***≡C–H bond bending vibrations:*** Exhibits a strong band between 700 – 610 cm^{-1} and a weak overtone band between 1300 – 1220 cm^{-1}.
- ***Other C–H bands of alkyl hydrocarbons if present in the alkyne compounds are also observed.***

Example: ***1–Hexyne*** (IR spectrum given in Figure 2.35).

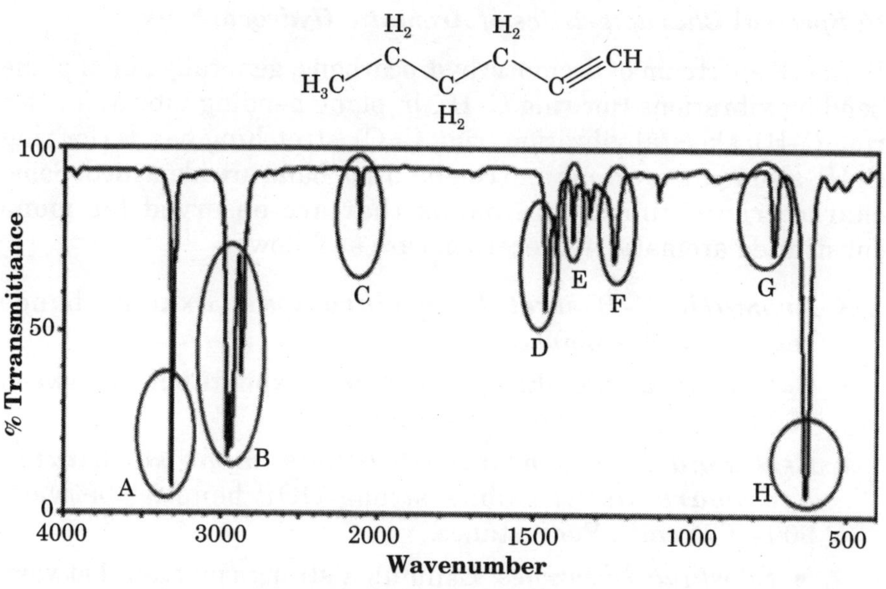

Figure 2.35: IR spectrum of 1-hexyne.

Spectral Interpretation

Region A : 3324 cm^{-1} (≡C–H Stretch).

Region B : 2971 cm^{-1} (C–H asymmetrical stretch of CH_3), 2879 cm^{-1} (C–H symmetrical stretch of CH_3), 2926 cm^{-1} (C–H asymmetrical stretch of CH_2), 2857 cm^{-1} (C–H symmetrical stretch of CH_2).

Region C : 2150 cm^{-1} (C≡C stretching).

Region D : 1465 cm^{-1} (C–H scissoring).

Region E : 1383 cm^{-1} (C–H methyl bending).

Region F : 1254 cm^{-1} (≡C–H bending overtone band).

Region G : 726 cm^{-1} (Methylene rocking a long chain band).

Region H : 630 cm^{-1} (≡C–H bending fundamental band).

2.5.6 Aromatic Hydrocarbons

IR Spectral Characteristics of Aromatic Hydrocarbons

In the IR spectrum of aromatic hydrocarbons, generally out of plane bending vibrations (for ring C–H), in plane bending vibrations (for ring C–H), skeletal vibration (ring C=C), stretching bands (for ring C–H), also weak overtone and combination band are identified. Some characteristic vibrational bands that are observed for mono substituted aromatic hydrocarbons are as follows:

- ***Aromatic C–H stretching vibrations:*** Exhibits bands between 3100 – 3000 cm^{-1}.
- ***C=C skeletal stretching vibrations:*** Exhibits bands between 1600 – 1585 cm^{-1} and 1500 – 1400 cm^{-1}.
- ***C–H bond OOP bending vibrations (mono substituted hydrocarbons):*** Exhibits strong OOP bending between 900 – 675 cm^{-1}. For instance,
  - *Substituted benzenes:* Exhibits a strong oop band between 600 – 400 cm^{-1}.

- ***C–H bond OOP bending vibrations (poly nuclear hydrocarbons:*** Exhibits strong out of plane bending between 900 – 675 cm^{-1}. These can be easily differentiated from mono substituted one as these exhibits more number bands than for example naphthalene exhibits three strong band between 900 – 675 cm^{-1} due to out of plane bending.
- ***Overtone and combination bands between 2000 – 1650 cm^{-1}.***
- ***C–H bond in plane bending vibrations:*** Exhibits in plane bending between 1300 – 1000 cm^{-1}.
- ***In case of alkylated aromatic compounds, other C–H bands of alkyl hydrocarbons if present in the aromatic compounds are also observed.***

Example: *Toluene* (IR spectrum given in Figure 2.36).

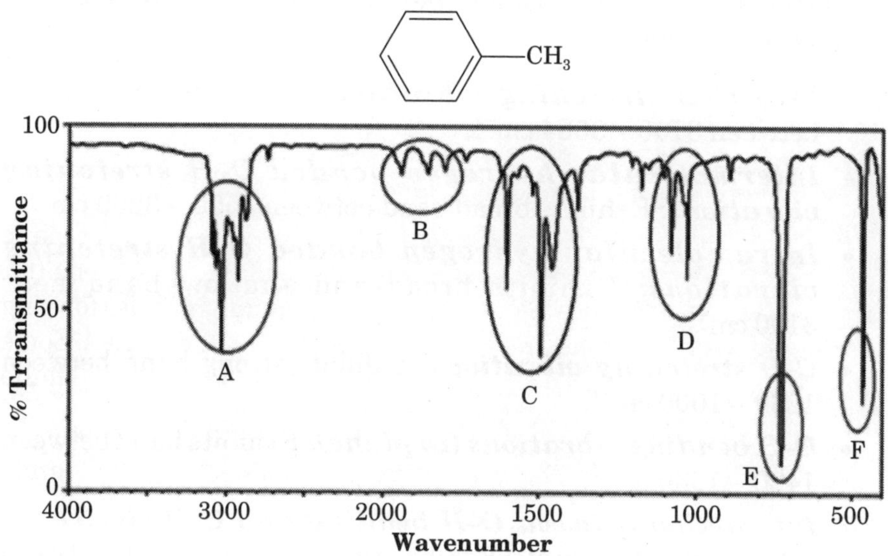

Figure 2.36: IR spectrum of toluene.

Spectral Interpretation

Region A : 3024 cm^{-1} (Aromatic C–H Stretch), 2965 cm^{-1} (C–H asymmetrical stretch of CH_3), and 2870 cm^{-1} (C–H symmetrical stretch of CH_3).

Region B : 2000 – 1650 cm^{-1} (overtone and combination bands).

Region C : 1614 cm^{-1}, 1506 cm^{-1} and 1465 cm^{-1} (C–C ring stretching).

Region D : 1086 cm^{-1} and 1035 cm^{-1} (in plane C–H bending).

Region E : 735 cm^{-1} (Out of plane C–H bending).

Region F : 438 cm^{-1} (Out of plane ring bending).

2.5.7 Alcohols and Phenols

IR Spectral Characteristics of Alcohols and Phenols

In the IR spectrum of alcohols and phenols-generally O–H stretching, C–O stretching and O–H bending bands are identified. Some characteristic vibrational bands that are observed for alcohols and phenols are as follows:

- ***Free O–H stretching vibrations:*** Exhibits sharp band between 3700 – 3584 cm^{-1}.
- ***Intermolecular hydrogen bonded O–H stretching vibrations:*** Exhibits broad band between 3550 – 3200 cm^{-1}.
- ***Intramolecular hydrogen bonded O–H stretching vibrations:*** Exhibits broad and shallow band near 3100 cm^{-1}.
- ***C–O stretching vibrations:*** Exhibits strong band between 1260 – 1000 cm^{-1}.
- ***O–H bending vibrations (in plane):*** Exhibits band between 1420 – 1330 cm^{-1}.
- ***Interaction between O–H bending and C–O stretching vibrations of phenols:*** Exhibits two additional bands between 1390 – 1330 cm^{-1} and 1260 – 1180 cm^{-1}.
- ***Interaction of O–H bending & C–H wagging of primary and secondary alcohols:*** Exhibits two additional bands near 1420 and 1330 cm^{-1}.
- ***O–H OOP bending of alcohols & phenols in liquid state:*** Exhibits strong band between 769 – 650 cm^{-1}.

- ***Other C–H bands of alkyl hydrocarbons if present in the compounds of alcohol or phenol compounds are also observed.***

Example: ***Hexanol*** (IR spectrum given in Figure 2.37).

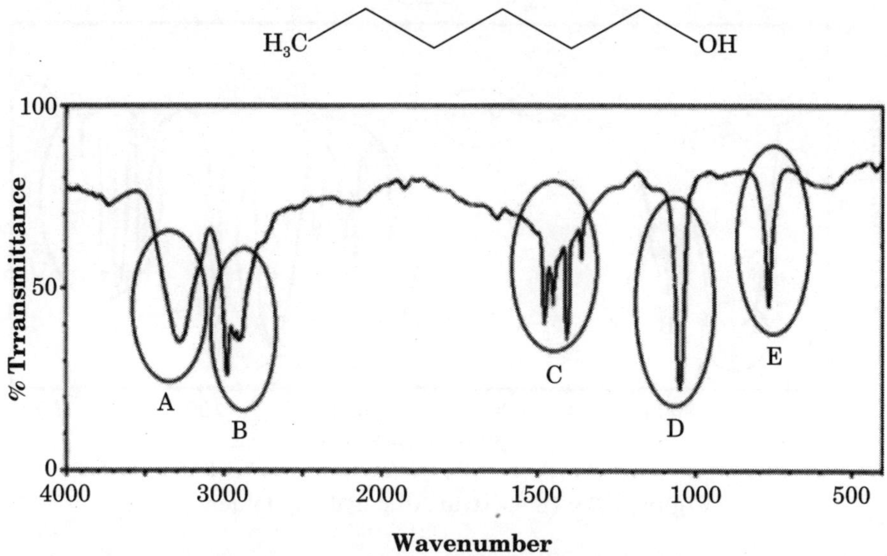

Figure 2.37: IR spectrum of 1-hexanol.

Spectral Interpretation

Region A : 3332 cm^{-1} (O–H Stretch intermolecular).

Region B : 2965 cm^{-1} (C–H asymmetrical stretch of CH_3), 2924 cm^{-1} (C–H asymmetrical stretch of CH_2), 2868 cm^{-1} (C–H symmetrical stretch of CH_2) and 2852 cm^{-1} (C–H symmetrical stretch of CH_2).

Region C : 1470 cm^{-1} (scissoring of CH_2), 1452 cm^{-1} (asymmetrical bending of CH_3), 1383 cm^{-1} (C-H bending of CH_3).

Region D : 1045 cm^{-1} (C–O stretching).

Region E : 725 cm^{-1} (methylene rocking, long chain band).

Example: *Para Hydroxy Toluene* (IR spectrum given in Figure 2.38).

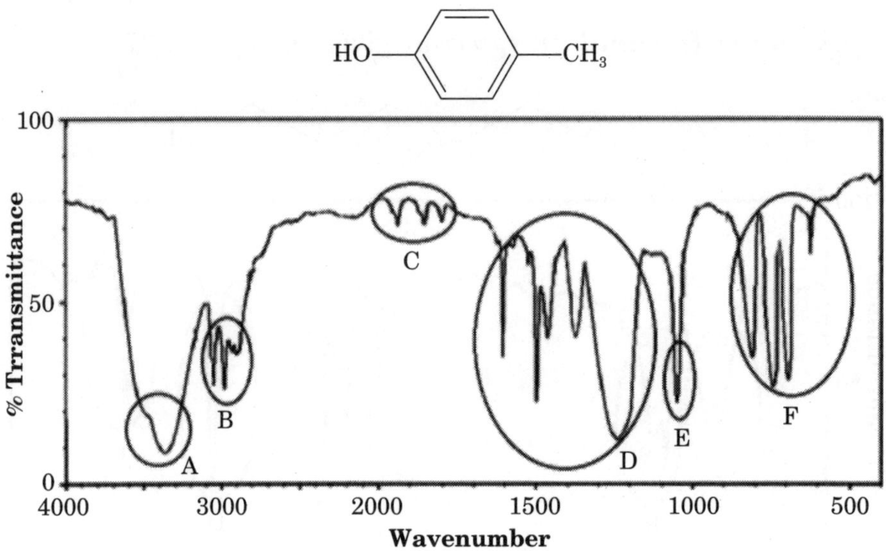

Figure 2.38: IR spectrum of p-hydroxy toluene.

Spectral Interpretation

Region A : 3372 cm^{-1} (O–H Stretch intermolecular).

Region B : 3045 cm^{-1} (Aromatic C–H stretch), 2968 cm^{-1} (C–H asymmetrical stretch of CH_3), and 2872 cm^{-1} (C–H symmetrical stretch of CH_3).

Region C : 2000 – 1650 cm^{-1} (overtone and combination bands).

Region D : 1608 cm^{-1}, 1504 cm^{-1} and 1470 cm^{-1} (C–C ring stretching); 1365 cm^{-1} (in plane O–H Bending), and 1224 cm^{-1} (C–O stretch).

Region E : 1085 cm^{-1} (in plane C–H bending).

Region F : 815 and 755 cm^{-1} (out of plane C–H bending); 690 cm^{-1} (OOP ring C–C bending); and 650 cm^{-1} (OOP O–H bending).

2.5.8 Ethers

IR Spectral Characteristics of Ethers

In the IR spectrum of ethers generally C–O stretching, C–H vibrational bands are identified. Some characteristic vibrational bands that are observed for ethers are as follows:

- ***C–O stretching vibrations in aliphatic ethers:*** Exhibits a strong asymmetrical band near 1125 cm^{-1}(generally between 1185 – 1085 cm^{-1}) and a symmetrical band in Raman spectrum.
- ***C–O stretching vibrations in cyclic ethers:*** Exhibits same bands as aliphatic ethers but asymmetrical band at lower and symmetrical band at higher wave number.
- ***C–O stretching vibrations in branched ethers (i.e., with branching on alpha carbon):*** Exhibits splitting of bands on higher wave number. For instance
    - Isopropyl ether exhibits a splitted (triplet) band from 1170 – 1114 cm^{-1}.
- ***C–O stretching vibrations in aryl alkyl ethers:*** Exhibits an asymmetrical band between 1275 – 1200 cm^{-1} and a symmetrical band between 1075 – 1020 cm^{-1}.
- ***C–O stretching vibrations in vinyl ethers:*** Exhibits an asymmetrical band between 1225 – 1200 cm^{-1} and a strong symmetrical band between 1075 – 1020 cm^{-1}.
- ***Other C–H bands of alkyl or aryl if hydrocarbons present in the ether compounds are also observed.***

Example: ***Di butyl ether*** (IR spectrum given in Figure 2.39).

Spectral Interpretation

Region A : 2962 cm^{-1} (C–H asymmetrical stretch of CH_3), 2865 (C–H symmetrical stretch of CH_3), 2926 cm^{-1} (C–H asymmetrical stretch of CH_2), 2853 cm^{-1} (C–H symmetrical stretch of CH_2).

Region B : 1470 cm^{-1} (C–H scissoring of CH_2), 1452 cm^{-1} (C–H asymmetrical bending of CH_3), 1383 cm^{-1} (C–H scissoring of CH_3).

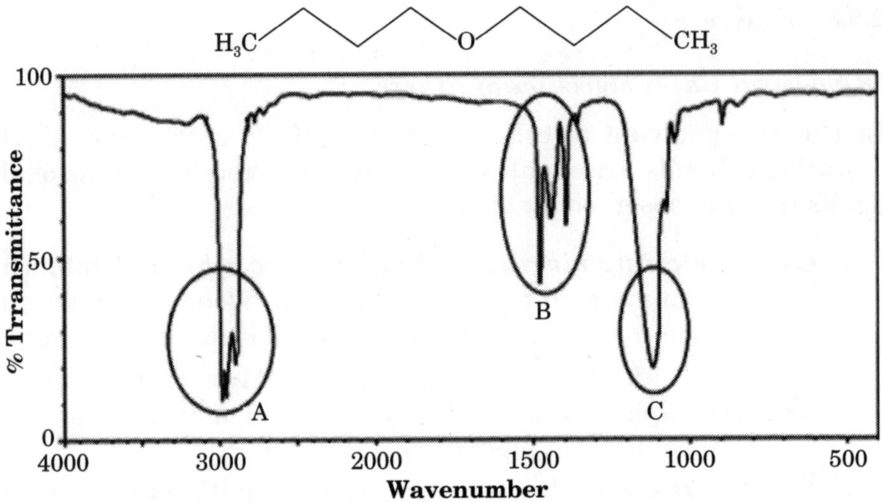

Figure 2.39: IR spectrum of di-butyl ether.

Region C : 1124 cm^{-1} (C–O stretching in C–O–C group).

Example: ***Phenyl methyl ether*** (IR spectrum given in Figure 2.40).

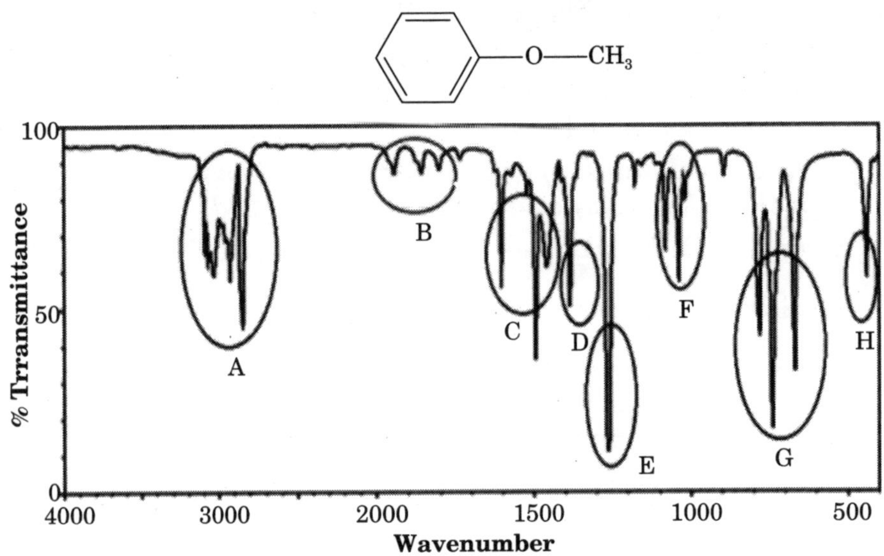

Figure 2.40: IR spectrum of Phenyl methyl ether.

Spectral Interpretation

Region A : 3056 cm^{-1} (Aromatic C–H Stretch), 2952 cm^{-1} (C–H asymmetrical stretch of CH_3), and 2838 cm^{-1} (C–H symmetrical stretch of CH_3).

Region B : 2000 – 1650 cm^{-1} (overtone and combination bands).

Region C : 1602 cm^{-1}, 1504 cm^{-1} and 1458 cm^{-1} (C–C ring stretching).

Region D : 1383 cm^{-1} (C-H bending of CH_3).

Region E : 1245 cm^{-1} (C–O stretching asymmetrical).

Region F : 1086 cm^{-1} (in plane C–H bending) and 1040 cm^{-1} (C–O stretching Symmetrical).

Region G : 785 & 755 cm^{-1} (out of plane C–H bending), and 690 cm^{-1} (OOP ring C–C bending).

Region H : 438 cm^{-1} (OOP ring bending).

2.5.9 Carbonyl Compounds (Ketones)

IR Spectral Characteristics of Ketones

In the IR spectrum of ketones generally C=O stretching, C–(C=O)–C stretching, C–(C=O)–C bending and C–H vibrational bands are identified. Some characteristic vibrational bands that are observed for ketones are as follows:

- ***C=O stretching vibrations in aliphatic ketones:*** Exhibits a strong band near 1715 cm^{-1}. Factors that affects C=O stretching of aliphatic ketones are as follows:
  - *Conjugation:* C=O stretching vibrations in ketones (conjugated) exhibits a strong band at lower wavenumber *i.e.*, between 1685 – 1666 cm^{-1} (when conjugated with alkene or phenyl group).
  - *Heteroatomic Substitution:* α-carbon replacement with heteroatom increases C=O stretching wavenumber (due to increase in force constant).

- *Solvent Effects:* Neat sample of acetone and cyclohexanone exhibits a strong band at 1715 cm^{-1}. C=O stretching wavenumber decreases and increases in polar solvent and non polar solvents respectively.
- *Steric Hindrance:* C=O stretching vibrations in sterically hindered aliphatic ketones (conjugated)—exhibits a strong band at higher wavenumber in comparison to normal conjugated ketones (due to sterical hindrance that causes reduction in effect of conjugation). For example s-cis α, β unsaturated ketone absorbs at 1699 cm^{-1} and s-trans α, β unsaturated ketone absorbs at 1674 cm^{-1}.
- *Hydrogen bonding:* C=O stretching vibrations in aliphatic ketones (intermolecular hydrogen bond with solvents like methanol) absorbs at lower wavenumber. Example, C=O stretching of neat 2-butanone absorbs at 1715 cm^{-1} and dilute solution of ketone in methanol absorbs at 1706 cm^{-1}.

- ***C=O stretching vibrations in β-Di ketones (mixtures of keto and enol forms):*** Enolic forms exhibits a very strong and broad band between 1640 – 1580 cm^{-1} (due to intramolecular hydrogen bonding). For example acetyl acetone in enolic form exhibits a strong band at 1613 cm^{-1}. Whereas the keto form and a small unbonded enol form exhibits two bands centred near 1725 cm^{-1}.
- ***C=O stretching vibrations in α-Di ketones:*** Exhibits single band as mono ketones do. For example, biacetyl absorbs at 1718 cm^{-1}.
- ***C=O stretching vibrations in quinones:*** Exhibits band between 1690 – 1655 cm^{-1}.
- ***C=O stretching vibrations in quinones with extended conjugation:*** Exhibits bands at lower frequency between 1655 – 1635 cm^{-1} than normal quinones.
- ***C=O stretching vibrations in acyclic halogenated (α-chloro) ketones:*** Exhibits band at 1745 cm^{-1} due to one conformer, which possess chlorine near to carbonyl group. Whereas a band is exhibited at 1725 cm^{-1} for another conformer, which possess chlorine away from carbonyl group. But in case of rigid molecules, for example in case of mono keto steroids only

halogen that are in equatorial position of carbonyl group tends to increase the wavenumber of C=O absorption. Whereas axial position of halogens does not mediates any effect.

- ***C=O stretching vibrations in cyclic ketones:*** Exhibits band to appear to move at higher frequency region when the ring size of cyclic ketone is reduced. This effect is due to the ring strain produced when bond angle reduces less than 120°. For example band at 1715 cm^{-1} (corresponds to cyclohexanone), 1751 cm^{-1} (corresponds to cyclo pentanone), and 1775 cm^{-1} (corresponds to cyclo butanone).
- ***C–C–C stretching and bending in C–(C=O)–C group of aliphatic ketones:*** Exhibits medium bands in 1230 – 1100 cm^{-1}.
- ***C–C–C stretching and bending in C–(C=O)–C group of aromatic ketones:*** Exhibits medium bands in 1300 – 1230 cm^{-1}.
- ***Other C–H bands of alkyl & aryl hydrocarbons present in ketones are also observed.***

Example: ***Ethyl methyl ketone*** (IR spectrum given in Figure 2.41).

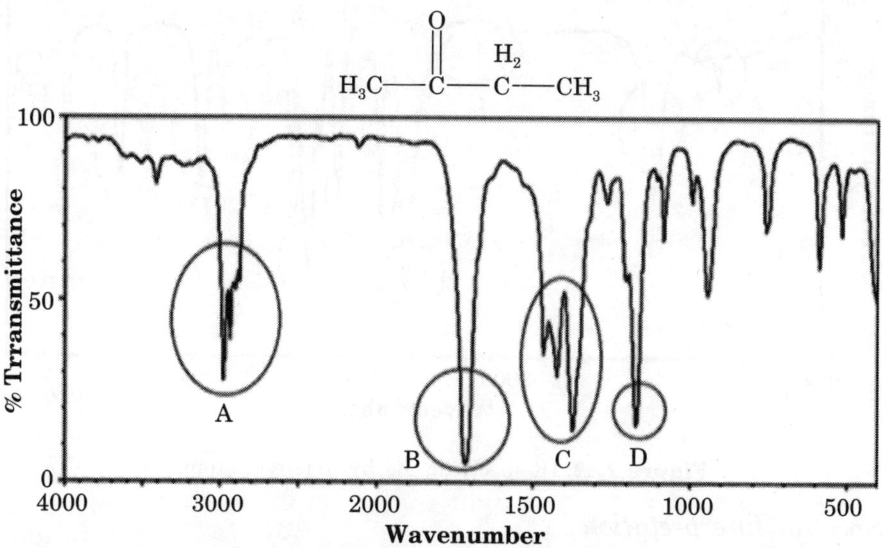

Figure 2.41: IR spectrum of ethyl methyl ketone.

Spectral Interpretation

Region A : 2971 cm^{-1} (C–H asymmetrical stretch of CH_3), 2863 cm^{-1} (C–H symmetrical stretch of CH_3), 2924 cm^{-1} (C–H asymmetrical stretch of CH_2), 2853 cm^{-1} (C–H symmetrical stretch of CH_2).

Region B : 1715 cm^{-1} (C=O stretching).

Region C : 1470 cm^{-1} (C–H scissoring of CH_2), 1452 cm^{-1} (C–H asymmetrical bending of CH_3), and 1383 cm^{-1} (C–H scissoring of CH_3 of CH_3CO group).

Region D : 1170 cm^{-1} (C–C–C stretching & bending in C–(C=O)–C group).

Example: ***Methyl phenyl ketone*** (IR spectrum given in Figure 2.42).

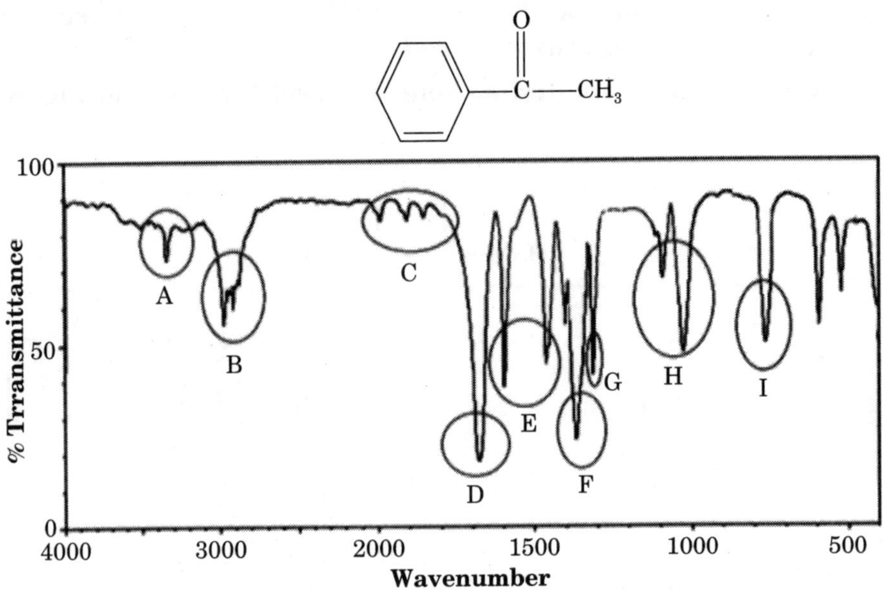

Figure 2.42: IR spectrum methyl phenyl ketone.

Spectral Interpretation

Region A : 3350 cm^{-1} (overtone of C=O stretching).

Region B : 3056 cm^{-1} (Aromatic C–H Stretch), 2952 cm^{-1} (C–H asymmetrical stretch of CH_3), and 2838 cm^{-1} (C–H symmetrical stretch of CH_3).

Region C : 2000 – 1700 cm^{-1} (overtone and combination bands).

Region D : 1685 cm^{-1} (C=O stretching).

Region E : 1602 cm^{-1}, and 1458 cm^{-1} (C–C ring stretching).

Region F : 1383 cm^{-1} (C–H scissoring of CH_3).

Region G : 1240 cm^{-1} (C–C–C stretching & bending in C–(C=O)–C group).

Region H : 1086 cm^{-1} and 1035 cm^{-1} (in plane C–H bending).

Region I : 735 cm^{-1} (Out of plane C–H bending).

2.5.10 Carbonyl Compounds (Aldehydes)

IR Spectral Characteristics of Aldehydes

In the IR spectrum of aldehydes generally C=O stretching, and C–H vibrational bands are identified. Some characteristic vibrational bands that are observed for aldehydes are as follows:

- ***C=O stretching vibrations in aliphatic aldehydes:*** Exhibits bands between 1740 – 1720 cm^{-1}, which is on a higher frequency than ketones. Factors affecting C=O stretching of aliphatic aldehydes are similar to ketones as follows:
  - *Heteroatomic Substitution:* α-carbon replacement with heteroatom increases C=O stretching wavenumber (due to increase in force constant). For example acetaldehyde exhibits band at 1730 cm^{-1}. Whereas trichloro acetaldehyde exhibits band at 1768 cm^{-1}.
  - *Conjugation:* C=O stretching vibrations in aldehydes (conjugated) exhibits a strong band at lower wavenumber *i.e.*, between 1710 – 1685 cm^{-1} (when conjugated with alkene or phenyl group).
  - *Hydrogen bonding:* In aldehydes hydrogen bonding (intramolecular) results in reduction in absorption frequency

to lower wavenumber. Example, C=O stretching of salicylaldehyde absorbs at 1666 cm^{-1}.

- ***C=O stretching vibrations in α-di aldehydes:*** Exhibits similar single band as mono aldehyde. For example, glyoxal exhibits only one C=O peak.
- ***C–H stretching vibration of aldehydic group:*** Exhibits two medium bands between 2830 – 2695 cm^{-1}, due to occurrence of Fermi resonance between fundamental band and overtone band (at 1390 cm^{-1}) of aldehydic C–H.
- ***C–H stretching vibration of aromatic aldehydes:*** Exhibits band on a higher frequency *i.e.*, 2900 cm^{-1}, when aromatic ring contains electronegative atom in ortho position of aldehyde.
- ***Other C–H bands of alkyl and aryl hydrocarbons if present in the aldehyde compounds are also observed.***

Example: ***2-Phenylpropanal*** (IR spectrum given in Figure 2.43).

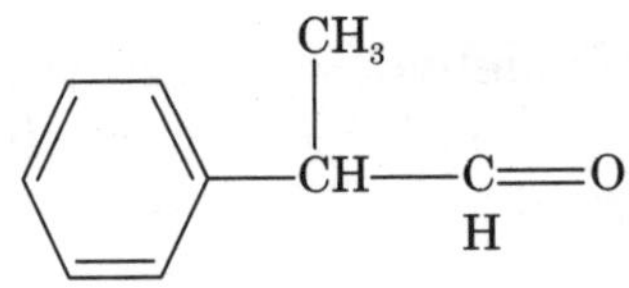

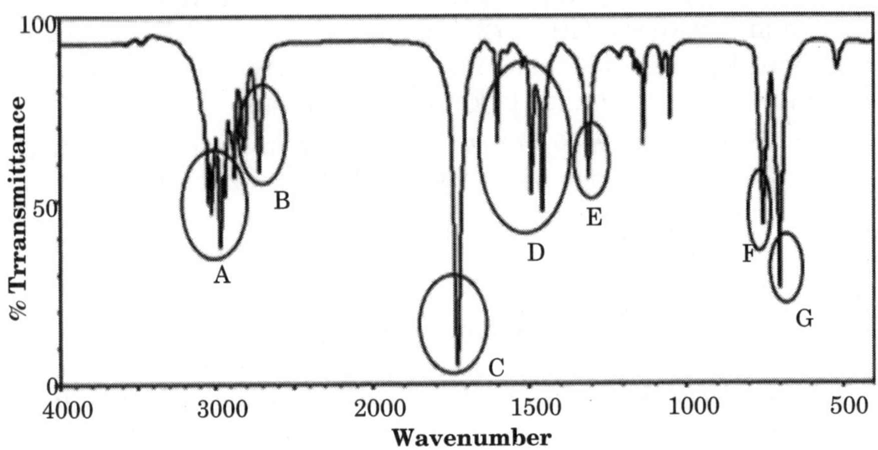

Figure 2.43: IR spectrum of 2-phenyl propanal.

Spectral Interpretation

Region A : 3065, 3026 cm^{-1} (Aromatic C–H Stretch); 2962 cm^{-1} (C–H asymmetrical stretch of CH_3); and 2875 cm^{-1} (C–H symmetrical stretch of CH_3).

Region B : 2824 cm^{-1} and 2720 cm^{-1} (aldehydic C–H stretch doublet a result of Fermi resonance with overtone of band at 1390 cm^{-1}).

Region C : 1724 cm^{-1} (C=O stretch of aldehyde).

Region D : 1602 cm^{-1}, 1493 cm^{-1} and 1458 cm^{-1} (C–C ring stretching).

Region E : 1390 cm^{-1} (aldehydic C–H bending).

Region F : 760 cm^{-1}(Out of plane C–H bending).

Region G : 702 cm^{-1} (Out of plane C–C ring bending).

2.5.11 Carbonyl Compounds (Carboxylic Acid)

IR Spectral Characteristics of Carboxylic Acids

In the IR spectrum of carboxylic acid generally O–H stretching, C=O stretching, C–O stretching and O–H bending vibrational bands are identified. Some characteristic vibrational bands that are observed for carboxylic acids are as follows:

- ***O–H stretching vibrations:*** Exhibits bands between 3300 – 2500 cm^{-1} (centers towards 3000 cm^{-1}). One of the factor that affects carboxylic acids O–H band stretching is
  - *Hydrogen bonding:* Solvents that can undergo inter-molecular hydrogen bonding such as dioxane, or tetrahydrofuran etc. causes bonded O–H absorption near 3100 cm^{-1}.
- ***C=O stretching vibrations in aliphatic carboxylic acids (saturated monomers):*** Exhibits bands near 1760 cm^{-1}. Factors affecting C=O stretching of carboxylic acid are as follows:

- *Dimerization:* Dimers of aliphatic carboxylic acids exhibits C=O stretching band on a lower wave number *i.e.*, between 1720 – 1706 cm^{-1}.
- *Hydrogen bonding:* Intra molecular hydrogen bonding exhibits C=O stretching band on a further lower wave number. For example, para hydroxy benzoic acid absorbs at 1680 cm^{-1}, whereas ortho hydroxy benzoic acid absorbs at 1665 cm^{-1}.
- *Unsaturation (conjugation):* Conjugation to the carbonyl of carboxylic acid causes reduction in absorption frequency to lower wave number. For example, dimers of unsaturated conjugated carboxylic acid and aryl carboxylic acids exhibits band between 1710 – 1680 cm^{-1}.
- *α-Halogen substitution:* Electronegative group substitution on α carbon of carboxylic acids exhibits C=O stretching band to appear on a lower wavenumber (*i.e.*, reduction by 10 – 20 cm^{-1}). For instance when a carboxylic acid is substituted with a halogen atom on α carbon it causes two conformations to occur. One is where chlorine atom is in close proximity and one in which it is away form carbonyl group. Hence, this halogens exhibits two C=O stretching bands one at higher frequency (corresponds to conformer in which chlorine atom is in close proximity to carbonyl group) and another at lower frequency.

- ***C–O stretching in carboxylic acids:*** Exhibits a band between 1320 – 1210 cm^{-1}.
- ***O–H bending vibrations in carboxylic acids:*** Exhibits a band between 1440 – 1395 cm^{-1}.
- ***Coupled vibrations:*** Interaction between C–O stretching and in plane C–O–H bending, generates two bands. One band occurs between 1315 – 1280 cm^{-1}, which appears as a doublet and corresponds to C–O stretching of long chain fatty acids. Another band occurs between 1440 – 1395 cm^{-1}, which corresponds to C–O–H bending.
- ***Other C–H bands of alkyl and aryl hydrocarbons if present in the carboxylic acid compounds are also observed.***

Example: *Hexanoic acid* (IR spectrum given in Figure 2.44).

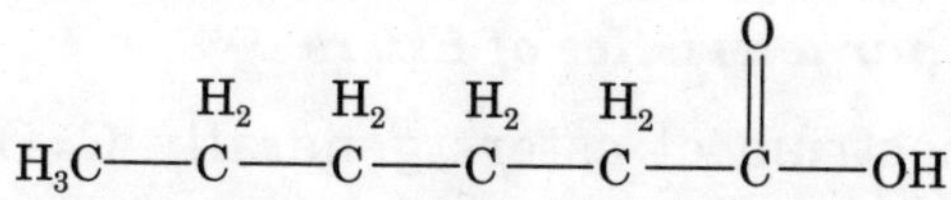

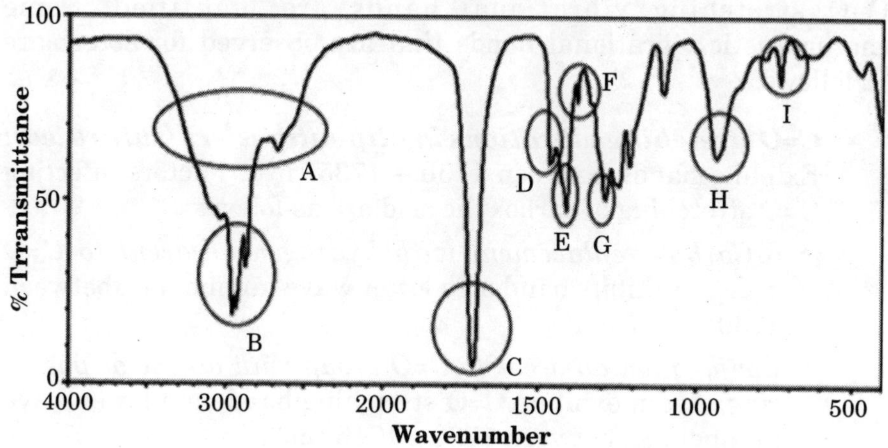

Figure 2.44: IR spectrum of 2, 2, 4-hexanoic acid.

Spectral Interpretation

Region A : 3300 – 2500 cm^{-1} (O–H stretch).

Region B : Super imposed bands in O–H band region *i.e.*, 2971 cm^{-1} (C–H asymmetrical stretch of CH_3), 2863 cm^{-1} (C–H symmetrical stretch of CH_3), 2924 cm^{-1} (C–H asymmetrical stretch of CH_2), 2853 cm^{-1} (C–H symmetrical stretch of CH_2).

Region C : 1721 cm^{-1} (C=O stretching).

Region D : 1470 cm^{-1} (C–H scissoring of CH_2).

Region E : 1419 cm^{-1} (C–O–H inplane bending).

Region F : 1383 cm^{-1} (C–H scissoring of CH_3).

Region G : 1296 cm^{-1} (C–O stretching).

Region H : 948 cm^{-1} (O–H out of plane bending).

Region I : 726 cm^{-1} (methylene rocking) long chain band.

2.5.12 Carbonyl Compounds (Esters)

IR Spectral Characteristics of Esters

In the IR spectrum of esters generally C=O stretching and C–O stretching vibrational bands are identified. Some characteristic vibrational bands that are observed for esters are as follows:

- ***C=O stretching vibrations in aliphatic esters (saturated):*** Exhibits bands between 1750 – 1735 cm^{-1}. Factors affecting C=O stretching of carboxylic acid are as follows:
  - *α-Carbon replacement with hydrogen adjacent to C=O group:* Exhibits band on a lower wave number *i.e.*, between 1730 – 1715 cm^{-1}.
  - *Conjugation adjacent to C=O group with alkene or phenyl ring:* Again exhibits C=O stretching band on a lower wave number *i.e.*, between 1730 – 1715 cm^{-1}.
  - *Conjugation adjacent to C–O bond with alkene or phenyl ring:* Exhibits C=O stretching band to appear on a higher wave number. For example vinyl acetate and phenyl acetate exhibits C=O stretching band at 1776 cm^{-1} and 1770 cm^{-1}.
  - *α-Halogen substitution adjacent to C=O group:* Exhibits band on a higher frequency. For example: ethyl chloro acetate exhibits a band at 1770 cm^{-1}.
  - *α-Carbon replacement with ketone group or carboxylic acid group adjacent to C=O group:* Exhibits C=O stretching band on same wavenumber as normal one. For example oxalates and α-di ketones exhibits a C=O band between 1755 – 1740 cm^{-1}.
- ***C=O stretching vibrations in six membered cyclic esters (δ-lactones):*** Exhibits band on same frequency as aliphatic esters do *i.e.*, 1750 – 1735 cm^{-1}. Factors affecting C=O stretching of six membered cyclic esters are as follows:
  - *Unsaturation (conjugation) α-Carbon adjacent to C=O group:* Exhibits a C=O band on a lower frequency. For example at 1720 cm^{-1}.

- *Unsaturation (conjugation) on α-Carbon adjacent to –O– group:* Exhibits a C=O band on a higher frequency. For example at 1760 cm^{-1}.
- *Double conjugation to α-Carbon adjacent to C=O group i.e., α-pyrones:* Exhibits a two C=O band on a higher frequency due to Fermi resonance *i.e.*, between 1775 – 1715 cm^{-1}.

- ***C=O stretching vibrations in five membered cyclic esters (γ-lactones):*** Exhibits band on higher frequency than aliphatic esters or δ-lactones *i.e.*, between 1795 – 1760 cm^{-1}. Factors affecting C=O stretching of six membered cyclic esters are as follows:
  - *Unsaturation (conjugation) on α-Carbon adjacent to C=O group:* Exhibits a C=O band on a lower frequency *i.e.*, 1750 cm^{-1}.
  - *Unsaturation (conjugation) on α-Carbon adjacent to –O– group:* Exhibits a C=O band on a higher frequency *i.e.*, 1800 cm^{-1}.

- ***C–O stretching vibrations in esters:*** Exhibits two asymmetrical bands between 1300 – 1000 cm^{-1}.
  - *C–C(=O)–O band appears on a higher frequency: These are of two types*

    (i) *Alcoholic acetates*: Saturated aliphatic esters appear at 1240 cm^{-1}. Vinyl and phenyl acetates appears at 1190 – 1140 cm^{-1}.

    (ii) *Acidic acetates*: Esters of α, β unsaturated acid exhibits multiple bands between 1300 – 1160. Where Esters of aromatic acid exhibits bands 1310 – 1250 cm^{-1}.
  - *O–C–C band appears on a lower frequency:* These are of three types

    (i) *Primary alcohol esters*: Exhibits band between 1165 – 1030 cm^{-1}.

    (ii) *Secondary alcohol esters*: Exhibits band near 1100 cm^{-1}.

    (iii) *Long chain fatty acid methyl esters*—exhibits three bands near 1250 cm^{-1}, 1205 cm^{-1}, and a strongest band at 1175 cm^{-1}.

- ***Other C–H bands of alkyl and aryl hydrocarbons if present in the ester compounds are also observed.***

Example: ***Methyl phenyl ester*** (IR spectrum given in Figure 2.45).

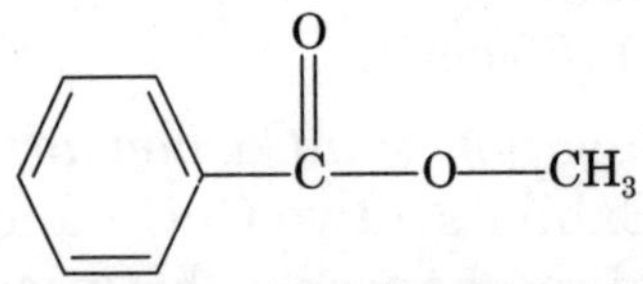

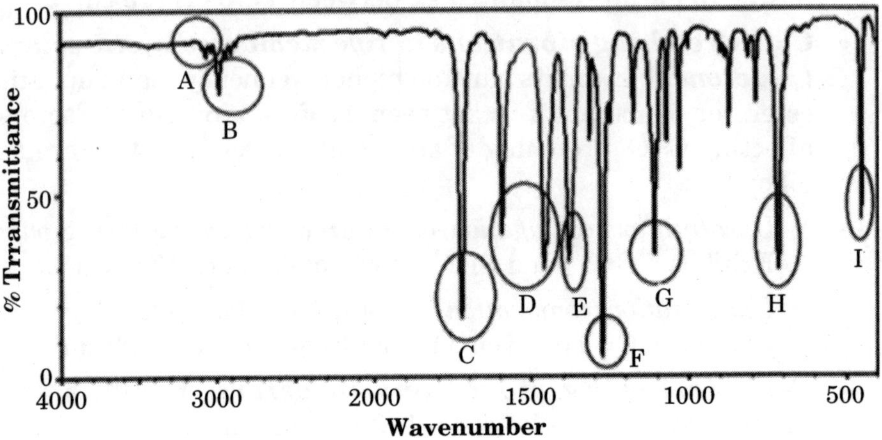

Figure 2.45: IR spectrum of phenyl acetate.

Spectral Interpretation

Region A : 3070 cm^{-1} and 3040 cm^{-1} (aromatic C–H stretching).

Region B : 2986 cm^{-1} (C–H asymmetrical stretch of CH_3), 2878 cm^{-1} (C–H symmetrical stretch).

Region C : 1726 cm^{-1} (C=O stretching); Note: if conjugation or phenyl ring would have been attached through C=O bond band would have appeared at 1765 cm^{-1}.

Region D : 1592 cm^{-1}, and 1458 cm^{-1} (C–C ring stretching).

Region E : 1383 cm^{-1} (C–H scissoring of CH_3).

Region F : 1286 cm^{-1} [C–C(=O)–O stretching in aromatic acid esters].

Region G : 1117 cm^{-1} (O–C–C stretching).

Region H : 730 cm^{-1} (Out of plane C–H bending).

Region I : 438 cm^{-1} (Out of plane ring bending).

2.5.13 Carbonyl Compounds (Acid Halides—*Acid Chlorides*)

IR Spectral Characteristics of Acid Chlorides

In the IR spectrum of acid chlorides generally C=O stretching and C–Cl stretching vibrational bands are identified. Some characteristic vibrational bands that are observed for acid chlorides are as follows:

- ***C=O stretching vibrations in unconjugated acid chlorides:*** Exhibits band between 1815 – 1785 cm^{-1}. Factors affecting C=O stretching of acid chlorides are as follows:
  - *Conjugation in acid chlorides:* Exhibits band on a lower frequency number *i.e.*, between 1800 – 1770 cm^{-1}.
  - *Aryl acid chloride:* Exhibits a weak band between 1750 – 1730 cm^{-1}.
- ***C–Cl stretching in acid chlorides:*** Exhibits band between 730 – 550 cm^{-1}.
- ***Other C–H bands of alkyl and aryl hydrocarbons if present in the acid chloride compounds are also observed.***

Example: *Benzoyl chloride* (IR spectrum given in Figure 2.46).

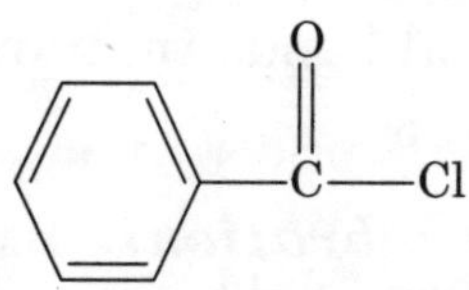

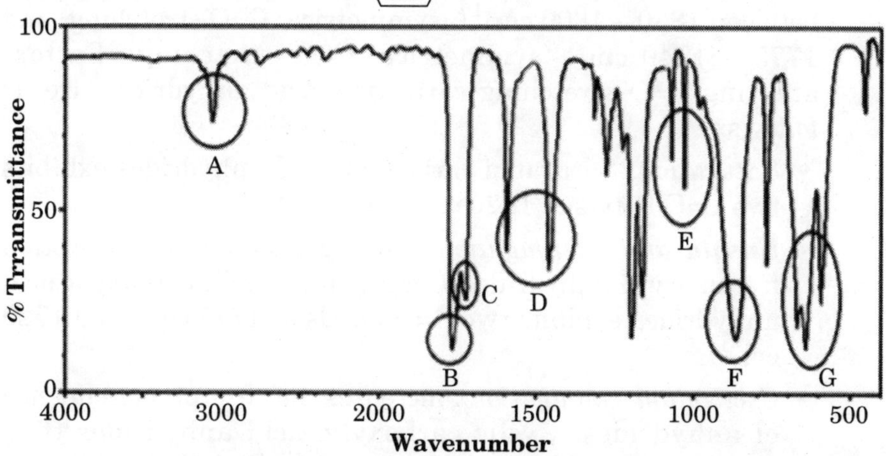

Figure 2.46: IR spectrum of benzoyl chloride.

Spectral Interpretation

Region A : 3065 cm^{-1} (aromatic C–H stretching).

Region B : 1772 cm^{-1} (C=O stretching).

Region C : 1730 cm^{-1} (Fermi resonance band due to C=O stretch and overtone of 870 cm^{-1}).

Region D : 1604 cm^{-1} and 1465 cm^{-1} (C=C ring stretching).

Region E : 1086 cm^{-1} and 1035 cm^{-1} (in plane C–H bending).

Region F : 870 cm^{-1} (Aromatic carbon stretch).

Region G : 692 – 635 cm^{-1} (C–Cl stretching).

2.5.14 Carbonyl Compounds (Carboxylic Acid Anhydrides)

IR spectral characteristics of Carboxylic acid Anhydrides

In the IR spectrum of carboxylic acid anhydrides generally C=O stretching and C–O stretching vibrational bands are identified. Some characteristic vibrational bands that are observed for anhydrides are as follows:

- ***C=O stretching vibrations:*** Exhibits two strong bands between 1830 – 1800 cm^{-1} (asymmetrical C=O stretching) and 1775 – 1740 cm^{-1} (symmetrical C=O stretching). Factors affecting C=O stretching carboxylic acid anhydrides are as follows:
  - *Saturation:* Saturated carboxylic acid anhydrides exhibits two C=O bands at 1820 cm^{-1} and 1750 cm^{-1}.
  - *Unsaturation (conjugation):* Conjugation reduces absorption frequency of anhydrides. Conjugated acyclic carboxylic acid anhydrides exhibits two C=O bands at 1775 cm^{-1} and 1720 cm^{-1}.
  - *Cyclization (conjugation):* increases the absorption frequency of anhydrides. Cyclic carboxylic acid anhydrides (five membered ring) for example succinic anhydride exhibits two C=O bands at 1865 cm^{-1} and 1782 cm^{-1}.

- ***C–O stretching vibrations of C–CO–O–CO–C:*** Exhibits strong multiple band from 1300 – 900 cm^{-1}. Factors affecting C–O stretching carboxylic acid anhydrides are as follows:
  - *Saturation:* Straight chain anhydrides exhibits C–O stretching band near 1047 cm^{-1}.
  - *Cyclization:* Cyclic anhydrides exhibits two bands between 952 – 909 cm^{-1} and 1300 – 1175 cm^{-1}.
- ***Other C–H bands of alkyl and aryl hydrocarbons if present in the carboxylic acid anhydride compounds are also observed.***

Example: *Butanoic anhydride* (IR spectrum given in Figure 2.47).

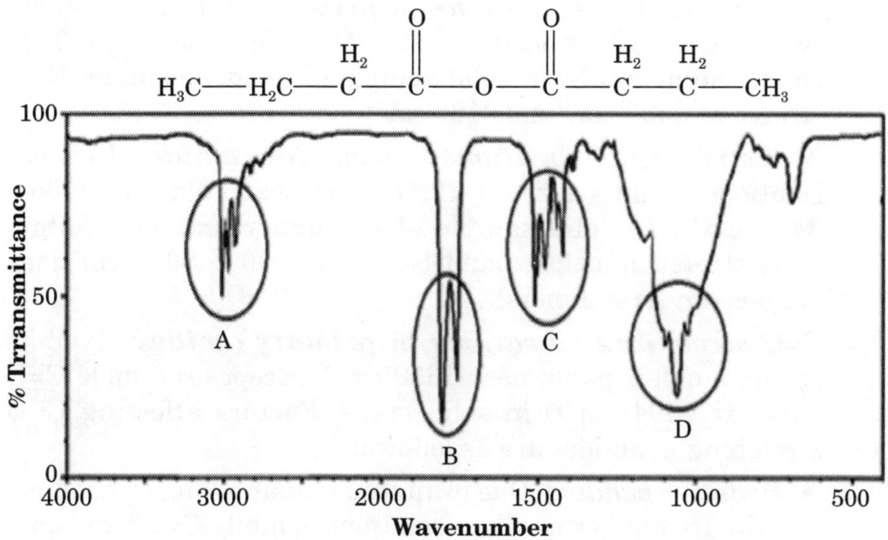

Figure 2.47: IR spectrum of butanoic anhydride.

Spectral Interpretation

Region A : 2992 cm^{-1} (C–H asymmetrical stretch of CH_3), 2872 cm^{-1} (C–H symmetrical stretch of CH_3), 2930 cm^{-1} (C–H asymmetrical stretch of CH_2), 2865 cm^{-1} (C–H symmetrical stretch of CH_2).

Region B : 1820 cm^{-1} and 1752 cm^{-1} (C=O Stretching).

Region C : 1470 cm^{-1} (C–H scissoring of CH_2), 1452 cm^{-1} (C–H asymmetrical bending of CH_3), 1383 cm^{-1} (C–H scissoring of CH_3).

Region D : 1045 cm^{-1} (C–O stretch of C–CO–O–CO–C).

2.5.15 Carbonyl Compounds (Amides)

IR Spectral Characteristics of Amides

In the IR spectrum of amides generally N–H stretching, C=O stretching (Amide I band), N–H bending (Amide II band) and C–N vibrational bands are identified. Some characteristic vibrational bands that are observed for amides are as follows:

- ***N–H stretching vibrations in primary amides:*** Exhibits two medium bands, near 3520 cm^{-1} (asymmetrical) and 3400 cm^{-1} (symmetrical). In solid sample of amides exhibits N–H stretch at 3350 cm^{-1} and 3180 cm^{-1}.
- ***N–H stretching vibrations in secondary amides:*** In dilute solutions exhibits a free N–H stretching band between 3500 – 3400 cm^{-1}. In solid sample of secondary amides exhibits N–H stretch multiple band between 3330 – 3060 cm^{-1} (as amides can form dimers).
- ***C=O stretching vibrations in primary amides:*** Exhibits strong amide I band, near 1650 cm^{-1} (except acetamide C=O band at 1694 cm^{-1}) in solid state. Factors affecting C=O stretching in amides are as follows:
  - *Hydrogen bonding:* Due to intermolecular hydrogen bonding, primary amides in dilute solutions exhibits C=O band on a higher frequency *i.e.*, near 1690 cm^{-1}.
  - *Electron withdrawing group near to nitrogen:* Exhibits C=O band on higher frequency.
- ***C=O stretching vibrations in secondary amides:*** Exhibits strong amide I band, near 1640 cm^{-1} in solid state. Factors affecting C=O stretching in amides are as follows:
  - *Hydrogen bonding:* Due to intermolecular hydrogen bonding, secondary amides in dilute solutions exhibits C=O band on a higher frequency *i.e.*, near 1690 cm^{-1}.

- *Electron withdrawing group near to nitrogen:* Exhibits C=O band on higher frequency.

- ***C=O stretching vibrations in tertiary amides:*** Exhibits strong amide I band, between 1680 – 1630 cm^{-1} in solid state. Factors affecting C=O stretching in amides are as follows:
  - *Solvent effects:* Tertiary amides in dilute solutions exhibits C=O band on a lower frequency *i.e.*, near 1615 cm^{-1}.
  - *Electron withdrawing group near to nitrogen:* Exhibits C=O band on higher frequency.
- ***N–H bending vibrations in primary amides:*** Exhibits medium amide II band between 1655 – 1620 cm^{-1}. Amides in dilute solution exhibits N–H bending on lower frequency, *i.e.*, between 1620 – 1590 cm^{-1}.
- ***N–H bending vibrations in secondary amides:*** Exhibits medium amide II band between 1570 – 1515 cm^{-1} (in solid state). Amides in dilute solution exhibits N–H bending on lower frequency, *i.e.*, near 1250 cm^{-1}.
- ***C–N stretching vibrations in primary amides:*** Exhibits band, near 1400 cm^{-1}.
- ***N–H wagging vibrations in secondary amides:*** Exhibits broad medium band between 800 – 666 cm^{-1}.
- ***Other C–H bands of alkyl and aryl hydrocarbons if present in the amido compounds are also observed.***

Example: *Propanamide* (IR spectrum given in Figure 2.48).

Spectral Interpretation

Region A	:	3350 cm^{-1} (N–H asymmetrical stretch) and 3175 (N–H symmetrical stretch).
Region B	:	2962 cm^{-1} (C–H asymmetrical stretch of CH_3), 2862 cm^{-1} (C–H symmetrical stretch of CH_3), 2926 cm^{-1} (C–H asymmetrical stretch of CH_2), 2852 cm^{-1} (C–H symmetrical stretch of CH_2).
Region C	:	1640 cm^{-1} (C=O Stretching, amide I band).
Region C′	:	1602 cm^{-1} (N–H bending, amide II band).

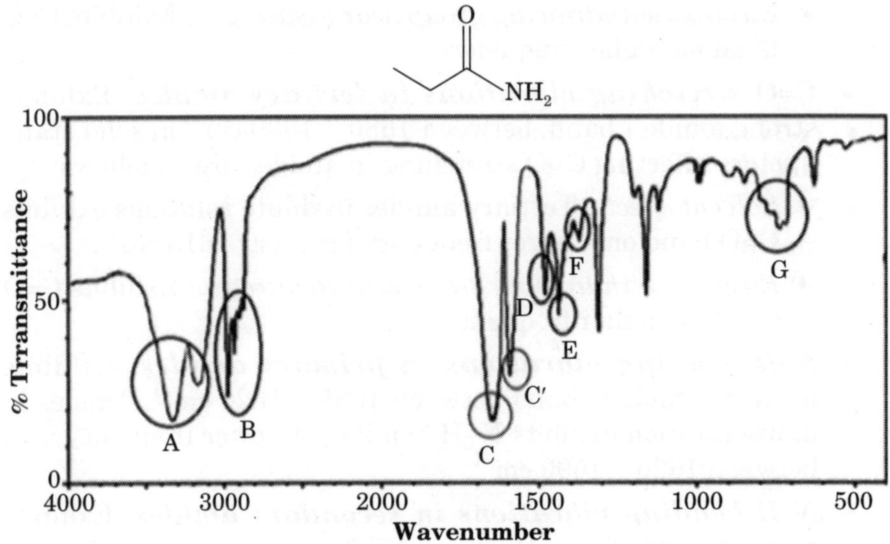

Figure 2.48: IR spectrum of propanamide.

Region D : 1470 cm^{-1} (C–H scissoring of CH_2).

Region E : 1425 cm^{-1} (C–N stretching).

Region F : 1383 cm^{-1} (C–H scissoring of CH_3).

Region G : 700 – 600 cm^{-1} (Broad band of N–H out of plane bending).

2.5.16 Amines

IR Spectral Characteristics of Amines

In the IR spectrum of amines generally N–H stretching, N–H bending and C–N stretching bands are identified. Some characteristic vibrational bands that are observed for amines are as follows:

- ***N-H stretching vibrations in primary amines:*** Exhibits two weak bands, between 3400 – 3300 cm^{-1} (asymmetrical) and 3330 – 3250 cm^{-1} (symmetrical). Factors affecting N–H stretching in primary amines are as follows:
  - *Aromaticity:* Aromatic primary amines exhibits N–H stretching on higher frequencies.

- *Solvent effect:* Primary amines in dilute solution exhibits N–H stretching on further higher frequencies *i.e.*, near 3500 cm^{-1} and 3400 cm^{-1}.

- ***N–H stretching vibrations in secondary amines:*** Exhibits single weak band between 3350 – 3310 cm^{-1}.
- ***N–H bending vibrations in primary amines:*** Exhibits medium to strong band between 1650 – 1580 cm^{-1}. Factors affecting N–H stretching in primary amines are as follows:
  - *Intermolecular bonding:* Aromatic primary amines as a dimer exhibits N-H stretching on higher frequencies.
  - *Liquid sample:* Exhibits a broad and strong band between 910 – 665 cm^{-1}.
- ***N–H bending vibrations in secondary amines:*** No band is observed for aliphatic secondary amines. But aromatic secondary amines exhibits a band near 1515 cm^{-1}.
- ***C–N stretching:*** Exhibits weak to medium band near 1250 – 1020 cm^{-1} in primary, secondary & tertiary amines. Aromatic amines exhibits band between 1340 – 1365 cm^{-1}.
- ***Other C–H bands of alkyl and aryl hydrocarbons if present in the amino compounds are also observed.***

Example: *Heptylamine* (IR spectrum given in Figure 2.49).

Spectral Interpretation

Region A : 3370 cm^{-1} (N–H asymmetrical stretch) and 3292 cm^{-1} (N–H symmetrical stretch).

Region B : 2962 cm^{-1} (C–H asymmetrical stretch of CH_3), 2862 cm^{-1} (C–H symmetrical stretch of CH_3), 2926 cm^{-1} (C–H asymmetrical stretch of CH_2), 2852 cm^{-1} (C–H symmetrical stretch of CH_2).

Region C : 1617 cm^{-1} (N–H bending).

Region D : 1470 cm^{-1} (C–H scissoring of CH_2), 1452 cm^{-1} (C–H asymmetrical bending of CH_3), 1383 cm^{-1} (C–H scissoring of CH_3).

Region E : 1073 cm^{-1} (C–N Stretching).

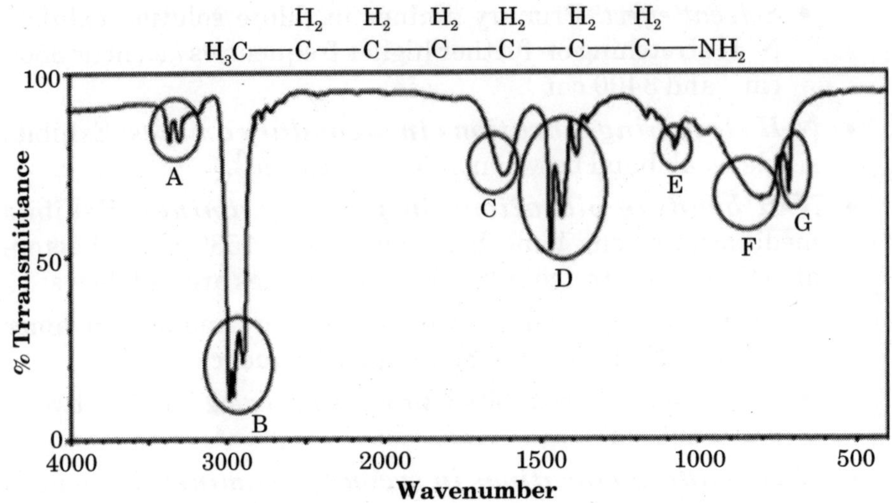

Figure 2.49: IR spectrum of heptyl amine.

Region F : 900 – 700 cm^{-1} (N–H wagging).

Region G : 726 cm^{-1} (Methylene rocking a long chain band).

2.5.17 Nitro Compounds

IR Spectral Characteristics of Nitro Compounds

In the IR spectrum of nitro compounds generally N–O stretching's are identified. Some characteristic vibrational bands that are observed for amines are as follows:

- ***N–O stretching vibrations in nitro compounds:*** Exhibits two strong bands, near 1550 cm^{-1} (asymmetrical) and 1350 cm^{-1} (symmetrical). For instance
  - *N–O stretching in Aliphatic nitro compounds:* Exhibits one strong asymmetrical N–O stretching between 1600 – 1530 cm^{-1} and one medium symmetrical N–O stretching between 1390 – 1300 cm^{-1}.
  - *N–O stretching in aromatic nitro compounds:* Exhibits one strong asymmetrical N–O stretching between 1550 – 1490 cm^{-1} and one medium symmetrical N–O stretching between 1355 – 1315 cm^{-1}.

- ***Other C–H bands of alkyl and aryl hydrocarbons if present in the nitrocompounds are also observed.***

Example: ***Nitro octane*** (IR spectrum given in Figure 2.50).

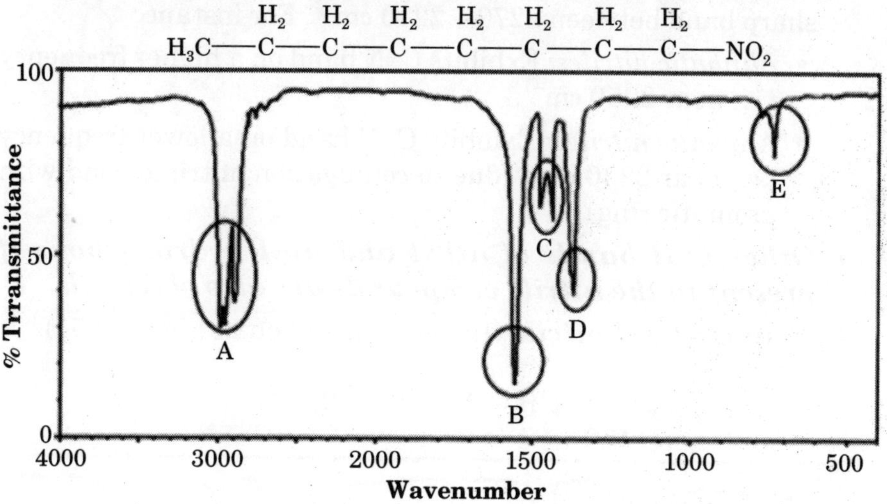

Figure 2.50: IR spectrum of nitro octane.

Spectral Interpretation

Region A	:	2962 cm^{-1} (C–H asymmetrical stretch of CH_3), 2862 cm^{-1} (C–H symmetrical stretch of CH_3), 2926 cm^{-1} (C–H asymmetrical stretch of CH_2), 2852 cm^{-1} (C–H symmetrical stretch of CH_2).
Region B	:	1550 cm^{-1} (N–O stretching asymmetrical).
Region C	:	1470 cm^{-1} (C–H scissoring of CH_2), 1452 cm^{-1} (C–H asymmetrical bending of CH_3).
Region D	:	1380 cm^{-1} (N–O stretching asymmetrical).
Region E	:	726 cm^{-1} (Methylene rocking—long chain band).

2.5.18 Nitriles

IR Spectral Characteristics of Nitriles

In the IR spectrum of nitriles compounds generally C≡N stretchings are identified. Some characteristic vibrational bands that are observed for nitriles are as follows:

- ***C≡N stretching vibrations in nitriles:*** Exhibits medium and sharp band between 2270 – 2210 cm^{-1}. For instance
    - *Aliphatic nitriles:* Exhibits C≡N band on a higher frequency *i.e.*, near 2250 cm^{-1}.
    - *Aromatic nitriles:* Exhibits C≡N band on a lower frequency *i.e.*, near 2230 cm^{-1} (due to conjugation of triple bond with aromatic ring).
- ***Other C–H bands of alkyl and aryl hydrocarbons if present in the nitrile compounds are also observed.***

Example: ***Hexyl nitrile*** (IR spectrum given in Figure 2.51).

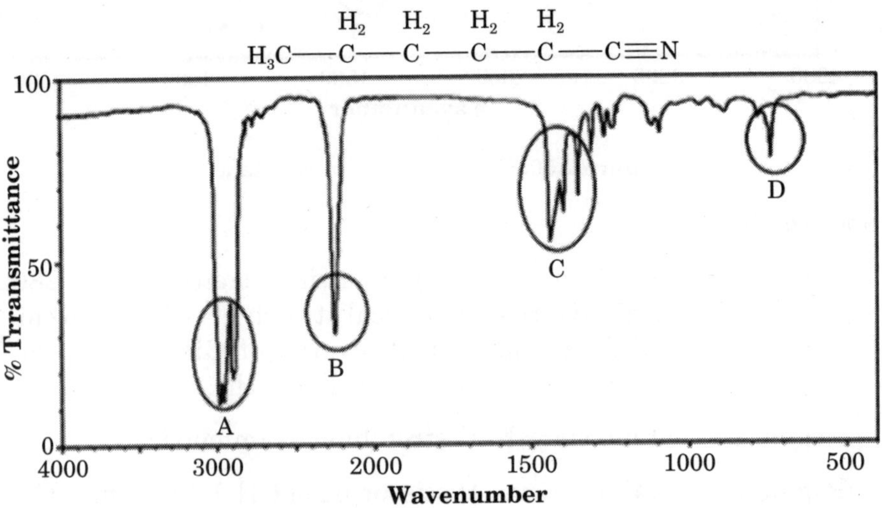

Figure 2.51: IR spectrum of hexyl nitrile.

Spectral Interpretation

Region A : 2962 cm^{-1} (C–H asymmetrical stretch of CH_3), 2862 cm^{-1} (C–H symmetrical stretch of CH_3), 2926 cm^{-1} (C–H asymmetrical stretch of CH_2), 2852 cm^{-1} (C–H symmetrical stretch of CH_2).

Region B : 2250 cm^{-1} (C≡N stretching).

Region C : 1470 cm^{-1} (C–H scissoring of CH_2), 1452 cm^{-1} (C–H asymmetrical bending of CH_3), 1383 cm^{-1} (C–H scissoring of CH_3).

Region D : 726 cm^{-1} (C–H rocking of CH_2) a long chain band.

2.5.19 Isocyanates

IR Spectral Characteristics of Isocyanates

In the IR spectrum of isocyanates generally –N=C=O stretching's are identified. Some characteristic vibrational bands that are observed for isocyanates are as follows:

- ***–N=C=O stretching vibrations in isocyanates:*** Exhibits broad and intense band between near 2270 cm^{-1}.
- ***Other C–H bands of alkyl and aryl hydrocarbons if present in the isocyanate compounds are also observed.***

Example: ***Hexyl isocyanate*** (IR spectrum given in Figure 2.52).

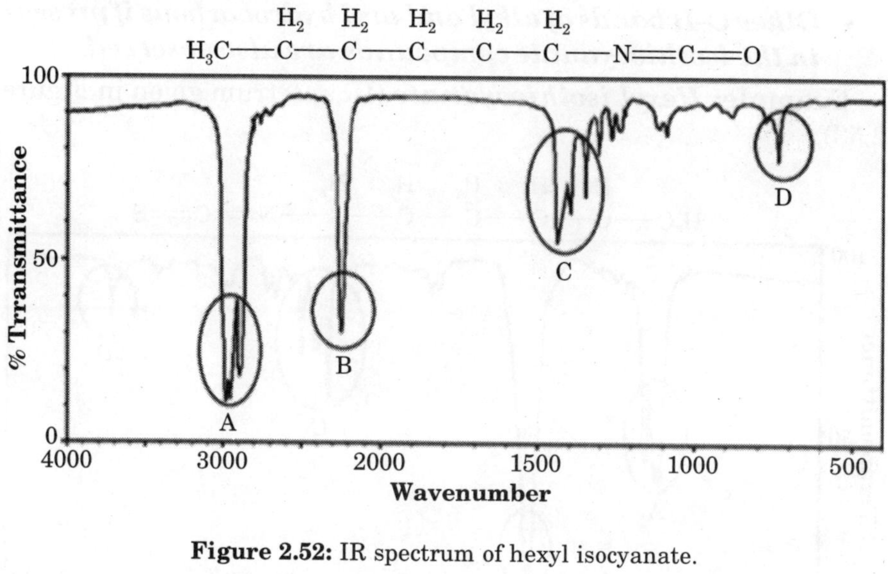

Figure 2.52: IR spectrum of hexyl isocyanate.

Spectral Interpretation

Region A : 2962 cm^{-1} (C–H asymmetrical stretch of CH_3), 2862 cm^{-1} (C–H symmetrical stretch of CH_3), 2926 cm^{-1}

(C–H asymmetrical stretch of CH_2), 2852 cm^{-1} (C–H symmetrical stretch of CH_2).

Region B : 2270 cm^{-1} (–N=C=O stretching).

Region C : 1470 cm^{-1} (C–H scissoring of CH_2), 1452 cm^{-1} (C–H asymmetrical bending of CH_3), 1383 cm^{-1} (C–H scissoring of CH_3).

Region D : 726 cm^{-1} (C–H rocking of CH_2) a long chain band.

2.5.20 Isothiocyanates

IR Spectral Characteristics of Isothiocyanates

In the IR spectrum of isothiocyanates generally –N=C=S stretchings are identified. Some characteristic vibrational bands that are observed for isothiocyanates are as follows:

- ***–N=C=S stretching vibrations in isothiocyanates:*** Exhibits broad and intense band between near 2125 cm^{-1}.
- ***Other C–H bands of alkyl and aryl hydrocarbons if present in the isothiocyanate compounds are also observed.***

Example: ***Hexyl isothiocyanate*** (IR spectrum given in Figure 2.53).

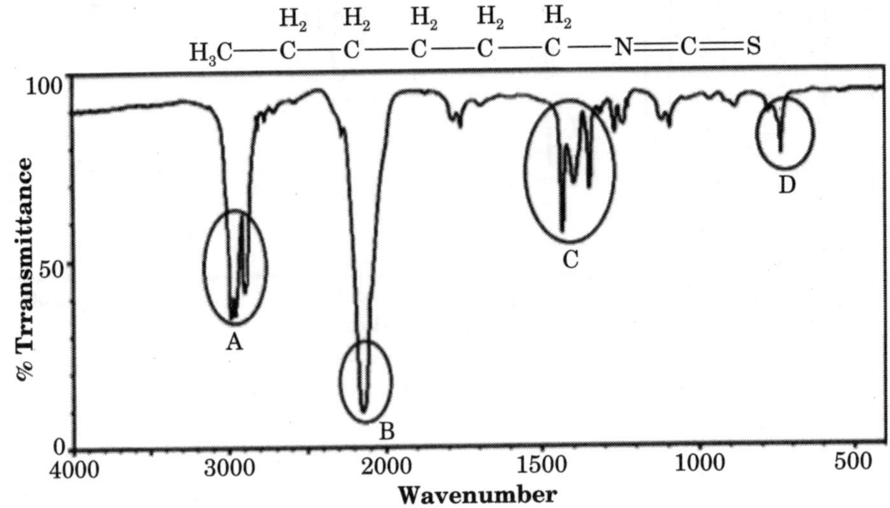

Figure 2.53: IR spectrum of hexyl isothiocyanate.

Spectral Interpretation

Region A : 2963 cm^{-1} (C–H asymmetrical stretch of CH_3), 2860 cm^{-1} (C–H symmetrical stretch of CH_3), 2924 cm^{-1} (C–H asymmetrical stretch of CH_2), 2853 cm^{-1} (C–H symmetrical stretch of CH_2).

Region B : 2124 cm^{-1} (–N=C=S stretching).

Region C : 1468 cm^{-1} (C–H scissoring of CH_2), 1454 cm^{-1} (C–H asymmetrical bending of CH_3), 1380 cm^{-1} (C–H scissoring of CH_3).

Region D : 726 cm^{-1} (C–H rocking of CH_2) a long chain band.

2.5.21 Carboxylate Group Ions and Amino Group Ions

A. IR Spectral Characteristics of Carboxylate Group Ions

In the IR spectrum of carboxylate ions generally C=O stretchings are identified. Some characteristic vibrational bands for carboxylate are as follows:

- ***C=O stretching vibrations in carboxylate ions:*** Exhibits two bands on a lower frequency than normal carboxylic acids *i.e.*, near 1600 cm^{-1} (strong asymmetrical band) and 1400 cm^{-1} (weaker symmetrical band).

B. IR Spectral Characteristics of Amino Group Ions

In the IR spectrum of amino ions generally N–H stretchings and N–H bendings are identified. Some characteristic vibrational bands for amino ions are as follows:

- ***N–H stretching vibrations:*** Free ammonium ions (NH_4^+) exhibits two strong bands between 3300 – 3030 cm^{-1} along with combinational bands in between 2000 – 1700 cm^{-1}. But in case of below mentioned compounds
  - *Primary amino ions* (NH_3^+): Exhibits broad/strong band between 3000 cm^{-1} and 2800 cm^{-1} along with combinational bands between 2800 – 2000 cm^{-1}.

- *Secondary amino ions* (NH_2^+): Exhibits broad/strong band between 3000 – 2700 cm^{-1} along with medium bands near 2000 cm^{-1}.
- *Tertiary amino ions* (NH^+): Exhibits broad/strong band between 2700 – 2250 cm^{-1}.
- *Quaternary amino ions* (N^+): Exhibits no N–H stretching band.

- ***N–H bending vibrations:*** Ammonium ions (NH_4^+) exhibits strong band near 1430 cm^{-1}. But in case of below mentioned compounds
  - *Primary amino ions:* Exhibits asymmetrical and symmetrical bending bands between 1600 – 1575 cm^{-1} and 1550 – 1505 cm^{-1}.
  - *Secondary amino ions:* Exhibits band between 1620 – 1560 cm^{-1}.
  - *Tertiary amino ions:* Exhibits very weak and unimportant bands.

Example: *Benzoic acid ammonium salt* (IR spectrum given in Figure 2.54).

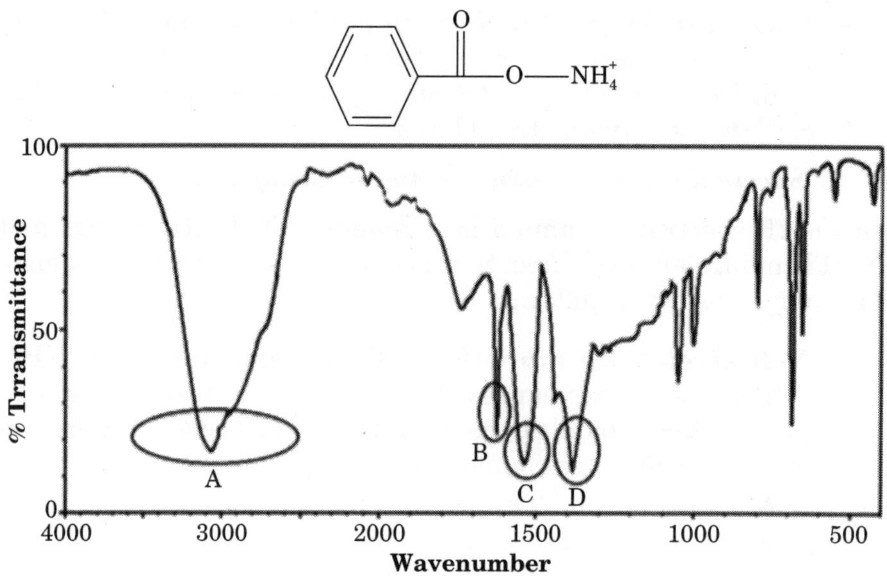

Figure 2.54: IR spectrum of benzoic acid ammonium salt.

Spectral Interpretation

Region A : 3600 – 2500 cm^{-1} (N–H and C–H stretch).

Region B : 1604 cm^{-1} (C=C ring stretching).

Region C : 1550 cm^{-1} (C=O stretch asymmetrical).

Region D : 1385 cm^{-1} (C–H scissoring of CH_3).

2.5.22 Amino Acid and Amino Acid Salts

IR Spectral Characteristics of Amino Acid and Amino Acid Salts

In the IR spectrum of amino acid generally NH_3^+ stretching, multiple combination/overtone bands, NH_3^+ bending, and C=O bands are identified. Some characteristic vibrational bands for amino acid salts are as follows:

- ***For NH_3^+ ion group***
  - *NH_3^+ stretching vibrations:* Exhibits band between 3100 – 2600 cm^{-1}.
  - *NH_3^+ bending vibrations:* Exhibits weak asymmetrical bending between 1660 – 1610 cm^{-1} and a symmetrical bending between 1550 – 1480 cm^{-1}.
  - Combination and overtone bands between 2220 – 2000 cm^{-1} due to combination of asymmetrical bending and torsional oscillation of NH_3^+.
  - *NH_3^+ torsion oscillations vibrations:* Exhibits band near 500 cm^{-1}.
- ***For Carboxyl ion group***
  - *C=O stretching vibration:* Exhibits two bands on a lower frequency than normal carboxylic acids *i.e.*, near 1600 cm^{-1} (strong asymmetrical band) and 1400 cm^{-1} (weaker symmetrical band).
- ***Other C–H bands if present in the amino acids and their salts are also observed.***

Example: ***Leucine*** (IR spectrum given in Figure 2.55).

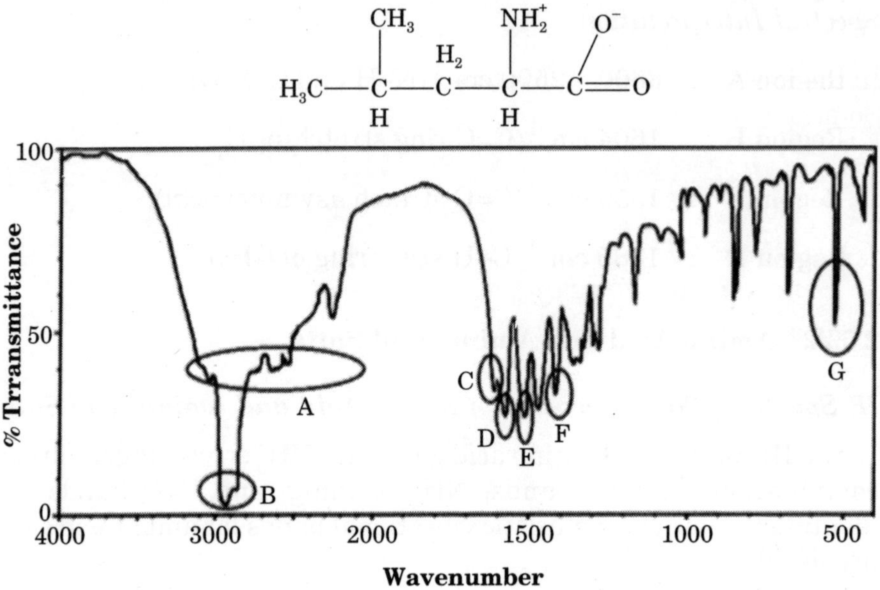

Figure 2.55: IR spectrum of leucine.

Spectral Interpretation

Region A	:	3100 – 2000 cm^{-1} (N–H stretch extended by combination band at 2140 cm^{-1} and other overtone bands).
Region B	:	2962 cm^{-1} (aliphatic C–H stretching).
Region C	:	1610 cm^{-1} (N–H bending asymmetrical).
Region D	:	1580 cm^{-1} (C=O stretching asymmetrical).
Region E	:	1505 cm^{-1} (N–H bending symmetrical).
Region F	:	1405 cm^{-1} (C=O stretching symmetrical).
Region G	:	525 cm^{-1} (N–H torsional oscillations).

2.5.23 Sulphur Compounds

2.5.23.1 *Mercaptans (Thiols)*

IR Spectral Characteristics of Mercaptans (Thiols)

In the IR spectrum of mercaptans generally S–H stretch is identified. Some characteristic vibrational bands for mercaptans are as follows:

- ***S–H stretching:*** Exhibits moderately weak band between 2600 – 2550 cm^{-1}.
- ***Other C–H bands of alkyl or aryl group if present in the mercaptan compounds are also observed.***

Example: *Hexane 1-Thiol* (IR spectrum given in Figure 2.56).

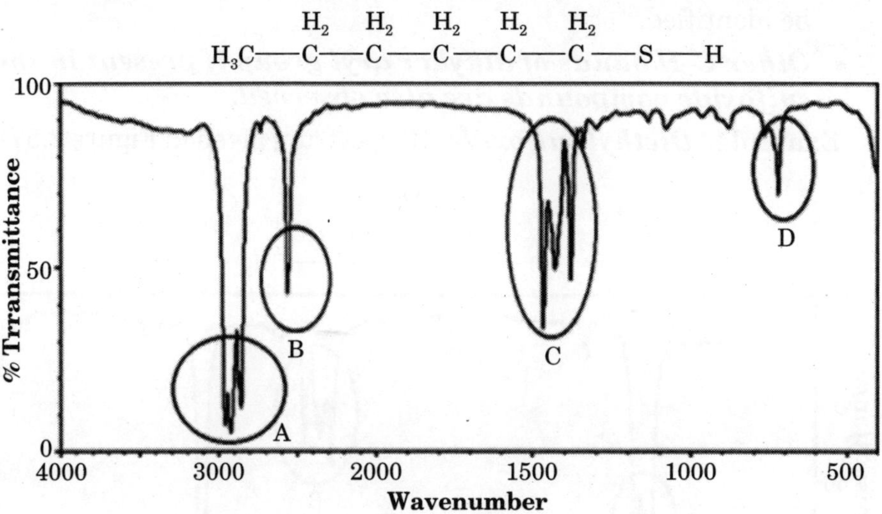

Figure 2.56: IR spectrum of hexane-1-thiol.

Spectral Interpretation

Region A : 2971 cm^{-1} (C–H asymmetrical stretch of CH_3), 2863 cm^{-1} (C–H symmetrical stretch of CH_3), 2924 cm^{-1} (C–H asymmetrical stretch of CH_2), 2853 cm^{-1} (C–H symmetrical stretch of CH_2).

Region B : 2670 cm^{-1} (S–H stretching).

Region C : 1470 cm^{-1} (C–H scissoring of CH_2), 1452 cm^{-1} (C–H asymmetrical bending of CH_3), 1383 cm^{-1} (C–H scissoring of CH_3).

Region D : 726 cm^{-1} (C–H rocking of CH_2) a long chain band.

2.5.23.2 *Sulfoxides*

IR Spectral Characteristics of Sulfoxides

In the IR spectrum of sulfoxides generally S=O stretch is identified. Characteristic vibrational bands for sulfoxides are as follows:

- ***S=O stretching:*** Exhibits a strong band near 1050 cm^{-1}. In straight chain aliphatic compounds it is easily recognised but in case of aromatic or some complex compounds it is difficult to be identified.
- ***Other C–H bands of alkyl or aryl group if present in the sulfoxide compounds are also observed.***

Example: ***Diethyl sulfoxide*** (IR spectrum given in Figure 2.57).

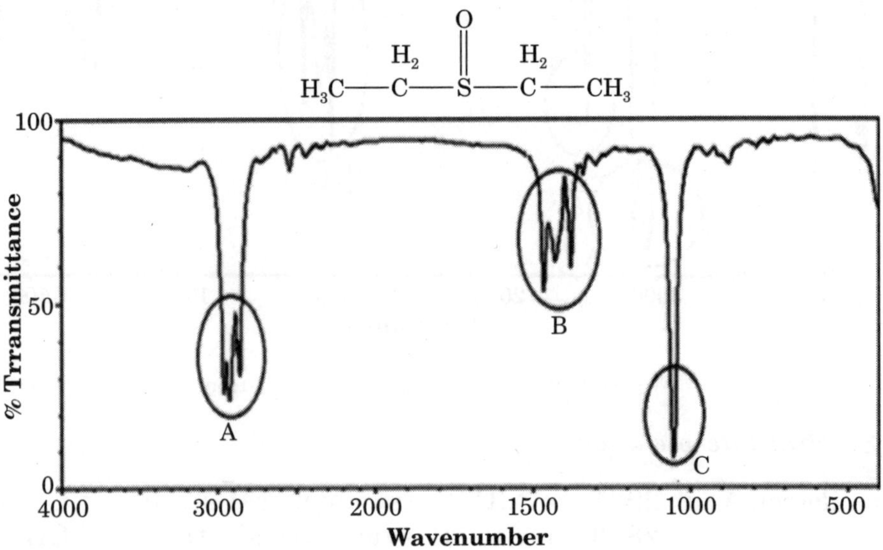

Figure 2.57: IR spectrum of diethyl sulfoxide.

Spectral Interpretation

Region A : 2971 cm^{-1} (C–H asymmetrical stretch of CH_3),
2863 cm^{-1} (C–H symmetrical stretch of CH_3),
2924 cm^{-1} (C–H asymmetrical stretch of CH_2),
2853 cm^{-1} (C–H symmetrical stretch of CH_2).

Region B : 1470 cm^{-1} (C–H scissoring of CH_2), 1452 cm^{-1} (C–H asymmetrical bending of CH_3), 1383 cm^{-1} (C–H scissoring of CH_3).

Region C : 1050 cm^{-1} (S=O stretching).

2.5.23.3 *Sulfones*

IR Spectral Characteristics of Sulfones

In the IR spectrum of sulfones generally S=O stretch is identified. Characteristic vibrational bands for sulfones are as follows:

- ***S=O stretching:*** Exhibits two strong band near 1300 cm^{-1} (asymmetrical stretching) and 1150 cm^{-1} (symmetrical stretching).
- ***Other C–H bands of alkyl or aryl group if present in the sulfone compounds are also observed.***

Example: ***Diethyl sulfone*** (IR spectrum given in Figure 2.58).

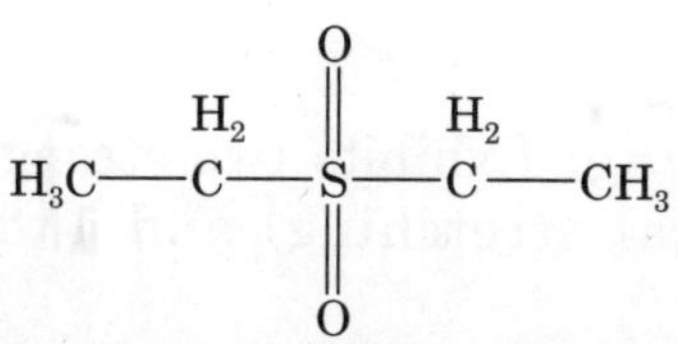

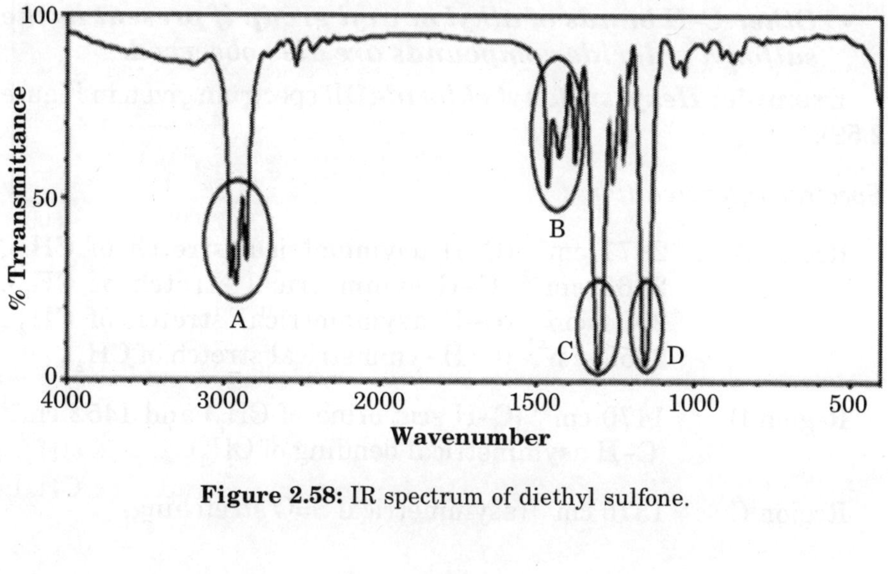

Figure 2.58: IR spectrum of diethyl sulfone.

Spectral Interpretation

Region A : 2971 cm^{-1} (C–H asymmetrical stretch of CH_3), 2863 cm^{-1} (C–H symmetrical stretch of CH_3), 2924 cm^{-1} (C–H asymmetrical stretch of CH_2), 2853 cm^{-1} (C–H symmetrical stretch of CH_2).

Region B : 1470 cm^{-1} (C–H scissoring of CH_2), 1452 cm^{-1} (C–H asymmetrical bending of CH_3), 1383 cm^{-1} (C–H scissoring of CH_3).

Region C : 1300 cm^{-1} (asymmetrical S=O stretching).

Region D : 1150 cm^{-1} (symmetrical S=O stretching).

2.5.23.4 *Sulfonyl Chlorides*

IR Spectral Characteristics of Sulfonyl Chlorides

In the IR spectrum of sulfonyl chlorides generally S=O stretch is identified. Characteristic vibrational bands for sulfonyl chlorides are as follows:

- ***S=O stretching:*** Exhibits two strong band near 1375 cm^{-1} (asymmetrical stretching) and 1185 cm^{-1} (symmetrical stretching).
- ***Other C–H bands of alkyl or aryl group if present in the sulfonyl chloride compounds are also observed.***

Example: ***Hexyl sulfonyl chloride*** (IR spectrum given in Figure 2.59).

Spectral Interpretation

Region A : 2971 cm^{-1} (C–H asymmetrical stretch of CH_3), 2863 cm^{-1} (C–H symmetrical stretch of CH_3), 2924 cm^{-1} (C–H asymmetrical stretch of CH_2), 2853 cm^{-1} (C–H symmetrical stretch of CH_2).

Region B : 1470 cm^{-1} (C–H scissoring of CH_2) and 1452 cm^{-1} (C–H asymmetrical bending of CH_3).

Region C : 1375 cm^{-1} (asymmetrical S=O stretching).

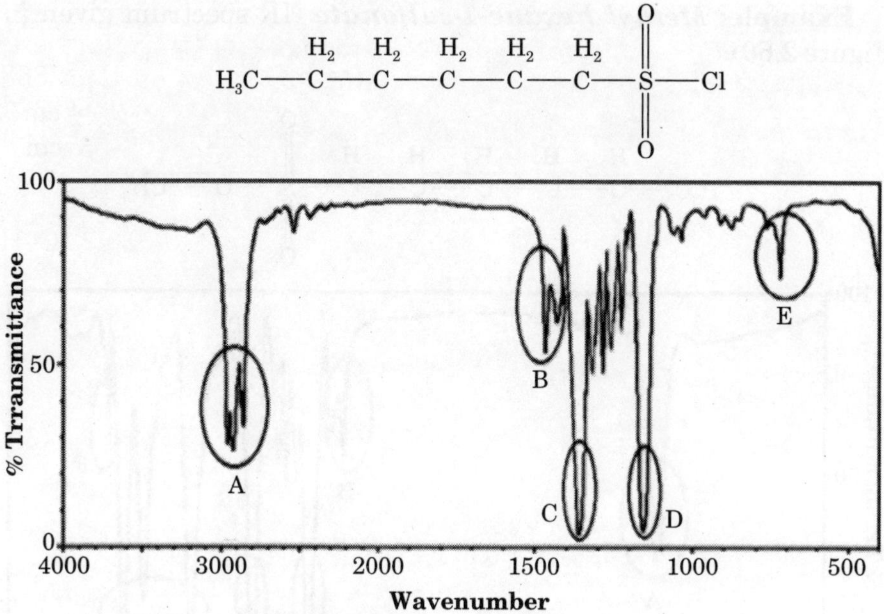

Figure 2.59: IR spectrum of hexyl sulfonyl chloride.

Region D : 1150 cm^{-1} (symmetrical S=O stretching).

Region E : 726 cm^{-1} (C–H rocking of CH_2) a long chain band.

2.5.23.5 *Sulfonates*

IR Spectral Characteristics of Sulfonates

In the IR spectrum of sulfonates generally S=O stretch and S–O stretch are identified. Characteristic vibrational bands for sulfonates are as follows:

- ***S=O stretching:*** Exhibits two strong bands near 1350 cm^{-1} (asymmetrical stretching) and 1175 cm^{-1} (symmetrical stretching).
- ***S–O stretching:*** Exhibits several strong bands between 1000 – 750 cm^{-1}.
- ***Other C–H bands of alkyl or aryl group if present in the sulfonate compounds are also observed.***

Example: *Methyl hexane-1-sulfonate* (IR spectrum given in Figure 2.60).

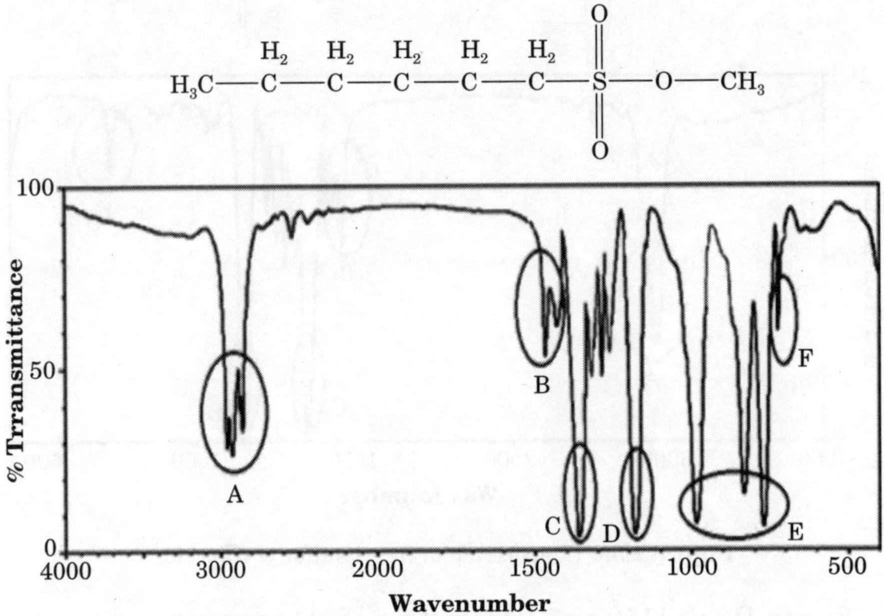

Figure 2.60: IR spectrum of methyl hexane-1-sulfonate.

Spectral Interpretation

Region A : 2971 cm^{-1} (C–H asymmetrical stretch of CH_3), 2863 cm^{-1} (C–H symmetrical stretch of CH_3), 2924 cm^{-1} (C–H asymmetrical stretch of CH_2), 2853 cm^{-1} (C–H symmetrical stretch of CH_2).

Region B : 1470 cm^{-1} (C–H scissoring of CH_2) and 1452 cm^{-1} (C–H asymmetrical bending of CH_3).

Region C : 1350 cm^{-1} (asymmetrical S=O stretching).

Region D : 1175 cm^{-1} (symmetrical S=O stretching).

Region E : 1000 – 750 cm^{-1} (several S–O stretching bands).

Region F : 726 cm^{-1} (C–H rocking of CH_2) a long chain band.

2.5.23.6 *Sulfonamides*

IR Spectral Characteristics of Sulfonamides

In the IR spectrum of sulfonamides generally S=O stretch, N–H stretch, and N–H bending are identified. Characteristic vibrational bands for sulfonamides are as follows:

- ***N–H stretching in primary amides:*** Exhibits two bands near 3350 cm^{-1} (asymmetrical stretching) and 3250 cm^{-1} (symmetrical stretching).
- ***N–H stretching in secondary amides:*** Exhibits band near 3250 cm^{-1}.
- ***N–H bending:*** Exhibits one band near 1550 cm^{-1}.
- ***S=O stretching:*** Exhibits two strong bands near 1325 cm^{-1} (asymmetrical stretching) and 1140 cm^{-1} (symmetrical stretching).
- ***Other C–H bands of alkyl or aryl group if present in the sulfonamide compounds are also observed.***

Example: ***Ethyl sulphonamide*** (IR spectrum given in Figure 2.61).

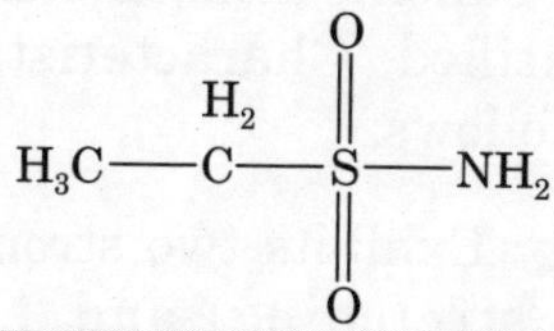

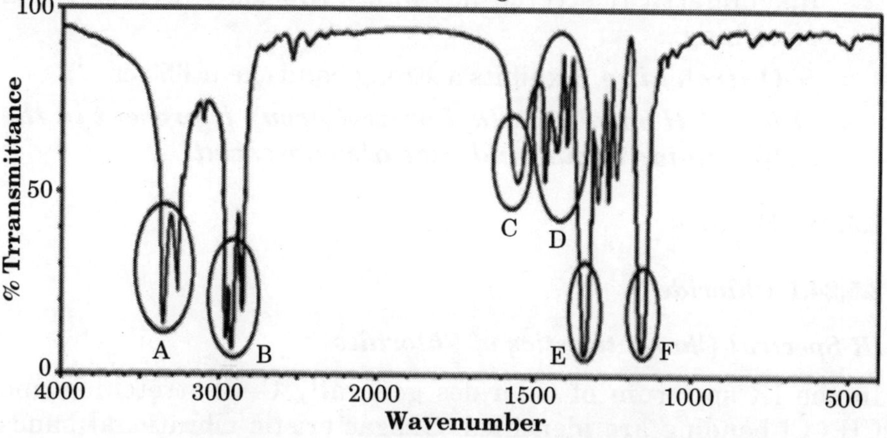

Figure 2.61: IR spectrum of ethyl sulfonamide.

Spectral Interpretation

Region A : 3350 cm^{-1} (asymmetrical N–H stretching) and 3250 cm^{-1} (symmetrical N–H stretching).

Region B : 2971 cm^{-1} (C–H asymmetrical stretch of CH_3), 2863 cm^{-1} (C–H symmetrical stretch of CH_3), 2924 cm^{-1} (C–H asymmetrical stretch of CH_2), 2853 cm^{-1} (C–H symmetrical stretch of CH_2).

Region C : 1550 cm^{-1} (asymmetrical N–H bending).

Region D : 1470 cm^{-1} (C–H scissoring of CH_2) and 1452 cm^{-1} (C–H asymmetrical bending of CH_3), and 1383 cm^{-1} (C–H scissoring of CH_3).

Region E : 1325 cm^{-1} (asymmetrical S=O stretching).

Region F : 1140 cm^{-1} (symmetrical S=O stretching).

2.5.23.7 *Sulfonic Acid*

IR Spectral Characteristics of Sulfonic Acid

In the IR spectrum of sulfonic acid generally S=O stretch, and S–O stretching's are identified. Characteristic vibrational bands for sulfonic acids are as follows:

- ***S=O stretching:*** Exhibits two strong bands near 1350 cm^{-1} (asymmetrical stretching) and 1150 cm^{-1} (symmetrical stretching).
- ***S–O stretching:*** Exhibits a strong bands near 650 cm^{-1}.
- ***Other C–H bands of alkyl or aryl group if present in the sulfonamide compounds are also observed.***

2.5.24 Halides

2.5.24.1 *Chloride*

IR Spectral Characteristics of Chlorides

In the IR spectrum of chlorides generally C–Cl stretching and CH_2–Cl bending are identified. Characteristic vibrational bands for chlorides are as follows:

- ***C–Cl stretching:*** Exhibits a strong band between 785 – 540 cm^{-1}. There can be more than one band in the given range attributed to different conformations of chlorine:
    - *C–Cl stretching in* CH_2Cl_2: Exhibits band near 739 cm^{-1}.
    - *C–Cl stretching in* $CHCl_3$: Exhibits band near 759 cm^{-1}.
    - *C–Cl stretching in* CCl_4: Exhibits band near 785 cm^{-1}.
- ***C–Cl bending in*** CH_2***–Cl:*** Exhibits bending (wagging) between 1300 – 1230 cm^{-1}.
- ***Other C–H bands alkyl or aryl group if present in the chloride compounds are also observed.***

Example: ***Tetra chloro methane*** (IR spectrum given in Figure 2.62).

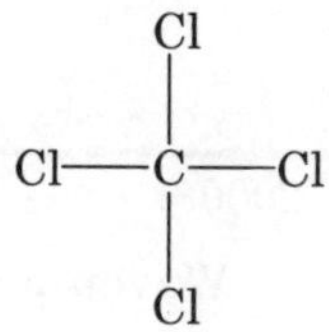

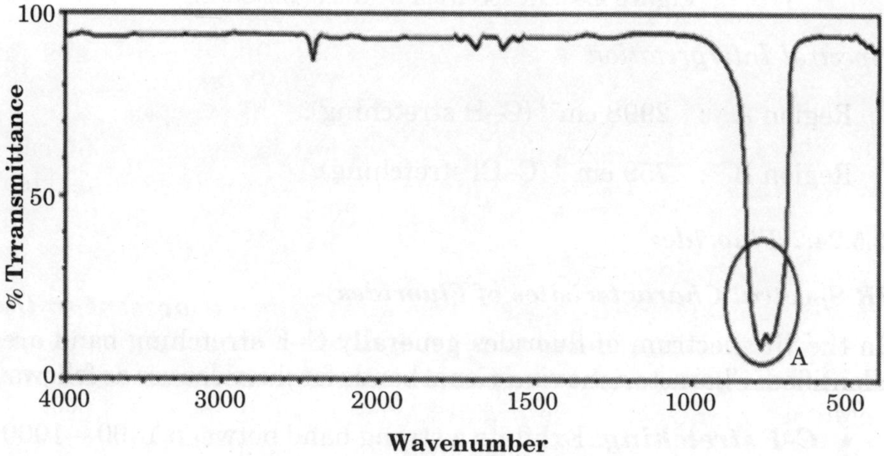

Figure 2.62: IR spectrum tetra chloro methane.

Spectral Interpretation

Region A : 785 cm^{-1} (C–Cl stretching).

Example: ***Tri chloro methane*** (IR spectrum given in Figure 2.63).

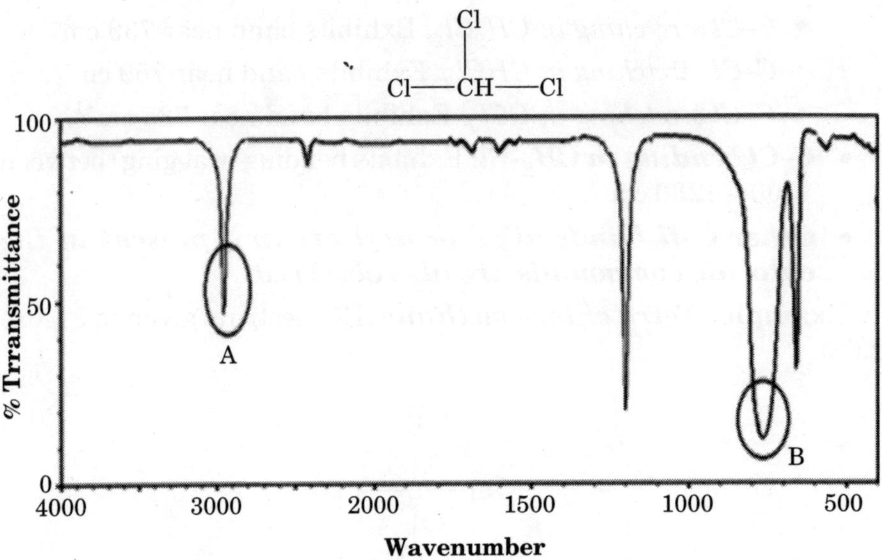

Figure 2.63: IR spectrum tri chloro methane.

Spectral Interpretation

Region A : 2998 cm^{-1} (C–H stretching).

Region B : 759 cm^{-1} (C–Cl stretching).

2.5.24.2 *Fluorides*

IR Spectral Characteristics of Fluorides

In the IR spectrum of fluorides generally C–F stretching band are identified. Characteristic vibrational bands for fluorides are as follows:

- ***C–F stretching:*** Exhibits a strong band between 1400 – 1000 cm^{-1}.
    - *C–F stretching in mono fluoro alkanes:* Exhibits one strong band between 1100 – 1000 cm^{-1}.
    - *C–F stretching in poly fluoro alkanes:* Exhibits one strong band between 1350 – 1100 cm^{-1}.

- ***C–F stretching aryl fluorides:*** Exhibits a strong band between 1250 – 1100 cm^{-1}.
- ***Other C–H bands alkyl or aryl group if present in the fluoride compounds are also observed.***

2.5.24.3 *Bromides*

IR Spectral Characteristics of Bromides

In the IR spectrum of bromides generally C–Br stretching and C–Br bending band are identified. Characteristic vibrational bands for bromides are as follows:

- ***C–Br stretching:*** Aliphatic bromides exhibits strong band between 650 – 510 cm^{-1}. Aromatic bromides exhibits strong band between 1075 – 1030 cm^{-1}.
- ***CH_2–Br bending:*** Exhibits band between 1250 – 1190 cm^{-1}.
- ***Other C–H bands alkyl or aryl group if present in the bromide compounds are also observed.***

2.5.23.4 *Iodides*

IR Spectral Characteristics of Iodides

In the IR spectrum of iodides generally C–I stretching and C–I bending band are identified. Characteristic vibrational bands for iodides are as follows:

- ***C–I stretching:*** Aliphatic iodides exhibits strong band between 600 – 485 cm^{-1}.
- ***CH_2–I bending:*** Exhibits band between 1200 – 1150 cm^{-1}.
- ***Other C–H bands alkyl or aryl group if present in the iodides compounds are also observed.***

2.5.25 Phosphorus Compounds

2.5.25.1 *Phosphines*

IR Spectral Characteristics of Phosphines

In the IR spectrum of phosphines generally P–H stretching, P–H bending, and P–C bands are identified. Characteristic vibrational bands for phosphines are as follows:

- ***P–H stretching:*** Exhibits two strong and sharp bands between 2320 – 2270 cm^{-1} (asymmetrical stretching) and 1150 cm^{-1} (symmetrical stretching).
- ***PH_2 bending:*** Exhibits two medium bands between 1090 – 1075 cm^{-1} and 840 – 810 cm^{-1}.
- ***P–H bending:*** Exhibits a medium band between 990 – 885 cm^{-1}.
- ***P–CH_3 bending:*** Exhibits medium band between 1450 – 1395 and 1346 – 1255 cm^{-1}.
- ***P–CH_2 bending:*** Exhibits medium band between 1440 – 1400 cm^{-1}.
- ***Other C–H bands of alkyl or aryl group if present in the phosphine compounds are also observed.***

2.5.25.2 *Phosphine Oxides*

IR Spectral Characteristics of Phosphine Oxides

In the IR spectrum of phosphine oxides generally P=O stretching band is identified. Characteristic vibrational bands for phosphine oxides are as follows:

- ***P=O stretching:*** Exhibits one strong band between 1210 – 1140 cm^{-1}
- ***Other C–H bands alkyl or aryl group if present in the phosphine oxide compounds are also observed.***

2.5.25.3 *Phosphate Esters*

IR Spectral Characteristics of Phosphate Esters

In the IR spectrum of phosphate esters generally P=O, P–O and C–O stretching band are identified. Characteristic vibrational bands for phosphate esters are as follows:

- ***P=O stretching:*** Exhibits one strong band between 1300 – 1240 cm^{-1}.

- ***P–O stretching:*** Exhibits one or more medium bands between 845 – 725 cm^{-1}.
- ***C–O stretching:*** Exhibits one strong band between 1088 – 920 cm^{-1}.
- ***Other C–H bands of alkyl or aryl group if present in the phosphate esters are also observed.***

2.5.26 Heterocyclic Compounds

Example: *Furan* (IR spectrum given in Figure 2.64).

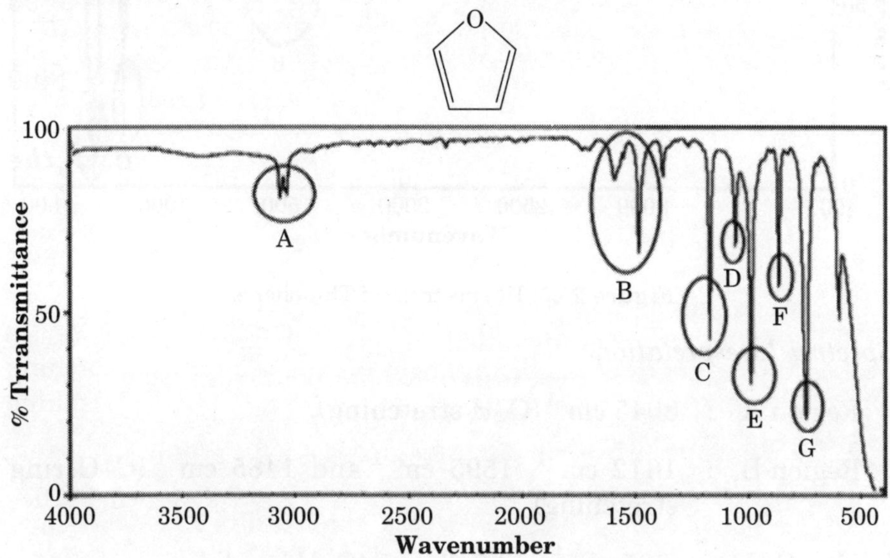

Figure 2.64: IR spectrum of furan.

Spectral Interpretation

Region A : 3086 cm^{-1} and 3025 cm^{-1} (C–H stretching).

Region B : 1592 cm^{-1} and 1485 cm^{-1} (C–C ring stretching).

Region C : 1172 cm^{-1} (C–O stretching asymmetrical).

Region D : 1060 cm^{-1} (C–O stretching symmetrical).

Region E : 991 cm^{-1} (C–H bending).

Region F&G : 869 cm^{-1} and 746 cm^{-1} (C–H bending).

Example: ***Thiophene*** (IR spectrum given in Figure 2.65).

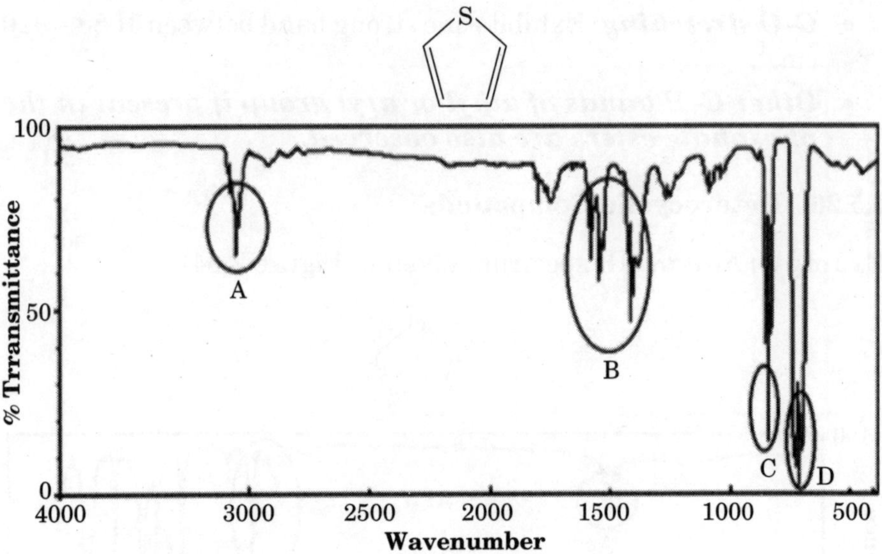

Figure 2.65: IR spectrum of Thiophene.

Spectral Interpretation

Region A : 3045 cm^{-1} (C–H stretching).

Region B : 1612 cm^{-1}, 1595 cm^{-1} and 1485 cm^{-1} (C–C ring stretching).

Region C&D : 865 cm^{-1} and 742 cm^{-1} (C–H bending).

Example: ***Indole*** (IR spectrum given in Figure 2.66).

Spectral Interpretation

Region A : 3458 cm^{-1} (N–H stretching).

Region B : 3064 cm^{-1} and 3032 cm^{-1} (C–H stretching).

Region C : 2000 – 1650 cm^{-1} (overtone and combination bands).

Region D : 1590 cm^{-1}, 1485 cm^{-1} and 1402 cm^{-1} (C–C ring stretching).

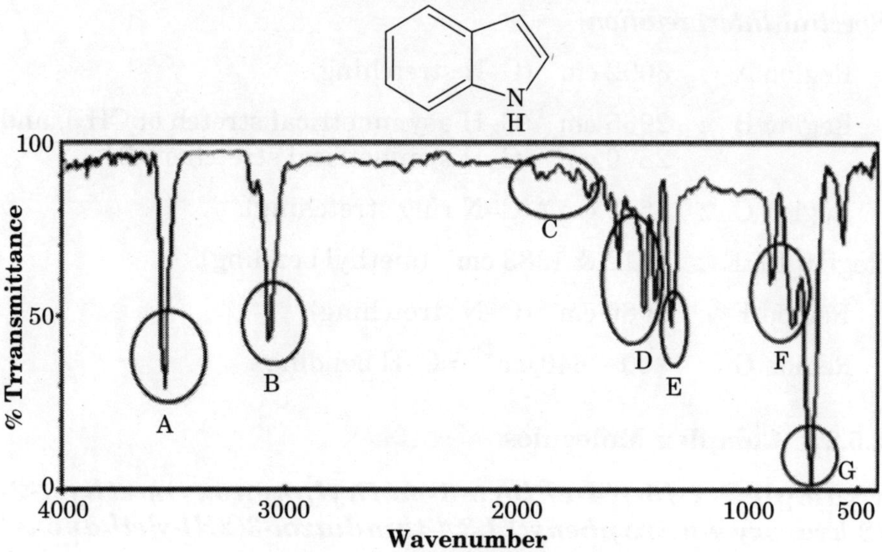

Figure 2.66: IR spectrum of Indole.

Region E : 1335 cm^{-1} (C–N stretching).

Region F&G : 912 cm^{-1}, 810 cm^{-1} and 720 cm^{-1} (C–H bending).

Example: ***1-methyl imidazole*** (IR spectrum given in Figure 2.67).

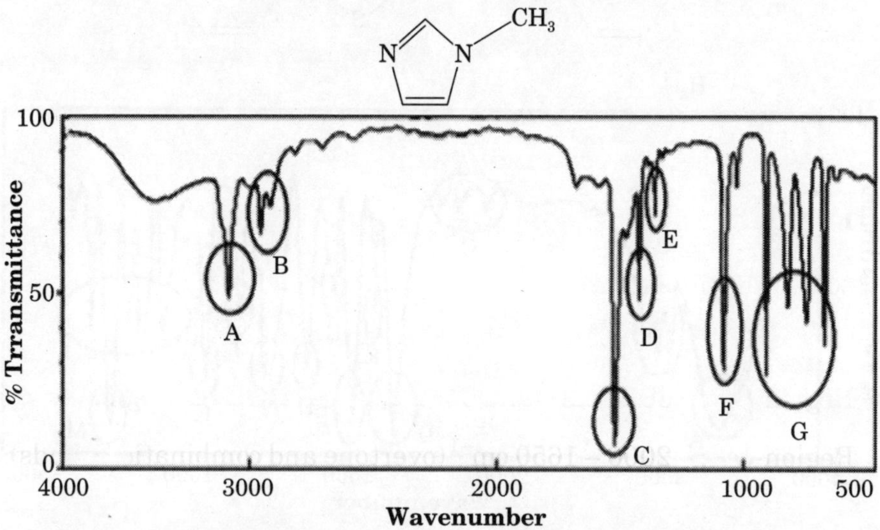

Figure 2.67: IR spectrum of 1-methyl imidazole.

Spectral Interpretation

Region A : 3092 cm^{-1} (C–H stretching).

Region B : 2965 cm^{-1} (C–H asymmetrical stretch of CH_3), and 2870 cm^{-1} (C–H symmetrical stretch of CH_3).

Region C : 1512 cm^{-1} (C=N ring stretching).

Region D&E : 1412 & 1383 cm^{-1} (methyl bending).

Region F : 1080 cm^{-1} (C–N stretching).

Region G : 900 – 640 cm^{-1} (=C–H bending).

2.5.27 Complex Molecules

Example: ***1-[5-{(4-chloro-3-methylphenoxy)methyl}-2-(2-hydroxy-4-nitro phenyl) 1,3,4-thiadiazol-3(2H)-ylethanone]*** (IR spectrum given in Figure 2.68).

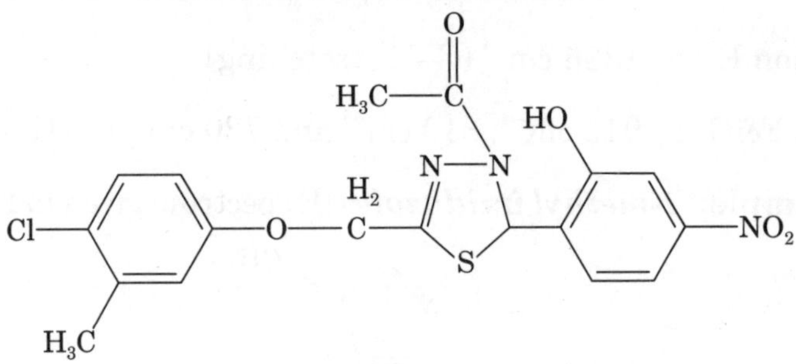

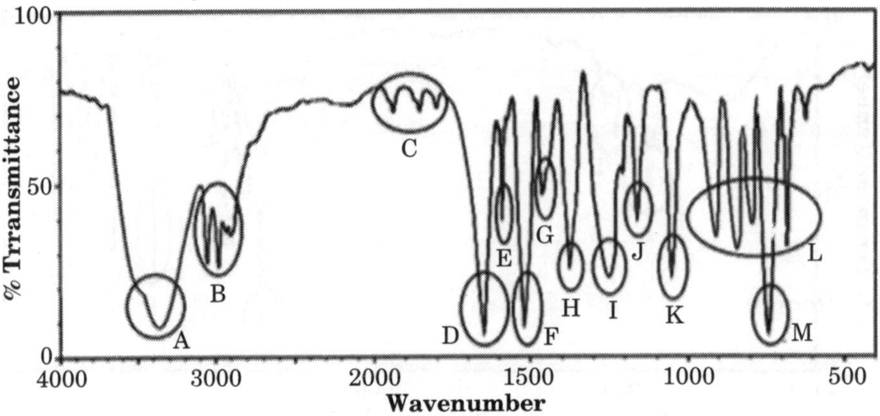

Figure 2.68: IR spectrum of 1-[5-{(4-chloro-3-methylphenoxy)methyl}-2-(2-hydroxy-4-nitrophenyl)-1, 3, 4-thiadiazol-3 (2H)-ylethanone].

Spectral Interpretation

Region A : 3372 cm^{-1} (O–H Stretch).

Region B : 3045 cm^{-1} (Aromatic C–H stretch), 2968 cm^{-1} (C–H symmetrical stretch of CH_3), & 2872 cm^{-1} (C–H symmetrical stretch of CH_3).

Region C : 2000 – 1650 cm^{-1} (overtone and combination bands).

Region D : 1640 cm^{-1} (C=O Stretching, tertiary amide).

Region E : 1598 cm^{-1} (C–C ring stretching).

Region F : 1515 cm^{-1} (N–O stretching asymmetrical).

Region G : 1470 cm^{-1} (C–H bending of CH_2).

Region H : 1380 cm^{-1} (N–O stretching symmetrical).

Region I : 1240 cm^{-1} (C–O stretching, asymmetrical of ether).

Region J : 1124 cm^{-1} (C–H bending of CH_2).

Region K : 1040 cm^{-1} (C–O bending of symmetrical of ether).

Region L : 900 – 640 cm^{-1} (=C–H bending).

Region M : C–Cl cm^{-1} (C–Cl stretching).

2.5.28 Pharmaceutical Compounds

Example: *Methyl salicylate* (IR spectrum given in Figure 2.69).

Spectral Interpretation

Region A : 3372 cm^{-1} (aromatic C–H stretching).

Region B : 3012 cm^{-1} (aromatic C–H stretching).

Region C : 2986 cm^{-1} (C–H asymmetrical stretch of CH_3), 2878 cm^{-1} (C–H symmetrical stretch).

Region D : 1726 cm^{-1} (C=O stretching).

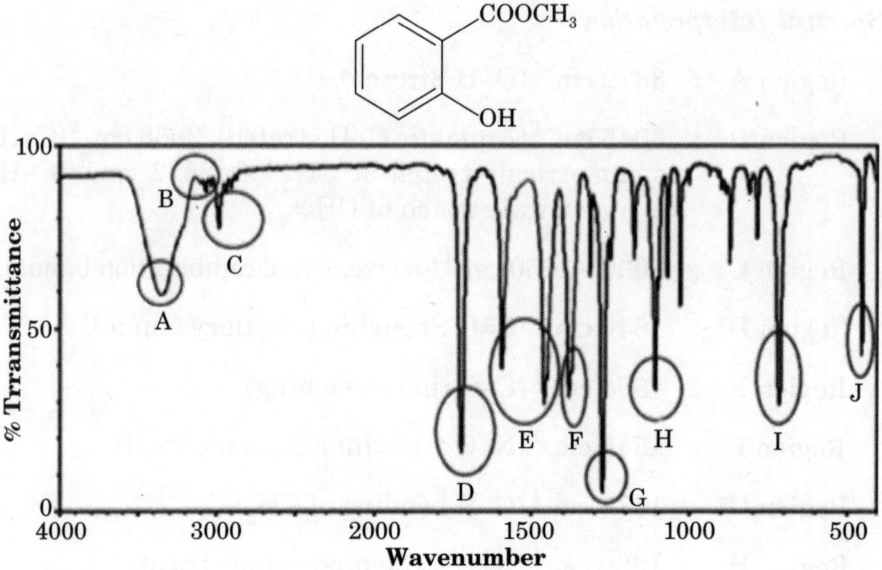

Figure 2.69: IR spectrum of methyl salicylate.

Region E : 1592 cm^{-1}, and 1458 cm^{-1} (C–C ring stretching).

Region F : 1383 cm^{-1} (C–H scissoring of CH_3).

Region G : 1286 cm^{-1} [C–C(=O)–C stretching in aromatic acid esters].

Region H : 1117 cm^{-1} (O–C–C stretching).

Region I : 730 cm^{-1} (Out of plane C–H bending).

Region J : 438 cm^{-1} (Out of plane ring bending).

Example: ***Isoniazid*** (IR spectrum given in Figure 2.70).

Spectral Interpretation

Region A : 3394 cm^{-1} (N–H stretching of NH group).

Region B : 3312 cm^{-1} (N–H asymmetrical stretch of NH_2 group).

Region C : 3114 cm^{-1} (N–H symmetrical stretch of NH_2 group).

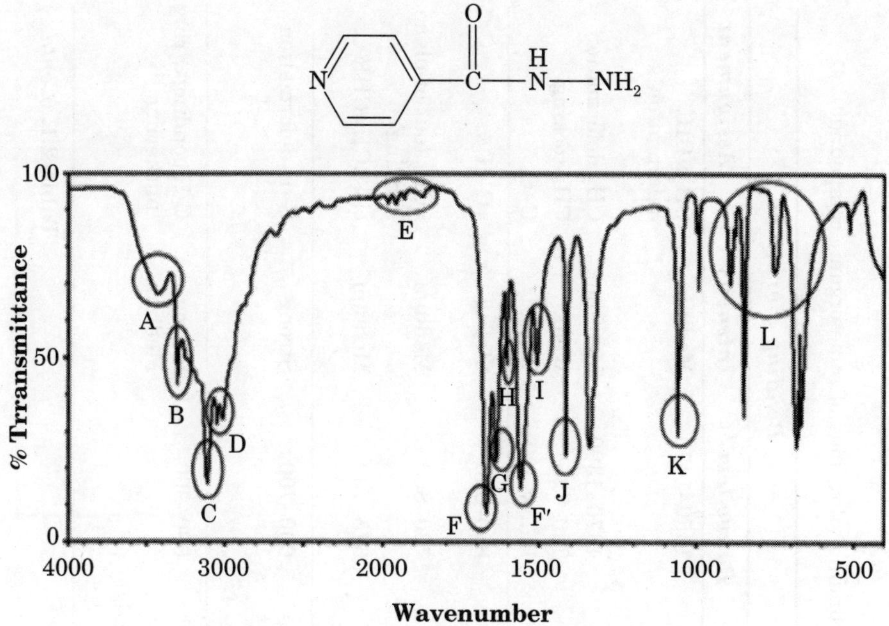

Figure 2.70: IR spectrum of isoniazid.

Region D : 3058 cm^{-1} (Aromatic C–H Stretch).

Region E : 2000 – 1800 cm^{-1} (overtone bands).

Region F : 1670 cm^{-1} (C=O Stretching, amide I band).

Region F′ : 1557 cm^{-1} (N–H bending, amide II band).

Region I : 1508 cm^{-1} (C=N ring stretching).

Region G,H,J : 1640 cm^{-1}, 1604 and 1412 cm^{-1} (C=C ring stretching).

Region K : 1070 cm^{-1} (C–N Stretching).

Region L : 900 – 675 cm^{-1} (=C–H bending out of plane).

For Further Assistance in IR Spectra Interpretation of Compounds Refer Table 2.1, Figure 2.71, Figure 2.72, Figure 2.73, and Figure 2.74.

Table 2.1. General IR absorption range, intensities and corresponding vibrations of some organic compounds.

Compounds	*Stretching vibrations*			*Bending vibrations*		
	Range (***cm^{-1}***)	***Intensity***	***Assignment***	***Range*** (***cm^{-1}***)	***Intensity***	***Assignment***
Alkanes	2850–3000	Strong	C–H Stretch CH_3, CH_2 & CH (2 or 3 bands)	1350–1470	Medium	CH_2&CH_3 deformation
				1370–1390	Medium	CH_3 deformation
				720–725	Weak	CH_2 rocking
	400–800	Weak	C–C	Below 500	Weak	C–C
Alkenes	3020–3100	Medium	=C-H & =CH_2 (usually sharp)	880–995	Strong	=C–H &
	1630–1680	Medium-Weak	C=C (symmetry reduces intensity)	780–850	Medium	=CH_2 (out-of-plane bending)
	1900–2000	Strong	C=C asymmetric stretch	675–730	Medium	cis-RCH=CHR
Alkynes	3300	Strong	C–H (usually sharp)	600–700	Strong	C–H deformation
	2100–2250	Varying	C≡C (symmetry reduces intensity)			
Aromatic compounds	3030	Varying	C–H (may beseveral bands)	690–900	Strong–Medium	C-H bending & ring puckering
	1600 & 1500	Medium-Weak	C=C (in ring) (2 bands) (3 if conjugated)			

Table 2.1. (*Contd...*)

Table 2.1. (*Contd...*)

Compounds	***Stretching vibrations***			***Bending vibrations***		
	Range (cm^{-1})	***Intensity***	***Assignment***	***Range*** (cm^{-1})	***Intensity***	***Assignment***
Alcohols & Phenols	3580–3650	Varying	O–H (free), usually sharp	1330–1430	Medium	O–H bending (in-plane)
	3200–3550	Strong	O–H (H-bonded), usually broad	650–770	Varying-Weak	O–H bend (out-of-plane)
	970–1250	Strong	C–O			
Amines	3400–3500 (dil. soln.)	Weak	N–H (1°-amines) 2 bands	1550–1650	Medium-Strong	NH_2 scissoring (1°-amines)
	3300–3400 (dil. soln.)	Weak	N–H (2°-amines)	660–900	var	NH_2 & N–H wagging (shifts on H-bonding)
	1000–1250	Medium	C–N			
Aldehydes & Ketones	2690–2840 (2 bands)	Medium	C–H (aldehyde C–H)	1350–1360	Strong	α-CH_3 bending
	1720–1740	Strong	C=O (saturated aldehyde)	1400–1450	Strong	α-CH_2 bending
	1710–1720	Strong	C=O (saturated ketone)	1100	Medium	C–C–C bending
	1690	Strong	aryl ketone			
	1675	Strong	α, β-unsaturation			
	1745	Strong	cyclopentanone			
	1780	Strong	cyclobutanone			

Table 2.1. (*Contd...*)

Table 2.1. (*Contd...*)

Compounds	***Stretching vibrations***			***Bending vibrations***		
	Range (cm^{-1})	***Intensity***	***Assignment***	***Range (cm^{-1})***	***Intensity***	***Assignment***
Carboxylic Acids & their Derivatives	2500–3300 (acids) overlap C–H	Strong	O–H (very broad)	1395–1440	Medium	C–O–H bending
	1705–1720 (acids)	Strong	C=O (H-bonded)	1590–1650	Medium	N–H (1°-amide) II band
	1210–1320 (acids)	Medium-Strong	O–C (sometimes 2-peaks)	1500–1560	Medium	N–H (2°-amide) II band
	1785–1815 (acyl halide)	Strong	C=O			
	1750 & 1820 (anhydrides)	Strong	C=O (2-bands)			
	1040–1100 (anhydrides)	Strong	O–C			
	1735–1750 (esters)	Strong	C=O			
	1000–1300 (esters)	Strong	O–C (2-bands)			
	1630–1695 (amides)	Strong	C=O (amide I band)			
Nitro Compound	Near 1550 & 1350	Strong	N–O (2- bands)			

Table 2.1. (*Contd...*)

Table 2.1. (*Contd...*)

Compounds	***Stretching vibrations***			***Bending vibrations***		
	Range (cm^{-1})	***Intensity***	***Assignment***	***Range (cm^{-1})***	***Intensity***	***Assignment***
Nitriles	2270–2210	Medium	C≡N (sharp)			
Isocyanates, Isothiocyanates, Diimides, Azides & Ketenes	2100–2270	Medium	–N=C=O, –N=C=S –N=C=N–, –N_3, C=C=O			
Imines & oximes	1690–1640	Weak & Strong	C=N			
Thiols	2600–2550	Weak	S–H			
Sulfoxides	Near 1050	Strong	S=O			
Sulfones	Near 1300	Strong	S=O			
Sulfonyl chlorides	Near 1375 & 1185	Strong	S=O (2 bands)			
Sulfonates	Near 1350 & 1175	Strong	S=O (2 bands)			
	100–750	Strong	S–O (several bands)			
Sulfonamide	Near 3350 & 3250	Medium	N–H (2-bands) 1° amide	1550	Medium	N–H
	Near 3250	Medium	N–H (2° amide)			

Table 2.1. (*Contd...*)

Table 2.1. (*Contd...*)

Compounds	***Stretching vibrations***			***Bending vibrations***		
	Range (cm^{-1})	***Intensity***	***Assignment***	***Range (cm^{-1})***	***Intensity***	***Assignment***
	Near 1325 & 1140	Strong	S=O (2-bands)			
	Near 1350 & 1150	Strong	S=O (2-bands)			
Sulfonic acid	Near 650	Strong	S-O (several bands)			
Phosphines	2320–2270 & near 1150	Strong	P–H (2-bands)	1090–1075 cm^{-1} & 840–810	Medium	PH_2 (2-bands)
				990–885	Medium	P–H
				1450–1395 & 1346–1255	Medium	P–CH_3 (2-bands)
				1440–1400	Medium	P-CH_2
Phosphine Oxides	1210–1140	Strong	P=O			
Phosphate esters	1300–1240	Strong	P=O			
	1088–920	Strong	C–O			
	845–725	Medium	P–O			
Halides	785–540	Strong	C–Cl			
	1400–1100	Strong	C–F			
	600–485	Strong	C–I			
	1200–1150	Strong	CH_2I			
	650–510	Strong	C–Br			

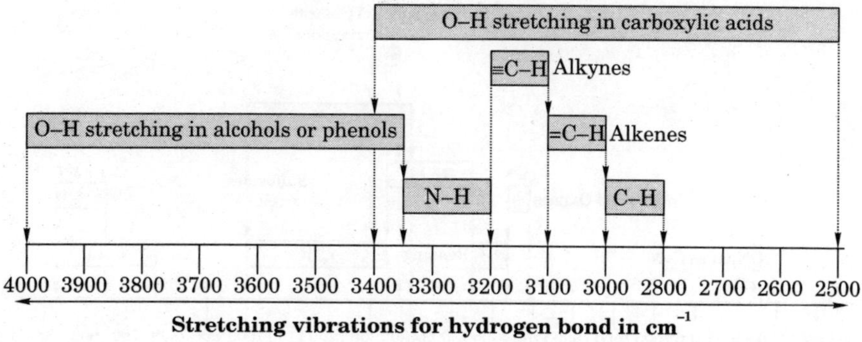

Figure 2.71: Chart representing stretching vibrations in IR region from 4000 – 2500 cm^{-1}.

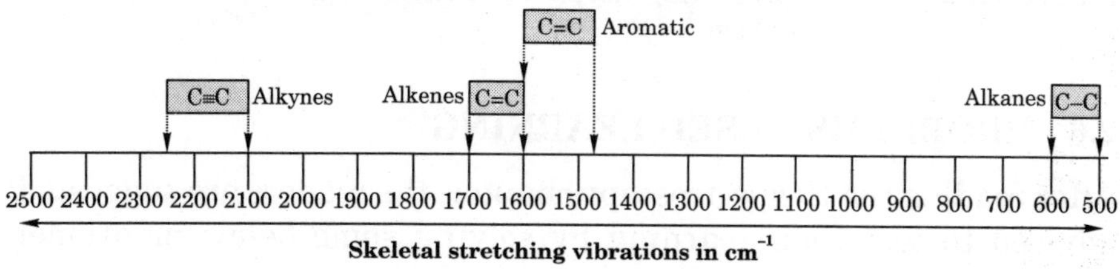

Figure 2.72: Chart representing skeletal stretching vibrations in IR region from 2500 – 500 cm^{-1}.

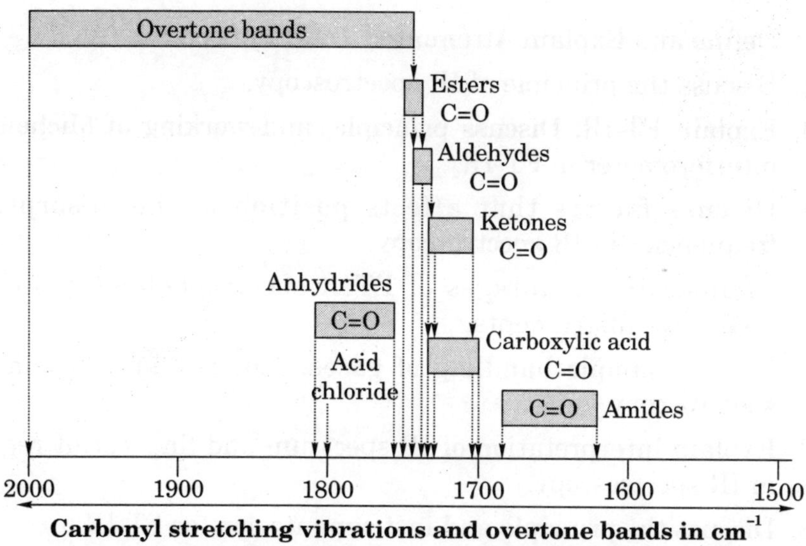

Figure 2.73: Chart representing carbonyl stretching vibrations and overtone bands in IR region from 2000 – 1500 cm^{-1}.

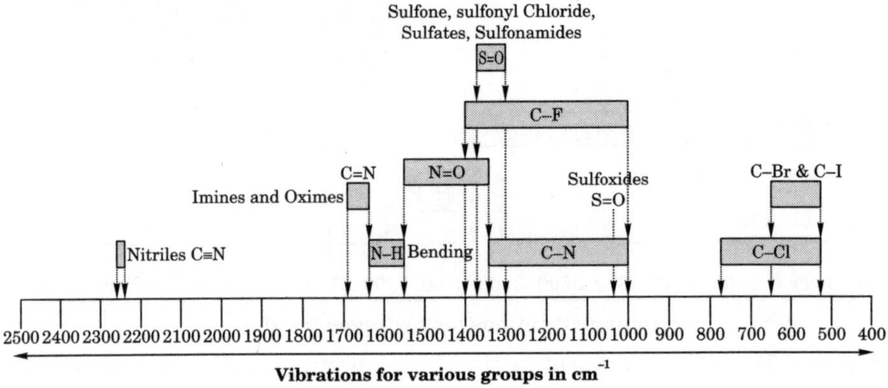

Figure 2.74: Chart representing vibrations of different groups in IR region from 2500 – 400 cm^{-1}.

2.6 PROBLEMS BASED LEARNING

After reading the IR spectroscopy chapter, the readers are suggested/ advised to test their learning by solving some below mentioned problems.

A. SUBJECTIVE PROBLEMS

1. Define and Explain Attenuated Total Reflectance (ATR).
2. Discuss the principle of IR spectroscopy.
3. Explain FT-IR. Discuss principle, and working of Michelson interferometer in FT-IR.
4. Discuss factors that affects position of the absorption frequencies in IR spectroscopy.
5. Enumerate advantages of FTIR spectrophotometers over dispersive instruments.
6. Discuss sample handling of gases, liquids and solids in IR spectroscopy.
7. Explain interpretation of IR spectrum and fingerprint region in IR spectroscopy.
8. Discuss the principle and instrumentation for FT-IR.
9. Explain modes of Vibrations in linear and non linear molecule.

10. Describe detectors used in infrared spectrophotometers.
11. Enlist the advantages of single beam IR over double beam IR instruments.
12. Describe various regions of IR radiations? Explain various types of stretching and bending vibrations with suitable example.
13. Discuss various types of stretching and bending vibrations in compounds IR spectra.

B. OBJECTIVE PROBLEMS

1. State True or False

(*i*) Position of these vibration bands in the IR spectrum of a molecule depends upon Bond strength and mass of atoms.

(*ii*) C=C always absorbs on higher frequency than C–C because of high bond strength.

(*iii*) O–H is always absorbed at higher frequency than C–C because of small reduced mass.

(*iv*) For a linear molecule of H_2O theoretically there can be four vibrational degree of freedoms.

(*v*) Absorbance is expressed as $A = \log_{10}$ (1/T).

(*vi*) Coupled vibrations occurs due to strong mechanical coupling between two bonds attached to the central atom.

(*vii*) Alkylation adjacent to carbonyl group causes C=O stretching to absorbs at lower frequency due to positive inductive effect.

(*viii*) Placement electronegative atom adjacent to carbonyl group causes C=O stretching to absorbs at higher frequency due to negative inductive effect.

(*ix*) Hydrogen bonding causes decrease in wavenumber of absorption of vibrational frequency.

(*x*) The number of rulings in a grating ranges from 15000 to 30000 grooves/lines per square inch or more.

Ans.: 1. (*i*) True (*ii*) True (*iii*) True (*iv*) True
(*v*) True (*vi*) True (*vii*) True (*viii*) True
(*ix*) True (*x*) True.

2. Fill up the blanks

(*i*) Practical IR region for organic compounds is from .... to .... cm^{-1}.

(*ii*) Transitions which are accompanied by a change in .... of a molecule are called as IR active transitions.

(*iii*) When the IR transitions occurs from a ground state ($v = 0$) to the first excited state ($v = 1$) it generates .... bands.

(*iv*) When the IR transitions occurs from a ground state ($v = 0$) to the second excited state ($v = 1$) it generates .... bands.

(*v*) Phenomenon that involves interaction between fundamental vibrations and overtone vibrations or combination-tone vibrations is called as ....

(*vi*) When lone pair of electron of atom contributes in extending conjugation with the double bond of a group, and thereby lowers vibrational frequency of its group this effect is called as .... effect.

(*vii*) Device that renders radiations parallel is called as ....

(*viii*) Device that makes a narrow optical image of radiation source is called as ....

(*ix*) Device that disperse the incident radiation is called as ....

(*x*) Device that isolates the desired spectral band by blocking the other dispersed radiations is called as ....

(*xi*) Technique that undergoes 7–10 internal reflections is called as .... reflectance.

(*xii*) Technique that undergoes more than 25 internal reflections is called as .... reflectance.

(*xiii*) When angle of reflection of reflected light from surface of sample becomes equals to angle of incidence then it mediates .... reflectance.

(*xiv*) The energy which penetrates one or more particles is reflected in all directions. This component is called as .... reflectance.

Ans.: 2. (*i*) 400 – 4000 (*ii*) Dipole moment (*iii*) Fundamental (*iv*) Overtone (*v*) Fermi resonance (*vi*) Mesomeric (*vii*) Collimator (*viii*) Entrance slit (*ix*) Monochromator (*x*) Exit slit (*xi*) Attenuated total (*xii*) Multiple Internal (*xiii*) Specular (*xiv*) Diffuse.

3. Match the following

(*i*) Symmetrical stretching	(*a*) Two atoms are equidistant from central atom.
(*ii*) Asymmetrical stretching	(*b*) Two atoms are non-equidistant differt from central atom.
(*iii*) Scissoring	(*c*) Two atoms are moves towards each other with respect to central atom.
(*iv*) Rocking	(*d*) Two atoms are moves in same direction with respect to central atom.
(*v*) Wagging	(*e*) Two atoms are moves either above or below the plane of central atom.
(*vi*) Twisting	(*f*) Among two atoms one move above and other below the plane of central atom.

Ans.: 3. (*i*) – (*a*), (*ii*) – (*b*), (*iii*) – (*c*), (*iv*) – (*d*), (*v*) – (*e*). (*vi*) – (*f*).

4. Match the following

(*i*) Nernst glower	(*a*) Made up of earth oxides.
(*ii*) Globar Source	(*b*) Made up of silicon carbide.
(*iii*) Incandescent Light bulb	(*c*) Contains a glass bulb with vacuum sealed tungsten filament.
(*iv*) Incandescent Wire	(*d*) Made up of Nichrome wire.
(*v*) Mercury arc	(*e*) Made up of quartz jacketed tube.

Ans.: 4. (*i*) – (*a*), (*ii*) – (*b*), (*iii*) – (*c*), (*iv*) – (*d*), (*v*) – (*e*).

5. Match the following

(*i*) Golay cell	(*a*) Composed of xenon gas filled metal cylinder whose one end is closed with a mirror diaphragm.
(*ii*) Thermocouple	(*b*) Assembly of two junctions formed by end to end welding of two different semi-conductor wires.
(*iii*) Bolometer	(*c*) Composed of a conductor metal strip of pt or Ni connected to wheat stone bridge.
(*iv*) Thermistor	(*d*) Contains strip of fused sintered oxide mixture connected to wheat stone bridge.
(*v*) Photovoltaic detector	(*e*) Made up of semiconductor IA, LTT or MCT.

Ans.: 5. (*i*) – (*a*), (*ii*) – (*b*), (*iii*) – (*c*), (*iv*) – (*d*), (*v*) – (*e*).

6. Match the following

(*i*) Solids sampling as solution	(*a*) Involves placement of solid solution drop on KBr disc evaporation to form a thin solid film.
(*ii*) Solids sampling as Mull	(*b*) Involves formation and placement of paste of solid in nujol oil.
(*iii*) Solids sampling as pellet	(*c*) Involves grounding and mixing of solid with KBr and placement into a press.
(*iv*) Solids sampling as film	(*d*) Used for amorphous solids.

Ans.: 6. (*i*) – (*a*), (*ii*) – (*b*), (*iii*) – (*c*), (*iv*) – (*d*).

C. DATA INTERPRETATION PROBLEMS

1. Predict the class of compound which fits into IR spectrum A.

   (*a*) Alkane compounds (*b*) Aromatic compounds

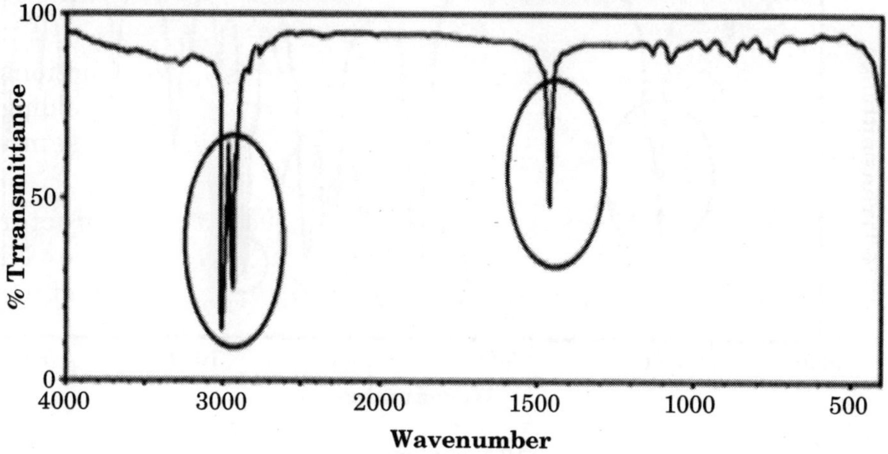

IR Spectrum–A

Ans.: *Alkane.*

2. Predict the compound which fits into IR spectrum B.

   (*a*) Hexanol (*b*) *p*-Hydroxy toluene

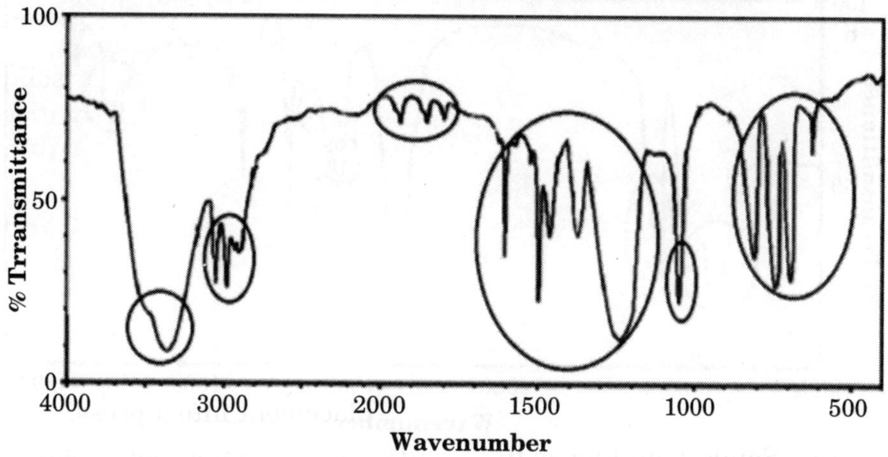

IR Spectrum–B

Ans.: *p-Hydroxy toluene.*

3. Predict the compound which fits into IR spectrum C.
   (*a*) Aliphatic ketone (*b*) Aliphatic Ether

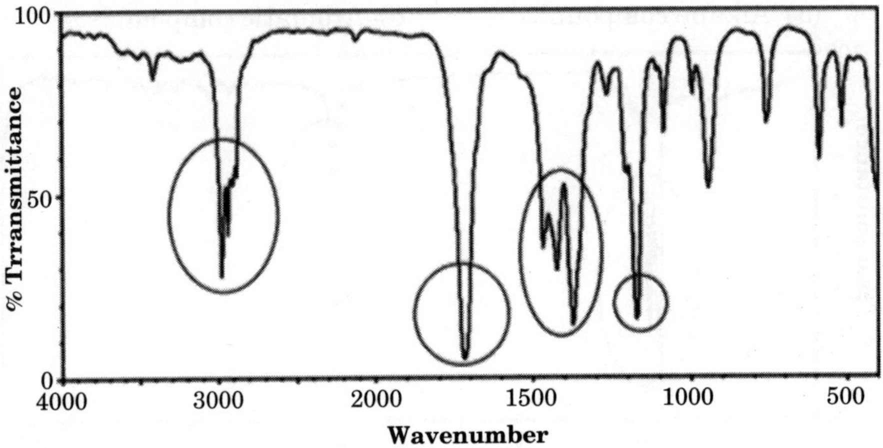

IR Spectrum–C

Ans.: *Aliphatic ketone.*

4. Predict the compound which fits into IR spectrum D.
   (*a*) Propanamide (*b*) Heptylamine

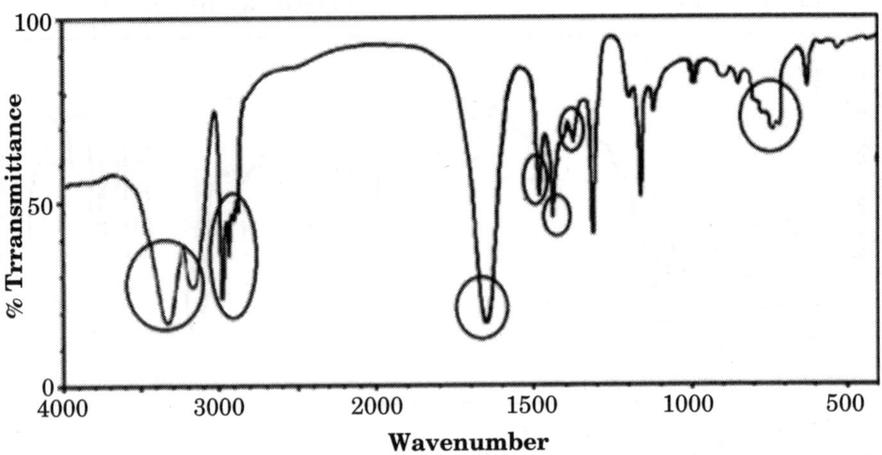

IR Spectrum–D

Ans.: *Propanamide.*

5. Predict the compound which fits best for IR spectrum E.
   (*a*) Hexanoic acid (*b*) Phenyl acetate

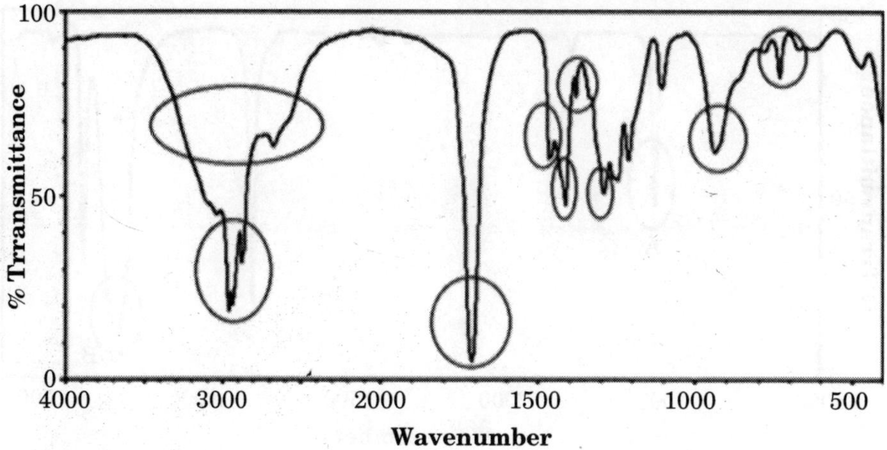

IR Spectrum–E

Ans.: *Hexanoic acid.*

6. Predict the compound which fits best for IR spectrum F.
   (*a*) Hexane 1-Thiol (*b*) Diethyl sulfoxide

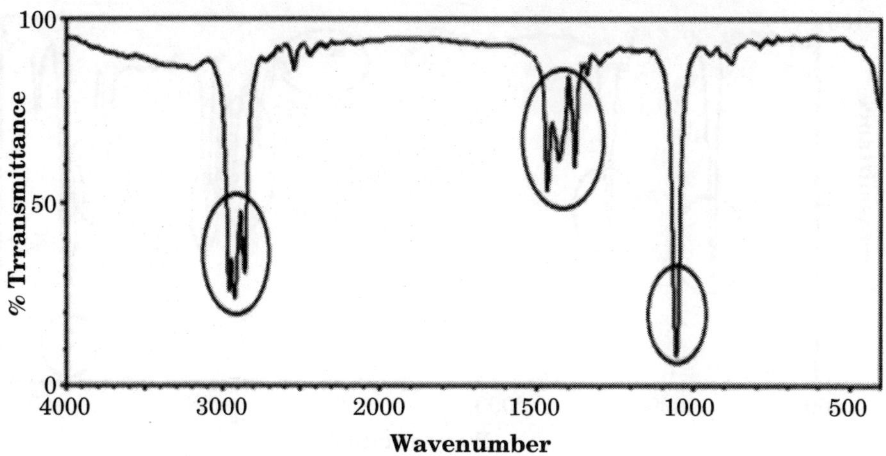

IR Spectrum–F

Ans.: *Diethyl Sulfoxide.*

7. Predict the compound which fits best for IR spectrum G.
   (*a*) Chloroform (*b*) Carbon Tetra Chloride

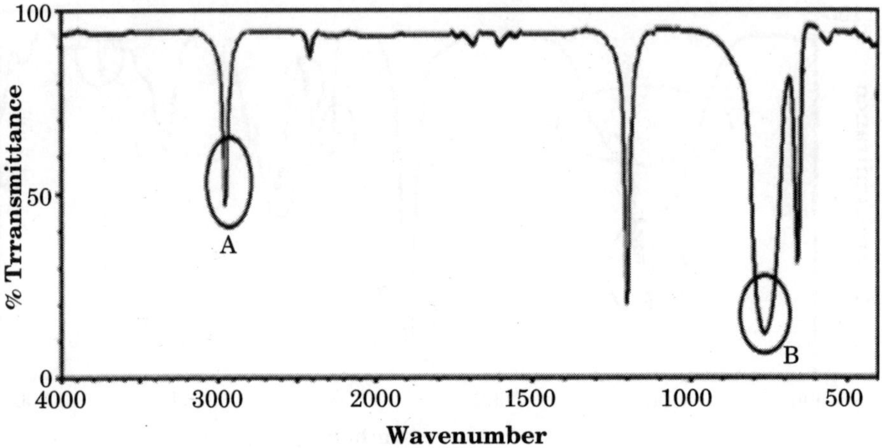

IR Spectrum–G

Ans.: *Chloroform.*

8. Predict the compound which fits best for IR spectrum H.
   (*a*) Indole (*b*) 1-Methyl Imidazole

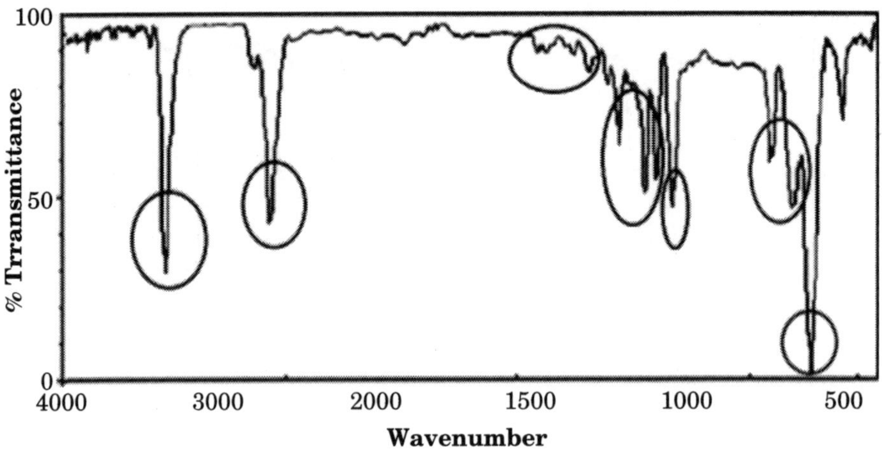

IR Spectrum–H

Ans.: *1-Methyl Imidazole.*

2.7 SUGGESTED READINGS

[1] Smith B.C., *Fundamentals of Fourier Transformed Infrared spectroscopy*, 2nd Edition, CRC press–Taylor & Francis, 2011.

[2] Pavia D.L., Lampman G.M., Kriz G.S., Vyvyan J.R., *Introduction to Spectroscopy*, 4th Edition, Brooks/Cole, 2009.

[3] Schradder B., *Infrared and Raman Spectroscopy*, N.Y., VCH, 2008.

[4] Skoog D.A., James Holler F., Crouch S.R., *Principles of Instrumental Analysis*, 6th edition, Brooks Cole, 2007.

[5] Silverstein R.M., Webster F.X., David J. Kiemle, *Spectrometric Identification of Organic Compounds*, 7th Edition, John Wiley, 2005.

[6] Kalsi P.S., *Spectroscopy of Organic Compounds*, 6th Edition, New Age International Publisher, 2004.

[7] Nyquist R.A., *Interpreting Infrared, Raman and Nuclear Magnetic Resonance Spectra*, Volume 1, Academic press, Elsevier, USA, 2001.

[8] Coleman P.B., *Practical Sampling Techniques for Infrared Analysis*, CRC Press, 1993.

[9] Vincent J.D., *Fundamentals of Infrared Detector Operation and Testing*, Wiley Interscience, 1990.

[10] Roeges N.P.G., *Guide to Interpretation of Infrared Spectra of Organic Structures*, Wiley, N.Y., 1984.

[11] Cooper J.W., *Spectroscopic Techniques for Organic Chemists*, 1st Edition, John Wiley, 1980.

[12] Nakanishi K., Solomon P.H., *Infrared Absorption Spectroscopy*, Holden Day, San Francisco, 1977.

[13] Bellamy L.J., *Infrared Spectra of Complex Molecules*, 3rd Edition, Springer, N.Y, 1975.

Chapter 3

Mass Spectroscopy

Objective

To make readers learn

- Basic Concepts, Theory and Underlying Principles of Mass Spectroscopy.
- Instrumentation of Mass Spectroscopy.
- Mass Spectra/Data Interpretation for Structural Determination of Compounds.

3.1 INTRODUCTION

Mass spectrometry is a technique to ionize the chemical compounds into charged molecules that are separated and quantified on the basis of their mass-to-charge ratios, so as to determine the masses, elemental composition, also structures of a sample or molecule. Generally, mass spectrometry, involves loading and vaporization of a sample molecule in the MS instrument. The vaporized sample molecules are ionized using different techniques (most commonly electron impact ionization). The generated ions are separated according to their mass-to-charge ratio in an analyzer (most commonly by electromagnetic fields). Lastly, the separated ions are detected using some quantitative method and converted into an ion signal on mass spectra. Commonly all mass spectrometers, comprises three basic elements those are ion source, mass analyzer, and a detector. To understand the mechanics of mass spectrometry let us go through series of events and components that come across during analysis of a sample in mass spectrometer.

3.2 INSTRUMENTATION

The components and mechanics of a mass spectrometer can be best understood with Figure 3.1, which shows the schematic diagram of a simple mass spectrometer.

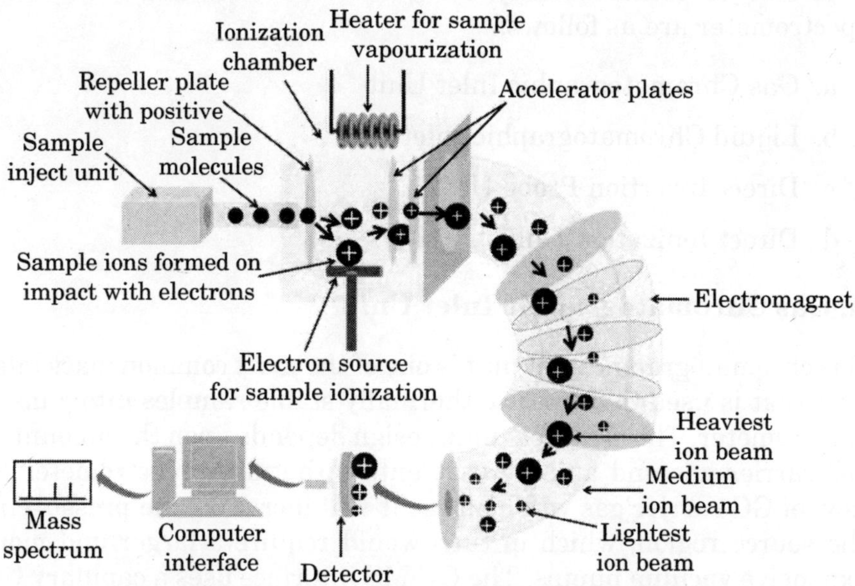

Figure 3.1: Schematic diagram of simple mass spectrometer.

The components of MASS spectrometer are as follows:

- Sample Inlet Unit
- Ionization Unit
- Mass Analyzer Unit
- Detector
- Computer Interface

3.2.1 Sample Inlet Unit

Sample introduction unit is one of the most crucial components of MS, which varies according to nature of sample and the sample matrix. Gaseous samples or samples with high vapour pressure are introduced directly into the ion source unit, because most of the MS ionization techniques are designed for gas phase molecules. Thermally stable liquids and solids sample are usually heated to increase the vapour pressure for analysis. Thermally labile sample must be directly ionized from the condensed phase.

A list of commercially used introduction units in mass spectrometer are as follows:

a. Gas Chromatographic Inlet Unit

b. Liquid Chromatographic Inlet Unit

c. Direct Insertion Probe Unit

d. Direct Ionization Unit

A. Gas Chromatographic Inlet Unit

Gas chromatographic inlet unit is one of the most common mass inlet unit that is used to introduce thermally stable samples into a mass spectrometer. The GC inlets unit design depends upon the amount of GC carrier gas and analyte that enters the mass spectrometer. If flow of GC carrier gas will be large it will increases the pressure in the source region, which in turn would require a larger and more expensive vacuum pumps. The GC/MS interface uses a capillary GC column. The end of the capillary is inserted directly into the source region of the mass spectrometer. As capillary columns are very common, so this inlet is widely used. Figure 3.2, shows the schematic diagram of GC inlet unit connected to a mass spectrometer.

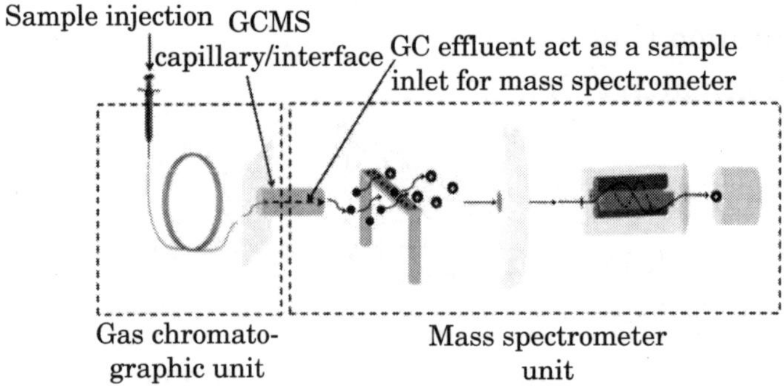

Figure 3.2: Schematic diagram of GC inlet unit connected to a mass spectrometer.

B. Liquid Chromatographic Inlet Unit

The Liquid chromatographic unit is used to introduce thermally labile compounds that face difficulty in separation by gas chromatography. As these inlets are used for temperature sensitive compounds, the sample is ionized directly from the condensed phase. Figure 3.3, shows the schematic diagram of HPLC inlet unit connected to a mass spectrometer.

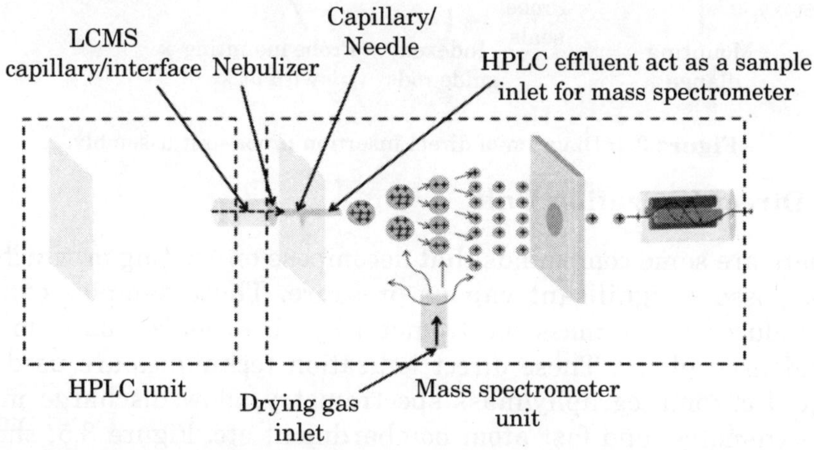

Figure 3.3: Schematic diagram of HPLC inlet unit connected to a mass spectrometer.

C. Direct Insertion Probe Unit

Direct insertion probe unit is widely used system for introduction of low vapor pressure liquids and solids into the mass spectrometer. In this unit, the sample is loaded into a short capillary tube at the end of a heated sleeve. This sleeve is then inserted through a vacuum lock to make the sample into the source region. After positioning the probe, the temperature of the capillary tube is increased to vaporize the sample. In addition, the sample is under vacuum and situated near to the source, as result lower temperatures are required for analysis. Direct insertion probe is more tedious than the direct vapour inlet. Figure 3.4, shows the diagram of direct insertion probe unit assembly.

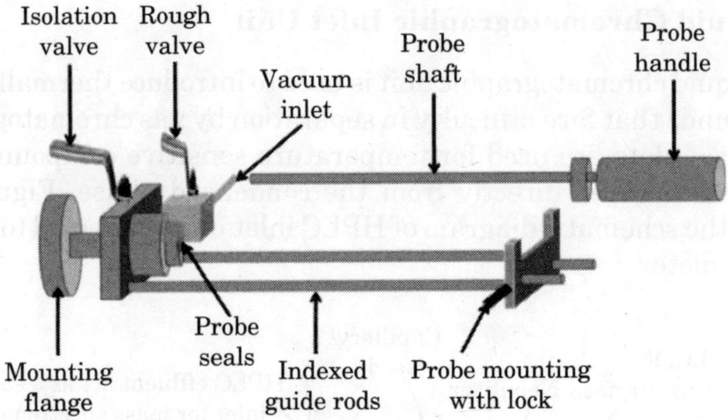

Figure 3.4: Diagram of direct insertion probe unit assembly.

D. Direct Ionization Unit

There are some compounds that decompose on heating or which do not possess significant vapour pressure. Those samples can be introduced to the mass spectrometer by direct ionization from the condensed phase. These direct ionization techniques are used for liquid chromatography/mass spectrometry, glow discharge mass spectrometry, and fast atom bombardment etc. Figure 3.5, shows the schematic diagram of direct ionization through fast atomic bombardment.

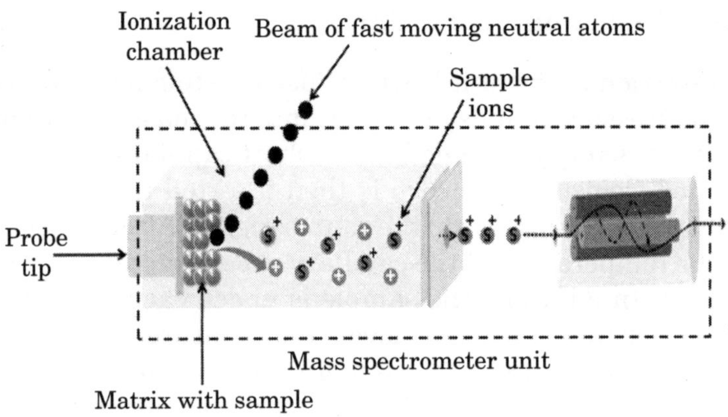

Figure 3.5: Schematic diagram of direct ionization through fast atomic bombardment.

3.2.2 Ionisation Unit

Ionization unit is meant for production of ions through different below mentioned techniques. This unit comprises a vacuum system so as to make the generated molecular ions free of air resistance. Ionization unit also contains a repeller plate, accelerating plates, and three slits. The positive ions are persuaded out into the rest of the chamber by the ion repeller plate (*a metal plate carrying a slight positive charge*). After the positive ions are repelled away from the very positive ionization chamber, these passes across the three slits, the final one of which is at 0 volts. The middle slit carries some intermediate voltage. Lastly, all the generated ions are accelerated into a finely focused beam. The whole ionization chamber is maintained at 10000 volts.

Commonly, a number of methods are available for ionization of sample molecules in the ionization unit. A list of commonly used ionization methods in mass spectrometry are as follows:

a. Electron Impact Method (EIMS)

b. Chemical Method (CIMS)

c. Field Method (FIMS & FDMS)

d. Laser Method (LIMA & LDMS MALDI)

e. Plasma Method (PDMS)

f. Ionic Impact Method (SIMS)

g. Atomic Method (FABMS)

h. Spray Method (ESIMS)

A. Electron Impact Method (EIMS—Electron Impact Mass Spectrometry)

EIMS is a technique of ionization in which a beam of electron is used for the ionization of the sample molecules. The vaporised sample passes into the ionization chamber. The metal coil (filament) is electrically heated to several thousand degrees, so as to produce electrons that are attracted by the electron trap (a positively charged metal plate). The sample molecules/atoms are bombarded with a stream of electrons, to remove one or more electrons out of the sample particles to generate positive ions. Generally, when an electron impacts on a molecule two conditions may occurs. Either the electron may be captured by the molecule to offer a radical anion or it may lead to displace an electron from the molecule to offer a radical cation.

$$M \rightarrow M^{-}$$
$$M \rightarrow M^{+}$$

An interesting fact is that the probability of formation of radical cation is generally 100 times more than the radical anion. In ionization chamber, sample molecules are continuously exposed to the beam of electrons. When the ionization potential of electron beam reaches to 10–15 ev, then the sample molecule gets converted into molecular ion. The generated molecular ion determines the molecular weight of any molecule. When the ionization potential of electron beam reaches up to 70ev in the ionization chamber, the generated molecular ions gets fragmented into daughter ions or fragment ions. These generated fragment ions determines the mode of synthesis of the molecule. Figure 3.6, shows the schematic diagram of EIMS.

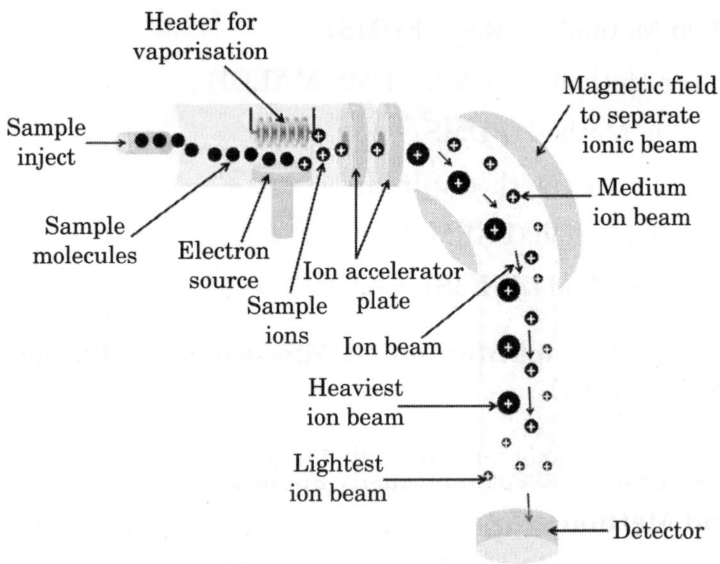

Figure 3.6: Schematic diagram of EIMS.

B. Chemical Method (CIMS)

Chemical Ionization Mass Spectrometry (CIMS) is a technique of ionization in which organic chemicals are used for the ionization of the sample molecules. In CIMS the organic sample is mixed with a stream of reactant gas like methane, isobutane, or ammonia. Further, the sample and gaseous mixture is subjected to electronic bombardment. The resultant effect is that firstly an electronic beam converts gas molecules into ionized form. Nextly, these generated molecular ions of gas further converts sample molecules into ionized form. Later on these ions are separated and analyzed into the analyzer. This technique is most suitable for ionization of hydrocarbons. Figure 3.7, shows the schematic diagram of CIMS.

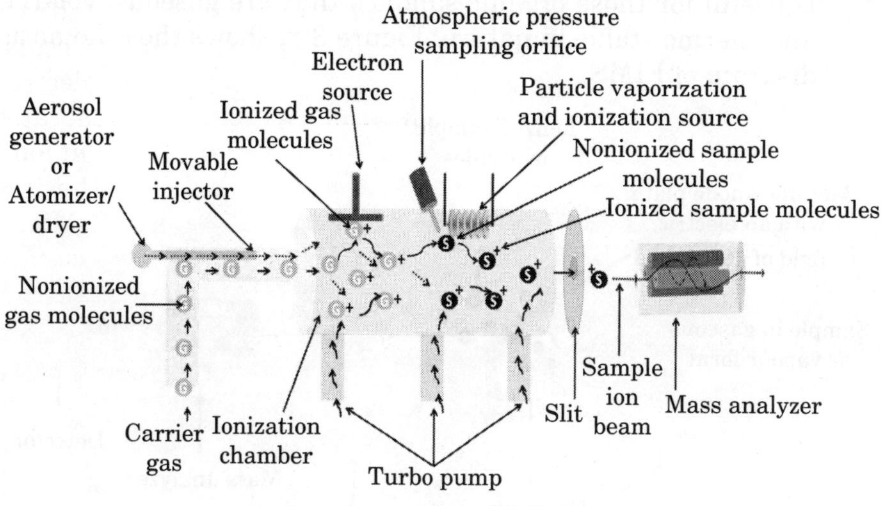

Figure 3.7: Schematic diagram of CIMS.

Merits

- All ionic peaks are more abundant than EIMS.
- Simple fragmentation pattern.
- GCMS interface can be easily applied.

C. Field Method

In this technique of ionization electric field is commonly used for the ionization of the sample molecules. In the field method of ionization based on the nature of organic sample to be ionized, two types of ionization techniques are most commonly used *viz.*, FIMS & FDMS.

(*i*) FIMS (*Field Ionization Mass Spectrometry*): In Field Ionization Mass Spectrometry (FIMS) technique, an organic sample is taken in the gaseous or vaporised form and introduced into the ionization chamber containing a metal plate as anode. Onto the metal plate an electric field of 10^{10} volt is applied. When the sample molecules in the gaseous or vaporised phase come in contact with the metallic anode plate they are converted into the ionized form. Ionization of sample molecules occurs as result of transfer of electrons from the sample molecules into the incomplete orbital of metallic anode plate. This technique is useful for those organic samples that are gaseous, volatile and thermo stable in nature. Figure 3.8, shows the schematic diagram of FIMS.

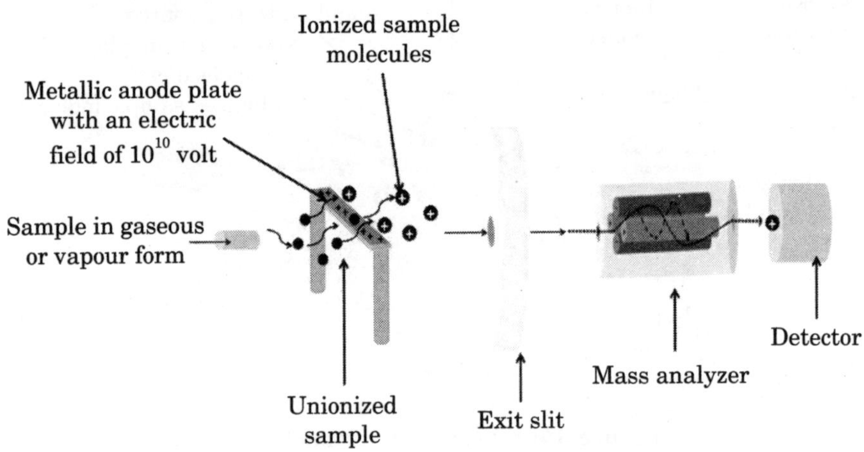

Figure 3.8: Schematic diagram of FIMS.

Merits

- High ionic peak abundance.
- Simple fragmentation pattern.

Demerits

- Low sensitivity.
- Low resolution.

(ii) **FDMS (*Field Desorption Mass Spectrometry*):** When the sample is of non-volatile or thermo labile nature then Field Desorption Mass Spectrometry (FDMS) technique is used. FDMS involves direct deposition of non-volatile organic sample on the surface of metallic anode plate of a pair of electrodes. Again onto this metal plate a high electric field is applied. This causes organic sample molecules to undergo desorption, and form molecular ions or quasi-molecular ion. Figure 3.9, shows schematic diagram of FDMS.

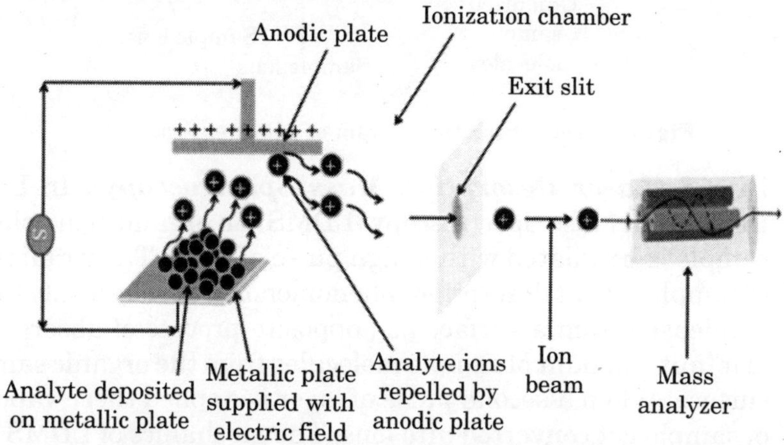

Figure 3.9: Schematic diagram of FDMS.

D. Laser Method

It is one another technique of ionization that makes use of laser for ionization of sample molecules. Based on the method of application of laser there are three types of laser method based ionization techniques such as LIMA, LDMS and MALDI.

(i) **LIMA (*Laser Ionization Mass Analysis*):** In laser ionization mass analysis (LIMA), the organic sample is taken in the gaseous or vaporised form and introduced into the ionization chamber where it is irradiated with pulse of laser of 10^5 Wcm^{-2} power. The irradiation of sample causes ionization of the sample molecules. Figure 3.10, shows the schematic diagram of LIMA.

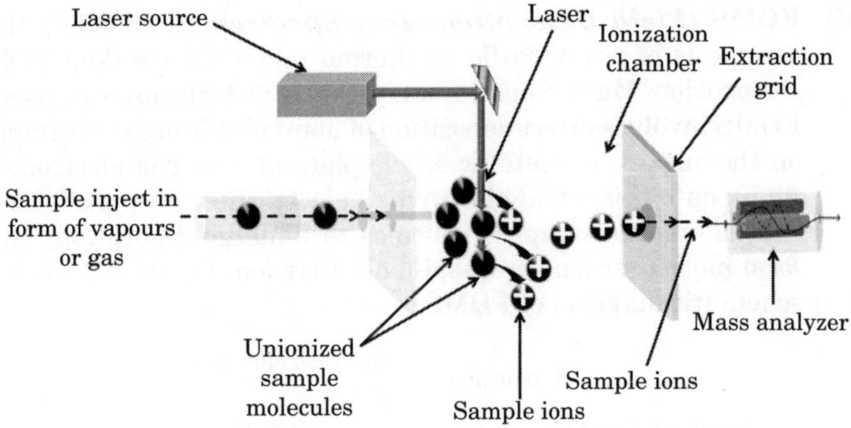

Figure 3.10: Schematic diagram of LIMA experiment.

(ii) **LDMS (*Laser Desorption Mass Spectroscopy*):** In Laser Desorption Mass Spectroscopy (LDMS) an organic non volatile sample is irradiated with a high pulse of laser. The irradiation of sample causes desorption (phenomenon whereby a substance is released from a surface *i.e.*, opposite process of absorption) of minute amount of sample molecules from the organic sample surface. When a second pulse of laser is applied these vapours of sample get converted into ions. The mechanics of LDMS can be best understood from the Figure 3.11. In this technique firstly

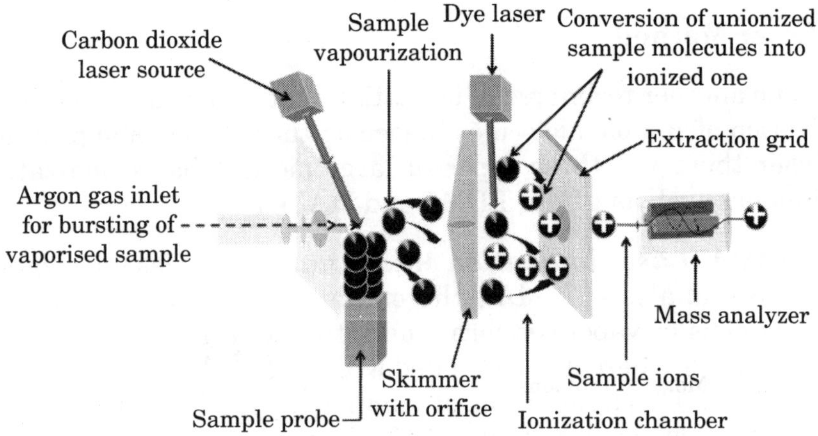

Figure 3.11: Schematic diagram of LDMS experiment.

a sample is placed on a probe which receives high energy laser pulses from a carbon dioxide laser. Secondly, an expanding argon gas stream propels the burst of vaporized sample into an accelerating region of mass spectrometer, through an orifice with the help of a skimmer. Here the vaporised molecules are again subjected to a high energy dye laser which produces ions through photo ionization process. Nextly, the formed ions are accelerated into a drift region of TOF mass analyzer.

(iii) **MALDI (*Matrix Assisted Laser Desorption Ionization*):** Matrix assisted laser desorption ionization is a modification of laser desorption method. In this technique the sample under investigation is dispersed on the surface of a matrix (can be glycerol or urea etc.). In the next step a laser beam is focused on the sample resting on the matrix. This makes sample to undergo desorption and ionization. Figure 3.12, shows the diagram of different stages of MALDI experiment.

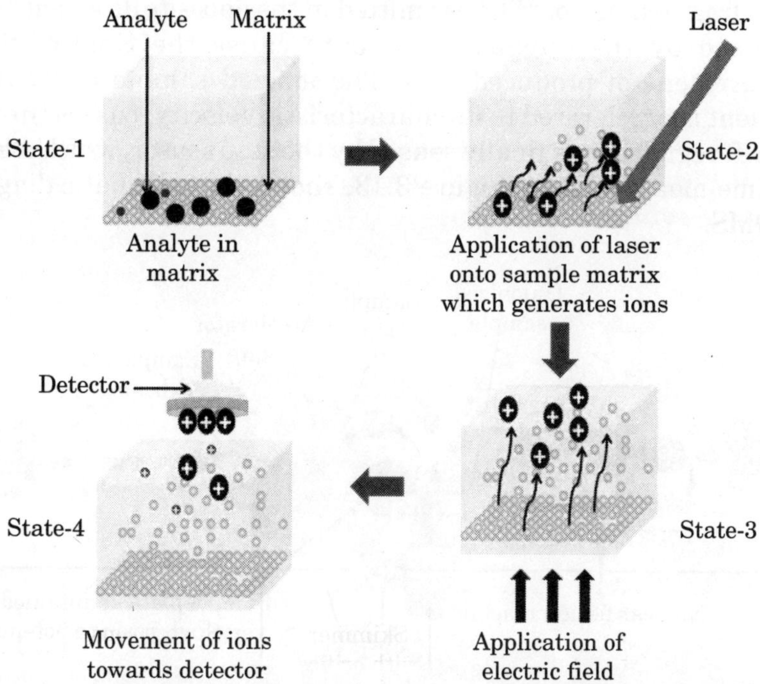

Figure 3.12: Schematic diagram of different stages of MALDI experiment.

E. Plasma Method (PDMS)

In Plasma Desorption Mass Spectrometry (PDMS), a plasma product is used for the ionization of the sample molecules and later on the produced sample ions are generally analyzed by the time of flight mass analyzer. PDMS experiment involves use of a fission fragment of ^{252}Cf (californium), which generates various ion pairs (almost 3%). The produced ionic pairs are so energetic that they also undergo decomposition to produce highly energetic ions such as $^{142}Ba^{18+}$ and $^{106}Te^{22+}$. The sample is placed on a thin sheet of aluminium or on a sheet of aluminized polyester film, which is connected to a very high positive potential. When fission occurs one of the fission fragment particle strikes over the sample molecule which produces about 10^9 W power and generates a localised plasma that possess a temperature of around 10000 Kelvin. As a result of such a highly elevated temperature the sample molecules undergoes ionization. Whereas the other fission fragment ion of ^{252}Cf, is emitted in the opposite direction which is sensed by the trigger detector to cause the time of flight measurement of produced ions. The ionized sample molecule or fragment is accelerated to its characteristic velocity, passes through the drift region and is finally sensed by the stop sensor, which arrests the time measurement. Figure 3.13, shows the schematic diagram of PDMS.

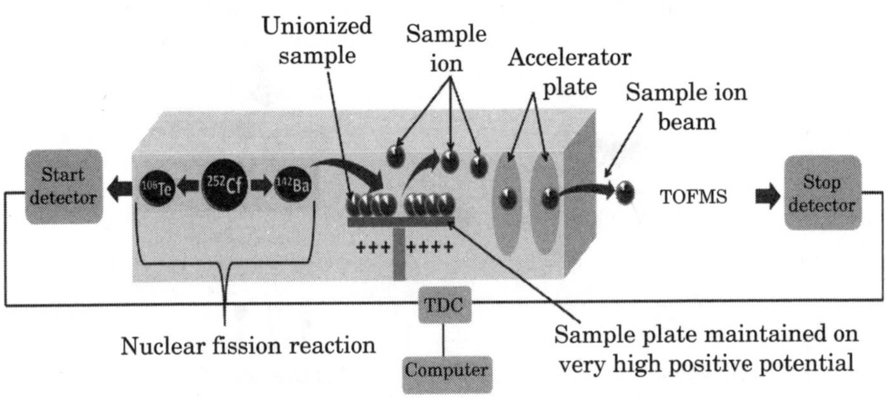

Figure 3.13: Schematic diagram of mechanics of PDMS experiment.

F. Ionic Impact Method (SIMS—Secondary Ion Mass Spectrometry)

In Secondary Ion Mass Spectrometry (SIMS), beam of primary ions are used for the ionization of the sample molecules which generates the secondary ions. Based on the application of secondary ions a technique called SIMS is used. SIMS is ionization involves generation of beam of ions from the inert gas after their introduction into the ionization unit. Argon or Xenon or Neon are the most commonly used gases for production of Ar^+ or Xe^+ or Ne^+ ions. For example, the generated beam of Ar^+ ion is allowed to strike onto the organic sample molecules. In this process Ar^+ ions transfer their energy into the sample molecules. As a result of which organic sample molecules undergo ionization. As in this technique a beam of inert gas primary ions is used to ionize the sample molecule and generates a secondary ions beam so this technique is called as Secondary Ion Mass Spectrometry (SIMS). Figure 3.14, shows the schematic diagram of SIMS.

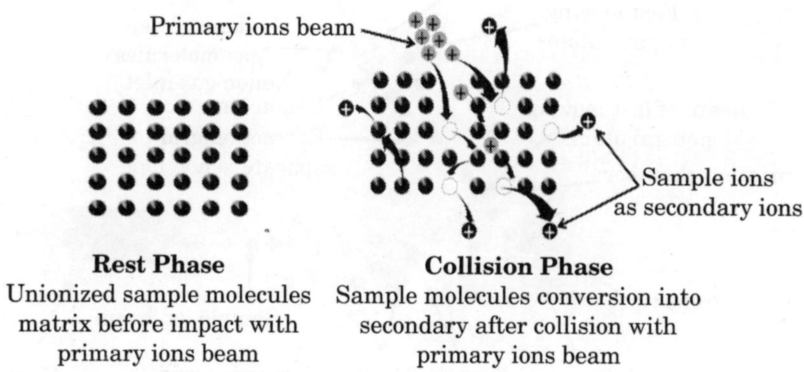

Figure 3.14: Schematic diagram of mechanics of SIMS experiment.

G. Atomic Method (FABMS—Fast Atomic Bombardment Mass Spectrometry)

Fast Atomic Bombardment Mass Spectrometry (FABMS) is a technique of ionization in which neutral atoms are commonly used for the ionization of the sample molecules. In FABMS firstly xenon gas molecules are ionized to generate xenon radical cations (xe^+). Secondly, these xe^+ cations are accelerated from 6000 to 10000 ev, which induces high translational energy into them. Thirdly, these highly energetic fast moving xe^+ cations, are passed into the xenon gas stream. At this stage, the highly energetic xe^+ cations withdraw electrons from the atoms of xenon gas stream to convert themselves into highly energized fast moving neutral xenon atoms. Fourthly, the left behind Xenon ions of gas stream are removed electro statically. Lastly, the highly energized fast moving neutral xenon atoms or fast atoms are allowed to bombard the organic sample molecules which ionize the sample molecules. Figure 3.15, shows schematic diagram of mechanics of FABMS.

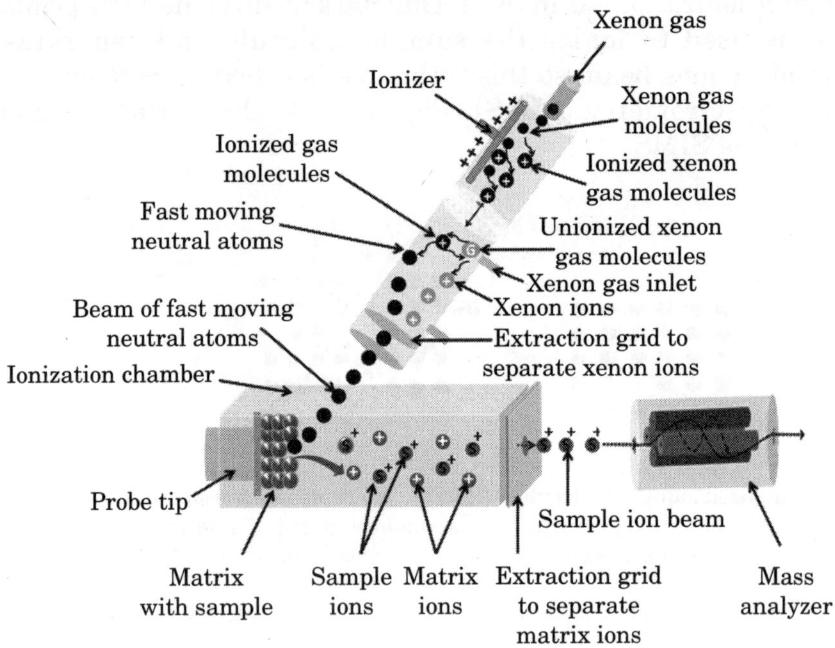

Figure 3.15: Schematic diagram of mechanics of FABMS experiment.

H. Spray Method (ESIMS)

Electron Spray Ionization Mass Spectroscopy (ESIMS) is a technique of ionization in which electron spray is commonly used for the ionization of the sample molecules. In this technique an outlet of HPLC is attached to a nebulizer. This nebulizer is attached to a needle (capillary) which is supplied with an ionizing voltage of several thousand volts. As the HPLC effluent or sample molecules come into the needle/capillary, due to high voltage the effluent or sample molecules gets ionized. Later the drying gas like nitrogen dries the sample ions mixed in solvent to make the sample ions free of solvent and totally dried. Figure 3.16, shows the schematic diagram of mechanics of ESIMS.

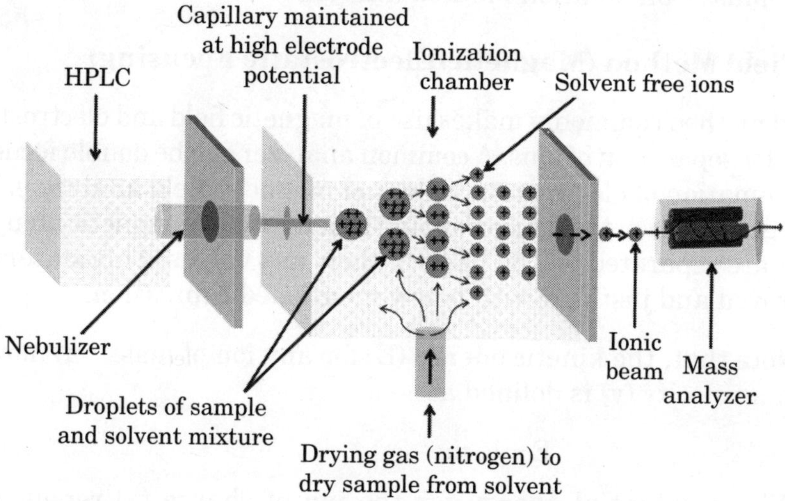

Figure 3.16: Schematic diagram of mechanics of ESIMS experiment.

Similar to the mechanics of ESI, one another technique is commonly used for ionization of sample in mass spectroscopy called thermal spray ionization (TSI). In this technique instead of the application of high electrostatic field on the capillary/needle attached to the HPLC, a high thermal filed is applied to ionize the sample effluent.

3.2.3 Ion Separation Unit or Mass Analyzer Unit

The ion separation units are meant to sort ions according to their mass-to-charge ratios or some other related property.

Commercially, a number of methods are available for separation of generated ions in ionization chamber. A list of methods for separation of ions are as follows:

a. Field Method (Magnetic & Electrostatic Focusing)
b. Velocity Method (TOFMS)
c. Pole Method (QMF& QIS)
d. Resonance Method (ICRMS & FTICRMS)
e. Mass Combination Method (MS/MS)

A. Field Method (Magnetic/Electrostatic Focusing)

Field method commonly makes use of magnetic field and electrostatic field for separation of ions. A common analyzer can be double focusing (combination of electrostatic field and magnetic field analyzer). The ions generated in the ionization unit when reaches magnetic analyzer they are separated on the basis of their m/z values. This statement is proved and justified with below mentioned expression.

Note that, the kinetic energy (E) for any ion of mass (m) moving with a velocity (v) is defined as

$$E = \tfrac{1}{2} mv^2 \qquad ...(3.1)$$

Where potential energy for the ion of charge (z) repelled by electrostatic filed of voltage (v) is defined as

$$v = z.v. \qquad ...(3.2)$$

It is obvious that when ion is repelled the potential energy is converted into kinetic energy.

Hence,
$$z \,.\, v = \tfrac{1}{2} mv^2 \qquad ...(3.3)$$
$$v^2 = 2\, z \,.\, v/m \qquad ...(3.4)$$

When ions enters into magnetic field (B) they undergoes circular motion (of radius r). Hence, at equilibrium centrifugal force becomes equal to centripetal force of magnetic field on the ions.

So,
$$mv^2/r = zBv \qquad ...(3.5)$$

Therefore, $v = z \cdot B \cdot r/m$ ...(3.6)

Putting, value of v in Eqn. (3.4)

we get $2z \cdot v/m = (z \cdot B \cdot r/m)^2$ ...(3.7)

Therefore, $m/z = B^2 \cdot r^2/2v$ ...(3.8)

Generally, it becomes difficult for a magnetic analyzer to differentiate between m^+ and $2m^{2+}$ ions if their m/z value of $B^2 \cdot r^2/2v$ remain same.

Hence, to differentiate these ions, their path can be changed by varying any of the B or v.

All the ions after repulsion from the accelerator plate do not possess similar kinetic energies.

Hence in double focusing mass analyzer a primary focusing is done by passing the ions through the two electro statically charged curved plates called electrostatic analyzer.

In this case centrifugal force

$$mv^2/r = z \cdot E \quad ...(3.9)$$

Hence combining zE with

$$zV = ½ mv^2$$

It shows that $r = 2v/E$ ...(3.10)

This is why the radial path of ion of a given velocity is independent of m/z value. Electrostatic analyzer focuses on ions of similar kinetic energy. Generally, the single focusing in mass spectrometry makes a resolution of 1 in 1000 or 1 in 7500. Whereas double focusing results in a resolution of 1 in 60000.

The Magnetic field bends the trajectories of ions into circular paths of radii that depend on the momentum-to-charge ratios of ions. Ions of larger m/z follow larger radius paths than ions of smaller m/z values, so ions of differing m/z values are dispersed in space. Changing the ion trajectories through variations of magnetic field strength makes the ions of different nominal mass-to-charge ratios to focus on detector and separate. Figure 3.17(*a*) and 3.17(*b*), shows schematic diagrams of single and double focussing mass spectrometers.

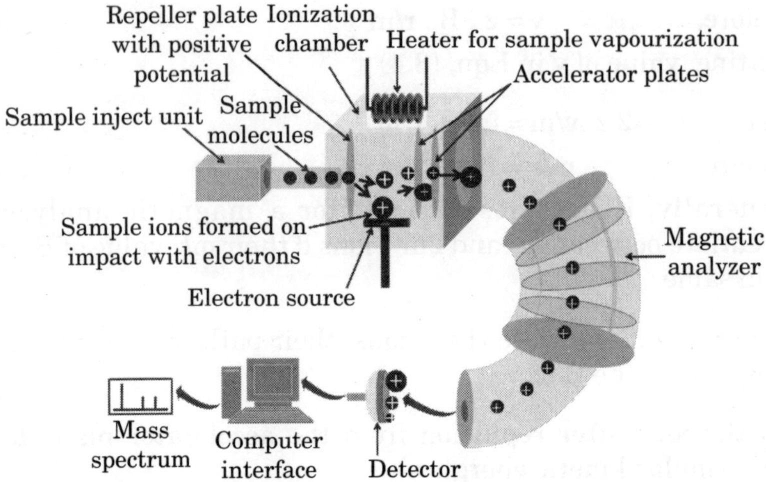

Figure 3.17(*a*): Schematic diagram of single focusing mass spectrometer.

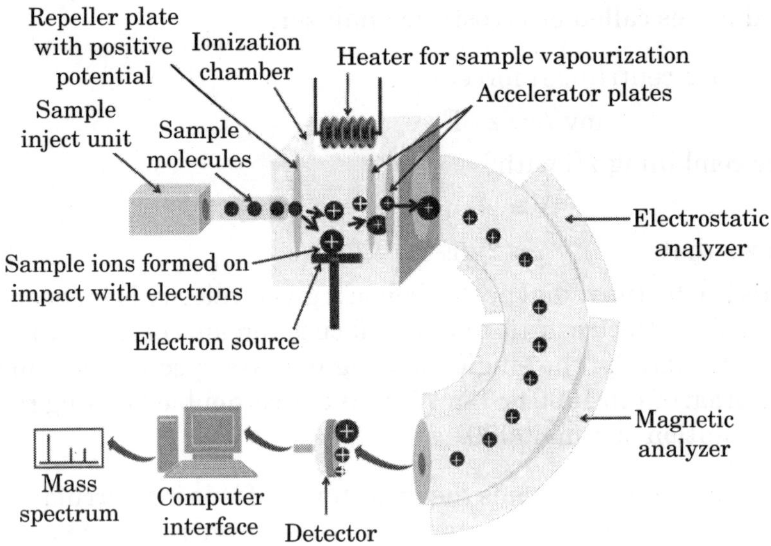

Figure 3.17(*b*): Schematic diagram of double focusing mass spectrometer.

B. Velocity Method (TOFMS)

In TOFMS, the ions are separated on the basis of their velocity instead of their m/z values. As this method encompasses measurement of time of flight so called as TOFMS (time of flight mass spectrometry). TOFMS contains a fly tube of approximately 1 metre of length. This tube is maintained under vacuum. In this tube firstly all of the generated ions are send in the short pulses and accelerated to a uniform kinetic energy. As a result the heavier ions travel slowly to the collector whereas the lighter ions move faster. The time difference between the ions is usually 10^{-8} seconds. The difference between ions pulses is 10^{-4} seconds. To observe the overall inside tube activity, this fly tube is attached with an oscilloscope. Figure 3.18, shows the schematic diagram of TOF mass analyzer.

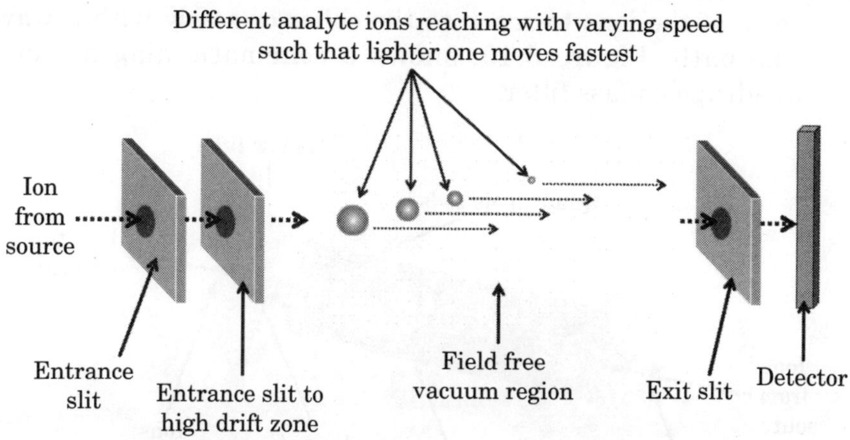

Figure 3.18: Schematic diagram of a TOF mass analyzer.

C. Pole Method

In this method the ions are separated on passage through four voltage carrying metallic rods enclosed in a tube called quadrupole. As pole method makes use of three different techniques for ions separation, hence this pole method is classified into two types *viz.*, QMF and QIS.

(*i*) **Quadruple Mass Filter (QMF):** This type of mass analyzer contains four pairs of voltage carrying metallic rods, in which opposite pairs are coupled together. These pairs of rods are enclosed in a tube. These rods are arranged symmetrically along the axis of tube. In these rods a complex electric field is generated by supplying a direct current voltage of few hundred volts. On this quadrupole electric field setup, a radiofrequency potential is applied. In this method firstly, the generated and accelerated +ve ions are allowed to enter the tube. Secondly, a constant ratio of DC voltage and RF is chosen and made to scan the quadruple from zero to maximum. This DC and RF ratio scan establishes a hyperbolic field among the four rods. There occurs a particular value of this hyperbolic field on which most of the +ve ions are deflected towards the rods to get discharged and does not reach the collector. But on the other hand this particular value of hyperbolic field will allow some of the ions to reach to the collector safely with a wave like path. Figure 3.19, shows a schematic diagram of a quadrupole mass filter.

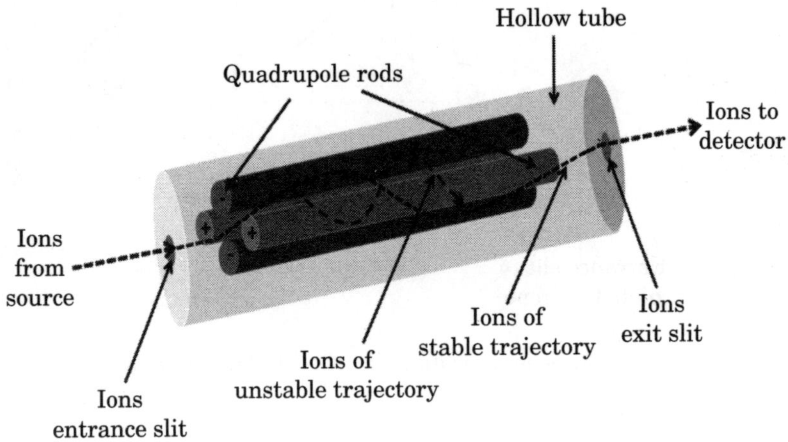

Figure 3.19: Schematic diagram of a Quadrupole mass filter.

(ii) **Quadrupole Ion Storage (QIS) or Ion Trap:** Quadrupole Ion Storage (QIS) mass analyzer contains a storage system for ions for temporarily storage, called quadrupole ion trap. The instrument contains a pair of end cap electrodes that contains a ring electrode in between them. In this circuit a DC and RF is applied. In this analyzer all the ions with different m/z values are trapped simultaneously in the storage system of QIS. When a RF is applied the ions with higher m/z value are ejected out of trap in axial direction towards detector. Figure 3.20, shows a schematic diagram of an ion trap.

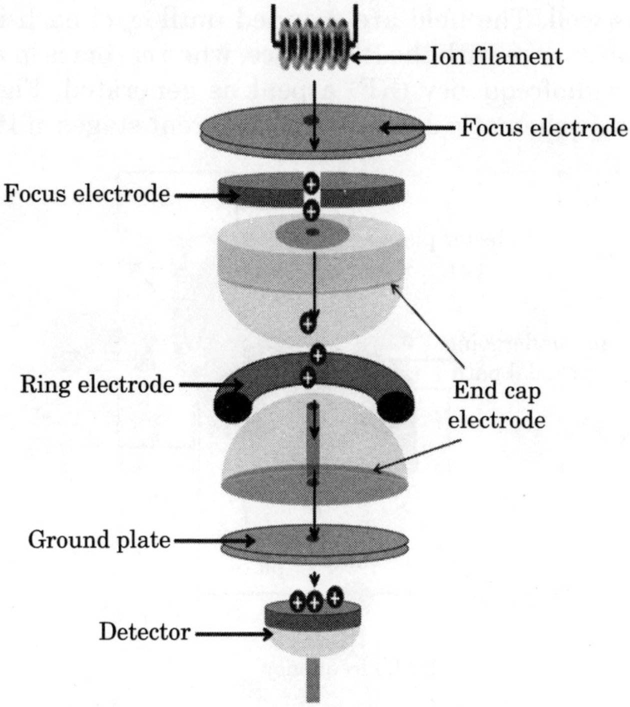

Figure 3.20: Schematic diagram of an ion Trap.

D. Resonance Method

In this method the ions are separated on the basis of their angular frequency. As ion resonance method makes use of two different techniques for ions separation, hence this method is classified into two types such as ICRMS and FTICRMS.

(i) **Ion Cyclotron Resonance Mass Spectrometry (ICRMS):** In Ion Cyclotron Resonance Mass Spectrometry (ICRMS) the ions generated in the ionization unit are drawn into the cubical cell analyzer. The cubical cell holds the ions with the help of electrical trapping and is supplied with constant magnetic field. The magnetic field application on the ions contained in cubical cell causes ions to undergo cycloidal motion. As a result of cycloidal motion of ions in the cubical cell, a cycloidal frequency or angular frequency persists among ions. This cyclotron frequency ω_c is a function of both electric field and m/z value of

ion as well. The field are scanned until ω_c of each ion comes into resonance with the RF. Hence, when ω_c comes in resonance with radiofrequency (RF) a peak is generated. Figure 3.21, shows a schematic diagram and different stages of ICRMS.

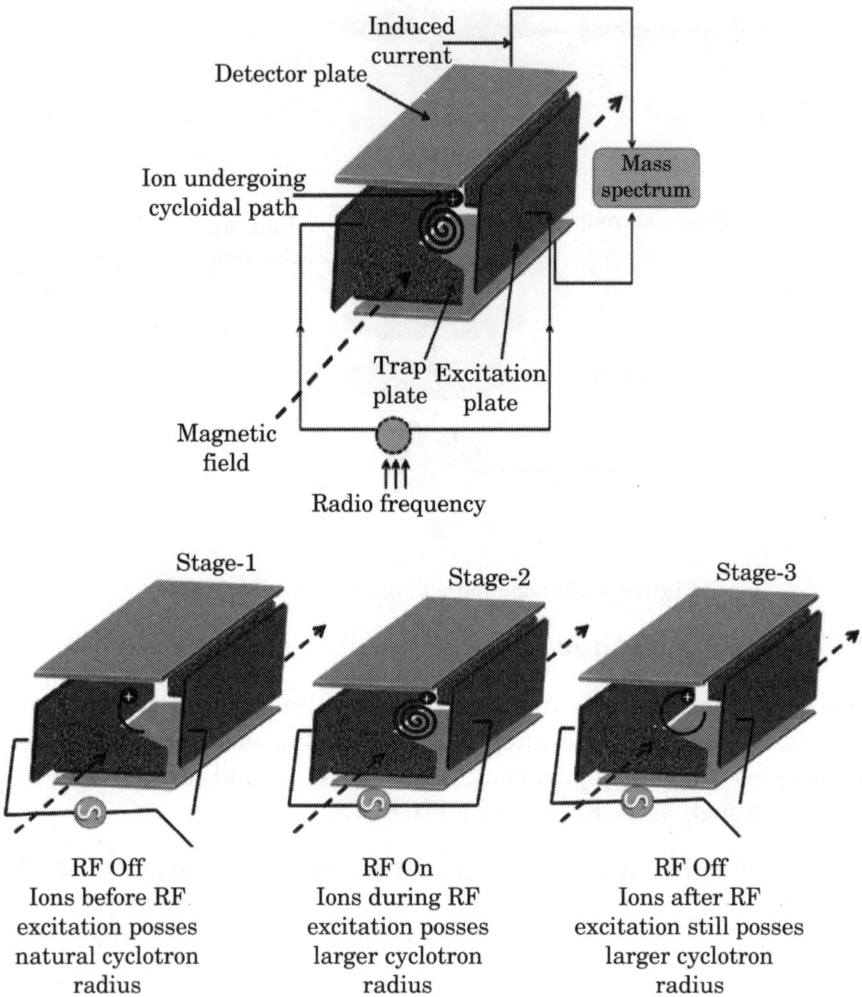

Figure 3.21: Schematic diagram of ICRMS and its stages.

(ii) **Fourier Transformed Ion Cyclotron Resonance Mass Spectrometry (FTICRMS):** Fourier transformed ion cyclotron resonance mass spectrometry is almost similar to the ICRMS.

But here instead of field scanning until cyclotron frequency of each ion comes in resonance with RF, directly a short pulse of RF is applied. This generates a current in receiver, which contains an interferogram that contains information on all of the ion frequencies. Lastly, mathematically this time domain curve is Fourier transformed into frequency domain curve. Figure 3.22, shows a schematic diagram FT ICRMS.

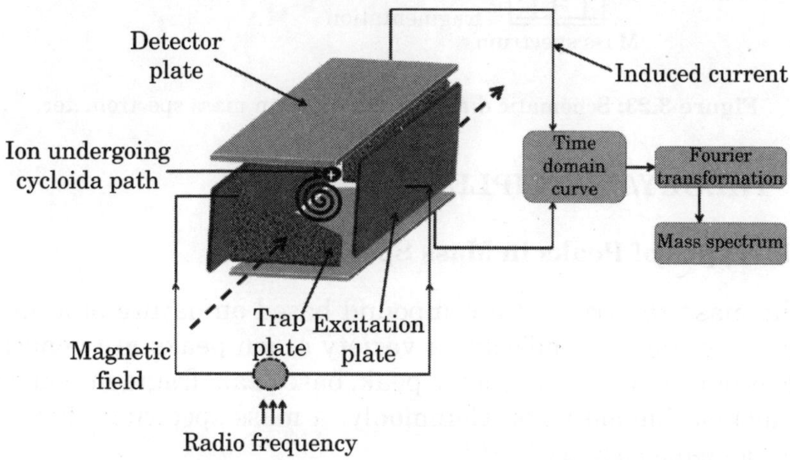

Figure 3.22: Schematic diagram of FTICRMS.

E. Mass Combination Method (MS/MS)

In mass combination method the ions generated in the ionization chamber are separated in two mass spectrometers that are connected through a filed free collision unit so called tandem mass spectrometer or MS/MS. In this technique in the first mass spectrometer the parent ions are separated and sequentially passed into the field free collision unit that possess helium gas. As a result in the collision unit containing stream of helium gas, the parent ions are fragmented into daughter ions. These generated fragment ions are further passed into second mass spectrometer to generate the fragment ion spectrum. The advantage of this technique is that the two different spectra for parent ions and fragment ions are generated from the same instrument. Figure 3.23, shows a schematic diagram tandem mass spectrometer.

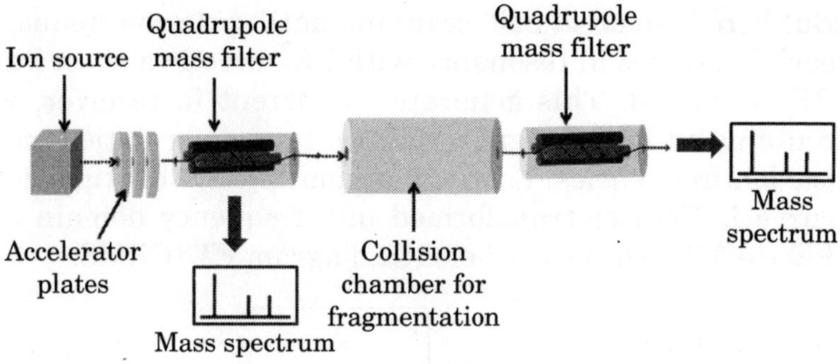

Figure 3.23: Schematic diagram of a Tandem mass spectrometer.

3.3 THEORY/PRINCIPLE

3.3.1 Types of Peaks in Mass Spectrum

In the mass spectrum of a compound based on nature of atoms or groups present in a molecule a variety of ion peaks are generated such as parent ion peak, isotope peak, base peak, fragment ion peak, and metastable ion peak. Commonly, a mass spectrum offers four types of peaks such as:

a. Molecular Ion (Parent Ion) Peak

b. Isotope Peak

c. Base Peak

d. Fragment Ion Peak

e. Metastable Ion Peak

A. Molecular Ion (Parent Ion) Peak

Molecular ion peak is defined as a peak of ionic species that is generated on a mass spectrum due to loss of one electron from the orbit of a molecule when exposed to an ionization potential of 10–15 ev energy in the ionization chamber. Generally, from the mass of parent ion the molecular mass of sample can be easily determined. The molecular ion is generally represented as M^+ radical cation. This shows that radical cation is formed due to loss of one electron from the neutral molecule. Usually, a molecular ion becomes unstable when it is exposed to an ionization potential up to 70ev and undergoes fragmentation to generate fragment ions.

Features of Molecular Ions

Abundance (Peak Intensity or Peak Height)

Generally, the peak abundance of any ion depends upon its stability in the mass spectrometer. Usually, molecular ion shows high abundance in case of aromatic system, hetero aromatic system, and triple bond system, double bond system and conjugate double bond system in the mass spectrum. Such ions are highly abundant due to principle of stability that is as a result of their capability of undergoing resonance and hyper conjugation. Elongation of carbon chain, chain branching and occurrence of C = O group in the molecular ion reduces its abundance in the mass spectrum.

Absence

Very high chain branching, occurrence of higher number of hydroxyl groups, and a long alkyl chain presence in the molecular assures its molecular ions absence in the mass spectrum.

Example: Mass spectrum of chloro butane representing M^+ ion peak (Figure 3.24).

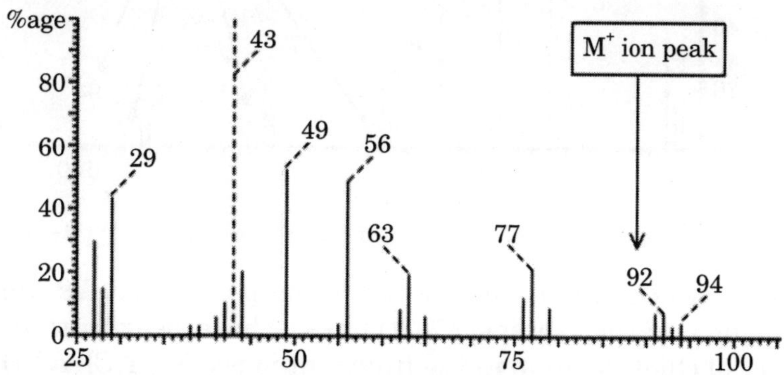

Figure 3.24: Mass spectrum of chloro butane representing only M^+ ion peak.

In the mass spectrum of chloro butane peak at 92 corresponds to molecular ion peak. This peak is due to loss of 1e- from the molecule of chlorobutane.

B. Isotope Peak

The isotope peak is defined as a peak of ionic species that is generated on mass spectrum due to presence of the less abundant isotopes in the molecule. An isotope peak generally appears adjacent to any of the molecular ion or fragment ion peak if they possess some isotopes (such as ^{12}C, ^{1}H, ^{15}N, ^{35}Cl, ^{19}F, ^{31}P, ^{79}Br).

Features of Isotope Peak

It is represented with M+1 or M+2 etc. Most commonly, the M+1 isotope peaks occurs when molecules contains some isotopes such as ^{13}C, ^{2}H, ^{15}N, and ^{17}O etc. Whereas M+2 isotope peaks occurs when molecule contains S, Si, Cl, and Br etc. M+2, M+4, M+6 and further higher isotope peaks are very common for those molecules that contain multiple halogens.

Example: Mass spectrum of chloro butane representing isotope peaks (Figure 3.25).

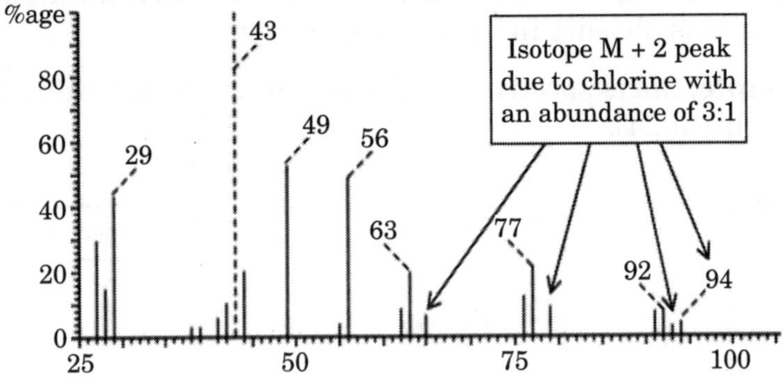

Figure 3.25: Mass spectrum of chloro butane representing isotope (M+2) peaks.

In the mass spectrum of chloro butane peak at 65, 79, and 94 corresponds to the isotope M+2 peaks. These peaks are due to presence of chlorine atom in the fragment ions of $C_2H_4Cl^+$, $C_3H_6Cl^+$, $C_4H_8Cl^+$ and $C_4H_9Cl^+$.

C. Base Peak

In a mass spectrum the ionic peak that shows the highest abundance (100%) is designated as a base peak. Generally, in a mass spectrum the base peak is adjusted to 100% and other peaks are signalled as a relative percentage of the base peak.

Features of Base Peak

The base peak occurs for that particular ion (*can be molecular ion or fragment ion*) that possess highest structural stability till it reaches to the detector or collector.

Example: Mass spectrum of chloro butane representing base peak (Figure 3.26).

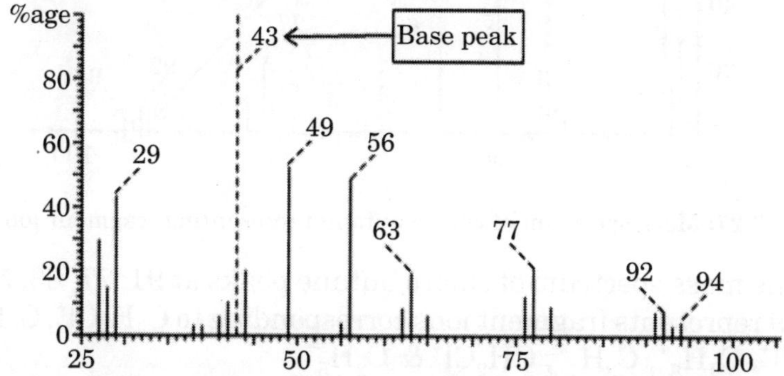

Figure 3.26: Mass spectrum of chloro butane representing base peak.

In the mass spectrum of chloro butane peak at 43 corresponds to the base peak. This peak is due to high stability of C3 carbocation.

D. Fragment Ion Peak

Fragment ion peak is defined as peak of ionic species that is generated on a mass spectrum due to fragmentation of molecular ion within the ionization chamber when exposed to an ionization potential of 70 ev energy.

Features of Fragment Ion Peak

A fragment ion is generally represented with m^+. Based on the successive loss of number of hydrogen, a molecule can produce a number of fragment ions such as M–1, M–2, and M–3. In any case a direct M–3, M–4 or M–5 fragment ion peak due to loss of 3, 4 or 5 hydrogen is rarely observed. M–14 and M–15 fragment ion peak may occur but with less abundance.

Example: Mass spectrum of chloro butane represents fragment ion peaks (Figure 3.27).

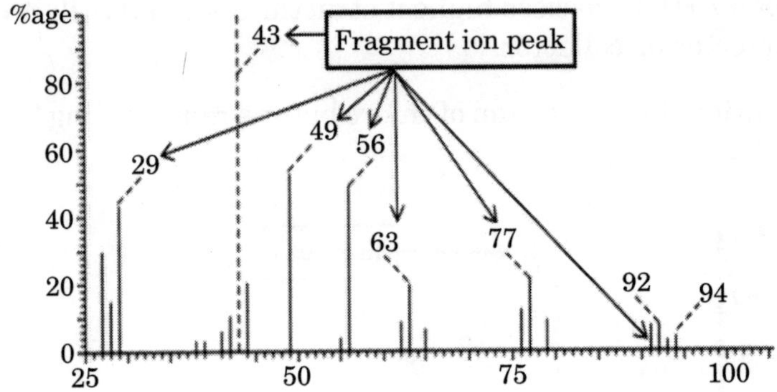

Figure 3.27: Mass spectrum of chloro butane representing fragment ion peak.

In the mass spectrum of chloro butane peaks at 91, 77, 63, 56, 49, 43 & 29 represents fragment ions corresponding to $C_4H_8Cl^+$, $C_3H_6Cl^+$, $C_2H_4Cl^+$, $C_4H_8^{\cdot+}$, $C_3H_7^+$, CH_2Cl^+ & $C_2H_5^+$.

E. Meta Stable Ion Peak

Metastable ion peak is defined as a peak of ionic species that is generated on a mass spectrum due to fragmentation of the molecular ion, which occurs out of the ionization chamber. Metastable ions are similar to the ordinary fragment ions that are produced with in the ionization chamber. But as the metastable ions are produced due to fragmentation of molecular ions especially out of the ionization chamber, so these ions possess different kinetic energy than the ordinary fragment ions. Hence these metastable ions reach to the collector in a slightly different fashion than the ordinary fragment ions that are produced with in the ionization chamber.

Features of Molecular Ions

- The metastable ion peaks usually occurs at non integer value for instance if an ordinary fragment ion peak of chlorobutane at 56, then metastable ion peak may occurs at non integer value for example at 56.2 (Figure 3.28). Likewise, there can be other metastable ion peaks in different compounds.

- The peaks are broader in size than ordinary peaks.
- The abundance of metastable ion is relatively low in comparison to the ordinary fragment ion peaks.

Example: Mass spectrum of chloro butane representing meta stable ion peak (Figure 3.28).

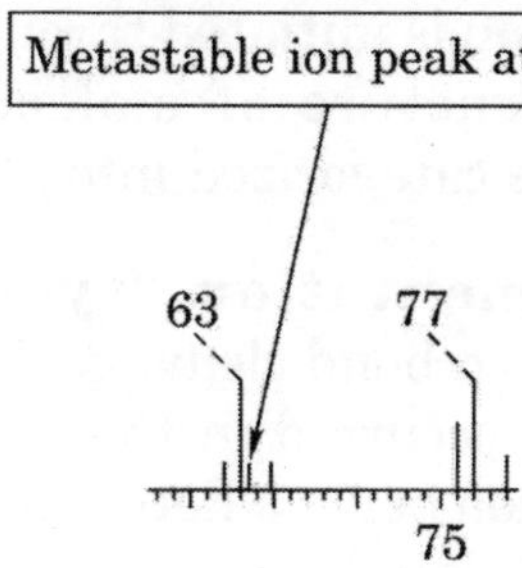

Figure 3.28: Mass spectrum of chloro butane representing meta stable ion peak.

In the given mass spectrum, peak at 63.5 represents metastable ion peak.

3.3.2 Mechanisms for Fragmentation of Molecules

A radical cation (molecular ion or parent ion) is composed of an unpaired valence electron that increases its reactivity. This reactivity of molecular ion causes its tendency to undergo new bond formation. The energy gained from new bond formation further makes the molecular ion highly energetic and unstable. Hence, to release this extra amount energy or to become stable molecular ion undergoes fragmentation by breaking some other bond in their structure. In mass spectrometry few mechanisms that mediate fragmentation of molecular species are as follows:

a. Homolytic Fragmentation (Type-I, Type-II, Type-III)

b. Heterolytic Fragmentation

c. Retro Diels Elder Fragmentation

d. Rearrangement (Cyclization, Elimination-ketene, Elimination-neutral)

e. McLafferty Rearrangement

f. Double McLafferty Rearrangement

A. Homolytic Fragmentation

A homolytic fragmentation is the one which occurs due to breaking of bonds between the two similar atoms (such as C–C). In a molecular ion, homolytic fragmentation is initiated through the radical region. Depending upon the structure of molecular radical cation homolytic fragmentation is categorized into three types:

(a) **Homolytic Fragmentation Type-I:** This type of fragmentation follows α bond cleavage. These fragmentations are most likely to be occurred in those molecules in which a molecular radical cation bears a heteroatom that is attached to a carbon with a single bond. In such case, firstly a parent ion will be formed (as the unpaired electron of hetero atom is involved in forming a double bond with carbon atom). Secondly, the energy gained in bond formation is utilised to break the bond between alpha and beta carbons. This breaking of bond or cleavage results in generation of one radical and one fragment ion. In this Fragmentation the +ve charge is always carried by the heteroatom of the heteroatom containing fragment.

List of some compounds that frequently undergoes homolytic fragmentation of Type-I are as follows:

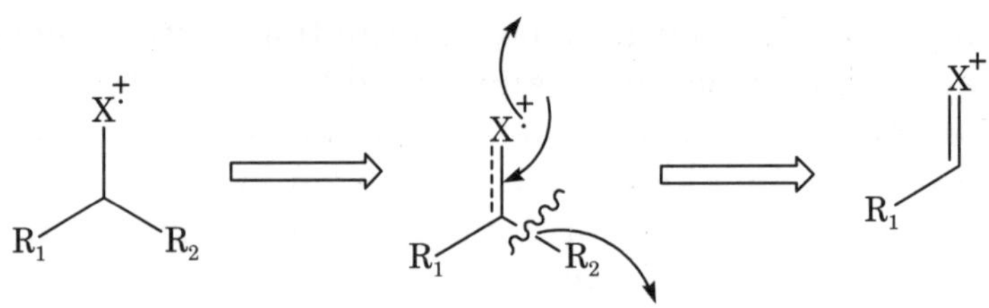

1. *Alcohols—(Example: 2-Butanol)*: The positive charge is carried by the heteroatom. Bond between alpha and beta carbon is broken and a new double bond is formed with heteroatom.

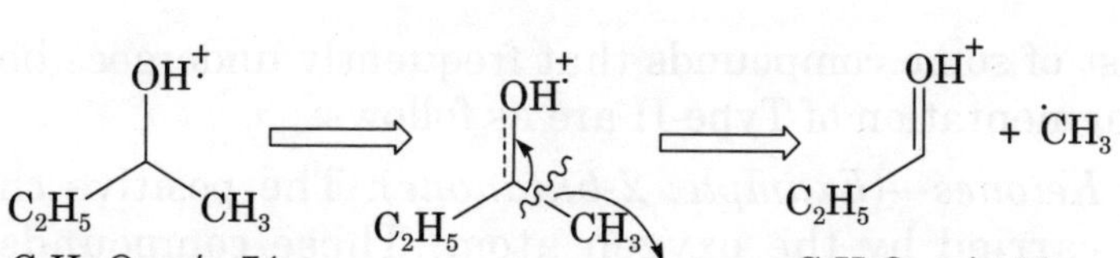

$C_4H_{10}O$, m/z: 74 $\qquad$ C_3H_7O, m/z: 59

2. *Ethers—(Example: Methyl ethyl ether)*: The positive charge is carried by the heteroatom. Bond between alpha and beta carbon is broken and a new double bond is formed with heteroatom.

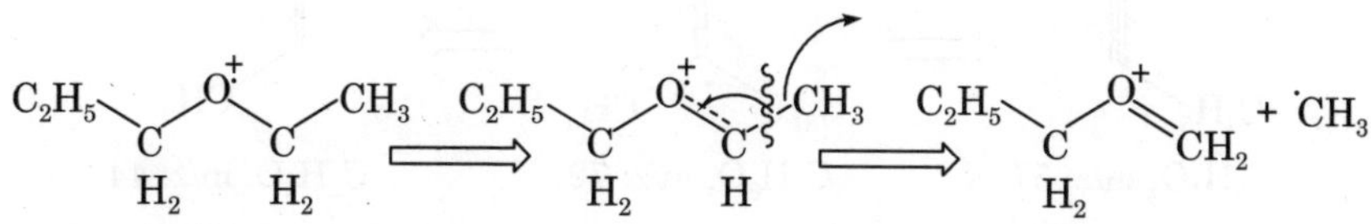

$C_5H_{12}O$, m/z: 88 $\qquad$ C_4H_9O, m/z: 73

3. *Amines—(Example: N-ethylbutan-2-amine)*: The positive charge is carried by the heteroatom. Bond between alpha and beta carbon is broken and a new double bond is formed with heteroatom.

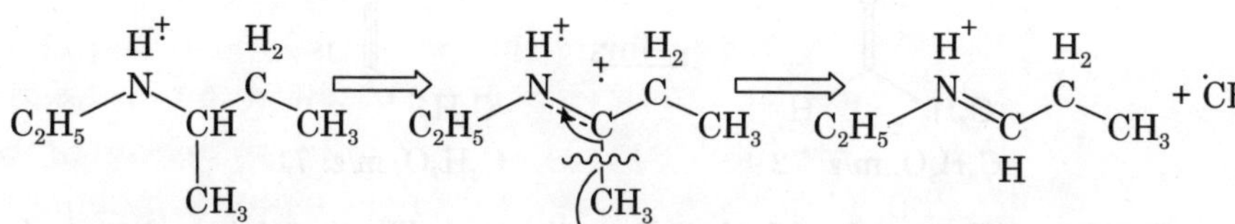

$C_6H_{15}N$, m/z: 101 $\qquad$ $C_5H_{12}N$, m/z: 86

(*b*) Homolytic Fragmentation Type-II: This type of fragmentation also follows α-bond cleavage. These fragmentations are most likely to be occurred in those molecules in which a molecular radical cation bears a heteroatom that is attached to a carbon with a double bond. In such case, firstly a parent ion will be formed (as the unpaired electron of hetero atom is involved in forming a triple bond with carbon atom). Secondly, the energy gained in bond formation is utilised to break the bond between α and β carbons. This breaking of bond or cleavage results in generation of one radical and one fragment ion. In this fragmentation the +ve charge is always carried by the heteroatom of the heteroatom containing fragment.

List of some compounds that frequently undergoes homolytic fragmentation of Type-II are as follows:

1. *Ketones—(Example: 2-butanone)*: The positive charge is carried by the oxygen atom. These compounds follow breakage of carbon-carbon bond that is alpha to carbonyl group. In unsymmetrical ketones two types of peaks are generated. In such fragmentation, elimination of bigger alkyl radical is commonly observed.

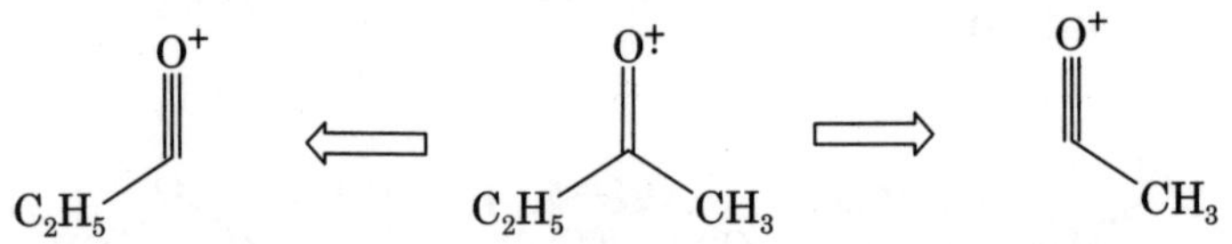

C_3H_5O, m/z: 57 C_4H_8O, m/z: 72 C_2H_4O, m/z: 44

2. *Aldehydes—(Example: 1-butanal)*: The positive charge is carried by the oxygen atom. These compounds follow breakage of carbon-hydrogen bond that is alpha to carbonyl group. In such fragmentation, elimination of bigger alkyl radical is commonly observed.

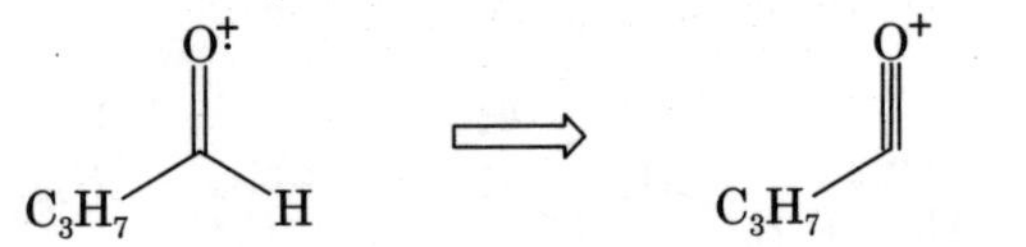

C_4H_8O, m/z: 72 C_4H_7O, m/z: 71

3. *Ester (Example: Methyl Propionate)*: The positive charge is carried by the oxygen atom. These compounds follow breakage of carbon-oxygen bond that is alpha to carbonyl group. In such fragmentation, elimination of bigger alkyl radical is commonly observed.

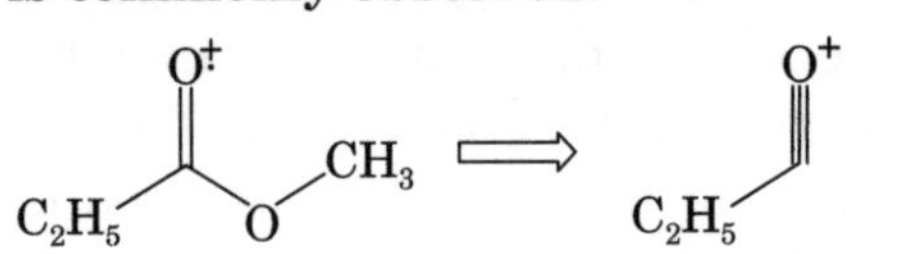

$C_4H_8O_2$, m/z: 88 C_3H_5O, m/z: 57

4. *Amides—(Example: propionamide)*: The positive charge is carried by the oxygen atom. These compounds follow breakage of carbon-nitrogen bond that is alpha to carbonyl group. In such fragmentation, elimination of bigger alkyl radical is commonly observed.

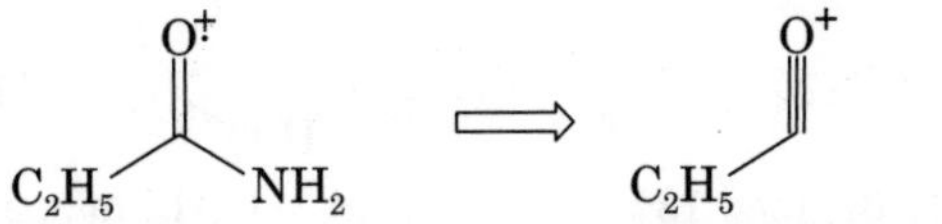
C_3H_7NO, m/z: 73 C_3H_5O, m/z: 57

(*c*) **Homolytic Fragmentation Type-III:** This type of fragmentation follows β bond cleavage. These fragmentations are most likely to be occurred in those molecules in which a molecular radical cation possesses a benzyl group in its molecular structure. In such case the carbon α to the benzene ring is involved in forming a bond with benzene to generate a tropilium ion. Whereas the carbon at β position from benzene ring undergoes cleavage. In this fragmentation the +ve charge is always carried by the tropilium fragment. Below mentioned is an example of compound that frequently undergoes homolytic fragmentation of Type-III are as follows:

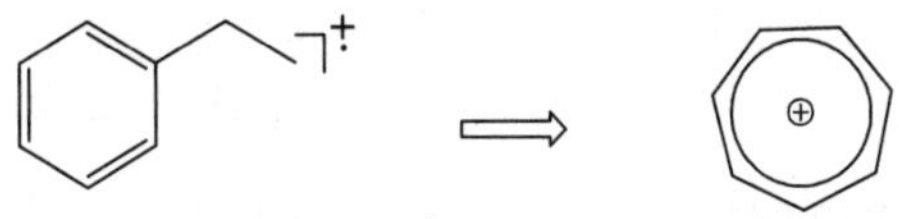
C_8H_{10}, m/z: 106 $C_7H_7^+$, m/z: 91

B. Heterolytic Fragmentation

A heterolytic fragmentation is the one which occurs due to breaking of bonds between the two different atoms (such as C–X, where X may be O, S, N, Cl). These fragmentations are most likely to be occurred in those molecules in which a molecular radical cation bears a heteroatom that is attached to a carbon with a single bond such as C–C–X. In such cases usually C–X bond is difficult to break when compared with C–C bond. In this fragmentation the +ve charge is not carried by the heteroatom containing fragment but is carried by the carbon containing fragment.

List of some compounds that frequently undergoes heterolytic fragmentation are as follows:

1. Halides

(*i*) *Example: Butyl Bromide*: In such case bromide bond is easily broken that results in a signal at m/z 57.

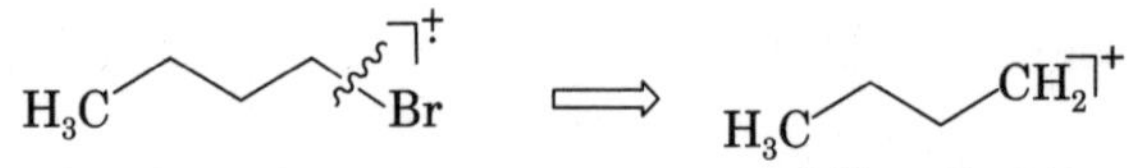

C_4H_9Br, m/z: 137 — C_4H_9, m/z: 57

(*ii*) *Example: Butyl iodide*: In such case Iodide bond is easily broken that results in a signal at m/z 57.

C_4H_9I, m/z: 184 — C_4H_9, m/z: 57

(*iii*) *Example: Butyl chloride*: In this case chloride bond is hard to be broken, such that it removes chlorine atom in form of HCl through rearrangement. This results in a signal at m/z 56.

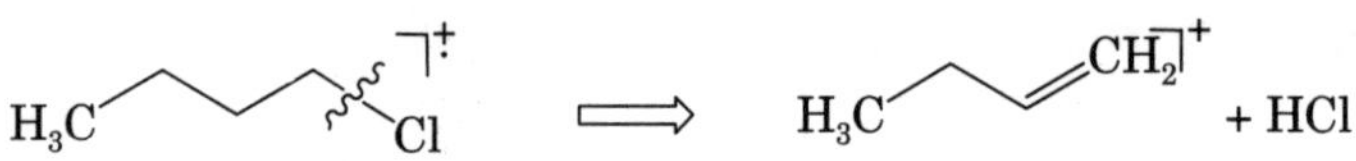

C_4H_9Cl, m/z: 92 — C_4H_8, m/z: 56

2. **Ethers (Example: Propyl ethyl ether):** In this case C–O bond is broken, such that it generates propyl ion and ethoxy radical fragment. This results in a signal at m/z 43.

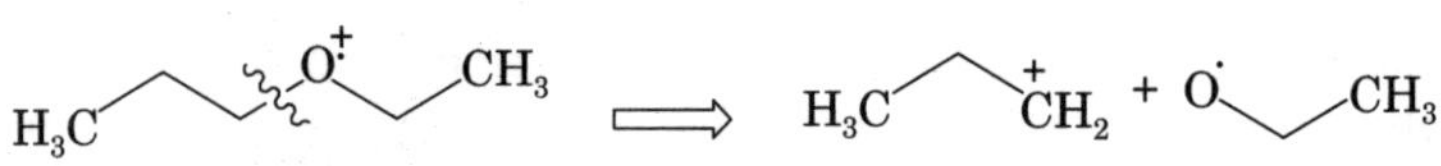

$C_5H_{12}O$, m/z: 88 — C_3H_7, m/z: 43

C. Retro Diels Elder Fragmentation

A Retro Diels Elder Fragmentation is the one which occurs generally in case of cyclic systems. In this fragmentation usually, if there occur a fragmentation of cyclic olefin molecular ion, a conjugated diene fragment ion is generated. But in case of cycloalkane molecular ion, an alkene fragment ion is generated. In such fragmentation process the +ve charge is generally carried by the fragment which is more substituted or more conjugated and possess a lower ionisation potential.

List of some compounds that frequently undergoes Retro Diels Elder fragmentation are as follows:

1. **Cycloalkene (Example: Cyclohexene):** In this case the cyclic ring is broken, such that it generates a dienyl ion and a neutral alkene. This results in a signal at m/z 54.

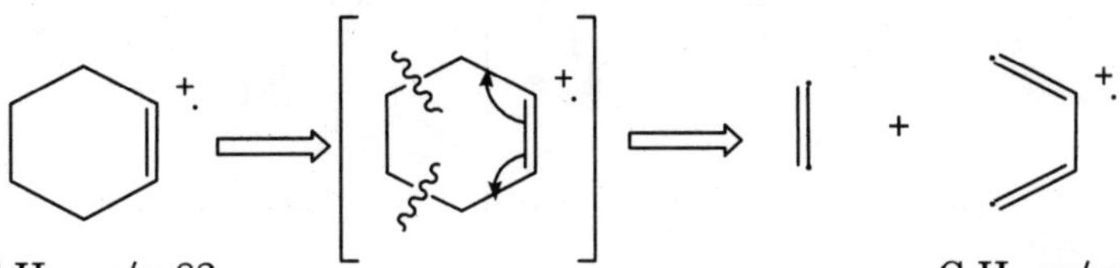

C_6H_{10}, m/z: 82 C_4H_6, m/z: 54

2. **Cycloalkane (Example: Cyclohexane):** In this case the cyclic ring is broken, such that it generates a enyl ion and a neutral alkene. This results in a signal at m/z 54.

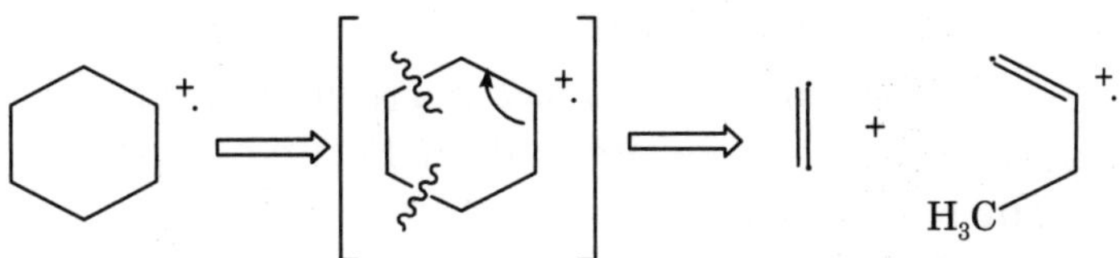

C_6H_{12}, m/z: 84 C_4H_7, m/z: 55

D. Rearrangement Fragmentation (Hydrogen Transfer)

This type of fragmentation mechanism is one of the most common intra-molecular rearrangement process of molecular radical ion structure, that leads to transfer of an atom (hydrogen) from one site of molecule to another. This as a result leads to cleavage or fragmentation of the molecule. Below mentioned are the different types of rearrangement fragmentations that occurs very frequently in the molecules.

(*a*) Cyclization Rearrangement: List of some compounds that frequently undergoes cyclization rearrangement are as follows:

(*i*) *Example: Alcohols* (*Propanol*): These undergo elimination of water such that positive charge retains on the carbon fragment

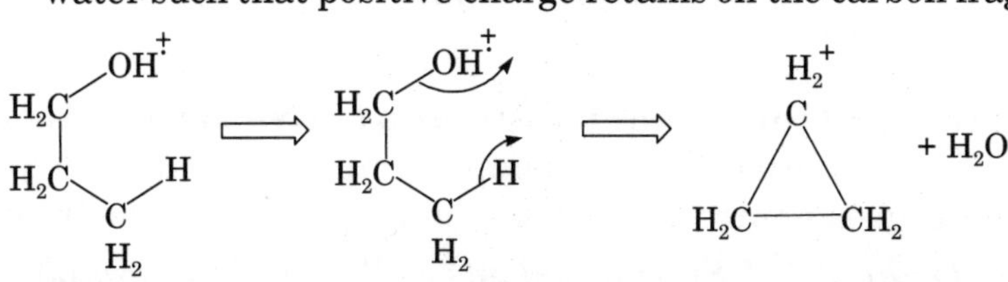

C_3H_8O, m/z: 60 C_3H_6, m/z: 42

(*ii*) *Example: Halides* (*Butyl chloride*): These undergo elimination of HCl such that positive charge retains on the carbon fragment.

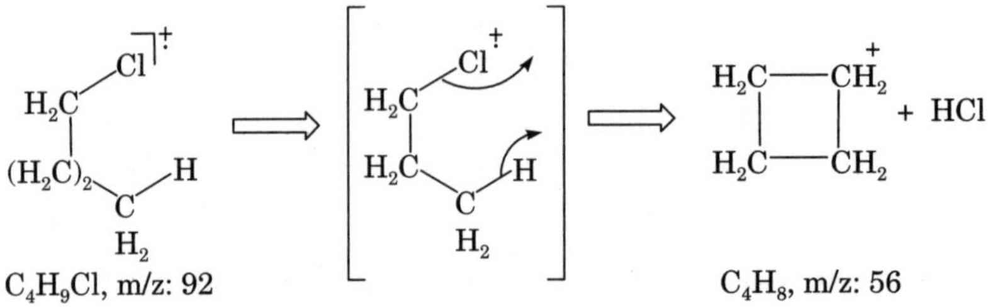

(*b*) Elimination Rearrangement (Ketene Elimination): List of some compounds that frequently undergoes ketene eliminations are as follows.

(*i*) *Amides* (*Example: Phenyl acetamide*): These involve characteristic ketene elimination. Positive charge is attained by the remaining amine fragment.

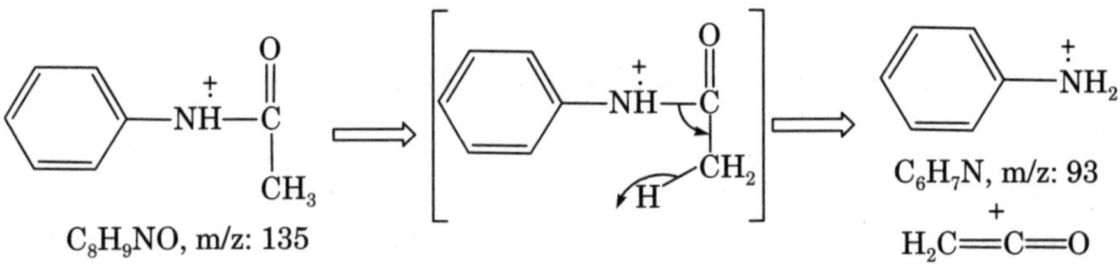

(*ii*) *Acetates* (*Example: Phenyl acetate*): These involve characteristic ketene elimination. Positive charge is attained by the remaining phenolic fragment.

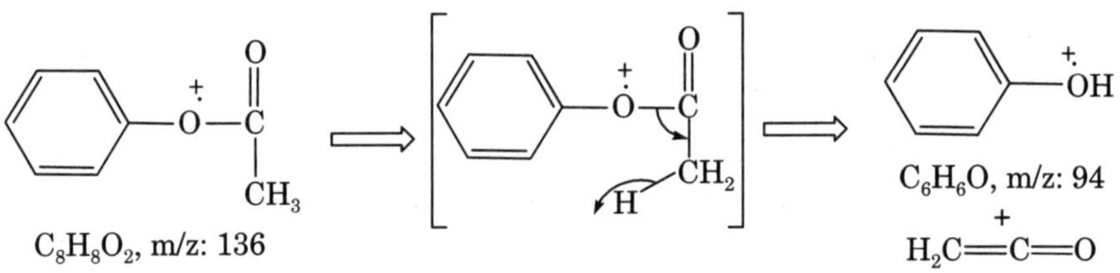

(*c*) Elimination Rearrangement (Neutral Molecule Elimination): List of some compounds that frequently undergoes neutral molecule eliminations are as follows:

(*i*) *Ortho Alkyl Substituted Alkyl Benzoates* (*Example: Ortho methyl 2-methylbenzoate*): These involve elimination of a neutral molecule of methanol and the positive charge is carried by the remaining fragment.

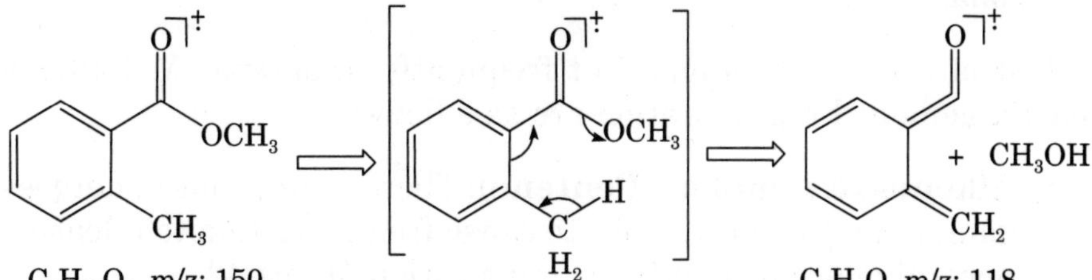

$C_9H_{10}O_2$, m/z: 150 C_8H_6O, m/z: 118

(*ii*) *Ortho Alkyl Substituted Benzyl Alcohol* (*Example: o-tolylmethanol*): These involve elimination of a neutral molecule of water and the positive charge is carried by the remaining fragment.

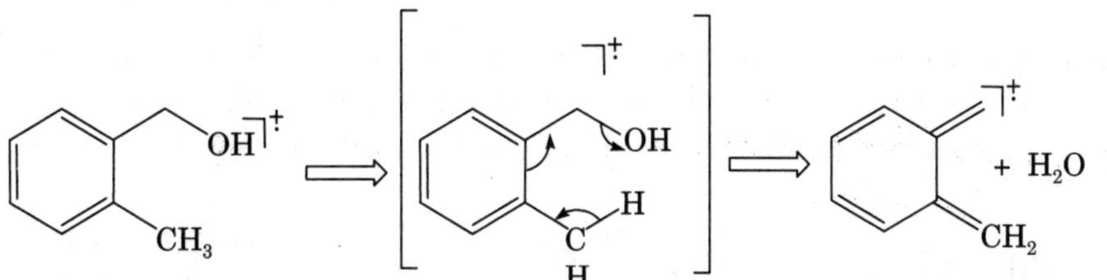

$C_8H_{10}O$, m/z: 122 C_8H_8, m/z: 104

(*iii*) *Alkyl Substituted Butenoate* (*Example: Methyl but-2-enoate*): These involve elimination of a neutral molecule of methanol and the positive charge is carried by the remaining fragment.

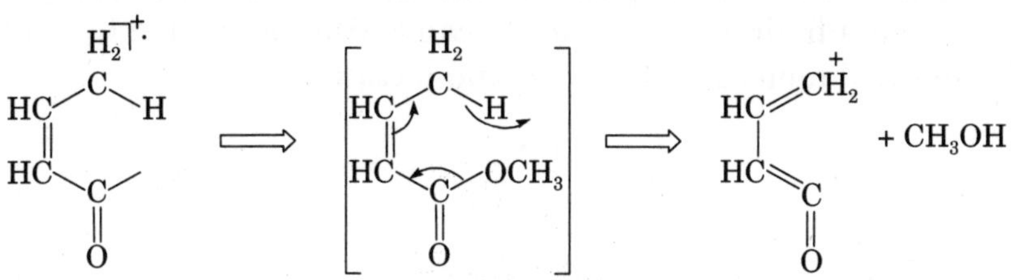

$C_5H_8O_2$, m/z: 100 C_4H_4O, m/z: 68

E. McLafferty Rearrangement Fragmentation

This type of fragmentation mechanism involves migration of gamma hydrogen and thereby elimination of a neutral molecule. This type of fragmentation is most commonly observed in the molecules of amines, unsaturated compounds, ketones and aldehydes etc. containing hydrogen at gamma position of their characteristic functional group.

List of some compounds that frequently undergoes McLafferty rearrangement Fragmentation are as follows:

(*a*) **Alkenes (Example: 4-Pentene):** These compounds undergoes gamma hydrogen transfer to cause fragmentation to release a neutral molecule of ethene and a cation at m/z 42.

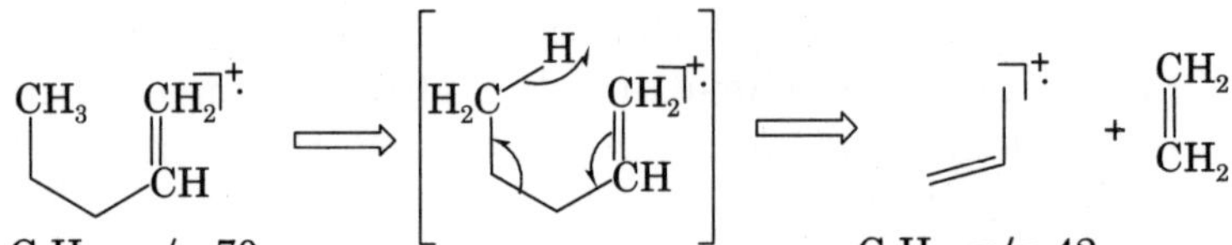

C_5H_{10}, m/z: 70 C_3H_6, m/z: 42

(*b*) **Ketones (Example: 2-Pentanone):** These compounds undergo gamma hydrogen transfer to cause fragmentation to release a neutral molecule of ethene and a cation at m/z 58.

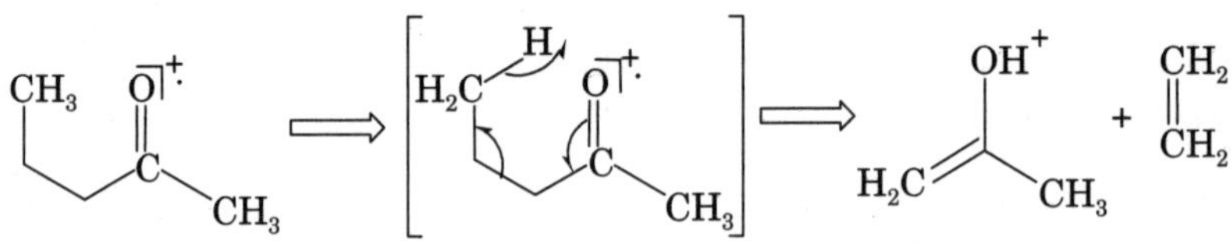

C_5H_{10}, m/z: 86 C_3H_6O, m/z: 58

(*c*) **Aldehydes (Example: Pentanal):** These compounds undergo gamma hydrogen transfer to cause fragmentation to release a neutral molecule of ethene and a cation at m/z 44.

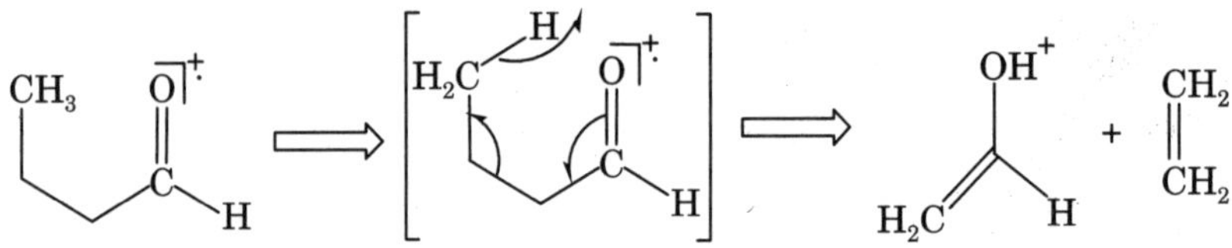

C_4H_8O, m/z: 72 C_2H_4O, m/z: 44

(*d*) **Aromatic Alkanes (Example: Propyl Benzene):** These compounds undergo gamma hydrogen transfer to cause fragmentation to release a neutral molecule of ethene and a cation at m/z 92.

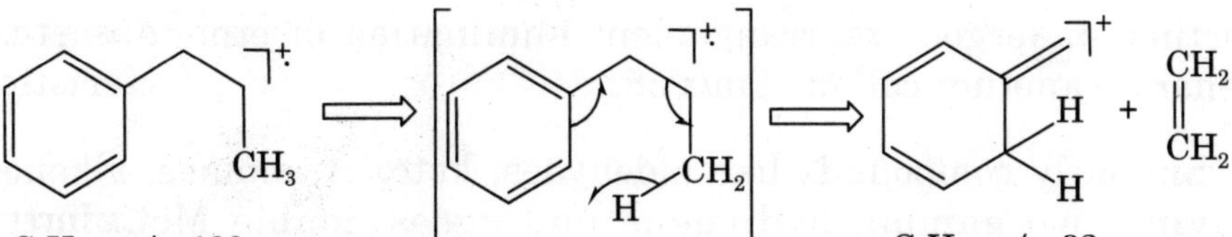

C_9H_{12}, m/z: 120 C_7H_8, m/z: 92

(*e*) **Amines (Example: Pentan-2-imine):** These compounds undergo gamma hydrogen transfer to cause fragmentation to release a neutral molecule of ethene and a cation at m/z 44.

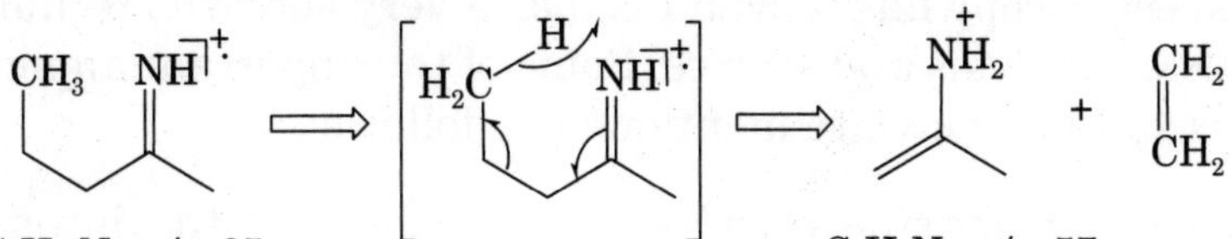

$C_5H_{11}N$, m/z: 85 C_3H_7N, m/z: 57

F. DOUBLE MCLAFFERTY REARRANGEMENT FRAGMENTATION

This type of fragmentation mechanism is same as McLafferty rearrangement but involves two times migration of gamma hydrogen and thereby eliminates two neutral molecules. This mechanism is common for those molecules that possess two gamma hydrogen positions.

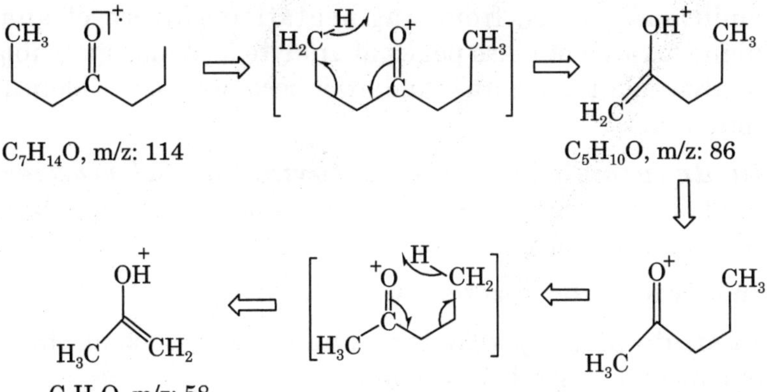

C_3H_6O, m/z: 58

For example in case of 4-heptanone, following gamma hydrogen transfer firstly a neutral molecule of ethene is eliminated to generate a fragment cation at m/z 86. Further this generated cation fragment further undergoes rearrangement elimination of ethane unit to generate another cation of m/z 58.

Similarly compounds like aldehydes, ketones, amines, alkenes having two gamma hydrogens undergoes double McLafferty rearrangements.

3.4 APPLICATIONS OF MASS SPECTROSCOPY

Mass spectroscopy has proven itself as a very versatile technique in the field of analysis and science. Some of the important applications of mass spectroscopy are mentioned as follows:

- ***In quantitative analysis*:** By comparing the intensity of fragment peaks of any given analyte under Mass investigation with some standard drug, the quantity of sample can be calculated.
- ***In establishment of structures*:** By making use of all principle fragmentation patterns, Nitrogen rule (molecule of odd/even nitrogen possess odd/even atomic weight respectively), Isotopic abundance and also peak matching the establishment of structure of both known and unknown compounds can be done.
- ***In detection of impurities*:** By identification of some additional peaks, from fragmentation pattern of analyte, highest value of mass peaks of analyte and also matching the analyte spectrum with standard it becomes easy to identify the impurities.
- ***In drug metabolism and toxicological studies*:** By analyzing the mass spectrum of metabolite and pure drug, the metabolism of drug can be easily investigated.
- **In clinical and forensic investigations.**
- **In study of gas phase ions chemistry**—such as chemistry of ions and neutrals in vacuum.
- In physical, chemical or biological properties investigations of compounds.
- Also in identification and sequencing of peptides and proteins.

3.5 INTERPRETATION

This section describes the mass spectral characteristics, spectral interpretation and fragmentation mode of various classes of compounds.

3.5.1 Alkanes

Mass Spectral Characteristics of Straight Chain Alkanes

- In the mass spectrum of straight chain alkanes the relative abundance of M^+ ion peak decreases on increasing homologation of alkanes.
- In the spectrum several bunch of fragment ion peaks (*i.e.*, peaks of C_nH_{2n+1}, C_nH_{2n}, and C_nH_{2n-1}) that are 14 units apart are commonly observed.
- In each bunch of fragment ion peak the order of abundance is $C_nH_{2n+1} > C_nH_{2n} > C_nH_{2n-1}$.
- Due to formation of carbocation in the mass spectrum of alkanes the most abundant peaks are of C3 and C4 ions.
- Fragmentation pattern of following example of octane is a key to explain the mass spectrum of other straight chain alkanes.

Example: *Octane* [Mass spectrum and fragmentation given in Figure 3.29(*a*) and (*b*) respectively].

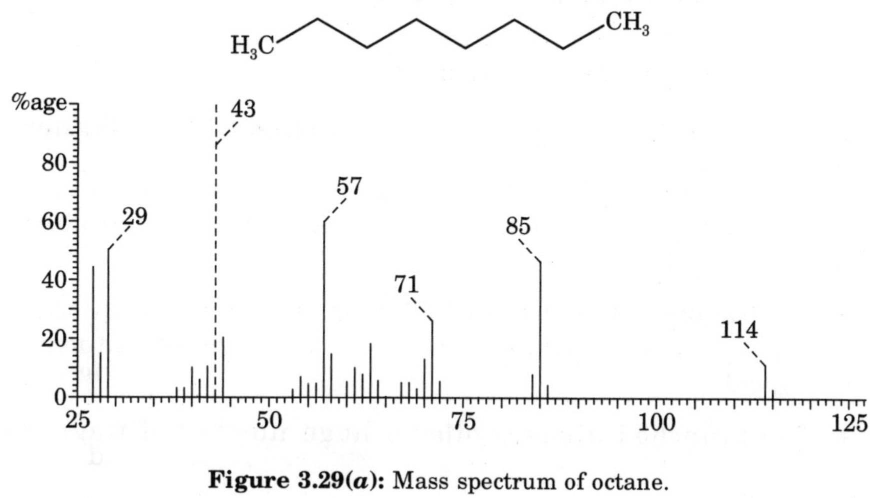

Figure 3.29(*a*): Mass spectrum of octane.

Spectral interpretation: Peaks observed at m/z 114 (corresponds to parent ion peak of $C_8H_{18}^{+}$ due to loss of 1e$^-$), 43 (base peak of $C_3H_7^+$ due to loss of $C_5H_{11}^{\bullet}$), 85 ($C_6H_{13}^+$ due to loss of $C_2H_5^{\bullet}$), 71 ($C_5H_{11}^+$ due to loss of $C_3H_7^{\bullet}$), 57 ($C_4H_9^+$ due to loss of $C_4H_9^{\bullet}$), and 29 ($C_2H_5^+$ due to loss of $C_6H_{13}^{\bullet}$).

Fragmentation mode

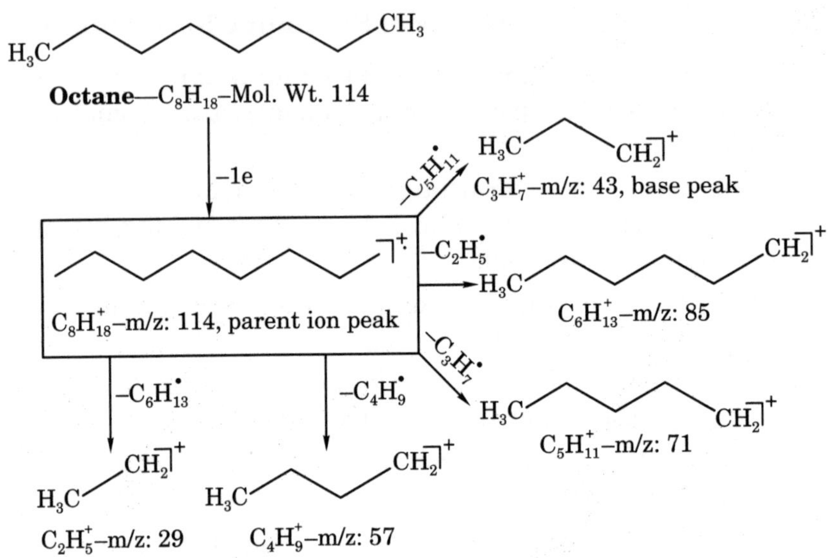

Figure 3.29(*b*): Mass fragmentation of octane.

3.5.2 Branched Chain Alkanes

Mass Spectral Characteristics of Branched Chain Alkanes

- In branched chain alkanes mass spectrum the relative abundance of parent ion peak is very less and generally absent in the spectrum.
- The fragmentation in branched chain alkanes generally occurs at the point of branching to eliminate the larger substituent as radical.
- The branched alkanes offer a huge number of fragment ions.

- C_nH_{2n+1} fragment ions generally appear in low abundance.
- If the organic compound contains longer chain of carbons then due to formation of carbocation the most abundant peaks are generated of C3 and C4 ions.
- Fragmentation pattern of following example of dimethyl pentane is a key to explain the mass spectrum of other branched chain alkanes.

Example: *2, 2-Dimethyl Pentane* (Mass spectrum and fragmentation given in Figure 3.30(*a*) and (*b*) respectively).

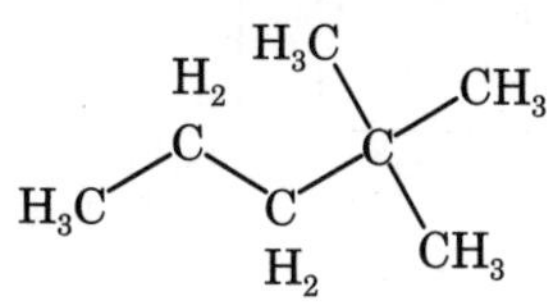

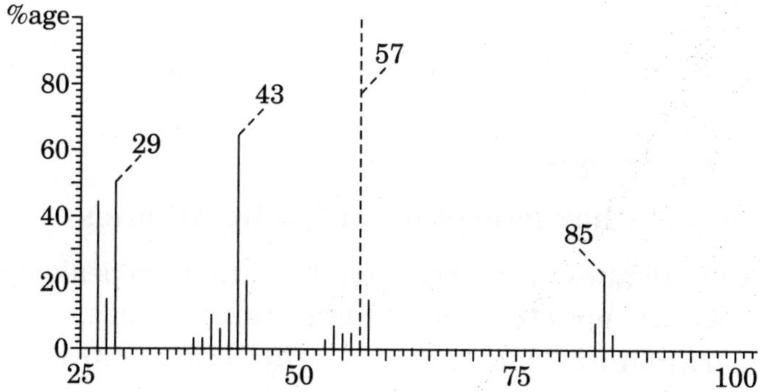

Figure 3.30(*a*): Mass spectrum of 2, 2-dimethyl pentane.

Spectral interpretation: Parent ion peak is absent due to branching. Peaks are observed at m/z 57 (corresponds to base peak of $C_4H_9^+$ due to loss of $C_3H_7^\bullet$), 85 ($C_6H_{13}^+$ due to loss of $CH_3^\bullet$), 43 ($C_3H_7^+$ due to loss of $C_4H_9^\bullet$) and 29 ($C_2H_5^+$ due to loss of $C_5H_{11}^\bullet$).

Fragmentation mode

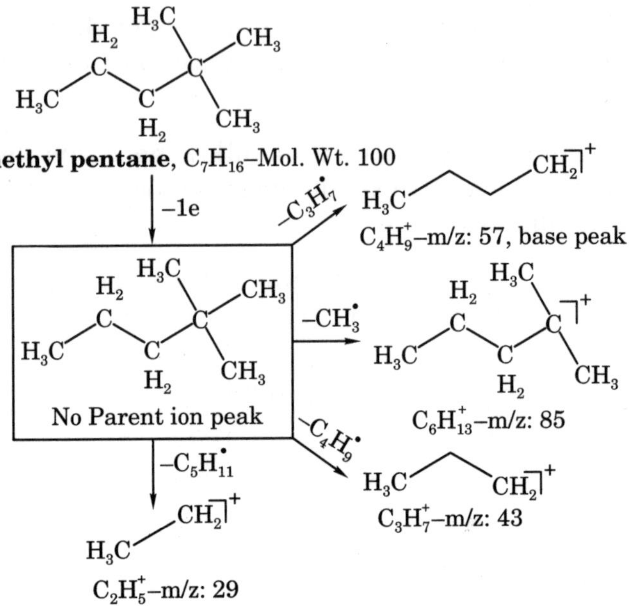

Figure 3.30(*b*): Mass fragmentation of 2, 2-dimethyl pentane.

3.5.3 Cyclic Alkanes

Mass Spectral Characteristics of Cyclic Alkanes

- In cyclic alkanes mass spectrum the parent ion peak abundance is generally more than of straight chain alkanes.
- The fragmentation in cyclic alkanes generally follows elimination of two carbon units.
- Fragment ions of cyclic alkanes are generally produced by loss of alkene as radical.
- Increasing the side chain carbons in cyclic alkenes causes formation of carbocation that result in abundant peaks of C3 and C4 ions.
- The cyclic alkanes commonly show a distinct fragment ion peak in the spectrum that is a result of Retro Diels Elder reaction.
- Fragmentation pattern of following example of methyl cyclohexane is a key to explain the mass spectrum of other cyclic alkanes.

Example: *Methyl cyclohexane* [Mass spectrum and fragmentation given in Figure 3.31(*a*) and (*b*) respectively].

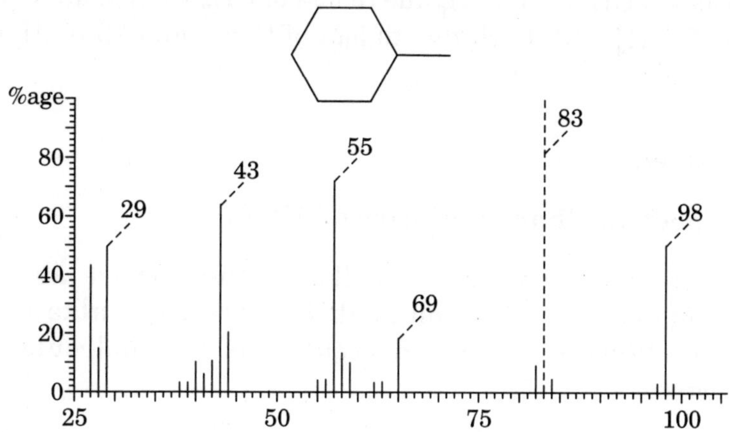

Figure 3.31(*a*): Mass spectrum of methyl cyclohexane.

Fragmentation mode

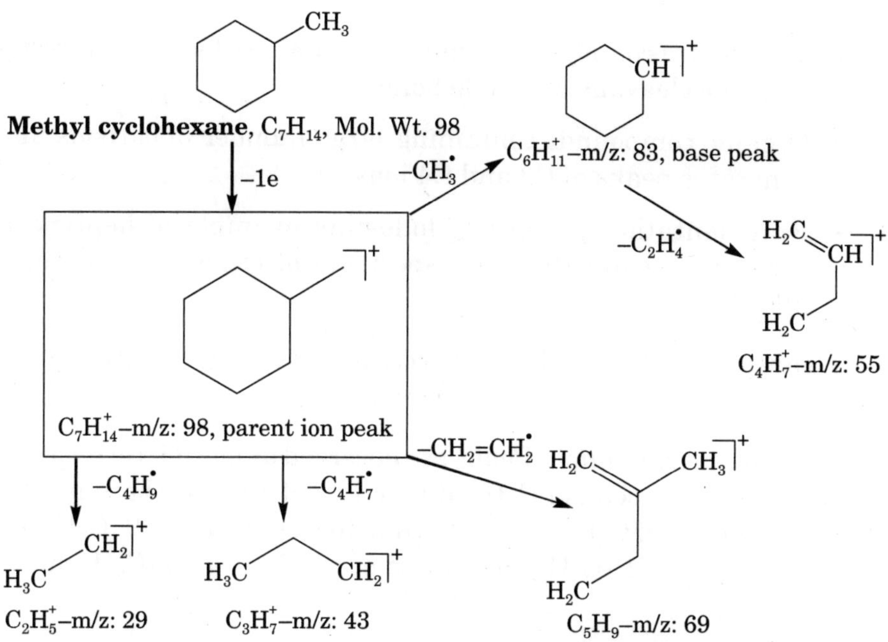

Figure 3.31(*b*): Mass fragmentation of methyl cyclohexane.

Spectral interpretation: Peaks are observed at m/z 98 (corresponds to parent ion peak of $C_7H_{14}^{+}$ due to loss of 1e$^-$), 83 (base peak of $C_6H_{11}^{+}$ due to loss of $CH_3^{\bullet}$), 69 (C_5H_9 due to loss of $CH_2 = CH_2^{\bullet}$), 55 ($C_4H_7^{+}$ due to loss of $C_2H_4^{\bullet}$), 43 ($C_3H_7^{+}$ due to loss of $C_4H_7^{\bullet}$) and 29 ($C_2H_5^{+}$ due to loss of $C_5H_9^{\bullet}$).

3.5.4 Alkenes

Mass Spectral Characteristics of Alkenes

- In the mass spectrum of alkenes the relative abundance of Molecular (M^+) ion peak is more than alkanes. The abundance of M^+ ion decreases as the molecular mass increases.
- If the gamma carbon containing hydrogen is available in alkene than it results in McLafferty rearrangement.
- In the mass spectrum of alkenes compounds the most abundant peaks are due to C_nH_{2n} ions.
- The spectra comprises fragment ions that are formed as a result of allylic cleavage of double bond.
- Organic compounds containing large number of carbons also generates peaks of C3 and C4 ions.
- Fragmentation pattern of following example of heptene is a key to explain the mass spectrum of other straight chain alkenes.

Example: *2-Heptene* [Mass spectrum and fragmentation given in Figure 3.32(*a*) and (*b*) respectively].

Spectral interpretation: Peaks are observed at m/z 98 (corresponds to the parent ion peak of $C_7H_{14}^{+}$ due to loss of 1e$^-$), 57 (base peak of $C_4H_9^{+}$ due to loss of $C_3H_5^{\bullet}$), 83 ($C_6H_{11}^{+}$ due to loss of $CH_3^{\bullet}$), 69 ($C_5H_9^{+}$ due to loss of $C_2H_5^{\bullet}$), 41 ($C_3H_5^{+}$ due to loss of $C_4H_9^{\bullet}$) and 29 ($C_2H_5^{+}$ due to loss of $C_5H_9^{\bullet}$).

Fragmentation mode

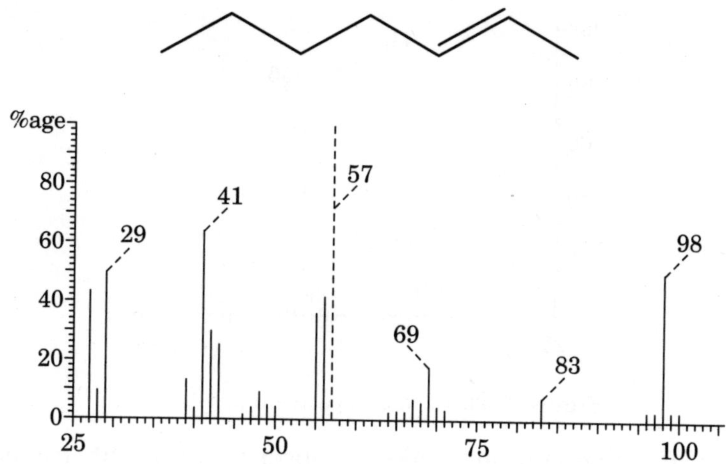

Figure 3.32(*a*): Mass spectrum of 2-heptene.

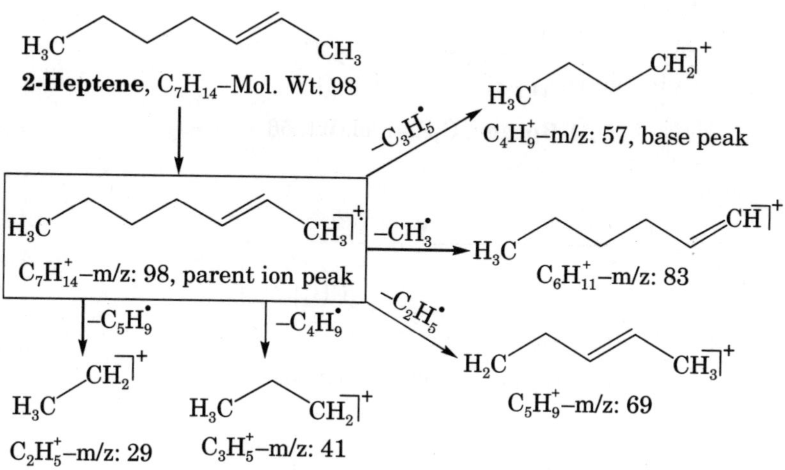

Figure 3.32(*b*): Mass fragmentation of 2-heptene.

Example: *1-Butene* [Mass spectrum and fragmentation given in Figure 3.33(*a*) and (*b*) respectively].

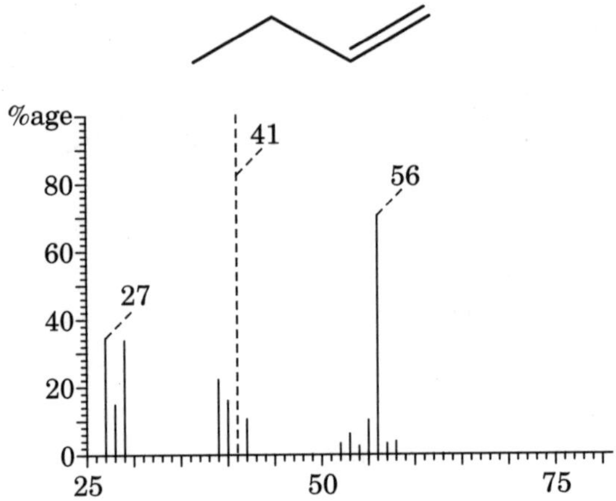

Figure 3.33(*a*): Mass spectrum of 1-butene.

Spectral interpretation: Peaks are observed at m/z 56 (corresponds to parent ion peak of $C_4H_8^{+\cdot}$ due to loss of $1e^-$), 41 (Base peak of $C_3H_5^+$ ion due to loss of $CH_3^{\bullet}$), and 27 ($C_2H_3^+$ ion due to loss of $C_2H_5^{\bullet}$).

Fragmentation mode

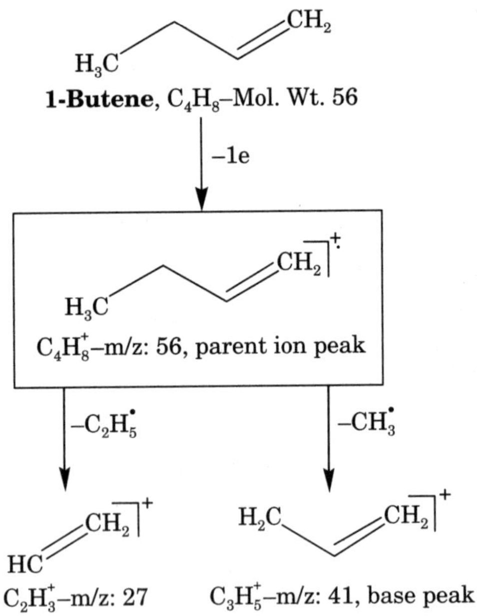

Figure 3.33(*b*): Mass fragmentation of 2-butene.

3.5.5 Cyclic Alkenes

Mass Spectral Characteristics of Cyclic Alkenes

- In the mass spectrum of cyclic alkenes, M^+ ion peak is more abundant than alkenes.
- In the spectrum of cyclic alkenes the most abundant peaks are due to C_nH_{2n} ions.
- The spectrum comprises fragment ions that are also 14 units apart.
- The cyclic alkenes commonly generate a distinct fragment ion that is result of Retro Diels Elder reaction.
- Fragmentation pattern of following example of cyclohexene is a key to explain the mass spectrum of other cyclic alkenes.

Example: *Cyclohexene* [Mass spectrum and fragmentation given in Figure 3.34(*a*) and (*b*) respectively].

Spectral interpretation: Peaks are observed at m/z 82 (corresponds to parent ion peak of $C_6H_{10}^{+\cdot}$ due to loss of 1e$^-$), 54 (base peak of $C_4H_6^{+\cdot}$ due to loss of $C_2H_4^{\bullet}$), 68 ($C_5H_8^{+\cdot}$ due to loss of $CH_2^{\bullet}$), 67 ($C_5H_7^{+\cdot}$ due to loss of $H^{\bullet}$), 41 ($C_3H_5^{+\cdot}$ due to McLafferty rearrangement) and 27 ($C_2H_3^{+}$ due to loss of $C_4H_7^{\bullet}$).

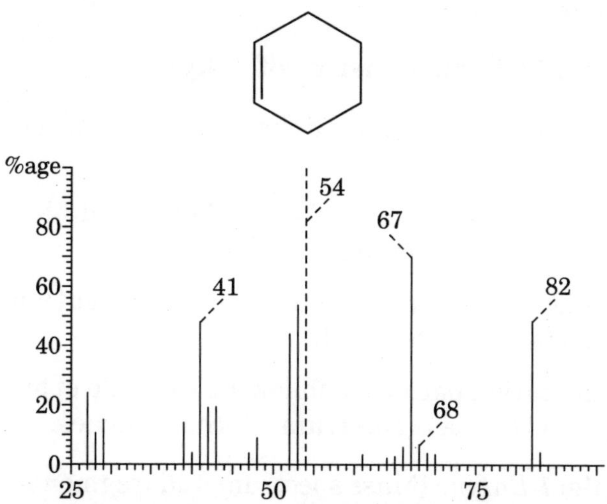

Figure 3.34(*a*): Mass spectrum of cyclohexene.

Fragmentation mode

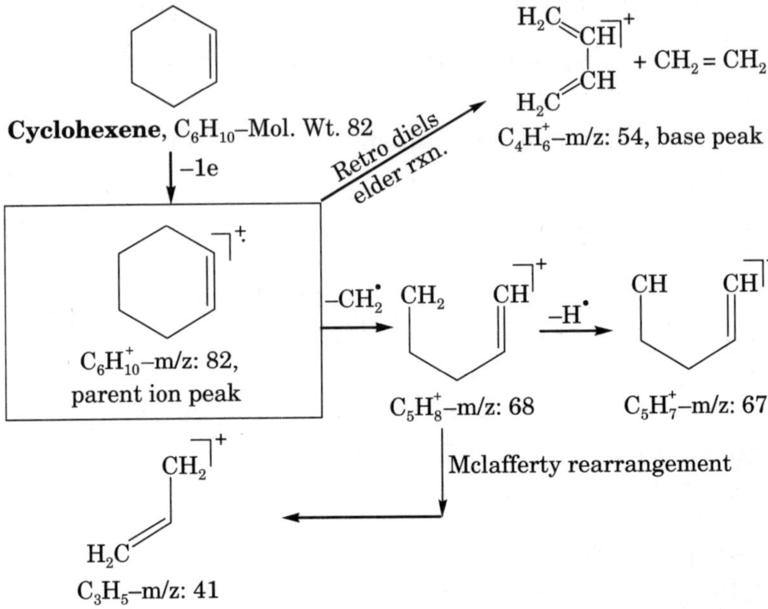

Figure 3.34(*b*): Mass fragmentation of cyclohexene.

3.5.6 Alkynes

Mass Spectral Characteristics of Alkynes

- In the mass spectrum of aliphatic alkynes, M^+ ion peak is the base peak.
- In the mass spectrum of alkynes M–15 and M–29 peaks are most commonly observed.
- The spectrum comprises distinct M–1 fragment ions peaks that are the result of loss of 1H.
- Fragmentation pattern of following example of butyne is a key to explain the mass spectrum of other alkynes.

Example: *1-Butyne* [Mass spectrum and fragmentation given in Figure 3.35(*a*) and (*b*) respectively].

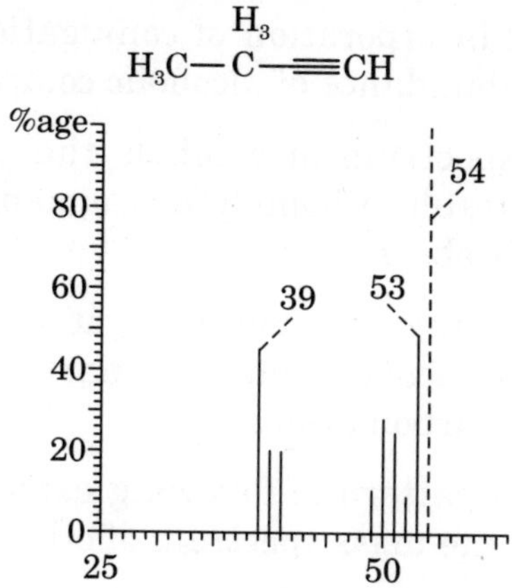

Figure 3.35(*a*): Mass spectrum of 1-butyne.

Spectral interpretation: Peaks are observed at m/z 54 (corresponds to parent ion and base peak as well of $C_4H_6^{+\cdot}$ due to loss of $1e^-$), 53 ($C_4H_5^+$ due to loss of $H^\bullet$), and 39 ($C_3H_3^+$ due to loss of $CH_3^\bullet$).

Fragmentation mode

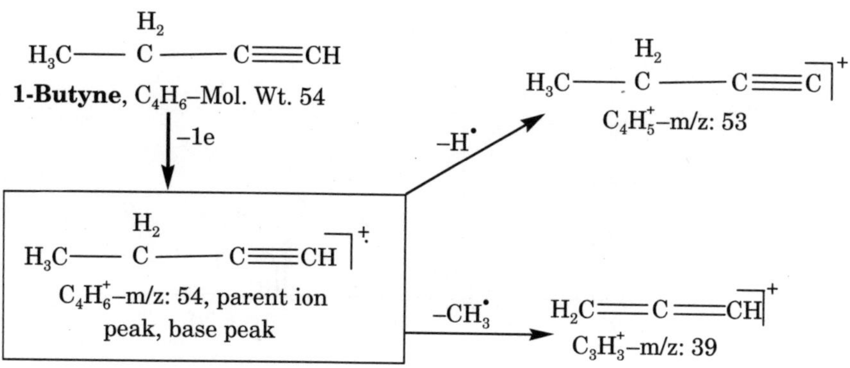

Figure 3.35(*b*): Mass fragmentation of 1-butyne.

3.5.7 Alcohols

Mass Spectral Characteristics of Alcohols

- In the mass spectrum of aliphatic alcohols M^+ ion peak is generally very low in abundance and sometimes absent in the spectrum. But incorporation of conjugation or aromatic ring increases the abundance of alcoholic compounds.

- In the mass spectrum of alcohols the peaks due to alpha cleavage as a result of homolytic fragmentation of type-I, are most commonly observed.
- The number of fragment ion peaks in the spectrum depends upon the type of alcohol such as primary, secondary or tertiary and number of carbon atoms.
- Fragmentation pattern of following example of pentanol is a key to explain the mass spectrum of other alcohols.

Example: *2-pentanol* [Mass spectrum and fragmentation given in figure 3.36(*a*) and (*b*) respectively].

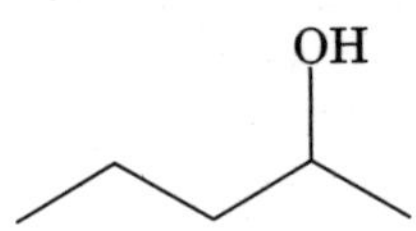

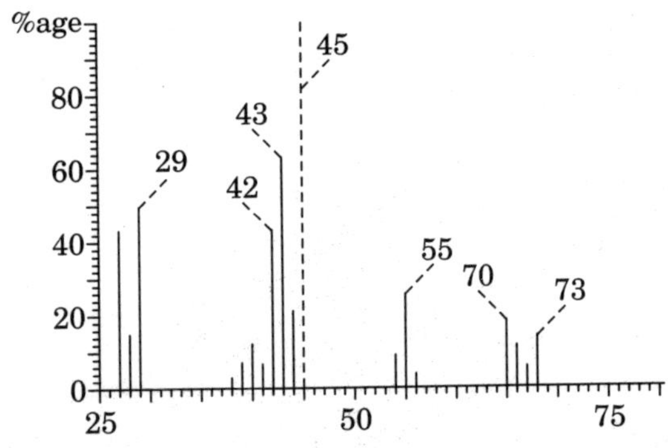

Figure 3.36(*a*): Mass spectrum of 2-pentanol.

Spectral interpretation: The parent ion peak is absent. Peaks are observed at m/z 45 (corresponds to base peak of $C_2H_5O^+$ due to loss of $C_3H_7^\bullet$), 73 ($C_4H_9O^+$ due to loss of $CH_3^\bullet$), 70 ($C_5H_{10}^+$ due to loss of $H_2O^\bullet$), 55 ($C_4H_7^+$ due to loss of $CH_3^\bullet$), 43 ($C_3H_7^+$ due to loss of $C_2H_5O^\bullet$), 42 ($C_3H_6^+$ due to loss of $C_2H_4^\bullet$), and 29 ($C_2H_5^+$ due to loss of $C_3H_7O^\bullet$).

Fragmentation mode

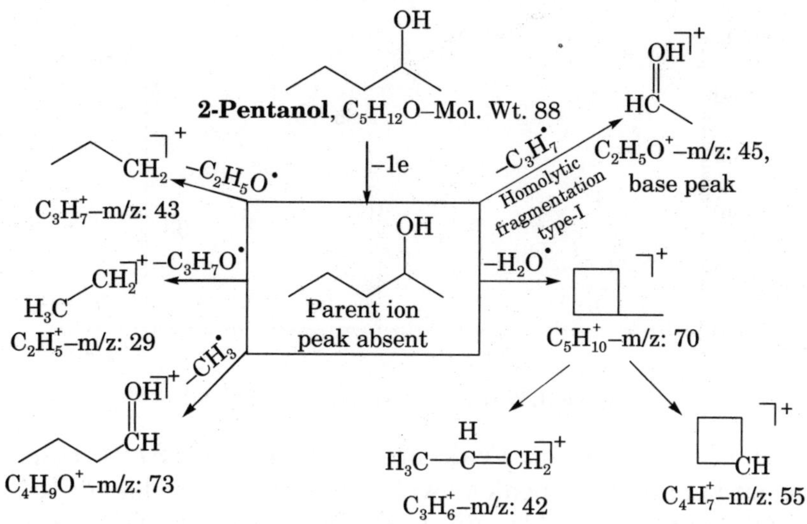

Figure 3.36(*b*): Mass fragmentation of 2-pentanol.

3.5.8 Phenols

General Mass Spectral Characteristics of Phenols

- In the mass spectrum of phenol, the M^+ ion peak is generally of high abundance.
- In phenol the phenoxy ion peak *i.e.*, peak due to loss of 1H from M^+ ion is usually of low abundance. But in case of cresol M–H peak is of high abundance.
- The number of fragment ion peaks in the spectrum depends upon the type of phenol.
- Homologation in the benzene causes generation of $C3^+$ and $C4^+$ peaks.
- Fragmentation pattern of following example of phenol is a key to explain the mass spectrum of other phenols.

Example: *Phenol* [Mass spectrum and fragmentation given in Figure 3.37(*a*) and (*b*) respectively].

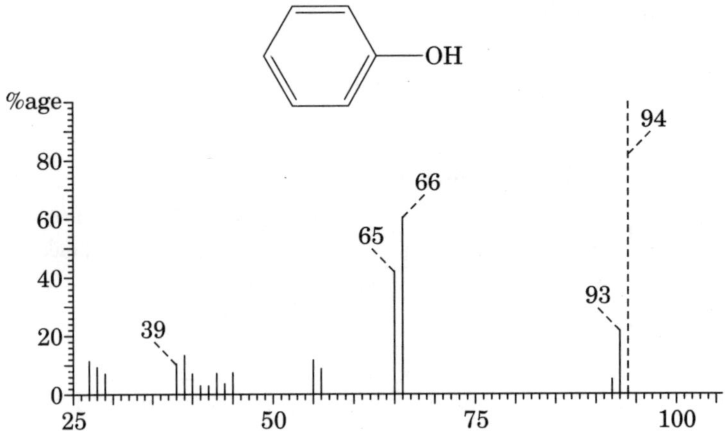

Figure 3.37(*a*): Mass spectrum of phenol.

Spectral interpretation: Peaks are observed at m/z 94 (corresponds to parent ion and base peak as well of $C_6H_6O^+$ due to loss of $1e^-$), 93 ($C_6H_5O^+$ due to loss of 1H), 66 ($C_5H_6^+$ due to loss of $CO^•$), 65 ($C_5H_5^+$ due to loss of $H^•$), and 39 ($C_3H_3^+$).

Fragmentation mode

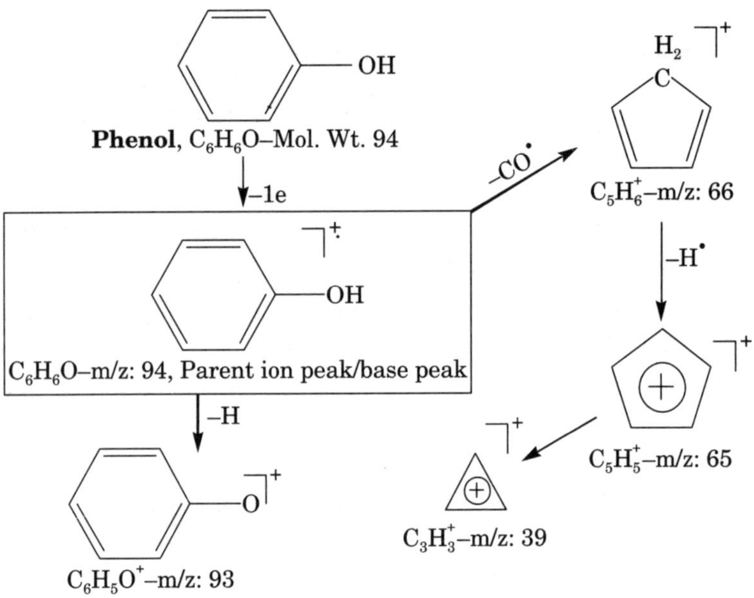

Figure 3.37(*b*): Mass fragmentation of phenol.

3.5.9 Ethers

Mass Spectral Characteristics of Ethers

- In the mass spectrum of ethers, the M^+ ion peak is generally of low abundance. But incorporation of any aromatic ring increases its abundance.
- In the mass spectrum of ethers the peaks due to alpha cleavage as a result of homolytic fragmentation of type-I, are most commonly observed.
- The mass spectrum of ethers also commonly involves peaks due to loss of alkyloxy radicals and carbocations.
- Homologation in the ethers causes generation of $C3^+$ and $C4^+$ peaks.
- Fragmentation pattern of following example of diethyl ether is a key to explain the mass spectrum of other ethers.

Example: *Diethyl ether* [Mass spectrum and fragmentation given in Figure 3.38(*a*) and (*b*) respectively].

Spectral interpretation: Peaks are observed at m/z 74 (corresponds to parent ion peak of $C_4H_{10}O^+$ due to loss of $1e^-$), 45 (base peak of $C_2H_5O^+$), 59 ($C_3H_7O^+$ due to loss of $CH_3^\bullet$), 31 (CH_3O^+ due to loss of $CH_2^\bullet$), and 29 (corresponding to $C_2H_5^+$ due to loss of $C_2H_5O^\bullet$).

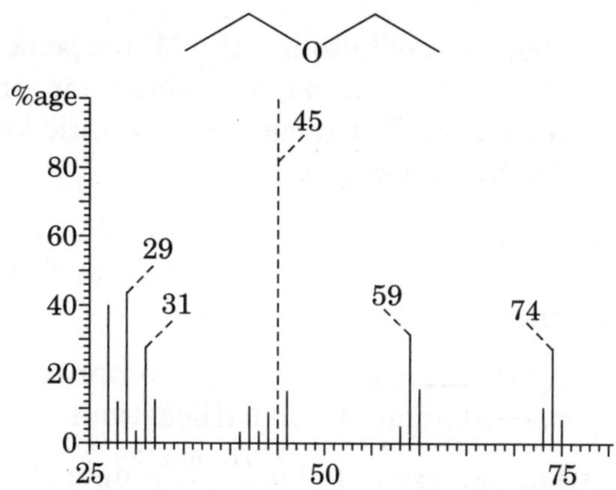

Figure 3.38(*a*): Mass spectrum of diethyl ether.

Fragmentation mode

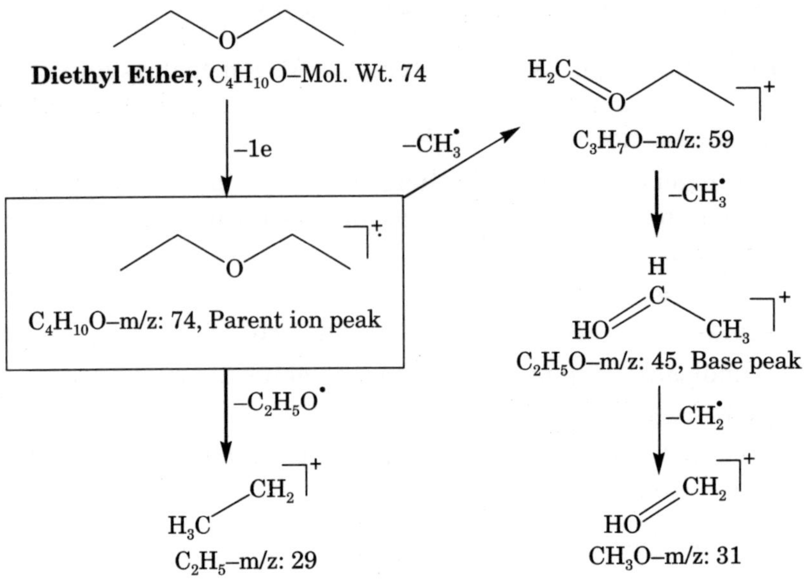

Figure 3.38(*b*): Mass fragmentation of diethyl ether.

3.5.10 Ketones

Mass Spectral Characteristics of Ketones

- In the mass spectrum of ketones, the M^+ ion peak are generally abundant in nature but as homologation increases the abundance decreases. But in case of aromatic ketones the M^+ ion peak are highly abundant.
- In the mass spectrum of ketones the peaks due to alpha cleavage and beta cleavage as a result of homolytic fragmentation of type-II, are most commonly observed.
- The mass spectrum of ketones also comprises peaks due to McLafferty rearrangement and carbocations.
- Fragmentation pattern of following example of pentanone is a key to explain the mass spectrum of other ketones.

Example: *2-pentanone* [Mass spectrum and fragmentation given in Figure 3.39(*a*) and (*b*) respectively].

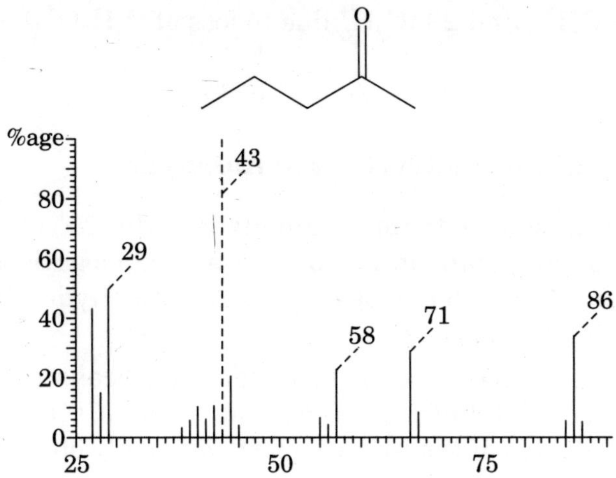

Figure 3.39(*a*): Mass spectrum of 2-pentanone.

Fragmentation mode

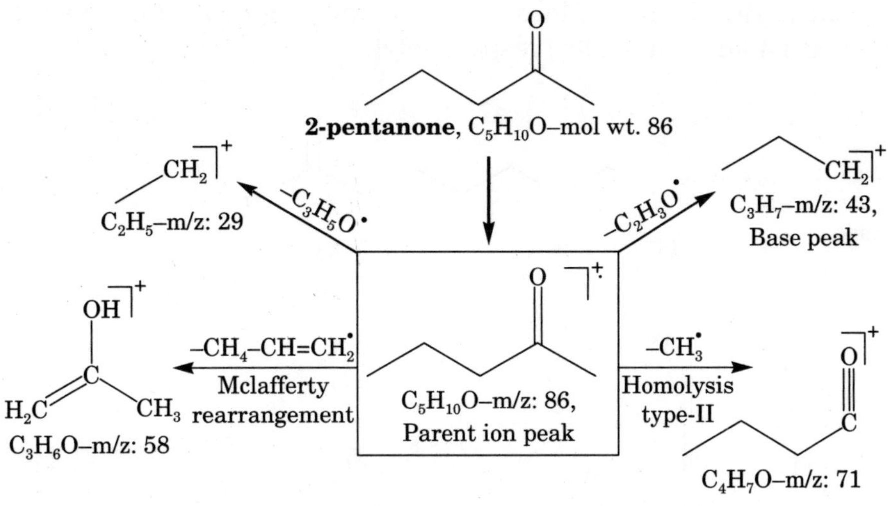

Figure 3.39(*b*): Mass fragmentation of 2-pentanone.

Spectral interpretation: Peaks are observed at m/z 86 (corresponds to parent ion peak $C_5H_{10}O^+$ due to loss of $1e^-$), 43 ($C_3H_7^+$ due to loss of $C_2H_3O^\bullet$), 71 ($C_4H_7O^+$ due to loss of $CH_3^\bullet$), 58 ($C_3H_6O^+$ due to loss of $CH_3 - CH = CH_2^\bullet$), and 29 ($C_2H_5^+$ due to loss of $C_3H_5O^\bullet$).

3.5.11 Aldehydes

Mass Spectral Characteristics of Aldehydes

- In the mass spectrum of aldehydes, the M^+ ion peak are generally abundant in nature but as homologation increases the abundance decreases. Aromatic aldehyde shows high abundant M^+ ion peaks.
- In the mass spectrum of aldehydes the peaks due to alpha cleavage and beta cleavage as a result of homolytic fragmentation of type-II, are most commonly observed.
- The mass spectrum of aldehydes also comprises peaks due to McLafferty rearrangement and carbocations.
- Fragmentation pattern of following example of octanal is a key to explain the mass spectrum of other aldehydes.

Example: *Octanal* [Mass spectrum and fragmentation given in Figure 3.40(*a*) and 3.40(*b*) respectively].

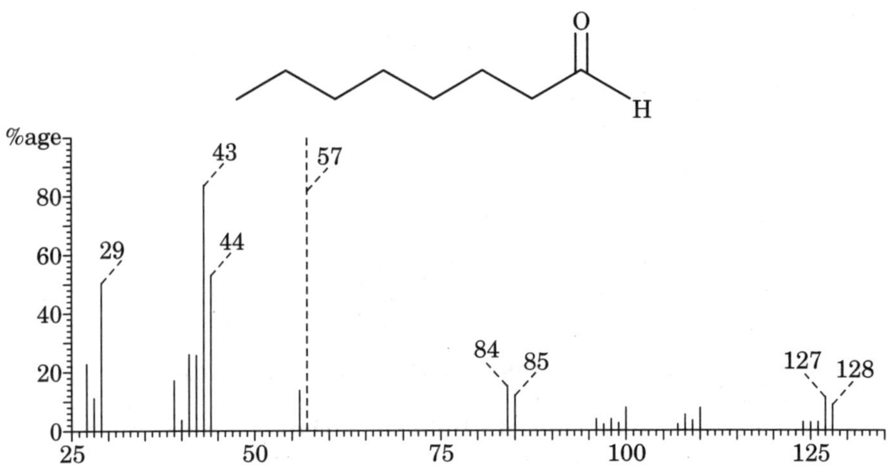

Figure 3.40(*a*): Mass spectrum of octanal.

Spectral interpretation: Peaks are observed at m/z 128 (corresponds to parent ion peak of $C_8H_{16}O^+$ due to loss of $1e^-$), 57 (base peak of $C_4H_9^+$ due to loss of $C_4H_7O^{\bullet}$), 127 ($C_8H_{15}O^+$ due to loss of $1H^{\bullet}$), 100 (fragment ion with loss of C_2H_5), 85 ($C_6H_{13}^+$ due to loss of $CH_2–CO^{\bullet}$), 84 ($C_6H_{12}^+$ due to loss of $CH_2{=}CH–OH^{\bullet}$), 44 ($C_2H_4O^+$ McLafferty product), 43 ($C_3H_7^+$ due to loss of $C_5H_9O^{\bullet}$), and 29 ($C_2H_5^+$ due to loss of $–C_7H_{15}^{\bullet}$).

Fragmentation mode

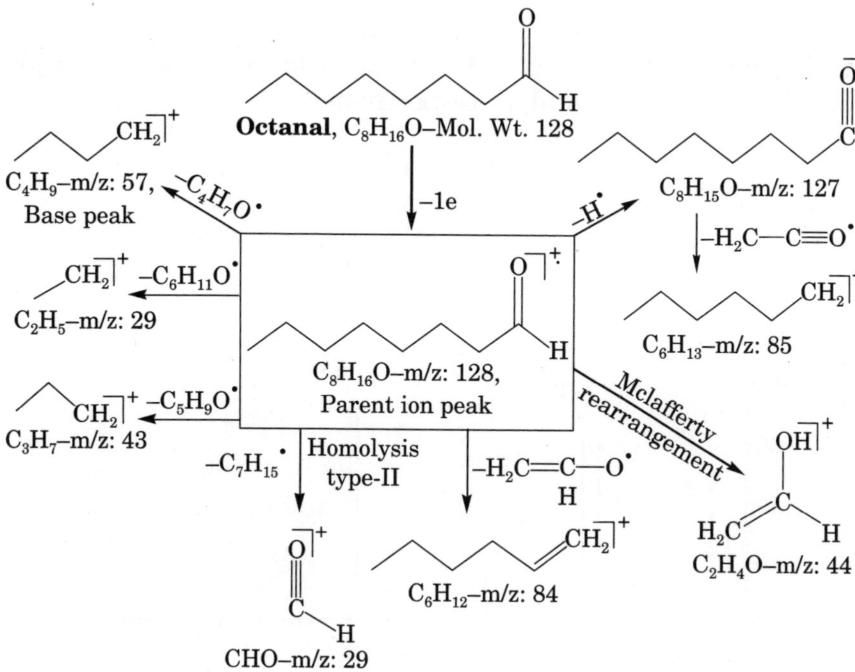

Figure 3.40(*b*): Mass fragmentation of octanal.

3.5.12 Carboxylic Acids

Mass Spectral Characteristics of Carboxylic Acids

- In the mass spectrum of aliphatic carboxylic acid, the M^+ ion peak is generally of low abundance, but in case of aromatic carboxylic acid abundance is higher.

- In the mass spectrum of carboxylic acid the most commonly observed fragment ion peaks are due to alpha cleavage and beta cleavage as a result of Homolytic type-II fragmentation.
- The mass spectrum of carboxylic acid also comprises the peaks that are due to McLafferty rearrangement and carbocations.
- Fragmentation pattern of following example of pentanoic acid is a key to explain the mass spectrum of other carboxylic acids.

Example: *Pentanoic acid* [Mass spectrum and fragmentation given in Figure 3.41(*a*) and (*b*) respectively].

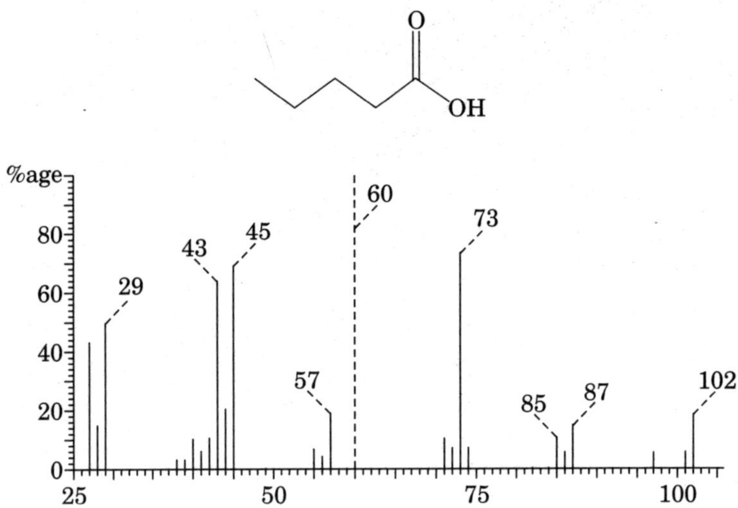

Figure 3.41(*a*): Mass spectrum of pentanoic acid.

Spectral interpretation: Peaks are observed at m/z 102 (corresponds to parent ion peak of $C_5H_{10}O_2^{+\cdot}$ due to loss of 1e$^-$), 60 (base peak of $C_2H_4O_2^{+\cdot}$ due to McLafferty rearrangement), 87 ($C_4H_7O_2^+$ due to loss of $CH_3^\bullet$), 85 ($C_5H_9O^+$ due to loss of $OH^\bullet$), 73 ($C_3H_5O_2^+$ due to loss of $C_2H_5^\bullet$), 57 ($C_4H_9^+$ due to loss of $COOH^\bullet$), 45 ($COOH^+$ due to loss of $C_4H_9^\bullet$), 43 ($C_3H_7^+$ due to loss of $C_2H_3O_2^\bullet$), and 29 (corresponds to $C_2H_5^+$).

Fragmentation mode

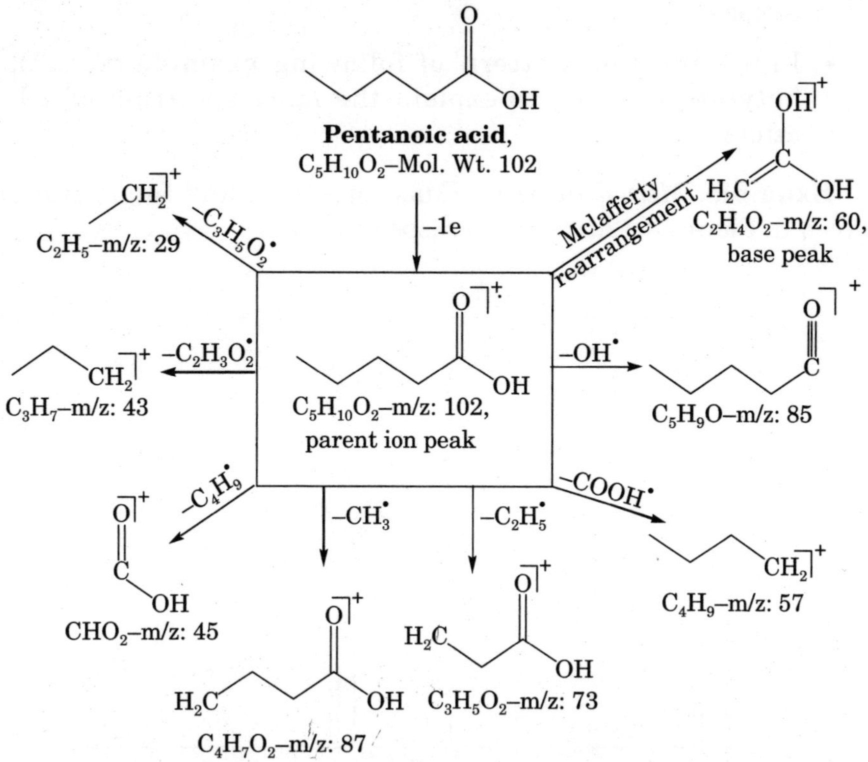

Figure 3.41(*b*): Mass fragmentation of pentanoic acid.

3.5.13 Ester

Mass Spectral Characteristics of Esters

- In the mass spectrum of ester, the M^+ ion peak is generally of low abundance, but in case of aromatic esters M^+ ion abundance is higher.
- In the mass spectrum of ester the most commonly observed fragment ion peaks are due to alpha cleavage as a result of Homolytic type-II fragmentation.

- The mass spectrum of ester also comprises the peaks that are due to McLafferty rearrangement and carbocations.
- Fragmentation pattern of following example of methyl butyrate is a key to explain the mass spectrum of other esters.

Example: *Methyl butyrate* [Mass spectrum and fragmentation given in Figure 3.42(*a*) and (*b*) respectively].

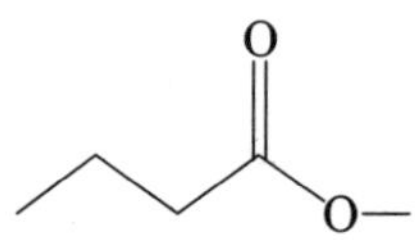

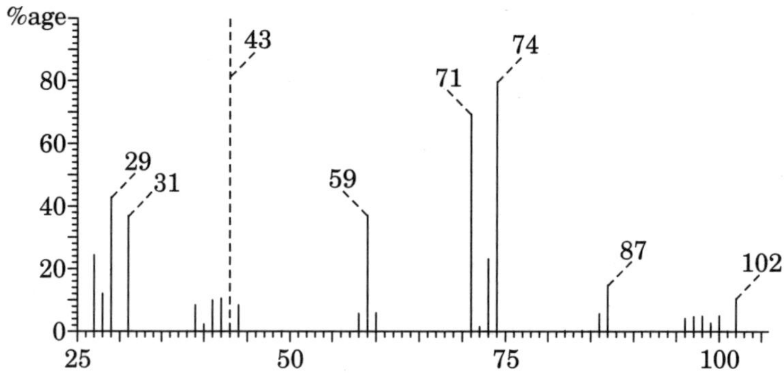

Figure 3.42(*a*): Mass spectrum of methyl butyrate.

Spectral interpretation: Peaks are observed at m/z 102 (corresponds to parent ion peak of $C_5H_{10}O_2^{\dot{+}}$ due to loss of 1e$^-$), 43 (base peak of $C_3H_7^{\dot{+}}$), 87 ($C_4H_7O_2^{\dot{+}}$ due to loss of $CH_3^{\bullet}$), 74 ($C_3H_6O_2^{\dot{+}}$ due to McLafferty rearrangement), 71 ($C_4H_7O^+$ due to loss of $OCH_3^{\bullet}$), 59 ($C_2H_3O_2^+$ due to loss of $C_3H_7^{\bullet}$),31 (O–$CH_3^{\dot{+}}$), and 29 ($C_2H_5^{\dot{+}}$ due to loss of $C_3H_5O_2^{\bullet}$).

Fragmentation mode

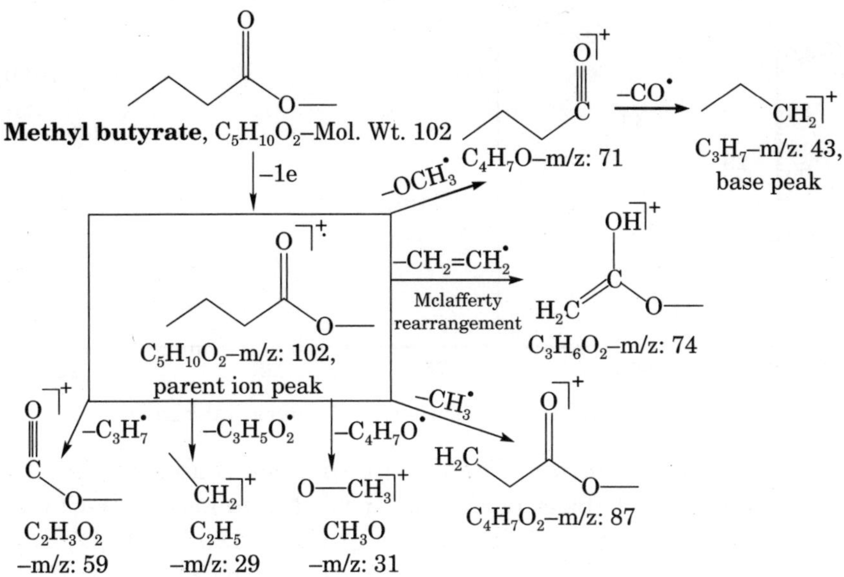

Figure 3.42(*b*): Mass fragmentation of methyl butyrate.

3.5.14 Anhydride

Mass Spectral Characteristics of Anhydride

- In the mass spectrum of anhydrides, the M^+ ion peak is generally very less abundant and sometime absent.
- In the mass spectrum of anhydrides the most commonly observed fragment ion peaks occurs due to alpha cleavage which is a result of Homolytic type-II fragmentation.
- The mass spectrum of anhydrides also comprises the peaks that are due to McLafferty rearrangement and carbocations.
- Fragmentation pattern of below mentioned example is a key to explain the mass spectrum of other anhydrides.

Example: *Butyric anhydride* [Mass spectrum and fragmentation given in Figure 3.43(*a*) and (*b*) respectively].

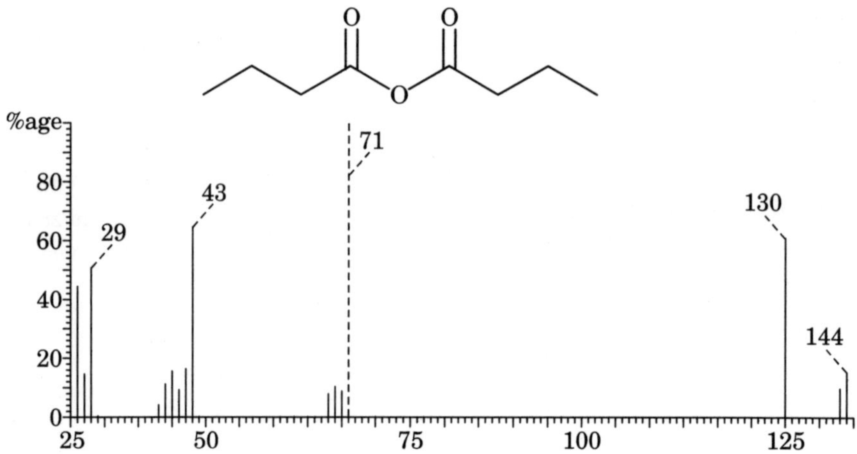

Figure 3.43(*a*): Mass spectrum of butyric anhydride.

Spectral interpretation: The parent ion peak is absent. Peaks are observed at m/z 71 (base peak of $C_4H_7O^+$ due to loss of O–CO–CH_2 CH_2–$CH_3^\bullet$), 144 ($C_7H_{12}O_3^+$ due to loss of $CH_3^\bullet$), 130 ($C_6H_{10}O_3^+$ due to McLafferty rearrangement), 43 ($C_3H_7^+$ due to loss of CO), and 29 ($C_2H_5^+$).

Fragmentation mode

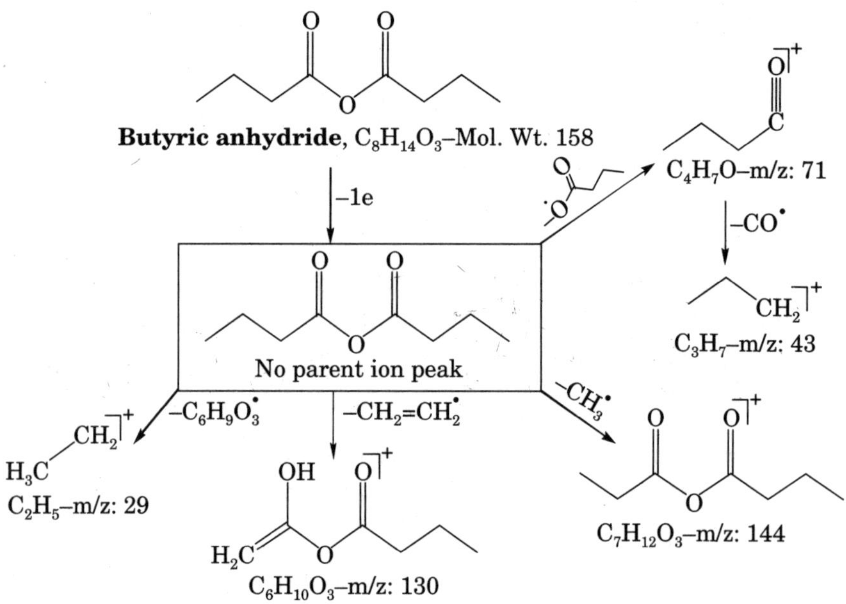

Figure 3.43(*b*): Mass fragmentation of butyric anhydride.

3.5.15 Halides

Mass Spectral Characteristics of Halides

- In the mass spectrum of halides, the abundance of M^+ ion peak generally increases when electro negativity of halogen substituent decreases. Whereas abundance of M^+ ion peak decreases when molecule is more branched and bears higher number of carbons.
- Depending upon number of halogen atoms the mass spectrum of halides contains distinct M+2, M+4, M+6 and M+8 isotope peaks. The abundance, number, and ratio of intensity of isotope peaks in comparison to M^+ ion peak depend upon number of halogens and combination of halogen atoms.
- In the mass spectrum of halides the most commonly observed fragment ion peaks are due to alpha cleavage as a result of heterolytic fragmentation.
- The mass spectrum of halides also comprises the peaks that are due to hydrogen transfer rearrangement, elimination, McLafferty rearrangement and carbocation.
- Fragmentation pattern of following example of chlorobutane is a key to explain the mass spectrum of other halides.

Example: *1-Chloro butane* [Mass spectrum and fragmentation given in Figure 3.44(*a*) and (*b*) respectively].

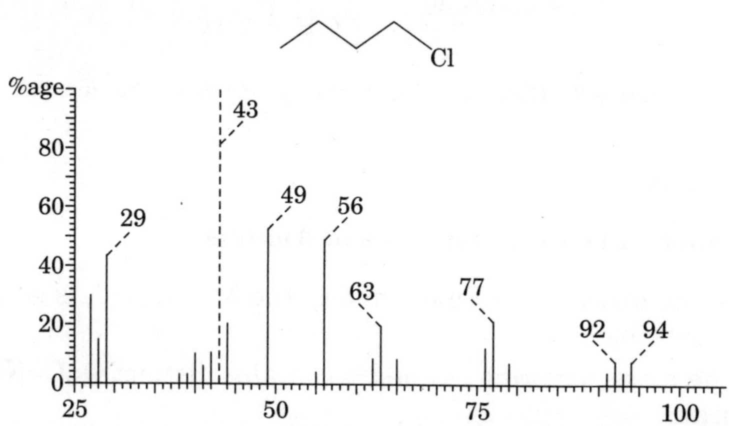

Figure 3.44(*a*): Mass spectrum of 1-chloro butane.

Spectral interpretation: Peaks are observed at m/z 92 (corresponds to parent ion peak of $C_4H_9Cl^+$ due to loss of $1e^-$), 94 (M+2 peak due to presence of Cl in the parent ion), 43 (base peak of $C_3H_7^+$), 91 ($C_4H_8Cl^+$ due to loss of $H^\bullet$), 77 ($C_3H_6Cl^+$ due to loss of $CH_3^\bullet$), 79 (M+2 peak due to presence of Cl in the fragment ion $C_3H_6Cl^+$), 63 ($C_2H_4Cl^+$ due to loss of $C_2H_5^\bullet$), 65 (M+2 peak due to presence of Cl in the fragment ion $C_2H_4Cl^+$), 56 ($C_4H_8^+$ due to loss of $HCl^\bullet$), 49 ($CH_2{=}Cl^+$ due to loss of $C_3H_7^\bullet$), and 29 (corresponds to $C_2H_5^+$).

Fragmentation mode

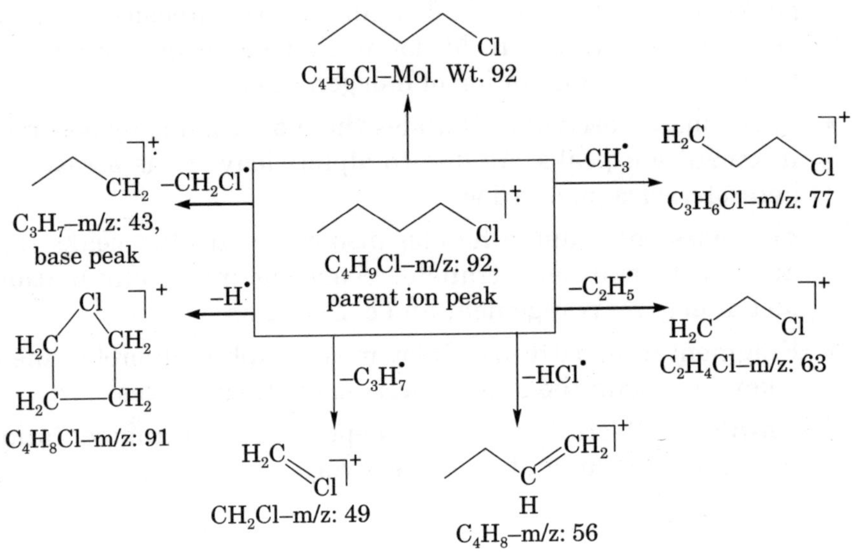

Figure 3.44(*b*): Mass fragmentation of 1-chloro butane.

3.5.16 Amides

Mass Spectral Characteristics of Amides

- In the mass spectrum of amides, the M^+ ion peak is generally observable.
- The mass spectrum of amides contains distinct $O{=}C{=}NH_2^+$ ion peak.

- In the mass spectrum of amides the most commonly observed fragment ion peaks are due to alpha cleavage as a result of Homolytic type-II fragmentation.
- The mass spectrum of amides also comprises the high abundant peaks that are due to McLafferty rearrangement and carbocation.
- Fragmentation pattern of following example of butyramide is a key to explain the mass spectrum of other amides.

Example: *Butyramide* [Mass spectrum and fragmentation given in Figure 3.45(*a*) and (*b*) respectively].

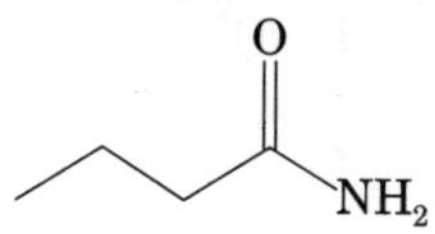

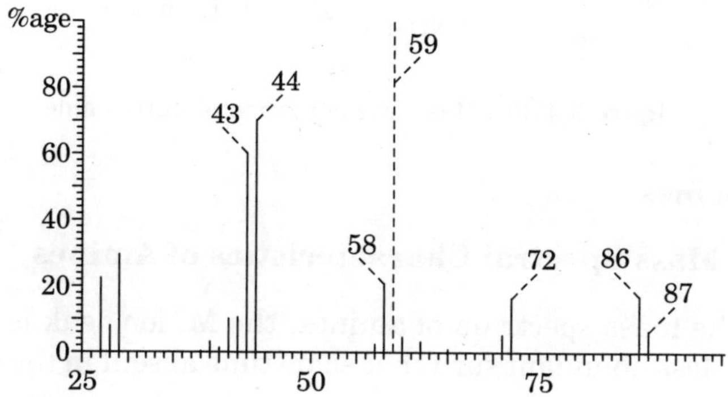

Figure 3.45(*a*): Mass spectrum of butyramide

Spectral interpretation: Peaks are observed at m/z 87 (corresponds to parent ion peak of $C_4H_9NO^+$ due to loss of $1e^-$), 59 (Base peak of C_2H_5NO due to McLafferty rearrangement), 86 ($C_4H_8NO^+$ due to loss of $H^\bullet$), 72 ($C_3H_6NO^+$ due to loss of $CH_3^\bullet$), 58 ($C_2H_4NO^+$ due to loss of $C_2H_5^\bullet$), 44 (CH_2NO^+ due to loss of $C_3H_7^\bullet$), 43 ($C_3H_7^+$ due to loss of $CONH_2^\bullet$), and 29 ($C_2H_5^+$).

Fragmentation mode

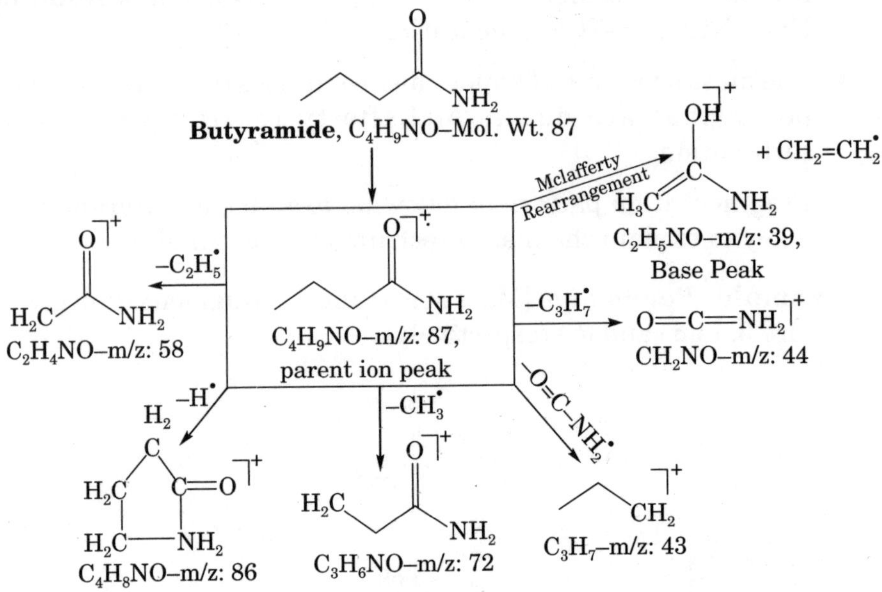

Figure 3.45(*b*): Mass fragmentation of butyramide.

3.5.17 Amines

General Mass Spectral Characteristics of Amines

- In the mass spectrum of amines, the M^+ ion peak is generally very less abundant such that some time absent in the spectrum.
- The mass spectrum of amines contains distinct CH_2=NH_2^+ ion peak.
- In the mass spectrum of amines the most commonly observed fragment ion peaks are due to alpha cleavage as a result of Homolytic type-I fragmentation.
- The mass spectrum of amines also comprises the high abundant peaks that are due to carbocations.
- Fragmentation pattern of following example of butylamine is a key to explain the mass spectrum of other amines.

Example: *1-Butylamine* [Mass spectrum and fragmentation given in Figure 3.46(*a*) and (*b*) respectively].

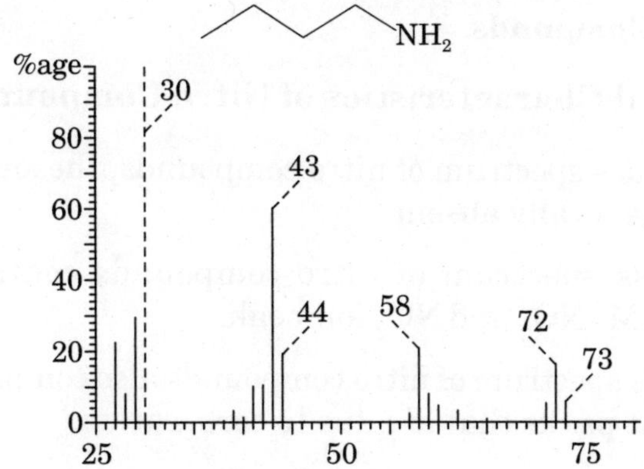

Figure 3.46(*a*): Mass spectrum of 1-butylamine.

Spectral interpretation: Peaks are observed at m/z 73 (corresponds to parent ion peak of $C_4H_{11}N^+$ due to loss of $1e^-$), 30 (base peak of CH_4N due to loss of $C_3H_7^•$) 72 ($C_4H_{10}N^+$), 58 ($C_3H_8N^+$ due to loss of $CH_3^•$), 44 ($C_2H_6N^+$ due to loss of $C_2H_5^•$), 43 ($C_3H_7^+$ due to loss of $CH_2{=}NH_2^•$), and 29 ($C_2H_5^+$).

Fragmentation mode

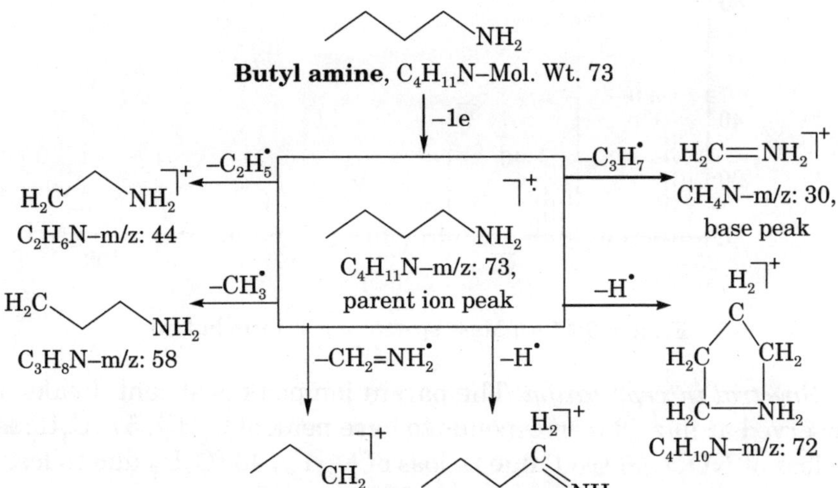

Figure 3.46(*b*): Mass fragmentation of 1-butylamine.

3.5.18 Nitro Componds

Mass Spectral Characteristics of Nitro Compounds

- In the mass spectrum of nitro compounds, the signal of M^+ ion peak is generally absent.
- The mass spectrum of nitro compounds contains distinct M–NO^+, M–NO_2^+ and NO_2^+ ion peak.
- The mass spectrum of nitro compounds also comprises the high abundant peaks that are due to carbocation.
- Fragmentation pattern of following example of nitro butane is a key to explain the mass spectrum of other nitro compounds.

Example: *1-Nitro butane* [Mass spectrum and fragmentation given in Figure 3.47(*a*) and (*b*) respectively].

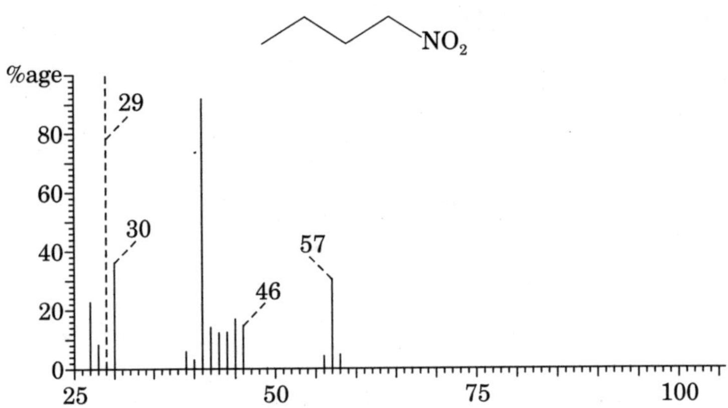

Figure 3.47(*a*): Mass spectrum of 1-nitro butane.

Spectral interpretation: The parent ion peak is absent. Peaks are observed at m/z 29 (corresponds to base peak of $C_2H_5^+$), 57 ($C_4H_9^+$ due to loss of $NO_2^\bullet$), 46 (NO_2^+ due to loss of $C_4H_9^\bullet$), 43 ($C_3H_7^+$ due to loss of CH_2NO_2), and 30 (corresponds to NO^+).

Fragmentation mode

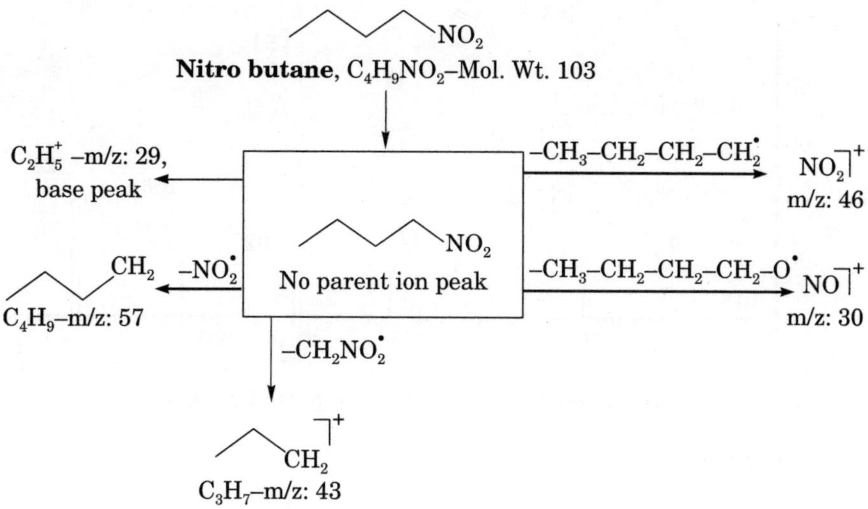

Figure 3.47(*b*): Mass fragmentation of 1-nitro butane.

3.5.19 Aromatic Compounds

Mass Spectral Characteristics of Aromatic Compounds

- In the mass spectrum of aromatic compounds, the M^+ ion peak is generally of high abundance.
- The mass spectrum of benzyl compounds contains distinct tropilium ion peak due to homolytic Type-III fragmentation.
- The mass spectrum of aromatic compounds shows significant abundant peaks of different aromatic rings that possess 3 to 6 carbons fragment ions.
- Fragmentation pattern of following example of phenyl propane, 2-isobutyl 2-hydroxybenzoate, nitrobenzene and 3-ethoxy-4-methylbenzene are the keys to explain the mass spectrum of other aromatic compounds.

Example: *Phenyl propane* [Mass spectrum and fragmentation given in Figure 3.48(*a*) and (*b*) respectively].

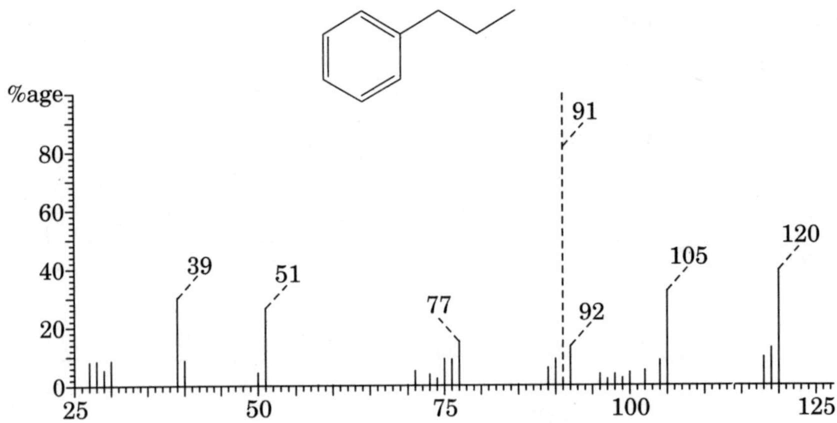

Figure 3.48(*a*): Mass spectrum of phenyl propane.

Fragmentation mode

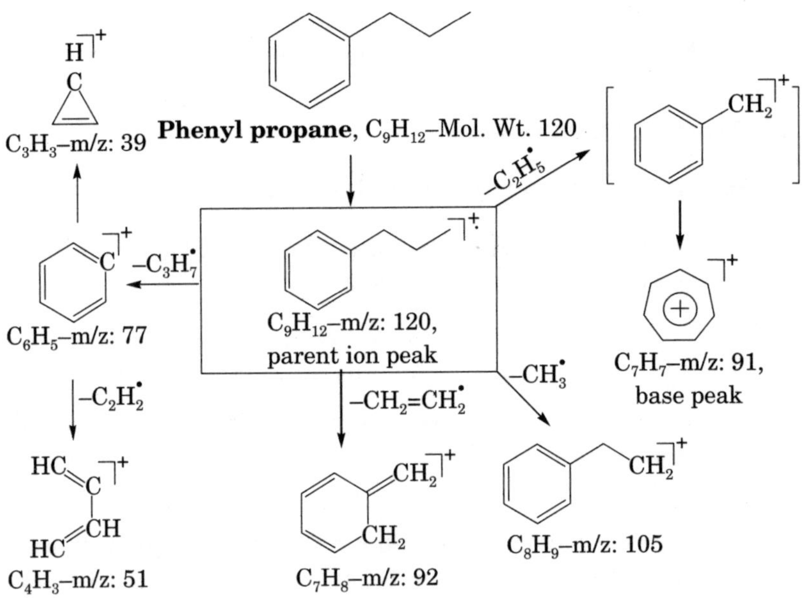

Figure 3.48(*b*): Mass fragmentation of phenyl propane.

Spectral interpretation: Peaks are observed at m/z 120 (corresponds to parent ion peak of C_9H_{12} due to loss of $1e^-$), 91 (corresponds to base peak of tropilium$^+$), 105 ($C_8H_9^+$ due to loss of

$CH_3^\bullet$), 92 ($C_7H_8^+$ due to loss of $CH_2{=}CH_2^\bullet$), 77 ($C_6H_5^+$ due to loss of $C_3H_7^\bullet$), 51 ($C_4H_3^+$), and 39 ($C_3H_3^+$).

Example: *2-isobutyl 2-hydroxybenzoate* [Mass spectrum and fragmentation given in Figure 3.49(*a*) and (*b*) respectively].

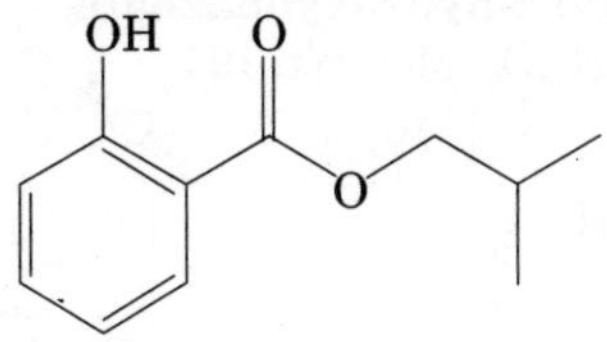

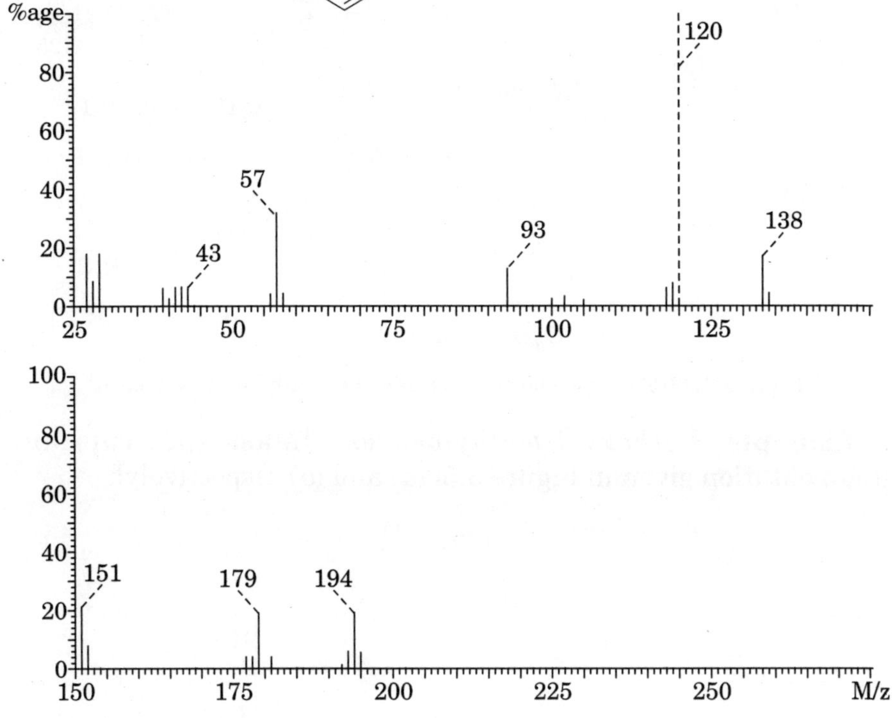

Figure 3.49(*a*): Mass spectrum of 2-isobutyl 2-hydroxybenzoate.

Spectral interpretation: Peaks are observed at m/z 194 (corresponds to parent ion peak of $C_{11}H_{14}O_3$ due to loss of 1e$^-$), 120 (base peak of $C_7H_4O_2^+$ due to loss of $C_4H_{10}O^\bullet$), 179 ($C_{10}H_{11}O_3^+$ due to loss of $CH_3^\bullet$), 151 ($C_8H_7O_3^+$ due to loss of $C_3H_7^\bullet$), 138 ($C_7H_6O_3^+$ due to Mclafferty rearrangement), 93 ($C_6H_5O^+$ due to loss of $C_5H_9O_2^\bullet$), 57 ($C_4H_9^+$), and 43 ($C_3H_7^+$).

Fragmentation mode

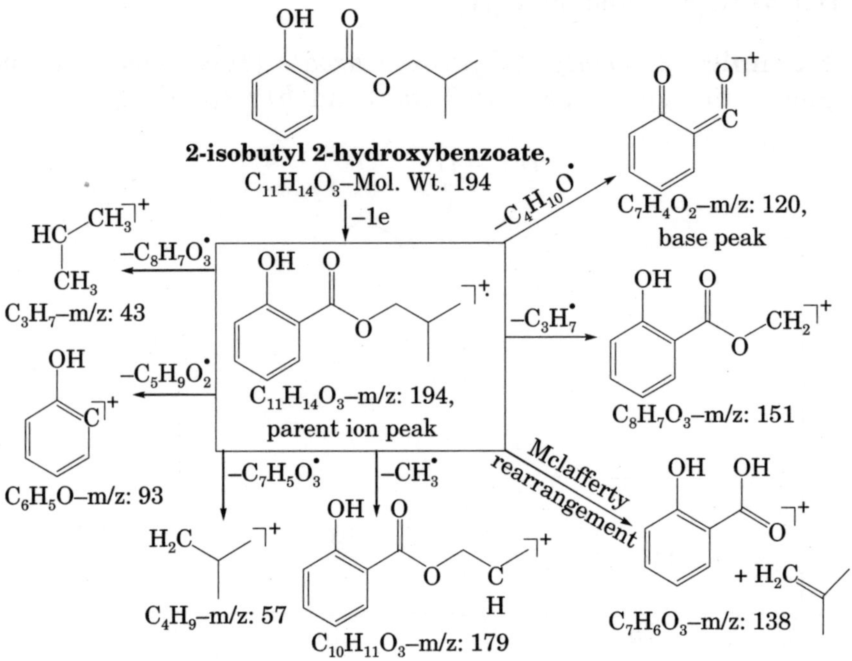

Figure 3.49(*b*): Mass spectrum of 2-isobutyl 2-hydroxybenzoate.

Example: *3-ethoxy-4-methylbenzene* [Mass spectrum and fragmentation given in Figure 3.50(*a*) and (*b*) respectively].

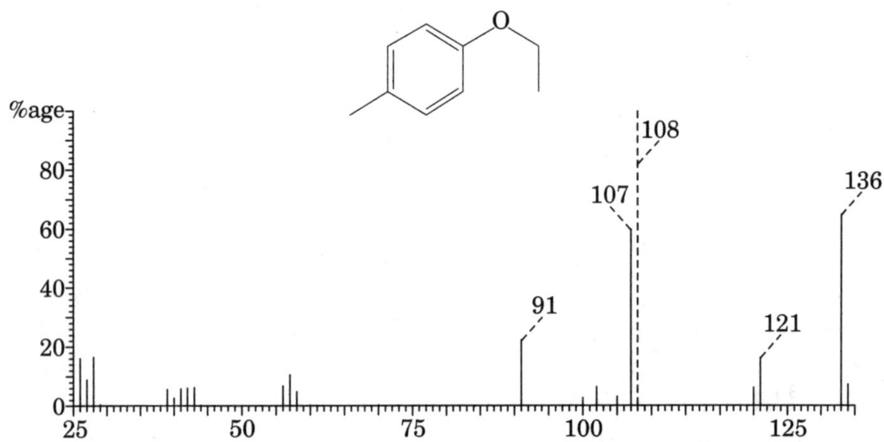

Figure 3.50(*a*): Mass spectrum of 3-ethoxy-4-methylbenzene.

Spectral interpretation: Peaks are observed at m/z 136 (corresponds to parent ion peak of $C_9H_{12}O$ due to loss of $1e^-$), 108 (base peak of $C_7H_8O^+$ due to Mclafferty rearrangement), 121 ($C_8H_9O^+$ due to loss of $CH_3^\bullet$), and 107 ($C_7H_7O^+$ due to loss of $C_2H_5^\bullet$).

Fragmentation mode

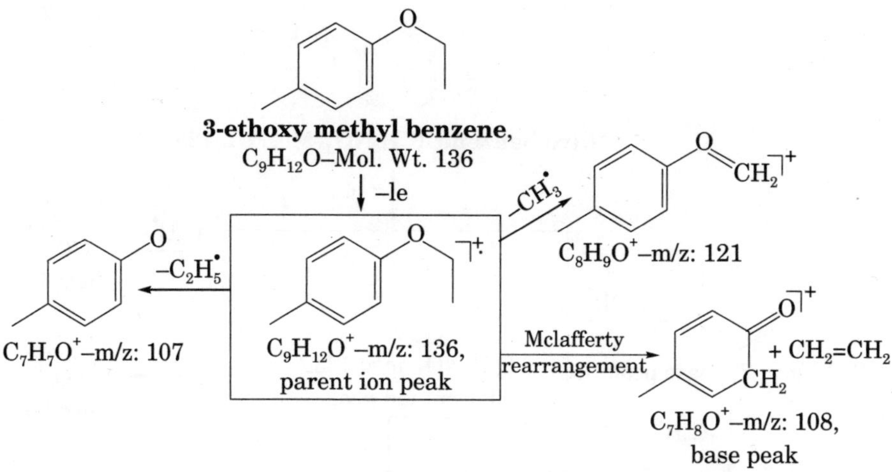

Figure 3.50(*b*): Mass spectrum of 2-isobutyl 2-hydroxybenzoate.

Example: *Nitrobenzene* [Mass spectrum and fragmentation given in Figure 3.51(*a*) and (*b*)].

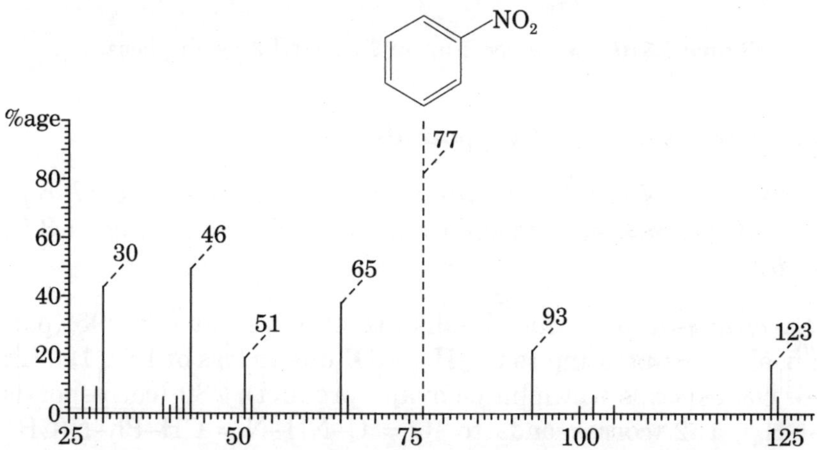

Figure 3.51(*a*): Mass spectrum of nitro benzene.

Spectral interpretation: Peaks are observed at m/z 123 (corresponds to parent ion peak of $C_6H_5NO_2$ due to loss of $1e^-$), 77 (base peak of Phenyl ion), 93 (corresponds to phenoxy ion), 65 ($C_5H_5^+$ ion), 51 ($C_4H_3^+$ ion), 46 (NO_2^+ ion), and 30 (NO^+ ion).

Fragmentation mode

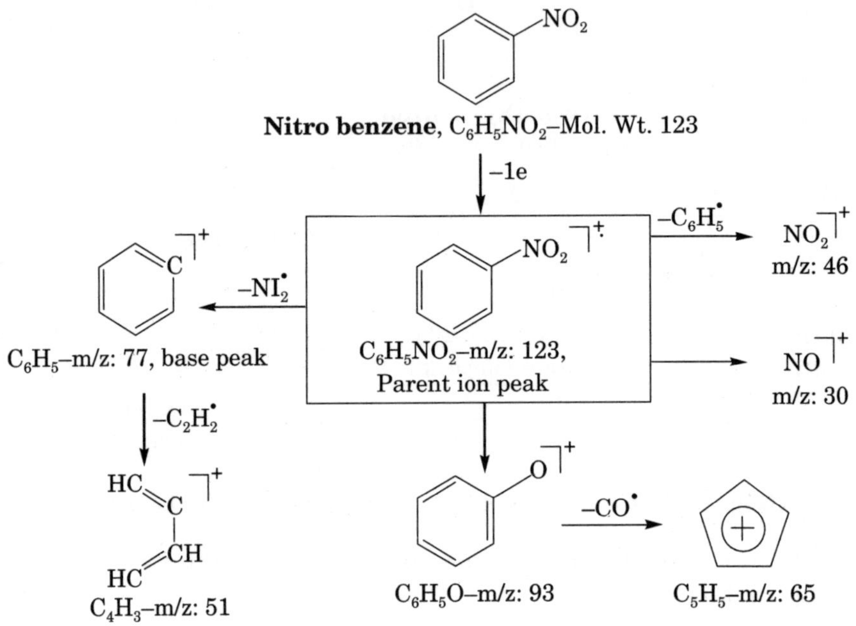

Figure 3.51(*b*): Mass spectrum of 2-isobutyl 2-hydroxybenzoate.

3.5.20 Some Complex Compounds

Example: *N-{4-(Dimethylamino) benzylidene}-3-phenylpropane hydrazide* [Mass spectrum and fragmentation given in Figure 3.52(*a*) and (*b*)].

Spectral interpretation: Peaks are observed at m/z 295 (parent ion peak corresponding to $C_{18}H_{21}N_3O^+$ due to loss of $1e^-$), 190 (Base peak corresponds to alpha cleavage product), 280 (corresponds to M–CH_3), 162 (corresponds to O = C–NH–N = CH–Ph–N$(CH_3)_2^+$),

148 (Corresponds to Ph–CH_2–CH_2–C(=O)–NH^+), 147 (Corresponds to N = CH–Ph–$N(CH_3)_2^+$), 120 (Corresponds to Ph–$N(CH_3)_2^+$), 105 (Ph–CH_2–CH_2^+), 91 (corresponds to tropilium ion $C_7H_7^+$) and 65 (corresponds to $C_5H_5^+$).

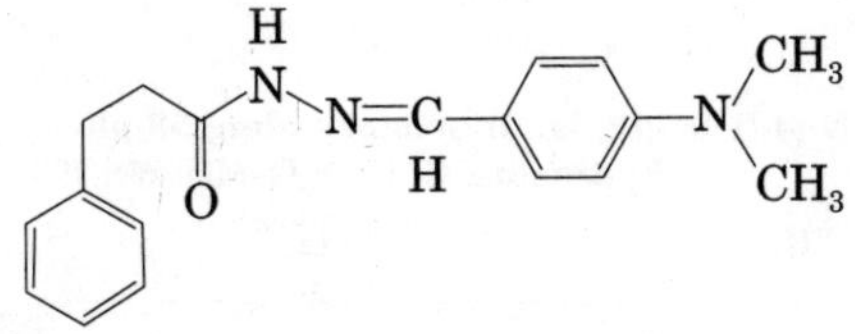

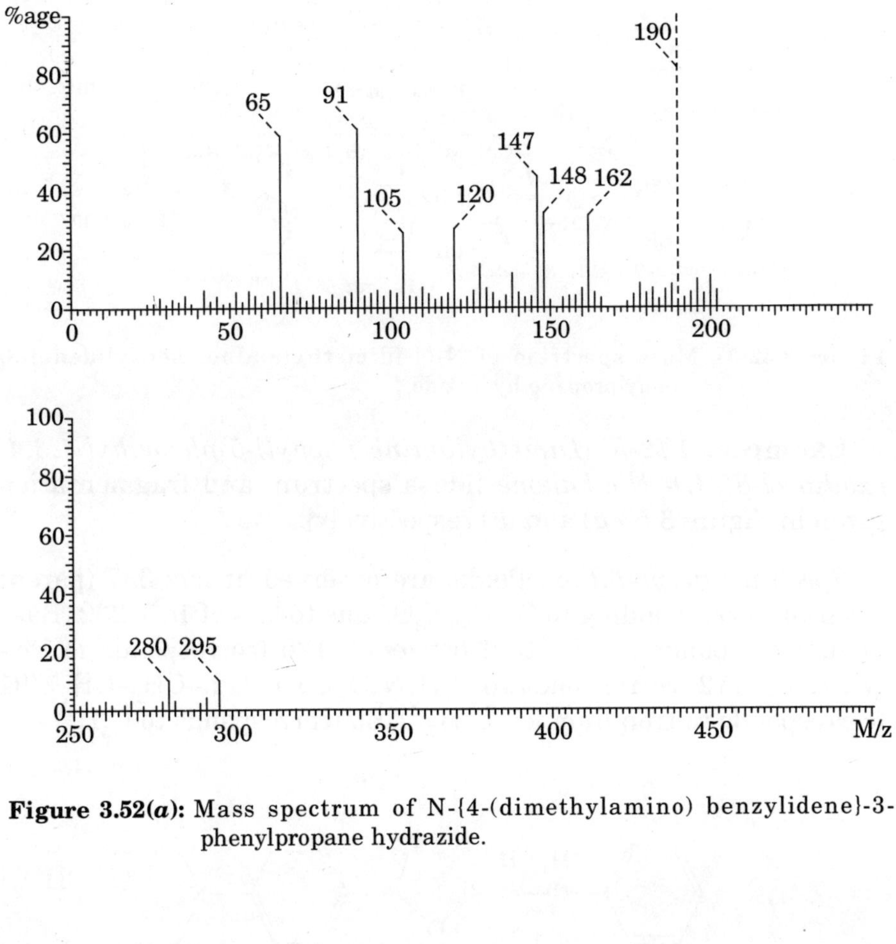

Figure 3.52(*a*): Mass spectrum of N-{4-(dimethylamino) benzylidene}-3-phenylpropane hydrazide.

Fragmentation mode

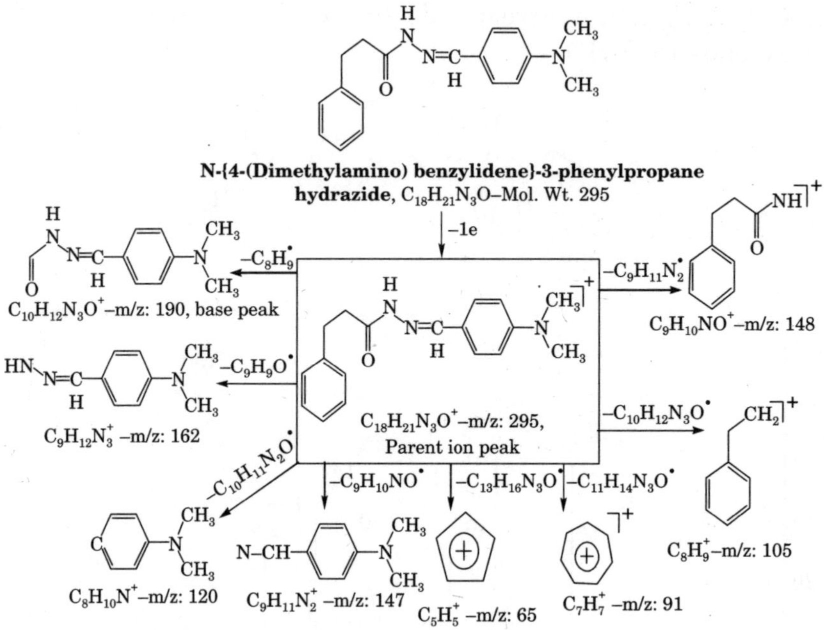

Figure 3.52(*b*): Mass spectrum of N-{4-(dimethylamino) benzylidene}-3-phenylpropane hydrazide.

Example: *1-[2-{4-(Dimethylamino)phenyl}-5-phenethyl-1,3,4-oxadiazol-3(2H)-yl] ethanone* [Mass spectrum and fragmentation given in Figure 3.53(*a*) and (*b*) respectively].

Spectral interpretation: Peaks are observed at m/z 337 (parent ion peak corresponding to $C_{20}H_{23}N_3O_2^+$ due to loss of $1e^-$), 232 (Base peak corresponds to M–Ethyl benzene), 120 (corresponds to Ph–$N(CH_3)_2^+$), 112 (corresponds to $C_4H_4N_2O_2^+$), 105 (Ph–CH_2–CH_2^+), 91 (corresponds to tropilium ion $C_7H_7^{+\bullet}$), 65 (corresponds to $C_5H_5^+$).

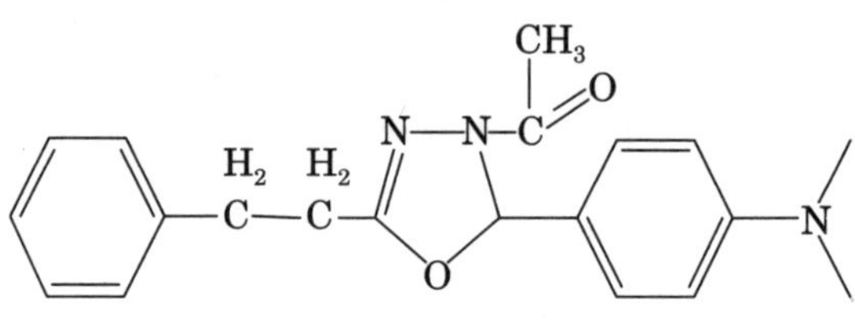

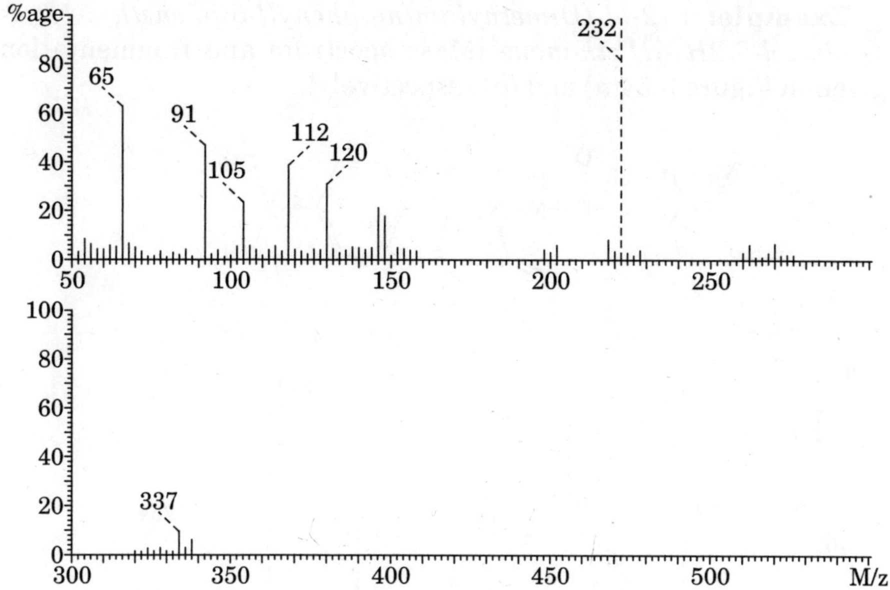

Figure 3.53(*a*): Mass spectrum of 1-[2-{4-dimethylamino)phenyl}-5-phenethyl-1,3,4-oxadiazol-3(2H)-yl] ethanone.

Fragmentation mode

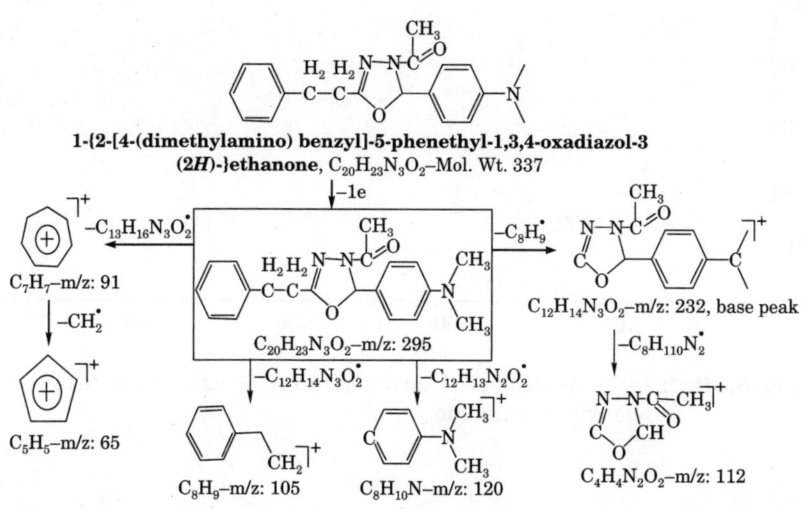

Figure 3.53(*b*): Fragmentation of 1-[2-{4-dimethylamino)phenyl}-5-phenethyl-1,3,4-oxadiazol-3(2H)-yl] ethanone.

Example: *1-[2-{4-(Dimethylamino)phenyl}-5-phenethyl-1,3,4-oxadiazol-3(2H)-yl] ethanone* [Mass spectrum and fragmentation given in Figure 3.54(*a*) and (*b*) respectively].

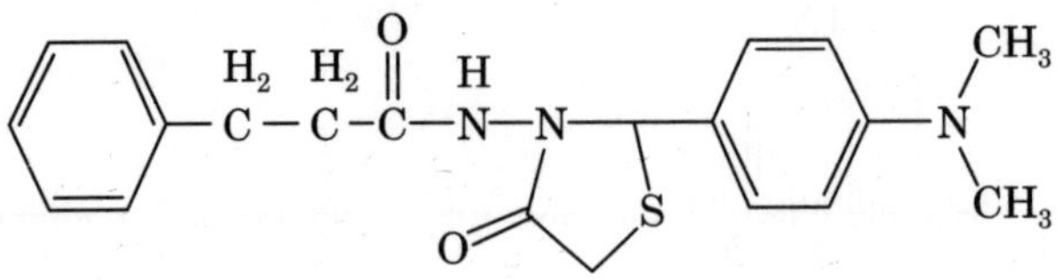

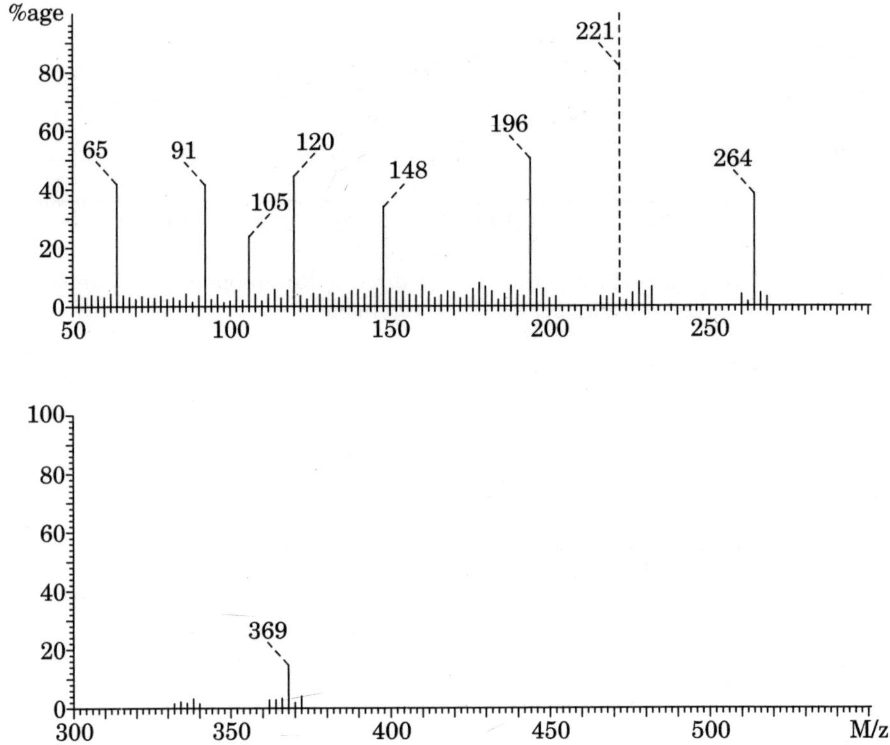

Figure 3.54(*a*): N-(2-(4-(dimethylamino)phenyl)-4-oxothiazolidin-3-yl)-3-phenylpropanamide.

Fragmentation mode

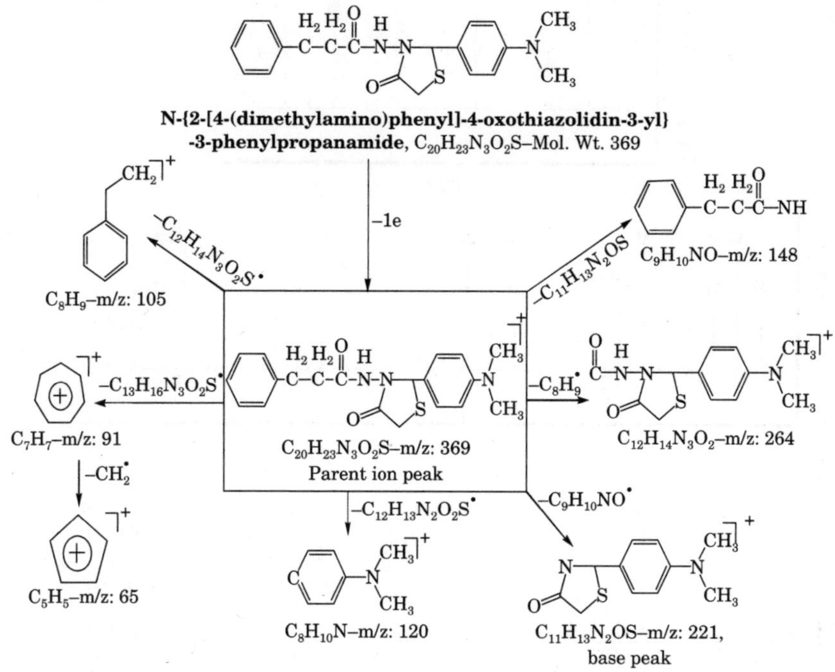

Figure 3.54(*b*): Fragmentation of 1-[2-{4-Dimethylamino)phenyl}-5-phenethyl-1,4-oxadiazol-3(2H)-yl] ethanone.

Spectral interpretation: Peaks are observed at m/z 369 (parent ion peak corresponding to $C_{20}H_{23}N_3O_2S^+$ due to loss of $1e^-$), 221 (Base peak corresponds to alpha cleavage product $C_{11}H_{13}N_2OS^+$), 264 (corresponds to M–Ethyl benzene$^+$), 148 (corresponds to $C_9H_{11}NO$), 120 [Corresponds to Ph–N$(CH_3)_2^{+}$], 105 (Ph–CH_2–CH_2^+), 91 (corresponds to tropilium ion $C_7H_7^{+\bullet}$), 65 (corresponds to $C_5H_5^+$).

For Further Assistance in Mass Spectra Interpretation of Compounds Refer, Figure 3.55.

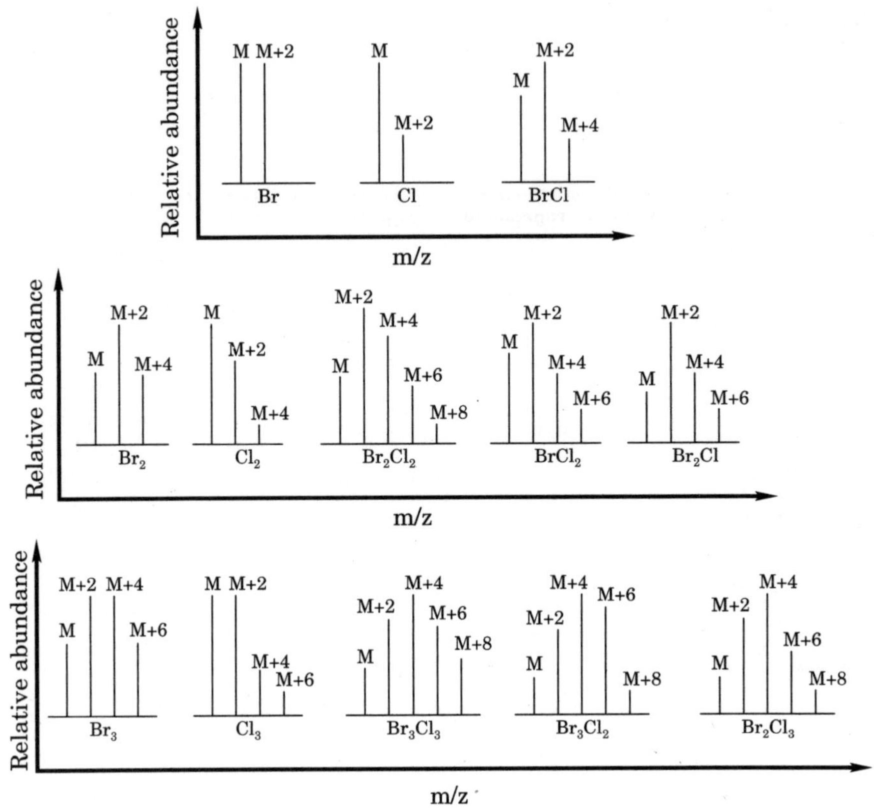

Figure 3.55: Chart representing molecular ion and isotope peaks for chlorine, bromine and their combinations.

3.6 PROBLEMS BASED LEARNING

After reading the MASS spectroscopy chapter, the readers are suggested/advised to test their learning by solving some below mentioned problems.

A. SUBJECTIVE PROBLEMS

1. Explain the principle of Mass spectroscopy. Describe Time of Flight analyzer with diagram.
2. Enumerate Various types of Mass analyzer. Discuss quadrupole mass analyzer with appropriate diagram.

3. Give brief account about different techniques of ion generation in mass spectroscopy.
4. Write a note on Mc Lafferty rearrangement and MALDI.
5. Explain Principle and functioning of quadrupole mass analyzer.
6. Explain chemical ionization Technique with its advantages and disadvantages.
7. Describe field desorption and MALDI techniques in detail.
8. Discuss McLafferty rearrangement with a suitable example.
9. Explain fragmentation pattern of alcohol and alkanes in mass spectroscopy.
10. Write different types of ions produced in a mass spectrometer.
11. Enlist general rules for fragmentation patterns in molecules?
12. Explain the principles of the double focusing and time of flight spectrophotometers.
13. Define and explain metastable peaks with a suitable example.

B. OBJECTIVE PROBLEMS

1. State True or False

(*i*) High chain branching, higher number of hydroxyl groups and longer alkyl chain makes M^+ ion peak absent in mass spectrum.

(*ii*) Fragmentation of M^+ ion occurs at 70ev.

(*iii*) Ketones, aldehydes, esters and amides undergoes homolytic fragmentation type-II.

(*iv*) Alcohols, ethers and amines undergoes homolytic fragmentation type-I.

(*v*) benzyl group bearing compounds commonly undergoes homolytic fragmentation type-III.

(*vi*) Halides and ethers commonly undergoes heterolysis.

(*vii*) Cyclo alkenes and cyclo alkanes commonly undergoes Retrodiels elder fragmentation to remove a neutral alkene.

(*viii*) Alcohols and halides commonly undergoes cyclization rearrangement.

(*ix*) Phenyl acetamide and phenyl acetate commonly undergoes ketene elimination.

Ans.: 1. (*i*) True (*ii*) True (*iii*) True (*iv*) True
(*v*) True (*vi*) True (*vii*) True (*viii*) True
(*ix*) True.

2. Fill up the blanks

(*i*) Peak of ionic species that is generated on mass spectrum due to presence of the less abundant isotopes in the molecule is called as .... peak.

(*ii*) When fragmentation of M^+ ion occurs out of ionization chamber such fragmented ion peak is called as .... peak.

(*iii*) Relative abundance of M^+ ion peak of alkanes decreases on increasing ....

(*iv*) 4-Pentene, 2-Pentanone, Pentanal, Propyl Benzene, pentan-2-imine undergoes McLafferty rearrangement due to availability of .... hydrogens.

(*v*) Ionic peak that shows highest abundance (100%) is designated as a .... peak.

(*vi*) In alkanes fragment ion peaks appears .... units apart.

Ans.: 2. (*i*) isotope peak (*ii*) metastable ion (*iii*) homologation
(*iv*) gamma (*v*) base (*vi*) Fourteen.

3. Match the following ionization methods

(*i*) Field Method	(*a*) FIMS & FDMS
(*ii*) Laser Method	(*b*) LIMA, LDMS & MALDI
(*iii*) Plasma Method	(*c*) PDMS
(*iv*) Ionic Impact Method	(*d*) SIMS
(*v*) Atomic Method	(*e*) FABMS
(*vi*) Spray Method	(*f*) ESIMS

Ans.: 3. (*i*) – (*a*), (*ii*) – (*b*), (*iii*) – (*c*), (*iv*) – (*d*), (*v*) – (*e*), (*vi*) – (*f*).

4. Match the following ion separation methods

(*i*) Field Method	(*a*) Magnetic/Electrostatic focusing
(*ii*) Velocity Method	(*b*) TOFMS

(*iii*) Pole Method (*c*) QMF & QIS

(*iv*) Resonance Method (*d*) ICRMS & FTICRMS

(*v*) Mass Combination Method (*e*) Tandem MS

Ans.: 4. (*i*) – (*a*), (*ii*) – (*b*), (*iii*) – (*c*), (*iv*) – (*d*), (*v*) – (*e*).

C. DATA INTERPRETATION PROBLEMS

1. Predict the compound which fits into IR spectrum A.

   (*a*) Methyl cyclo hexane (*b*) Cyclohexene

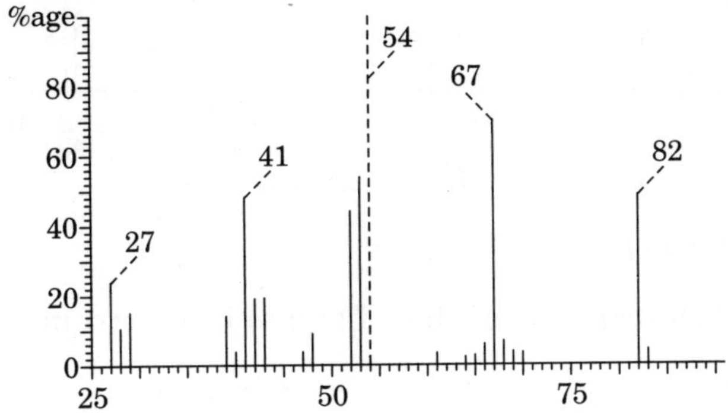

IR Spectrum–A

Ans.: *Cyclohexene.*

2. Predict the compound which fits into IR spectrum B.

   (*a*) 1-Butyne (*b*) 1-Butene

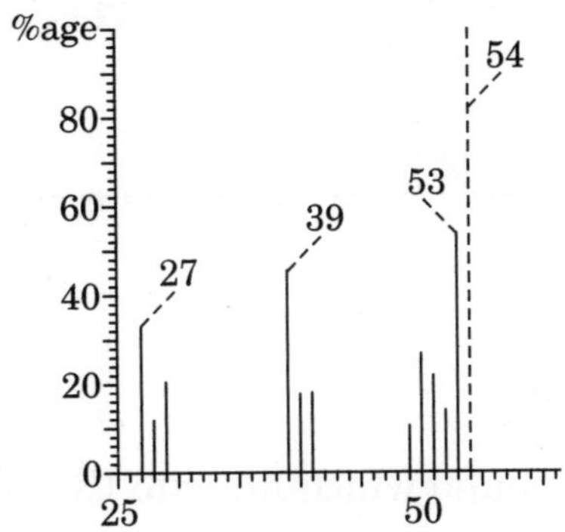

IR Spectrum–B

Ans.: *1-Butyne.*

3. Predict the compound which fits into IR spectrum C.

(*a*) Phenol (*b*) 2-pentanol

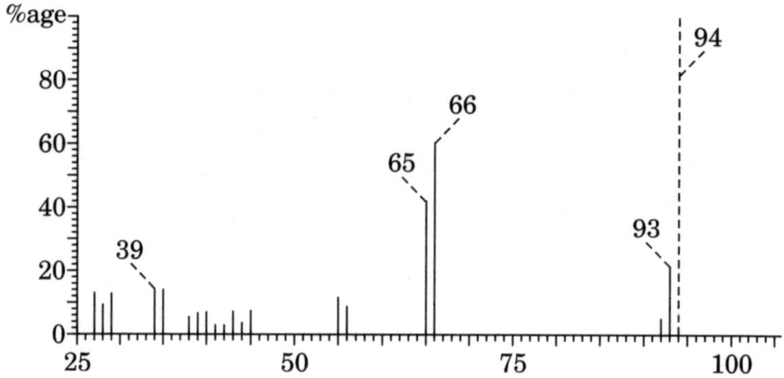

IR Spectrum–C

Ans.: *Phenol.*

4. Predict the compound which fits into IR spectrum D.

(*a*) 2-Pentanone (*b*) 2-Octanal

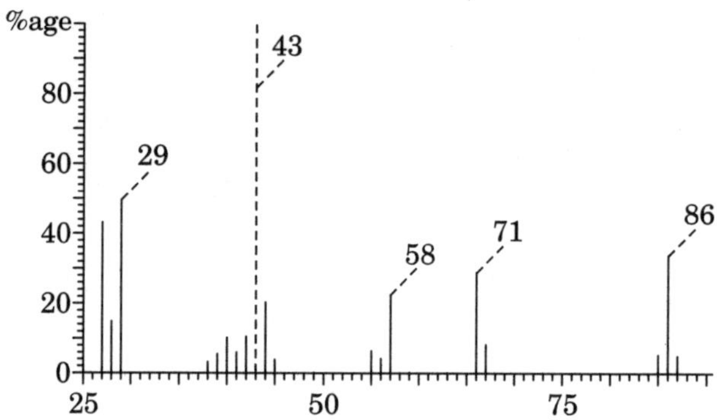

IR Spectrum–D

Ans.: *2-Pentanone.*

5. Predict the compound which fits into IR spectrum E.

(*a*) Pentanoic acid (*b*) Methyl butyrate

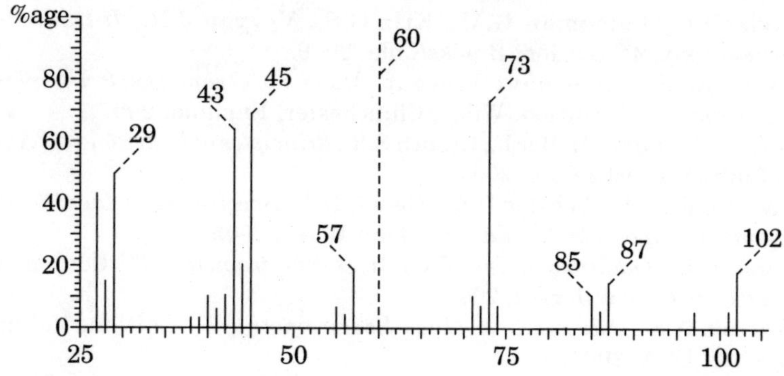

IR Spectrum–E

Ans.: *Pentanoic acid.*

6. Predict the compound which fits into IR spectrum F.

   (*a*) Methyl butyrate (*b*) Butyric anhydride

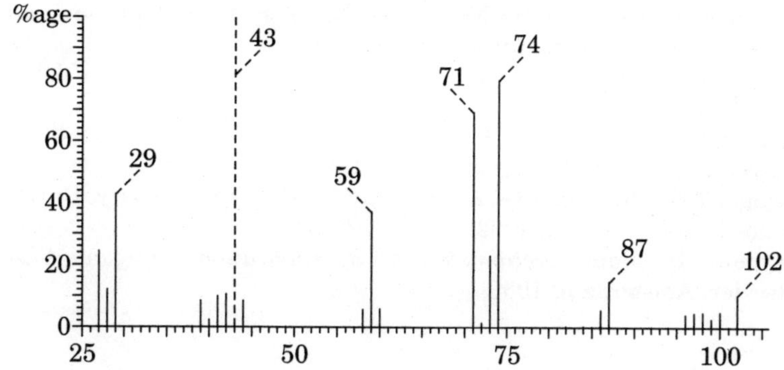

IR Spectrum–F

Ans.: *Pentonoic acid.*

3.7 SUGGESTED READINGS

[1] Duckworth H.E., Barber R.C., Venkatasubramanian V.S., *Mass Spectroscopy*, 2nd Edition, Cambridge University Press, England, 2010.

[2] Cole R.B., *Electrospray and MALDI Mass Spectrometry*, 1st Edition, John Wiley and Sons, Hoboken, New Jersey, 2010.

[3] Watson J.T., Sparkman O.D., *Mass spectrometry*, 4th Edition, John Willey, England, 2009.

[4] Pretsch E., Buhlmann P., Badertscher M., *Structure Determination of Organic Compounds: Tables of Spectral Data*, 4th Edition, Springer, 2009.

[5] Pavia D.L., Lampman G.M., Kriz G.S., Vyvyan J.R., *Introduction to Spectroscopy*, 4th Edition, Brooks/Cole, 2009.

[6] Hoffmann E., Stroobant Vincent, *Mass spectrometry Principles and Applications*, 3rd Edition, Wiley, Chinchester, England, 2007.

[7] Skoog D.A., James Holler F., Crouch S.R., *Principles of Instrumental Analysis*, 6th Edition, Brooks Cole, 2007.

[8] Silverstein R.M., Webster F.X., Kiemle D.J., *Spectrometric Identification of Organic Compounds*, 7th Edition, John Wiley, 2005.

[9] March R.E., *Quadrupole Ion Trap Mass Spectrometry,* 2nd Edition, Wiley-Interscience, New Jersey, 2005.

[10] Ashcroft A.E., *Advances in Mass Spectrometry*, 1st Edition, Volume 16, Elsevier, USA, 2004.

[11] Smith R.M., *Understanding Mass Spectra: A Basic Approach*, 2nd Edition, Wiley Interscience, New Jersey, 2004.

[12] Robinson J.W., Frame E.M.S., Frame II G.M., *Undergraduate Instrumental Analysis*, 6th Edition, CRC Press, 2004.

[13] Schalley C.A., *Modern Mass Spectrometry*, Springer, 2003.

[14] Ashcroft A.E., *Ionization Methods in Organic Mass Spectrom*etry, Royal Society of Chemistry, UK, 1997.

[15] Kitson F.G., Larsen B.S., McEwen C.N., *Gas chromatography and Mass Spectroscopy*, Academic press, London, 1996.

[16] Chapman J.R., *Practical organic Mass spectrometry*, 2nd Edition, John Wiley, New York, May 1994.

[17] McLafferty F.W., Tureek F., *Interpretation of Mass spectra*, 4th Edition, University Science Books, California 1993.

[18] Beynon J.H., Brenton A.G., *Introduction to Mass spectrometry*, University of Wales Press, Swansea, 1982.

[19] Beynon J.H., *Mass Spectrometry and its Application to Organic Chemistry*, Elsevier, Amsterdam, 1960.

Chapter 4

^{1}H-Nuclear Magnetic Resonance (NMR) Spectroscopy

Objective

To make readers learn

- Basic Concepts, Theory and Underlying Principles of ^{1}H-NMR Spectroscopy.
- Instrumentation of ^{1}H-NMR Spectroscopy.
- ^{1}H-NMR Spectra/Data Interpretation for Structural Determination of Compounds.

4.1 INTRODUCTION

Nuclear magnetic resonance (NMR) spectrometry is a form of absorption spectroscopy which involves placement of molecules (atomic nucleus) in an external magnetic field that causes degeneration of spin state into two spin states of unequal energy. The nuclei with the spin state of lower energy or aligned nuclei are induced to absorb energy of radiofrequency (RF) range to change their spin orientation with respect to applied magnetic field. The absorption frequencies peaks are plotted against peak intensities which generates an NMR spectrum. NMR experiments are used to identify the structure of compounds. NMR is a powerful technique that provides detailed information on the topology, dynamics and three-dimensional structure of molecules in solution and the solid state. **In ^{1}H-NMR spectroscopy, the experimental conditions are so designed that spin transitions of only hydrogen nuclei are noticed.** To use ^{1}H-NMR as an analytical tool, it is necessary to understand the basic principles of NMR spectroscopy.

4.2 THEORY/BASIC PRINCIPLES

4.2.1 Nuclear Spin

All nuclei carries a charge, which spins on nuclear axis and thereby generates a magnetic dipole along nuclear axis. The spinning charge creates an angular momentum which is described as spin quantum number (I). Most of the atomic nuclei possess property of spin. Net spins of some nuclei are given in Table 4.1.

Table 4.1: Net spin for different nuclei.

Nuclei	*Unpaired protons*	*Unpaired neutrons*	*Net spin*
^{1}H	1	0	1/2
^{2}H	1	1	1
^{13}C	0	1	1/2
^{31}P	1	0	1/2
^{23}Na	1	2	3/2
^{14}N	1	1	1
^{19}F	1	0	1/2

- **If any nuclei has odd atomic mass, odd atomic number or both as odd, will possess quantized spin angular momentum and magnetic moment.** *[Odd mass nuclei (odd number of nucleons) have fractional spins eg. I = 1/2 (^{1}H, ^{13}C, ^{19}F), I = 3/2 (^{11}B) & I = 5/2 (^{17}O). Spectra of ^{1}H, ^{3}H, ^{13}C, ^{15}N, ^{19}F, ^{31}P are easily obtained as they have I= ½ and uniform spherical charge distribution. The most commonly studied nuclei are ^{1}H and ^{13}C, although nuclei from isotopes of many other elements such as ^{2}H, ^{6}Li, ^{10}B, ^{11}B, ^{14}N, ^{15}N, ^{17}O, ^{19}F, ^{23}Na, ^{29}Si, ^{31}P, ^{35}Cl, ^{113}Cd, ^{129}Xe, ^{195}Pt have been studied by high-field NMR.]*
- **Isotopes which contain an odd number of protons or neutrons have an intrinsic magnetic moment and angular momentum, possess nonzero spin.** Even mass nuclei (*odd number of protons and neutrons*) have integral spins *e.g.*, I = 1 (^{2}H, ^{14}N).
- **If any nucleus has even number of protons and/or neutrons, will possess a zero spin.** [*Even mass nuclei (even*

number of protons and neutrons) have zero spin ($I = 0$) e.g., ^{12}C, & ^{16}O)].

Allowed spin states for a nucleus = $2I + 1$

Note:

Isotopes—Elements which can have different neutrons numbers can have isotopes.

Magnetic Moment—a moment of electrons and nuclear particles caused by the intrinsic spin of a particle and the orbital motion of a particle, as around a nucleus or a measure of the strength of a localized source of magnetic field).

Angular Momentum—rotational momentum.

Spin—it is a type of angular momentum.

Spin State—Magnetic moment of spin with higher energy is anticlockwise called –1/2 (β state) which opposes the magnetic field and the magnetic moment of the spin with lower energy is clockwise called +1/2 (α state), which aligns with the magnetic field.

Spin Quantum Number (I)—it can be determined from atomic mass and atomic number (can be 0, ½, 1, 3/2, etc.), I = 1 means non-spherical charge distribution.

Charge—Property or force of a particle with which it affects the other particles.

Example 1: Hydrogen Spin States ($I = ½$)—There are two spin states [2(½) + 1 = 2] for hydrogen nucleus: *i.e.*, clockwise (+½) or anticlockwise (–½) spin. In the absence of an external magnetic field, these magnetic moments are randomly oriented in all directions. If an external magnetic field is applied, the nuclear spins of the hydrogen nuclei orients themselves either in alignment with or in opposition to the field (Figure 4.1).

The two spin states for hydrogen nuclei normally degenerate. However, in the presence of a magnetic field, the two spin states become unequal in energy. Those nuclei whose spins are in alignment with the external magnetic field are lower in energy than those nuclei whose spin are in opposition to the external field.

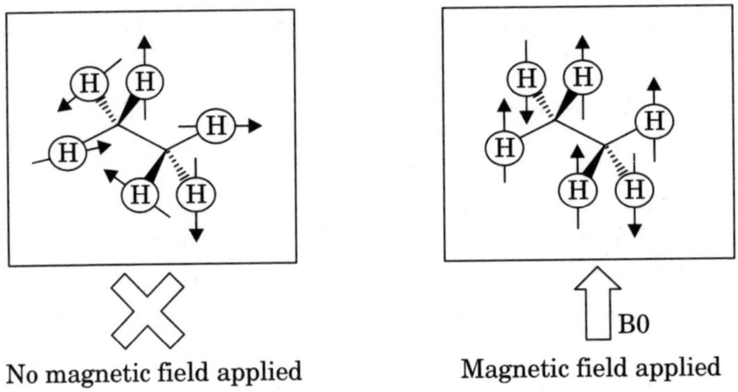

Figure 4.1: 'H' Nuclear spin in no external magnetic field (left) and an applied magnetic field (right).

The energy difference between the two spin states in a magnetic field of 7.0586 Tesla is about 1.20×10^{-4} kJ/mole (Figure 4.2).

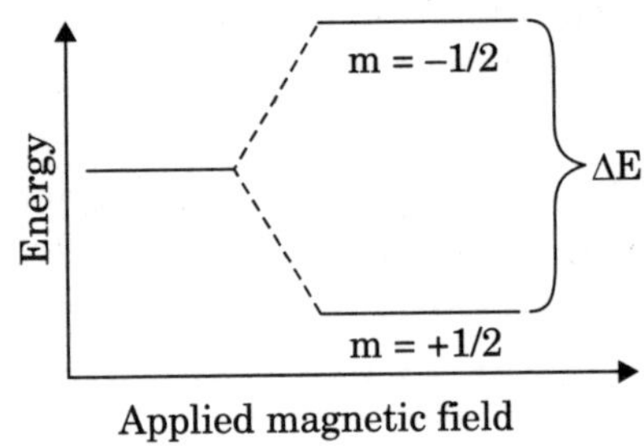

Figure 4.2: Energy differences between hydrogen nuclei state.

Radiation with a frequency of about 300 MHz, which lies in the radio frequency region of the electromagnetic spectrum, corresponds to the given energy difference.

$$\Delta E = 2.675 \times 10^{8}\ s^{-1}T^{-1} * (6.626 \times 10^{-34}/2p) * 7.0586\ T \quad ...(4.1)$$

$$\Delta E = 1.20 \times 10^{-4}\ kJ/mol \quad ...(4.2)$$

4.2.2 Transition Energy

As the nucleus carries a positive charge and is spinning, due to which it generates a small magnetic field. The nucleus therefore possesses a magnetic moment, m, which is proportional to its spin *I*.

$$\mu = \frac{\gamma . I . h}{2\pi} \quad ...(4.3)$$

The constant, γ, is called the *magnetogyric ratio* and is a fundamental nuclear constant which has a different value for every nucleus. h is Plancks constant. The energy of a particular energy level is given by

$$E = \frac{\gamma . h}{2\pi} . m . B \quad ...(4.4)$$

Where B is the strength of the magnetic field at the nucleus.

Therefore transition energy is expressed as

$$\Delta E = \frac{\gamma . h . B}{2\pi} = h . v \quad ...(4.5)$$

Hence, if the magnetic field, B, is increased, and therefore ΔE also increases. This also indicates that when a nucleus has a relatively large magnetogyric ratio, then ΔE also becomes correspondingly large.

Therefore, $$v = \frac{\gamma}{2\pi} . B \quad ...(4.6)$$

ABSORPTION OF RADIATION BY A NUCLEUS IN A MAGNETIC FIELD

Behaviour of a nucleus in a magnetic field can be well understood from Figure 4.3.

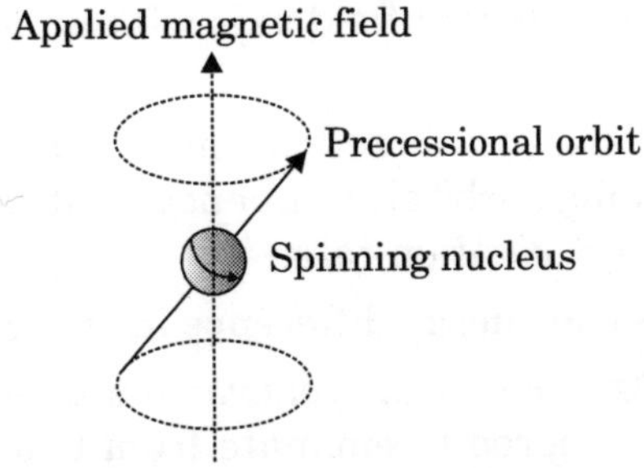

Figure 4.3: Nucleus behaviour in magnetic field.

Imagine a nucleus (of spin 1/2) in a magnetic field. This nucleus is in the lower energy level (*i.e., its magnetic moment does not oppose the applied field*). **The nucleus is spinning on its axis. In the presence of a magnetic field, this axis of rotation will precess around the magnetic field. The frequency of precession is called as larmor frequency.**

The potential energy of precessing nucleus is

$$E = -\mu \,.\, B \,.\, \cos\theta \qquad ...(4.7)$$

Where, θ is angle between direction of applied field and axis of nuclear rotation.

If energy is absorbed by the nucleus, then the angle of precession (θ), will change. For a nucleus of spin 1/2, absorption of radiation "flips" the magnetic moment so that it opposes the applied field (the higher energy state).

It is important to note that only a small portion of 'target' nuclei are in the lower energy state (and can absorb radiation). **There is the possibility that by exciting these nuclei, the populations of the higher and lower energy levels will become equal. If such condition occurs, then there will be no further absorption of radiation. Therefore spin system will be considered as saturated.** The possibility of saturation means that we must be aware of the relaxation processes which return nuclei to the lower energy state.

Note:

Larmor frequency (*precessional rate*)—angular momentum vector precesses about external field axis with an angular frequency is Larmor frequency.

Angular frequency (ω) also referred to angular speed, radial frequency, circular frequency, orbital, frequency, and radian frequency is a scalar measure of rotation rate.

Transition energy—The energy difference in energy between levels.

Magnetic Dipole—An object that generates a magnetic field in which the field is considered to emanate from two opposite poles. The magnitude of generated dipole is expressed in terms of nuclear magnetic moment (μ).

4.2.3 Spin States Population

Spin states population is defined as the population of small excess nuclei in the lower spin state that is the only responsible for generation of NMR signal when undergoes transition from lower state to higher. The two spin states possess different nuclei population. There is small excess populations of nuclei in the lower energy spin state *i.e.*, out of 2 million nuclei in two states there are only nine more nuclei are in lower spin state at 1.41 Tesla so called excess nuclei population. The nuclei in the lower spin state can be excited into the upper spin state by absorption of energy of about 60 MHz (resonance). Table 4.2. Represents variation in spin population with operating frequency.

Table 4.2: Variation in spin population with operating frequency.

Frequency in MHz	*Nuclei Population*
20	3
40	6
60	9
80	12
100	16
200	32
300	48
600	96

This produces a signal which provides the NMR spectrum. Hence, by raising the operating frequency of NMR instrument, the excess population of nuclei in lower state is increased; thereby the transition of nuclei between two states is increased. If two spin states become equal in nuclei population, then no net spin transitions occurs and hence no signal is produced and such condition is called as saturation. Saturation occurs when too powerful radio frequency pulse is used. Hence, operating frequency should be increased but without increasing its power, so as to increase the energy difference thereby increase the excess nuclei population and hence sensitivity of instrument is increased.

4.2.4 Resonance

Phenomenon by virtue of which a proton absorbs radiofrequency and gets promoted to a less favorable high energy state is called as resonance. As nucleus is a charged species, so exhibits the two spin states (of unequal energy) when placed in applied magnetic field. The movement of nuclear charge generates its own magnetic field, and nucleus possesses a magnetic moment μ (Figure 4.4).

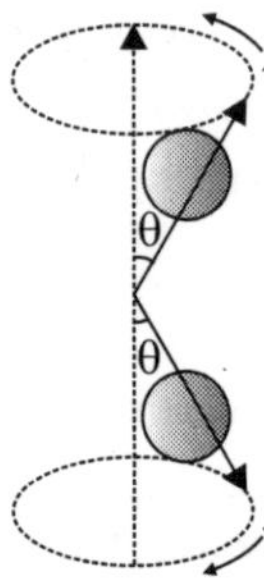

Figure 4.4: Spin states of nuclei.

Key Concepts

- **The magnetic axis of proton precesses about 'z' axis of stationary magnetic field B_0. During this process spinning top precess** (*rotates like a gyroscope*) **off perpendicularly under gravity.** Therefore all the equivalent protons precessing in random phase around *z* axis (*i.e.*, in direction of magnetic field B_0) possess a net magnetization M_0 along *z* axis instead of *x* and *y* axis. A schematic diagram for nuclear mechanics of magnetization is given in Figure 4.5.
- **When applied RF (ν_1) becomes equal to precessional frequency of protons resonance occurs (*i.e.*, RF flips the spin of nuclei from lower energy state to higher energy state).**
- Here motive is to tip M_0 in x and y axis of cartesian frame of reference and measure the resultant component of magnetization in that plane. RF (ν_1) is applied in such a way that its magnetic component B_1 is perpendicular to B_0 that rotates with precessing proton assemblage M_0.

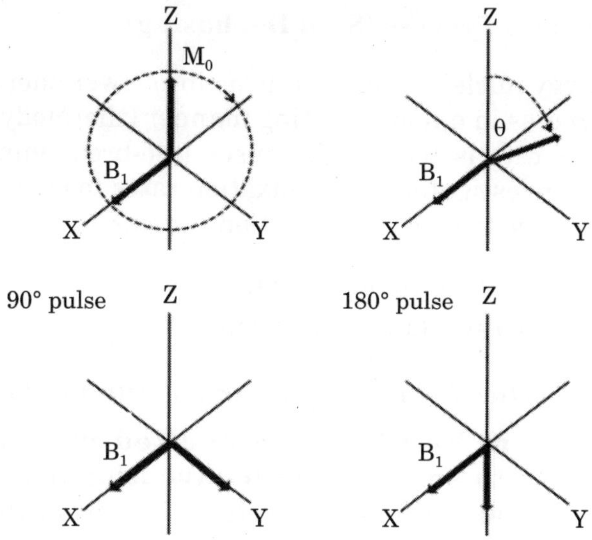

Figure 4.5: Schematic diagram of nuclear mechanics of magnetization.

- Such condition can be maintained by using RF oscillator along x axis *i.e.*, perpendicular to axis of B_0. Hence, it generates CW (continuous wave), oscillating, magnetic field B_1 in x axis. This B_1 is possess two components rotating in opposite directions of which one is direction of precessional orbit of protons and the other is in opposite of it.

Note:

Gyroscope motion—A slow, periodic conical motion of the rotation axis of a spinning body.

Precession—A change in the orientation of the rotational axis of a rotating body.

Diamagnetism—Property of nuclei that causes it to create a magnetic field in opposition to an externally applied magnetic field.

Paramagnetism—Property of nuclei that causes it to create a magnetic field aligned to externally applied magnetic field.

Spin-Spin Splitting—Coupling of nuclear spin of one nucleus via bonding electrons to other nucleus (which experiences two different micro environments), thereby resonates at two slightly different frequencies). This results in splitting of resonance line into two lines (*e.g.*, doublet).

4.2.5 Relaxation Process (Spin Dephasing)

The high energy nuclei (excited) returns into lower energy state by relaxation process in a non-radiating manner (thermodynamics). If the relaxation rate is too fast it causes line-broadening in NMR spectrum. So, investigators like relaxation rates to be fast—but not too fast. There are two major relaxation processes.

- Spin—lattice (longitudinal) relaxation
- Spin—spin (transverse) relaxation

4.2.5.1 *Spin-Lattice-Relaxation (T_1) or Longitudinal Relaxation*

The mechanism by which an excited magnetization vector returns to equilibrium (conventionally along the z axis) is called as Spin-lattice-relaxation or T_1 Relaxation. A schematic diagram that represents T_1 phase is given in Figure 4.6.

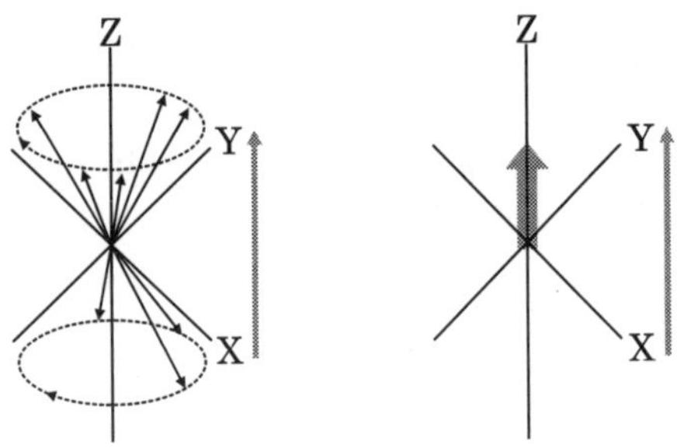

Figure 4.6: Schematic diagram for T_1 relaxation.

Key Concepts

- T_1 Phase is slightly slower than Transverse relaxation (T_2).
- **In T_1 Phase excited protons transfers their energy to the surrounding protons with appropriate tumbling frequency. As a result the net magnetization returns from X and Y axis to Z axis.**
- As nuclei in an NMR experiment are contained in a sample. **The sample in which nuclei are held is called the lattice. Nuclei in the lattice are in vibrational and rotational**

motion, which creates a complex magnetic field. The magnetic field caused by motion of nuclei with in the lattice is called lattice field. This lattice field has many components. Some of these components will be equal in frequency and phase to the Larmor frequency of the nuclei of interest. **Components of the lattice field can interact with nuclei in the higher energy state, and cause them to loose energy** (returning to the lower state).

- **The energy that a nucleus looses, increases the amount of vibration and rotation within the lattice** (resulting in a tiny rise in the temperature of the sample).
- **The relaxation time, T_1** (the average lifetime of nuclei in the higher energy state) **is dependent on the magnetogyric ratio of the nucleus and the mobility of the lattice**.
- As mobility increases, the vibrational and rotational frequencies increase, making it more likely for a component of the lattice field to be able to interact with excited nuclei. However, at extremely high mobilities, the probability of a component of the lattice field being able to interact with excited nuclei decreases.

4.2.5.2 *Spin-Spin Relaxation (T_2) or Transverse Relaxation*

The mechanism by which the excited magnetization vector (conventionally in the x–y plane) decays is called as Transverse relaxation or T_2 relaxation. A schematic diagram that represents T_2 phase is given in Figure 4.7.

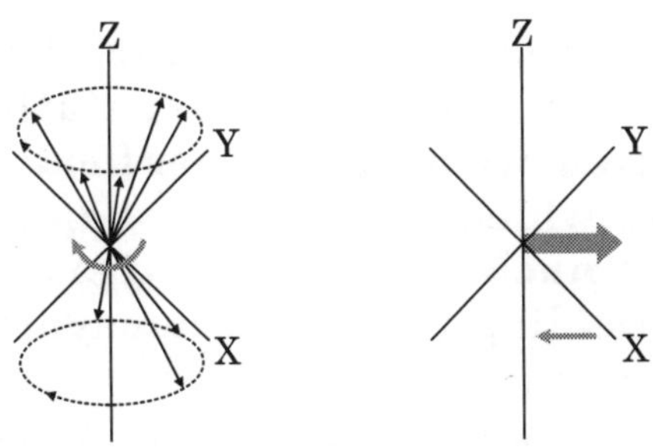

Figure 4.7: Schematic diagram for T_2 relaxation.

Key Concepts

- **In T_2 Phase energy is transferred among two precessing protons.** This is slightly faster than longitudinal relaxation.
- Spin-spin relaxation describes the interaction between neighboring nuclei with identical precessional frequencies but differing magnetic quantum states.
- In this situation, the nuclei can exchange quantum states; a nucleus in the lower energy level will be excited, while the excited nucleus relaxes to the lower energy state.
- There is no net change in the populations of the energy states, but the average lifetime of a nucleus in the excited state will decrease. This can result in line-broadening.

4.2.6 Chemical Shift

The difference in signal position of any nuclei under investigation in comparison to the signal position of reference nuclei (*e.g., TMS nuclei*) on a δ value/τ scale is called as chemical shift.

Key Concepts

- Protons in different chemical environment shows different chemical shift and also protons in similar chemical environment show peak on same positions on a δ value/τ scale.
- Usually, protons in a molecule are shielded by surrounding electronic cloud and this electronic cloud density depends upon the nuclei chemical environment.
- **The electrons surrounding nuclei in presence of magnetic field undergoes circulation and due to electronic circulation one another field (secondary field) is generated which opposes the applied field (RF) to resonate the nuclei under investigation, this effect is called as shielding effect or diamagnetism.** Higher the shielding, greater is energy required to resonate nuclei under investigation or lower δ value. Lower δ value (*higher τ value*) implies nuclei under investigation shows a signal near (*or towards*) δ value of standard, this is also called as upfield region.

- **If molecule contains some other atomic nuclei which reduce or hamper the electron cloud density of nuclei under investigation, this consequently reduces the secondary field and thereby easily allows (does not oppose) the applied field (RF) to resonate the nuclei under investigation, this effect is called as deshielding effect or paramagnetism.** Higher the deshielding effect, the lesser is energy required to resonate nuclei under investigation and higher δ value (*lower τ value*). Such that nuclei under investigation shows a signal away from δ value of standard this is also called as downfield effect.

4.2.6.1 *Factors Affecting Chemical Shift*

The shielding and deshielding effects depends upon the electron cloud density surround nuclei, and this density of electron cloud depends upon various factors such as:

a. Inductive effect

b. Vander Waal's deshielding

c. Anisotropy and

d. Hydrogen bonding

A. Inductive Effect

When a molecule contains electronegative atoms (*like halogen, oxygen and nitrogen*) ***near to the nuclei under investigation, it poses an electron withdrawing effect. As a result the electronic cloud surrounding nuclei under investigation is reduced or removed, that leads to downfield effect, such an effect is called as inductive effect.***

Therefore inductive effect causes deshielding effect on nuclei under investigation. The greater the distance between the nuclei under investigation and electronegative atom the lesser the inductive effect occurs.

Influence of Inductive effect on the chemical shift can be understood by taking examples of propane (*with no inductive effect*) and propyl chloride molecule (*with profound inductive effect*).

- **In case of simple molecule of propane** (Figure 4.8), without inductive effective two signals appears on a ^{1}H-NMR, spectrum on a δ value scale.

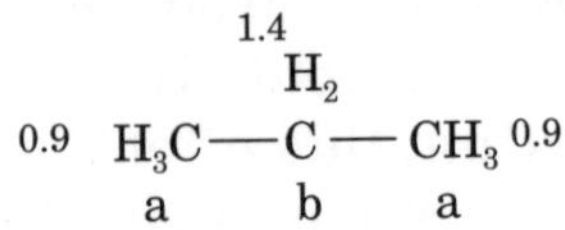

Figure 4.8: Propane.

  - One of signal appears at δ value of 0.9 ppm, in upfield region (*shielded*) in form of a triplet for 'a' protons of CH_3 group's protons with very high intensity. Both methyl hydrogens appear on same δ value as possess same chemical environment.
  - Other signal appears at δ value of 1.4 ppm, in downfield region (*deshielded*) in form of a septet or multiplet for 'b' protons of CH_2 group with low intensity.

- **But, in case of propyl chloride** (Figure 4.9), three signals appears on a ^{1}H-NMR, spectrum on a δ value scale.

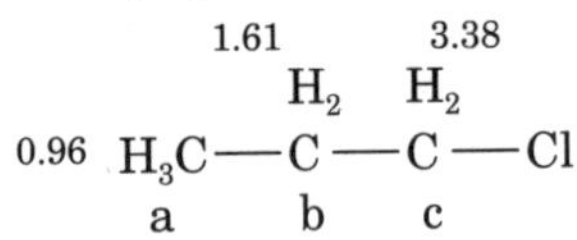

Figure 4.9: Propyl chloride.

  - One of signal appears at δ value of 0.9 ppm, in upfield region (*shielded*) in form of a triplet for 'a' protons of CH_3 group protons with very high intensity.
  - Other signal appears at δ value of 1.61 ppm, in upfield region (*shielded*) in the form of a sextet or multiplet which accounts for 'b' protons of CH_2 group protons with low intensity.
  - Another signal appears at δ value of 3.38 ppm, in downfield region (*deshielded*) in form of a sextet or multiplet for 'c' protons of CH_2 group which is directly attached to chlorine atom.

- **Among δ value of methylene group of both propane and propyl chloride molecules**, the protons of methylene group of propyl chloride are very much deshielded. This is because the electron withdrawing effect of chlorine atom, which removes the electron cloud/electronic shield around hydrogens of methylene group. As a consequence, it becomes easy for applied frequency to resonate methylene hydrogen nuclei to resonate at lower frequency. So, inductive effect of electronegative atom causes nuclei under investigation to absorb downfield or becomes more deshielded.

B. Vander Waal's Deshielding

When the electron cloud of nuclei under investigation is sterically hindered due to some bulky groups in an overcrowded molecule, this causes nuclei to appear in downfield region *(deshielded)* ***on a δ value scale, such an effect is called as Vander Waal's deshielding.*** Vander Waal's deshielding causes nuclei under investigation to absorb at higher δ value.

Influence of Vander Waal's Deshielding on chemical shift can be best understood from examples of 2-butene (*without Vander Waal's Deshielding*) and 1, 2-Diphenylethene (*with profound Vander Waal's deshielding*) *molecules*.

- **In case of 2-butene molecule** (Figure 4.10), two signals appears on a ^{1}H-NMR, spectrum on a δ value scale.

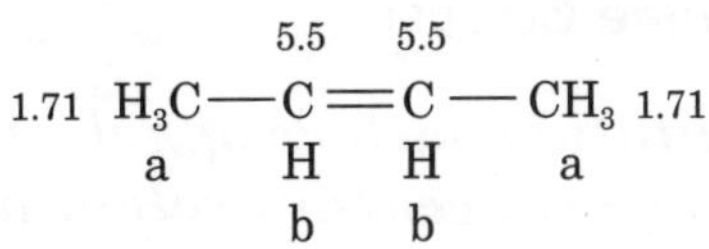

Figure 4.10: 2-Butene.

  - One of signal appears at δ value of 1.71 ppm, in upfield region in form of a doublet for 'a' protons of CH_3 group protons with very high intensity.
  - Other signal appears at δ value of 5.5 ppm, in downfield region (*deshielded*) in the form of a quartet which accounts for 'b' protons of two =CH group hydrogens with very low intensity as contributed by protons of two hydrogens.

- **But in case of 1, 2-Diphenylethene** (Figure 4.11), two types of signals appears on a δ value scale.

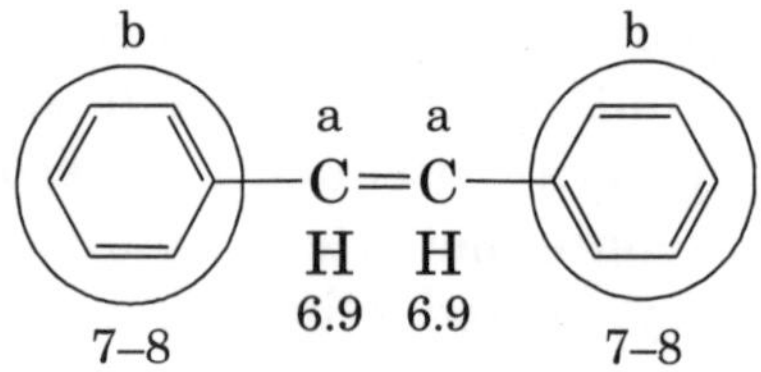

Figure 4.11: 1, 2-Diphenylethene.

- One of signal appears at δ value of 6.9 ppm, in upfield region (*shielded*) in form of singlet for 'a' protons of =CH group protons.
- Other multiple signal appears between 7–8 ppm, in more downfield region (*deshielded*) for 'b' aromatic protons of two benzene rings.

- **Among δ value of =CH group of both 2-butene and 1, 2-Diphenylethene molecules**, protons of =CH group of 1, 2-diphenylethene are much deshielded. This is because presence of two bulky groups causes repulsion of the electron cloud/ electronic shield around hydrogens of =CH group. As a result, it becomes easy for applied frequency to resonate methylene hydrogen nuclei at lower frequency. So, Vander Waal's force causes nuclei to absorb downfield or become more deshielded.

C. Anisotropy (Space Effect)

In some cases circulation pattern of π electrons under applied field or orientation of molecular protons affects the chemical shift of protons of molecules. Such an effect is called as anisotropic effect. That is why protons that are attached to C=C experiences more deshielding effect than due to inductive effect. In case of aldehyde and aromatic compounds protons appears on higher δ value than alkynes proton.

- **In case of Alkenes** (Figure 4.12), plane of double bond is at right angle to the applied field and the induced circulation of π electrons generates induced magnetic field which is diamagnetic around carbon atom region and paramagnetic in the protons region. As a result protons feel higher field strength and thereby resonate at lower applied field.

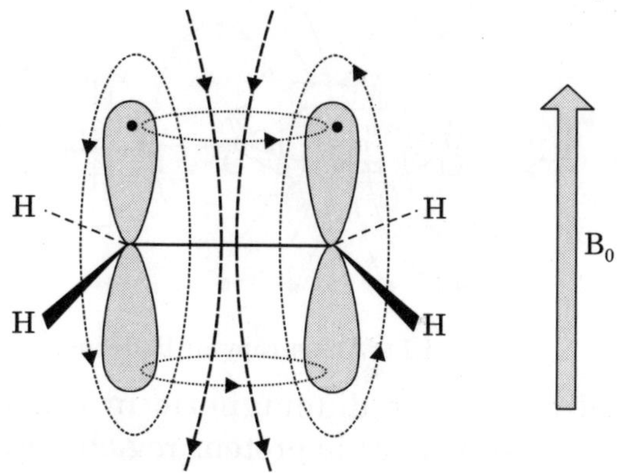

Figure 4.12: Anisotropy in alkenes.

For example in case of ethene anisotropic effect causes protons of hydrogens of $=CH_2$ group to resonates at 5.25 δ value, a highly deshielded region.

- **In case of Alkynes** (Figure 4.13), plane of triple bond is parallel to the applied field and the induced circulation of π electrons generates induced magnetic field which is paramagnetic around carbon atom region and diamagnetic in the protons region. As a result protons feel lower field strength and thereby resonate at lower applied field.

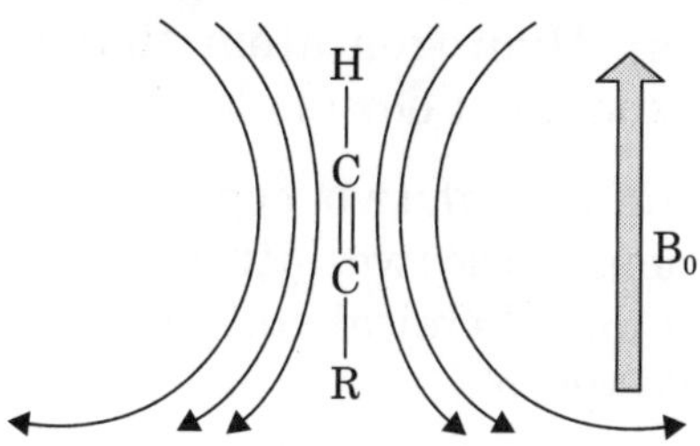

Figure 4.13: Anisotropy in alkynes.

For example in case of ethyne molecule (CH≡CH) the protons of hydrogens of ≡CH group resonates at 1.80 δ value, a highly shielded region.

- **In case of Benzene** (Figure 4.14) due to induced circulation of π electrons induced magnetic field called ring current is generated.

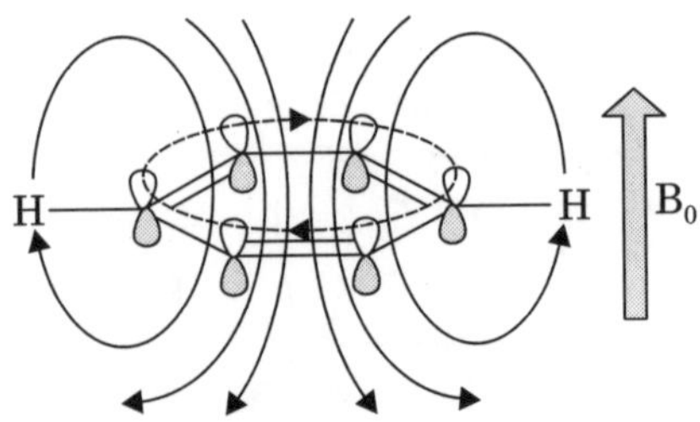

Figure 4.14: Anisotropy in Benzene.

The induced current is diamagnetic in centre of ring and paramagnetic outside ring *i.e.*, in protons region. As a result protons feel higher field strength and thereby resonate at lower applied field. The protons that are attached below and above plane of ring resonate at low d value (high applied field). This is the reason why protons of aromatic/benzene usually appear between 6–8 d values.

D. HYDROGEN BONDING

Hydrogen bonding is one another factor that contributes to chemical shift. *Hydrogen bonding causes a further deshielding of protons, and a further downfield shift of proton resonances. Hydrogen bonding effects are concentration and temperature dependent.*

In general, protons attached to oxygen and nitrogen resonates between 0.5–5 ppm. Hydrogen atom that undergoes hydrogen bonding tends to resonate at downfield.

Hydrogen bonding is of two type *i.e.*, intermolecular and intramolecular. Among these two, intramolecular hydrogen bonding does not undergoes any chemical shift on change in concentration but intermolecular does so.

For example in case of phenols that undergoes intermolecular hydrogen bonding the absorption shifts occurs from 6–8 δ down field region to upfield region when their concentration is reduced by addition of more of solvent.

In case of enols, that undergoes intramolecular hydrogen bonding a high chemical shift is resulted to greater downfield region *i.e.*, 10.5–12 δ. Amines, also shows similar chemical shift due to hydrogen bonding.

4.2.7 Spin-spin Coupling (Peak Splitting or Spin-spin Splitting)

Coupling is a phenomenon in which the nearby spin-active nuclei, *(nuclei having non-zero nuclear spin)*, **affects the local magnetic field of a nucleus under investigation.** Coupling implies physical interaction of the intrinsic angular momentum (spin) of different nuclei in a molecule. Figure 4.15, describes the splitting pattern of nuclei undergoing spin coupling.

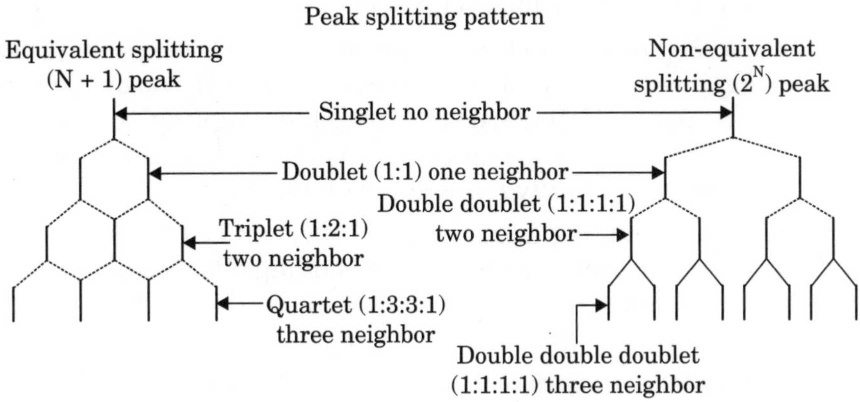

Figure 4.15: Splitting pattern of peaks while nuclear spins coupling.

Key Concepts

- Spin-spin splitting study supports in determining the number and types of hydrogens contained in a molecules.
    - In spin-spin splitting study hydrogens on adjacent carbons are commonly studied.
    - In ^{1}H-NMR experiment, homonuclear three bond couplings ($^3J_{H-H}$) causes peak splitting by N + 1 and 2^N pattern according to number of neighboring hydrogens.
- Coupling leads to additional splitting of the resonances for groups of coupled nuclei.
- Coupling is transmitted by the valence electrons of a molecule. The magnitude of the coupling is related to the number of bonds separating the coupled nuclei.
- These are of two types *i.e.*, equivalent and nonequivalent.

- Generally, a single peak should be obtained for each type of proton in a molecule. But **when a proton (H_A) under investigation is near to nonequivalent proton (H_B), then;**
  - In half of molecules H_A proton will be adjacent to an H_B (*which is aligned with field*) and in other half H_A proton will be adjacent to H_B (*which is against field*).
  - Thus, half of H_A's in sample feels a larger magnetic field than they would in absence of H_B and half would feel a slightly smaller magnetic field.
  - So, H_A proton exhibits two absorptions (*same case will happen for H_B*) and such splitting of the H_A resonance into two peaks is termed as "spin-spin coupling" or and the distance between two peaks (in Hz) is called as "coupling constant" (*usually represented by the symbol J*).
- Spin-spin coupling is transmitted through electrons in bonds. So, spin-spin coupling depends on bonding relationship between two hydrogens.
- Figure 4.16, describes the peak intensity/area ratio with respect to signals in a mnemonic manner called Pascal's triangle.

Peak intensity/Peak area ratio	Signal/Peak
1	Singlet
1 : 1	Doublet
1 : 2 : 1	Triplet
1 : 3 : 3 : 1	Quartet
1 : 4 : 6 : 4 : 1	Quintet
1 : 5 : 10 : 10 : 5 : 1	Sextet
1 : 6 : 15 : 20 : 15 : 6 : 1	Septet

Figure 4.16: Pascal's Triangle representing peak intensity/area ratio with respect to signals.

N + 1 Rule—Spin-spin splitting is best explained by N + 1 rule. According to N + 1 rule, each hydrogen (proton) of a carbon after resonance senses the number of equivalent hydrogens/protons (N) on the adjacent carbon and tends to split its peak by N + 1.

Example: *Ethyl chloride*

$$\underset{a}{H_3C}-\underset{b}{CH_2}-Cl$$

Ethyl chloride molecule possess different types of hydrogens shown as 'a' and 'b' (*chemically non equi-valent*). As a result these undergo $^3J_{H-H}$ couplings, so that a triplet for 'a' types of protons appears at 1.5 in the ratio of 1 : 2 : 1 and a quartet for 'b' types of protons appears at 3.4 in the ratio of 1 : 3 : 3 : 1.

Few examples of some compounds that undergo homonuclear three bond couplings ($^3J_{H-H}$) are given in Table 4.3.

Table 4.3: Description of types of protons, number of signals chemical shifts and peak area ratio for some organic compounds pursuing homonuclear three bond couplings ($^3J_{H-H}$).

Compound Structure and Name	***Types of Protons***	***Number of Signals***	***δ value in ppm/Peak Splitting (Peak Area Ratio)***
H_3C-CH_3 (a, a) Ethane	One	One	a = 0.8/Singlet
H_3C-CH_2-Cl (a, b) Ethyl Chloride	Two	Two	a = 1.5/Triplet (1:2:1) b = 3.4/quartet (1:3:3:1)
$H_3C-CH_2-CH_3$ (a, b, a) Propane	Two	Two	a = 0.9/Triplet (1:2:1) b = 1.3/septet (1:6:15:20:15:6:1)
$H_3C-C(CH_3)_2-NH_2$ (a: CH_3; b: NH_2) 2-Methylpropane-2-amine	Two	Two	a = 1.1/Triplet (1:2:1) b = 4.5/septet or multiplet (1:6:15:20:15:6:1)
$Br-CH_2-CH_2-Br$ (a, a) 1, 2-Dibromoethane	One	One	a = 3.8/singlet (1)

Table 4.3: (*Contd...*)

Table 4.3: (*Contd...*)

Compound Structure and Name	***Types of Protons***	***Number of Signals***	***δ value in ppm/Peak Splitting (Peak Area Ratio)***
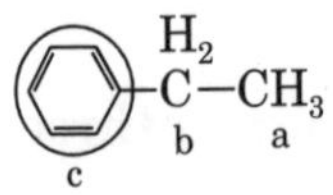 1-Ethyl benzene	Three	Three	a = 1.25/Triplet (1:2:1) b = 2.6/quartet (1:3:3:1) c = aromatic protons from 7–7.25/multiplet
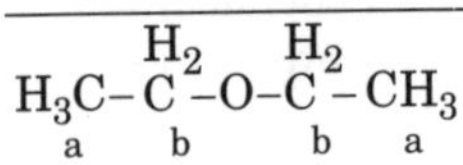 Diethyl ether	Two	Two	a = 1.1/Triplet (1:2:1) b = 3.4/quartet (1:3:3:1)
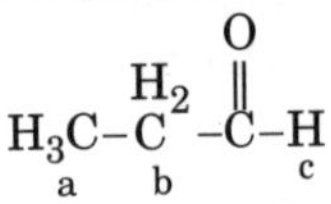 1-Propanal	Three	Three	a = 1.07/Triplet (1:2:1) b = 2.4/Quintet (1:4:6:4:1) c = 9.9/Triplet (1:2:1)
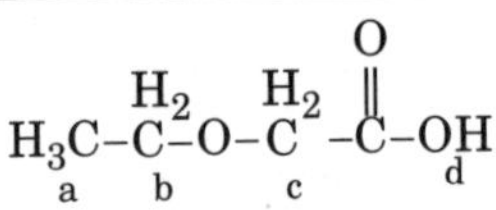 Ethoxy acetic acid	Four	Four	a = 1.1/Triplet (1:2:1) b = 3.4/quartet (1:3:3:1) c = 4.3/singlet (1) d = 11.0/singlet (1)
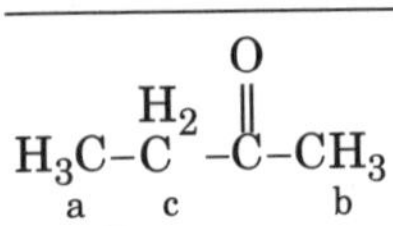 2-Butanone	Three	Three	a = 1.06/Triplet (1:2:1) b = 2.1/Singlet (1) c = 2.5/Quartet (1:3:3:1)

4.2.8 Coupling Constant

As the distance between centers of two adjacent peaks of a multiplet is generally constant, and is independent of external field, hence it is called as coupling constant. Coupling constant is denoted with symbol J. A value of J is measured in Hertz (Hz) or cycles per second (cps).

Key Concepts

- The J value is always constant for a multiplet peak and is independent of applied field.

- With the help of J value one can easily distinguish between two singlet's from a doublet. This is accomplished by recording a spectrum at two different radio frequencies. If a separation occurs between the two peaks it reflects presence of two singlet peaks and if no separation occurs it reflects presence of a doublet.
- In general the J value occurs between 0–20. Similarly, one can distinguish a quartet from doublet.
- For example in ^{1}H-NMR spectrum of ethyl chloride, for 'a' and 'b' protons there appears a triplet and quartet respectively. In case of triplet peak of three 'a' protons, the J_{ab} is same because 'a' protons undergoes coupling with the two equivalent 'b' protons. Moreover, in the case of quartet peak of two 'b' protons also, J_{ba} is same because 'b' protons undergoes coupling with the two equivalent 'a' protons.
- Similarly, J value of multiplet of a compound is always constant when the signal for one set of equivalent protons is splitted by another set of equivalent protons or vice versa.
- Generally, the J value for ^{1}H-NMR experiment is represented with $^3J_{H-H}$ to represent coupling between H–C–C–H (*i.e.*, two coupling of two hydrogens separated by three bonds). Usually, in ^{1}H-NMR experiment the $^3J_{H-H}$ value ranges from 0–20 Hz.
- The J value depends upon the number of covalent bonds through which protons may interact and also on the structural relationship between coupled protons.

Example 1: *Ethyl Chloride*

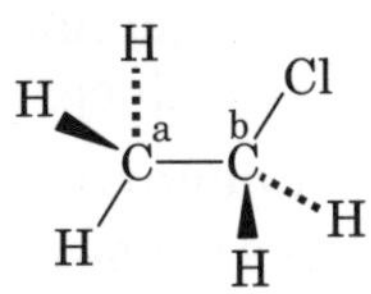

The NMR spectrum of ethyl chloride offers a triplet (for 'a' hydrogens) of low δ value (up field) and a quartet (for 'b' hydrogens) of high δ value (down field). The J value for both multiplets is same *i.e.*, 7.1 cps.

Example 2: *Isopropyl Iodide*

Isopropyl iodide NMR spectrum offers a doublet (for 6H of methyl groups) of low δ value and a septet (^{1}H) of high δ value. The J value for both multiplets is same *i.e.*, 6.7 cps.

The J value of multiplets of a compound will not be constant when signal is splitted due to two different sets of protons. For example in the NMR spectrum of propyl iodide, J_{ab} = 6.8 and J_{bc} = 7.3 cps.

4.2.9 Types of Coupling

Couplings are categorized into three types *viz.*,

- Geminal Coupling
- Vicinal Coupling
- Long range Coupling

4.2.9.1 *Geminal Coupling*

A coupling which occurs in geminal protons i.e., protons attached on same carbon but with different environments is called as geminal coupling. The geminal J value depends upon the bond angle of geminal protons. Usually the geminal J value is –25 cps at 105°, –12 cps at 109°, zero at 125° and is positive above 125° bond angle.

Key Concepts

- Geminal coupling becomes very strong when the two geminal protons are of different environment. Increasing the 's' character (bond angle) increases the J value of geminal protons. For example, in case of methane geminal J value is –12.4 cps where as in case of ethylene it is +2.5 cps.
- Increase in the electronegativity of atom or group that withdraws sigma electrons causes increase in geminal J value. For example, in case of methyl chloride it is –10.8 cps whereas in case of methyl fluoride it is –9.4.
- Increase in the electronegativity of atom or group that withdraws π electrons causes decrease in geminal J value. For example, in case of ethene it is +2.3 cps whereas in case of vinyl fluoride it is –3.3 cps.

- In mono substituted olefinic compounds geminal J value for cis is more than the trans.
- In isomeric olefinic compounds the geminal J value for trans is more than the cis.

4.2.9.2 *Vicinal Coupling*

A coupling which occurs in vicinal nuclei i.e., nuclei which are separated by three bonds and with different environments is known as vicinal coupling. In this case the vicinal J value depends upon the dihedral angle of vicinal protons.

Usually the vicinal J value is highly positive when the dihedral angle is 0° or 180° and is slightly negative when dihedral angle is 90°.

Key Concepts

- The vicinal J value for the proton of gauche form is 8 cps and for anti form is 5–12 cps.
- The vicinal J value for the cyclopentane cis proton (at zero dihedral angle) is 8 cps and for trans proton (at 80° dihedral angle) is zero cps.
- The vicinal J value in case of cyclohexane axial-axial proton (at 180° dihedral angle) is 8 cps. Whereas in case of axial-axial and equatorial-equatorial is 2 cps.
- In olefinic compounds the vicinal J value for trans is always more than the cis.

4.2.9.3 *Long Range Coupling*

When two nuclei are separated by more than three bonds then no coupling occurs. But if a coupling occurs between two nuclei that are separated by more than three bonds such coupling is termed as long range coupling. These are common in unsaturated and fluoro compounds. These involve couplings between nuclei that are separated by even more than 4–5 bonds.

4.3 NMR INSTRUMENTATION

The essential components that commonly an NMR instrument comprises are as follows:

- A Magnet (highly stable)
- A Sample probe
- A Radio frequency radiation source
- A detector (Phase sensitive)
- A Data processing unit

4.3.1 Magnet

Magnet is the most important component of NMR spectrometer, because its strength and quality only determines the sensitivity and resolution of any spectrometer. The higher the strength of magnet the better the resolution and sensitivity of NMR instrument, hence an NMR instrument is always operated at highest possible magnetic field strength. Apart from it, the field of a NMR spectrometer should be reproducible, homogeneous, and uniform. It is very essential that stability of a magnet is maintained at all costs. There are three types of magnets of use in NMR spectrometers such as:

a. Permanent Magnet
b. Electromagnet
c. Super Conducting Solenoid

A. Permanent Magnets

Permanent magnets possess strengths that require an oscillator frequency of less than 100 MHz for protons. These magnets are used in commercial continuous wave spectrometers. These are temperature sensitive and require extensive cooling and shielding. These suffer from field drift problem so these are not ideal for extended periods of data accumulation.

B. Electromagnets

The electromagnets were also supposed to be used in early 50's of 19th century for proton resonance in a range of 60 MHz, 80 MHz, and 100 MHz are now rarely used in the NMR instruments.

C. Superconducting Magnets (Solenoid)

Superconducting magnets are simple, small and produces high field strength besides having low operating cost. These are helium cooled.

Modern high resolution spectrometers above 100 MHz use superconducting magnets and operate in pulsed FT mode.

4.3.2 Sample Probe

This is another component of NMR spectrometer which possess several functions such as:

- Holds the sample in a fixed position in the magnetic field.
- Contains an air turbine to rotate (spin) the sample. Spinning of the sample is done to ensure that sample experiences a uniform field.

Sample probe houses a coil for generating radiofrequency for excitation and also for detection of the NMR signal. Old NMR spectrometers use to possess a separate transmitter and receiver coils which were perpendicular to each other for excitation of the pulse and signal detection. But now the modern spectrometers usually contain a single coil probe which is simple and more efficient.

4.3.3 Detector System

To detect NMR signal high frequency radio signal is converted to an audio frequency signal. An audio frequency signal is made up of two components, such as:

- A carrier signal of frequency of oscillator which is used to produce the signal and
- A superimposed NMR signal of analyte.

Analyte signal differs in frequency from standard by a few ppm. In a proton spectrum, chemical shifts are in a range of 1–10 ppm. So, proton magnetic resonance data generated by 400 MHz spectrometer would appear in the frequency range of 400,000,000 Hz to 400,004,000 Hz. It is impossible to digitise such a small difference of frequency. Therefore different signals are generally obtained which lies in audio frequency in kilo hertz range.

The modern spectrometers possess a phase sensitive detector (*quadrupole*) which can sense the sign of frequency difference. So, a detector differentiates between the frequency of standard and sample.

4.3.4 Sample Handling

Commonly high resolution NMR spectrometer needs clear transparent sample solution of 2 to 15% concentration, whereas non viscous or less viscous pure liquids can also be used.

Liquid samples are usually taken in NMR glass tube of about 5 mm outer diameter and 15–20 cm length with a capacity of 0.5 cm^3. Small amount samples, micro tubes can also be used.

In ^{1}H-NMR, aprotic solvents are commonly used. This is because hydrogen atom present in solvent may show its absorbance and can interfere with analyte absorption spectrum. As carbon tetrachloride does not contain any hydrogen, so considered as most ideal but due to its low solubility for many compounds its use is limited.

Hence, deuterated organic solvents for example deuterated chloroform ($CDCl_3$), deuterated benzene (C_6D_6), deuterated acetone (CD_3COCD_3) and deuterated dimethyl sulfoxide (CD_3SOCD_3) etc. are most commonly used in ^{1}HNMR spectroscopy.

4.3.5 Types of ^{1}H-NMR Instruments

On the basis of application and measurement of RF frequency in NMR spectrometry different NMR instruments are developed such as:

4.3.5.1 *Continuous Wave (CW)-NMR Spectrometer*

Continuous wave NMR experiment is defined as technique in which different nuclei are brought into resonance one by one with continuous variation in either magnetic field or oscillating frequency.

In NMR spectrometer sample under investigation is dissolved in a suitable and non-interfering solvent. In practice along with the sample an internal standard is also commonly taken in NMR tube (*cylindrical glass tube*) which is placed into the space of magnetic assembly of the NMR spectrometer.

Nextly, the sample is spun, so that, sample can experience uniform magnetic field throughout. In the magnetic assembly space there exists a housing of transmitter and receiver coils that are perpendicular to each other, for excitation of the pulse and signal detection. The transmission coil supplies 60 MHz radiofrequency to change the spin of sample nuclei. When RF energy is absorbed by the sample, it induces a change in orientation of nuclear spin. While reorientation of the nuclei spin an RF signal is induced in the detector coil, which is recorded as an NMR signal or a peak.

As per equation 4.6, there are two variable parameters, such as frequency (υ) and magnetic field (B). So, experimentally, there are two different ways to achieve resonance namely,

a. Frequency Sweep Method
b. Field Sweep Method

A. Frequency Sweep Method

Method of nuclei resonance which involves application of varying radio frequency under a fixed or constant magnetic field strength is called as frequency sweep method.

Key Concepts

- Sample is placed between poles of a strong magnet which generates two energy levels and tends the nucleus to undergo precessional motion.
- Irradiation of sample with a variable radiofrequency generated by an oscillator or transmitter.
- The resonance occurs when frequency of the radio frequency radiation matches with the Larmor precessional frequency of the sample nuclei.
- **As a result of resonance, nuclei of lower spin state absorb the radiation and moves on to the higher energy state. Such a transition is called as 'spin flipping'.**

B. Field Sweep Method

A nuclei resonance method which involves use of varying of magnetic field strength at a fixed/constant oscillator frequency is known as field sweep method. In field sweep spectrometers continuous variation of the magnetic field strength is achieved with the help of special coils present at the poles of the magnet.

Key Concepts

- When field strength is of low precessional frequency which is lower than the applied radiofrequency, then the two frequencies will not match and no energy will be absorbed.
- As magnetic field is increased, the precessional frequency starts increasing. At a particular magnetic field strength it matches exactly to the applied radio frequency thereby resonance occurs and radiation is absorbed.
- Transition occurs from lower spin state to higher spin state.

Among the two sweep methods field-sweep method is the most commonly used method for NMR. This is due to ease of varying magnetic field. Now a days, high resolution modern NMR instrument don't use CWNMR but these are still used in modern NMR instruments requiring low resolution.

4.3.5.2 *FT-NMR Spectrometer*

Sometime quantity of sample for analysis is too small, as in case of biological samples. Due to this Signal/Noise (S/N) ratio is too small and so it becomes difficult to distinguish the signals from background noise. A solution for such a condition is to 'scan' the sample number of times and add different scans using computer. With this strategy S/N ratio improves, but it is time consuming *i.e.* about an hour. Another strategy is to use one important process called *Fourier transformation*.

Fourier transform (FT) NMR spectrometry is defined as experiment in which all the sample nuclei are simultaneously excited by irradiating with brief RF pulses containing all frequencies. The resultant time domain FID signals so obtained are digitized and converted to frequency domain curve by Fourier transformation (*mathematical programme to convert a time domain curve into a frequency domain curve*). Fourier transform (FT) spectrometry is a general concept used to study very weak signals after isolating it from environmental noise.

Key Concepts

- In FT-NMR, sample nuclei in a magnetic field are subjected to very brief pulses of intense radiofrequency radiation.

- These brief pulses of duration of few microseconds contain all the frequencies in the range where the nuclei can absorb.
- This causes all the nuclei in the sample to absorb the radiofrequency and get excited simultaneously.
- After a delay of a few seconds another pulse is applied. The delay is so chosen that during the interval between the two pulses all the excited nuclei relax back to the ground state and the population difference is achieved back. This relaxation is characterized by free induction decay (FID).
- This FID can be detected with a radio receiver coil which is perpendicular to the static magnetic field. In fact, a single coil is frequently used both to provide pulse to the sample and also to detect the FID signal.
- The FID signal is digitised and stored in a computer for data processing. The process of giving a pulse and collecting the FID takes a few seconds, therefore a large number of such FIDs which are in a way equivalent to a scan can be collected in a reasonable amount of time.
- The time domain decay signals (FIDs) from numerous successive pulses are added to improve the signal to noise ratio. The resulting summed data are then converted to a frequency domain signal by Fourier transformation (Figure 4.17).

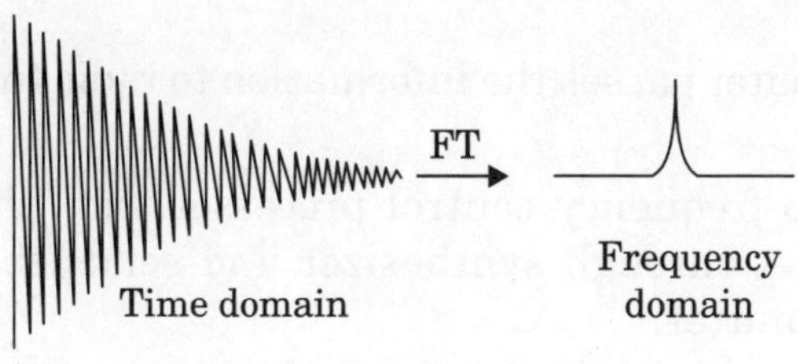

Figure 4.17: Conversion/fourier transformation of time domain signal in to frequency signal.

- Finally, digital filtering may be applied to data to further increase the signal to noise ratio. The resulting frequency domain spectrum is similar to the spectrum produced by a scanning continuous wave experiment.
- FT instruments possess high resolving power and wavelength reproducibility which make possible the analysis of complex spectra.

- Old NMR instruments use to possess a separate transmitter and receiver coils but now modern spectrometers contain a single coil probe. Modern instruments are simple and more efficient.

The Figure 4.18, describes and highlights sequential steps while obtaining an NMR spectrum through a high resolution spectrometer.

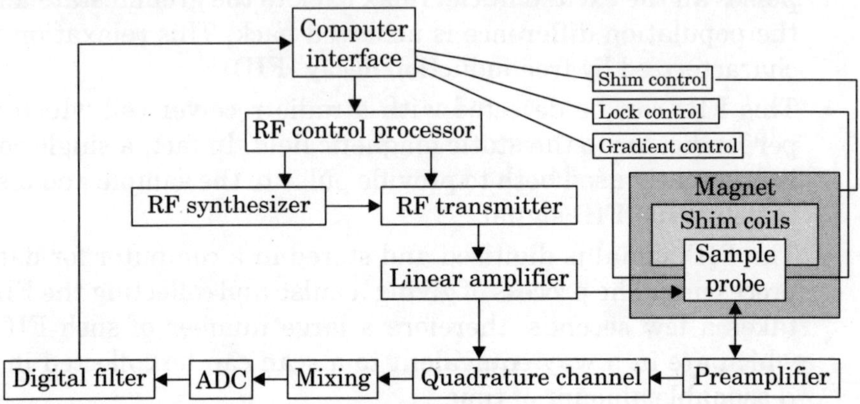

Figure 4.18: Sequential steps while obtaining NMR spectrum through a high resolution spectrometer.

The sequential steps involved in conducting FTNMR experiment are as follows:

- The computer passes the information to radio frequency control processor.
- The radio frequency control processor, sets the required RF frequencies through synthesizer and sends the *RF* pulses to the transmitter.
- Transmitter prepares the pulses with the required shapes, phases, durations, also power settings and lastly sends the pulses into different amplifiers or linear amplifiers (It is important for smooth functioning of NMR spectrometers to keep all the stages linear during transmission and amplification as this whole activity is done in microseconds).
- After linear amplifiers, the RF pulse reaches to the sample probe containing a coil which applies and receives the *RF* frequency to the sample situated on top of probe.

Generally, when no *RF* pulse is applied the amplifiers remains completely blank, and the spins of nuclei precession is free from *RF* disturbance.

- As same coil is used for transferring (application of *RF* pulses at the observed frequency) and receiving of very weak signal/ response from the spins, hence a preamplifier is used for the pulses at the observed frequency. After receiving of the NMR signal, the signals are picked up by the receiving coil and reaches to the preamplifier with minimal attenuation. The speciality of preamplifier is that, after the NMR signals are released from preamplifier, the signals still possess actual and sufficiently amplified NMR frequency with preserved the S/N ratio in the subsequent amplification and mixing stages.
- A special safety circuit is provided with preamplifier which protects the preamplifier from the very high voltage present during RF pulse and directs the pulse only towards the coil.
- Because a signal cannot be digitized at NMR frequency, so a high dynamic digitizer is used to digitize the signal by transforming it into a frequency range between 10–100 kHz. This reduction in frequency is done in different stages.
    - In the Ist stage of digitization, all of the NMR frequencies are transformed to the same intermediate frequency and the two quadrature channels are created.
    - In the IInd stage of digitization, mixing is done so as to reduce the signal frequencies of the quadrature channels to lower frequencies which can be digitized in the analog-to-digital converter (ADC—samples the signal at discrete time points and converts it into a series of numbers).
    - In the IIIrd stage of digitization, digital filters are used to perform digital filtration of ADC oversampled data (much larger frequency range than desired) for the selection of final spectral range. If digital filters are not available the corresponding analog filters must be used before the ADC.
- After digitization and filtration the signals are sent to the computer memory for summing up and processing of data of different scans.

4.3.6 Internal Standard

As in a NMR spectrometer, it is impossible to measure the chemical shifts of a sample without an internal standard. Hence, an internal standard is always used to determine the relative extent of shielding or deshielding of the external field in various chemical environments. For organic compounds tetra methyl silane (TMS) is found as the most suitable standard. The reasons behind choice of TMS as an internal standard are as follows:

- TMS is highly inert and does not interact with most of the organic compounds, so does not interferes in NMR measurements.
- As TMS possess low boiling point (*i.e.*, 27°C) and is soluble in most of the organic solvents, therefore it can be easily removed or separated from other organic solvents after the NMR experiment.
- At very low concentrations also all the protons of TMS displays a highly intense signal, because all the TMS protons are chemically and magnetically equivalent.
- As TMS displays a strong and sharp signal, so no chances of interference with all other protons signals of different organic molecules because of their absorbance at lower field relative to TMS.

4.3.7 Integral in ^{1}H-NMR

In ^{1}H-NMR experiment of a molecule, it is possible to characterise different types of protons and also the quantification of each type of protons. In a ^{1}H-NMR spectrum the area under each peak is directly proportional to the number of hydrogens correspondent to that particular peak.

In modern NMR instrument it is possible to electronically integrate the area under each peak by vertically raising a line after tracing each peak called integral. It is very important to note that height of integral line does not give exact number of hydrogens of any particular peak but it gives a relative number of hydrogen.

For example in ^{1}H-NMR spectrum of 1, 1 dichloro propane (Figure 4.19), three integrals are shown for 'a', 'b', and hydrogens 'c' hydrogens with division of 8, 5 and 3 respectively. The number of hydrogens for each peak can be easily calculated by dividing all the three integral division with least division of 3. As a result peak of a, b, c shows 2.6 = 3, 1.6 = 2, and 1 hydrogen respectively.

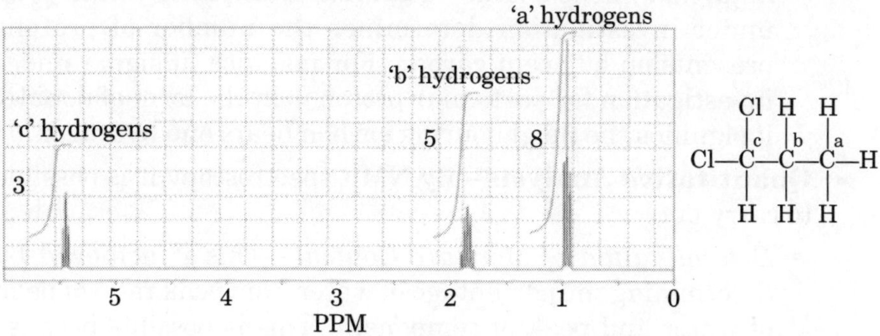

Figure 4.19: ^{1}H-NMR spectrum of 1, 1-dichloro propane.

4.4 APPLICATIONS OF ^{1}H-NMR SPECTROSCOPY

NMR spectroscopy has laid new standards in spectroscopy and science. Below mentioned are some of the important applications of ^{1}H-NMR spectroscopy.

- **In Establishment of structures**—By NMR spectroscopy one can determine
  - *Protons Types Count*—such that the number of peaks in a NMR spectrum represents the types of protons. For instance if an NMR spectrum contains five peaks it represents that molecule contains five different types of protons and if seen keenly peaks shows molecule contains seven types of protons.
  - *Protons Environment*—such that proton peak under investigation relative to other peak or reference peak is shielded/deshielded or upfield or downfield. For instance in a molecule of toluene methyl protons shows a signal near δ value of 1, that is shielded because near to standard peak *i.e.*, upfield, whereas aromatic protons produces a signal

between 6–8 δ value that is away from standard, that is downfield.

- *Count of Protons For Each Type*—such that if intensity of peaks (height/area) is more the higher the number of proton corresponds that peak whereas if intensity of peaks is low it signifies lesser protons available.
- *Adjacent Protons Counts*—such that multiplicity of each peak under investigation determines the number of protons presents on adjacent carbon. For instance if signal under investigation for particular proton is in the form of doublet it signifies the neighbouring carbon bears one hydrogen.

- **Quantitative Analysis**—By NMR spectroscopy it is possible to carry out
  - *Determination of Moisture Content*—This is achieved by determining the percentage of water from peak ratio of peak of water and peak of component. This is possible because water gives a characteristic peak.
  - *Single Component or Multi Components Estimation*—This is achieved by identifying specific peak for each component. Thereby the peak area/height ratio of component is determined from integral value using standard with which the quantity can be estimated.
  - *Determination of chain length of Surfactant*—This is achieved by determining the hydrogen nuclei proportion in poly-oxyethylene chain.
  - Also determination of percentage hydrogen and iodine value (measurement for double or triple bond).
- In determination of isomerism, keto-enol tautomerism, hydrogen bonding, molecular asymmetry and optical purity etc.
- In investigation of molecular interactions such as micelle and drug receptor interactions.

4.5 INTERPRETATION OF ^{1}H-NMR SPECTRA

This section describes the ^{1}H-proton NMR absorptions of various classes of compounds, along with their ^{1}H-NMR spectra and *J* values using some relevant examples.

4.5.1 Alkanes (Paraffins)

^{1}H-NMR Spectral Characteristics of Alkanes

In the ^{1}H-NMR spectrum of alkanes, the hydrogens of methyl, methylene and methine group of alkanes shows NMR signals on a δ value scale, in upfield region between 1–2 ppm. Some characteristic chemical shifts and J values of ^{1}H-NMR values observed for alkanes are as follows:

- **R–CH_3**—Exhibits sharp and tall peak between 0.75–1.3 (^3J = 7–8 Hz) ppm.
- **R–CH_2–R**—Exhibits overlapped peaks in long chain alkanes between 1.2–1.5 (^3J = 7–8 Hz) ppm.
- **R_3–CH**–Exhibits overlapped peaks in long chain alkanes between 1.5–2.0 (^3J = 7–8 Hz) ppm.
- **The order of shielding of hydrogens in methyl, methylene and methine groups relative to TMS is R–CH_3 > R–CH_2–R > R_3–CH, such that methyl group hydrogens are most shielded or appears upfield.**
- **Alkanes commonly exhibits ^3J couplings.**

Example: *Ethane* (^{1}H-NMR spectrum given in Figure 4.20).

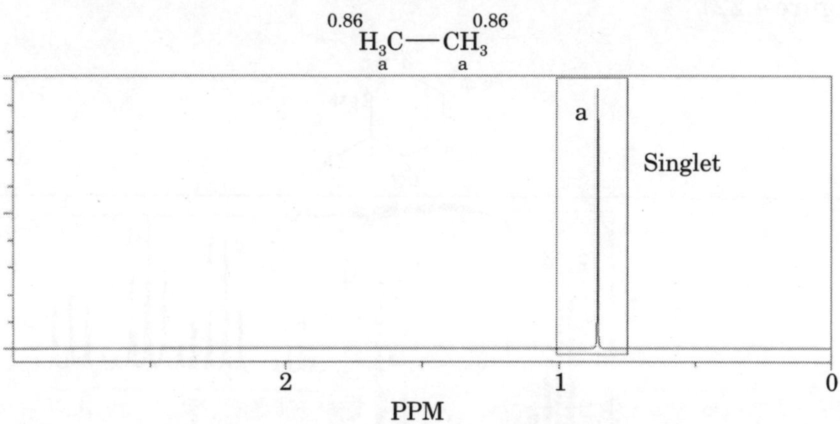

Figure 4.20: ^{1}H-NMR spectrum of ethane.

Interpretation

^{1}H-NMR δ (ppm): 0.86 (Singlet, 6H of CH_3).

Example: *2-Methyl Butane* (^{1}H-NMR spectrum given in Figure 4.21).

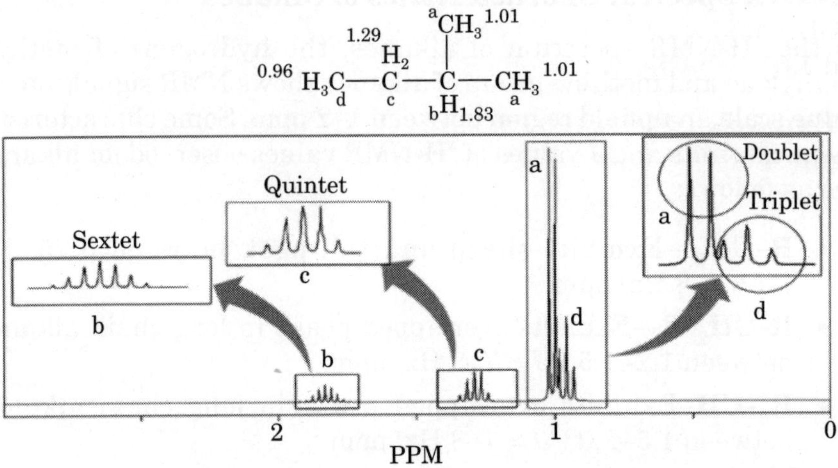

Figure 4.21: ^{1}H-NMR spectrum of 2-methylbutane.

Interpretation

^{1}H-NMR δ (ppm): 0.96 (triplet, 3H of CH_3–d), 1.01 (doublet, 6H of CH_3–a), 1.29 (quintet, 2H of CH_2–c), and 1.83 (sextet, 1H of CH–b).

Example: *Methyl Cyclohexane* (^{1}H-NMR spectrum given in Figure 4.22).

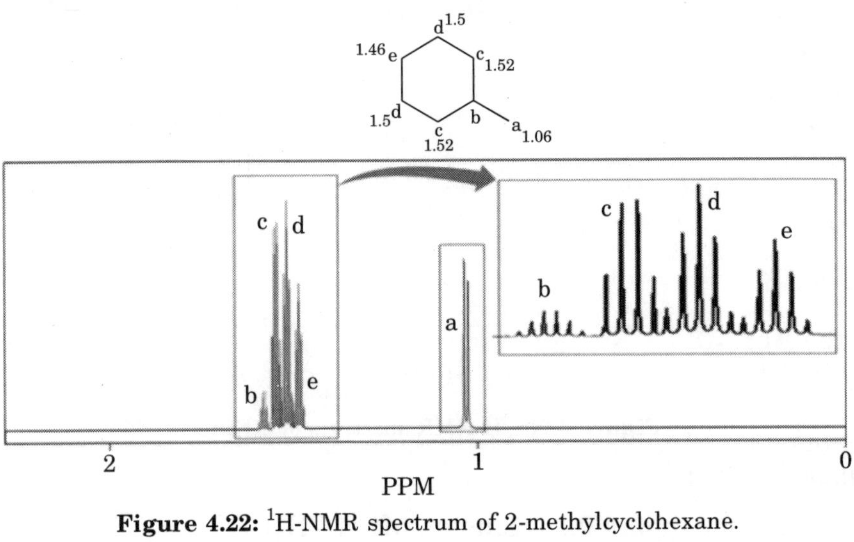

Figure 4.22: ^{1}H-NMR spectrum of 2-methylcyclohexane.

Interpretation

¹H-NMR δ (ppm): 1.06 (doublet, 3H of CH_3–a), 1.46 (quintet, 2H of CH_2–e), 1.5 (quintet, 4H of CH_2–d), 1.52 (quartet, 4H of CH_2–c), and 1.61 (octet, 1H of CH–b).

Example: *2-Methyl Propane* (¹H-NMR spectrum given in Figure 4.23).

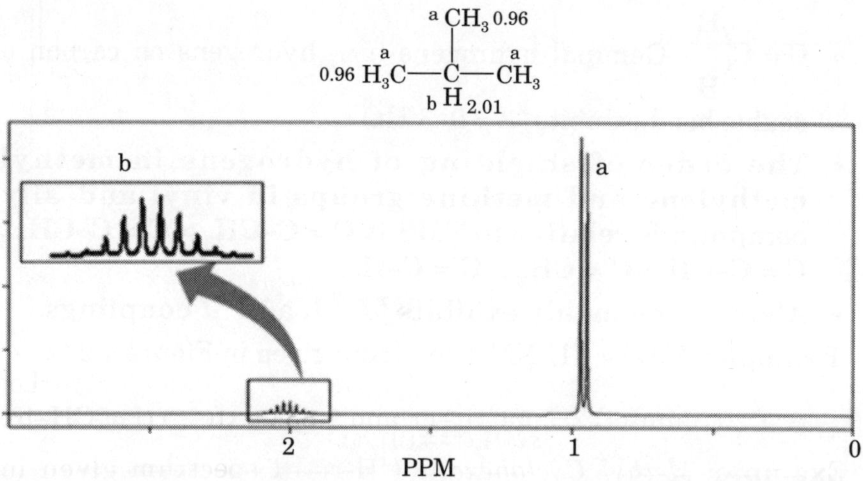

Figure 4.23: ¹H-NMR spectrum of 2-methylpropane.

Interpretation

¹H-NMR d (ppm): 0.96 (doublet, 9H of CH_3) and 2.01 (multiplet, 1H of CH).

4.5.2 Alkenes (Olefins)

¹H-NMR Spectral Characteristics of Alkenes

In the ¹H-NMR spectrum of alkenes, the allylic and vinylic hydrogens shows NMR signal in downfield region (due to deshielding caused by double bond) on a δ value scale. Some characteristic chemical shifts and *J* values of ¹H-NMR values observed for alkenes are as follows:

- **C = C–H**—Vinyl hydrogen (*i.e.*, hydrogen on a double bond) exhibits peak between 4.5–6.5 (^3J = 11–18 Hz for trans H–C = C–H, and ^3J = 6–15 Hz for cis H–C = C–H) ppm. This deshielding is due to anisotropy of alkenyl bond.
- **C = C–C–H**—Allylic hydrogens (*i.e.*, hydrogens on carbon adjacent to the double bond) exhibits a peak between 1.6–2.5 (^4J = 0–3 Hz for H–C = C–C–H) ppm. This smaller deshielding to nearness to alkenyl bond.
- $\mathbf{C = C{<}^{H}_{H}}$ —Geminal hydrogens (*i.e.*, hydrogens on carbon of double bond exhibits ^2J = 0–3 Hz).
- **The order of shielding of hydrogens in methyl, methylene and methine groups in vinyl and allyl compounds relative to TMS is C = C–CH$_3$ > C = C–CH$_2$ > C = C–CH > C = CH$_2$ > C = C–H.**
- **Alkenes commonly exhibits ^2J, ^3J, and ^4J couplings.**

Example: *Ethene* (^{1}H-NMR spectrum given in Figure 4.24).

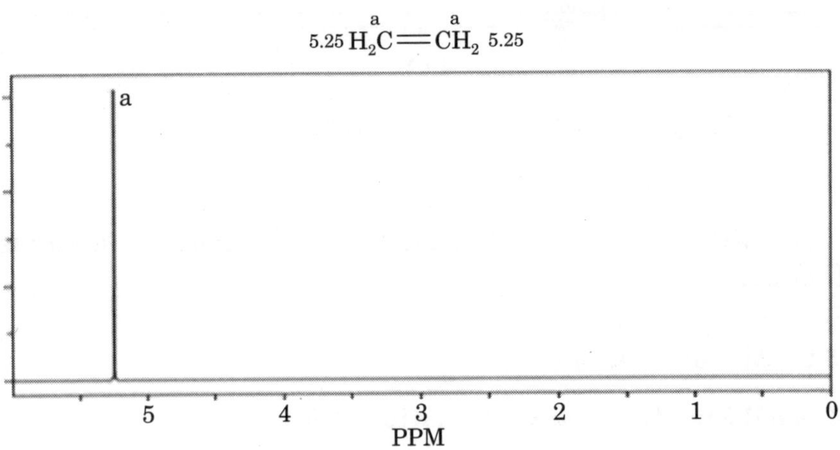

Figure 4.24: ^{1}H-NMR spectrum of ethene.

Interpretation

^{1}H-NMR δ (ppm): 5.25 (singlet, 4H of CH$_2$).

Example: *Cyclohexene* (¹H-NMR spectrum given in Figure 4.25).

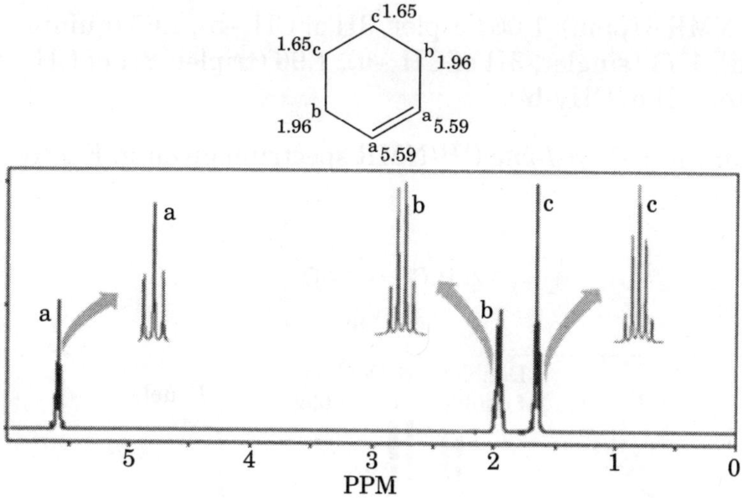

Figure 4.25: ¹H-NMR spectrum of cyclohexene.

Interpretation

¹H-NMR δ (ppm): 1.65 (quintet, 4H of CH_2–c), 1.96 (quartet, 4H of CH_2–b), 5.59 (triplet, 2H of = CH–a).

Example: *1-Ethyl-2-methylcyclohex-1-ene* (¹H-NMR spectrum given in Figure 4.26).

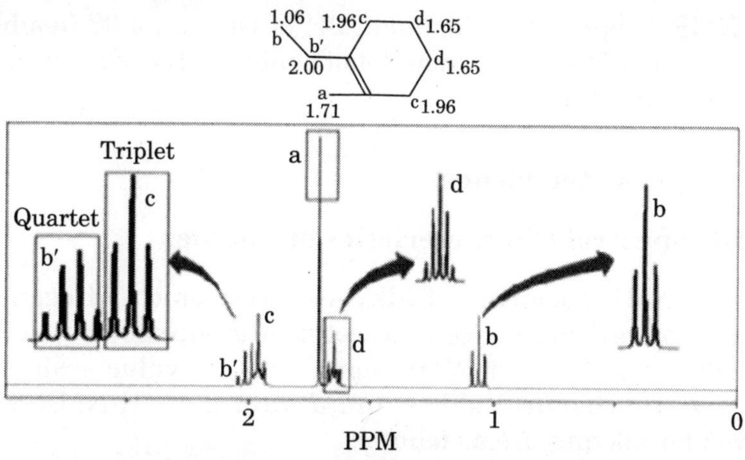

Figure 4.26: ¹H-NMR spectrum of 1-ethyl-2-methylcyclohex-1-ene.

Interpretation

^{1}H-NMR δ (ppm): 1.06 (triplet, 3H of CH_3–b), 1.65 (quintet, 2H of CH_2–d), 1.71 (singlet, 3H of CH_3–a), 1.96 (triplet, 2H of CH_2–c), 2.0 (quartet, 2H of CH_2–b′).

Example: *Prop-1-ene* (^{1}H-NMR spectrum given in Figure 4.27).

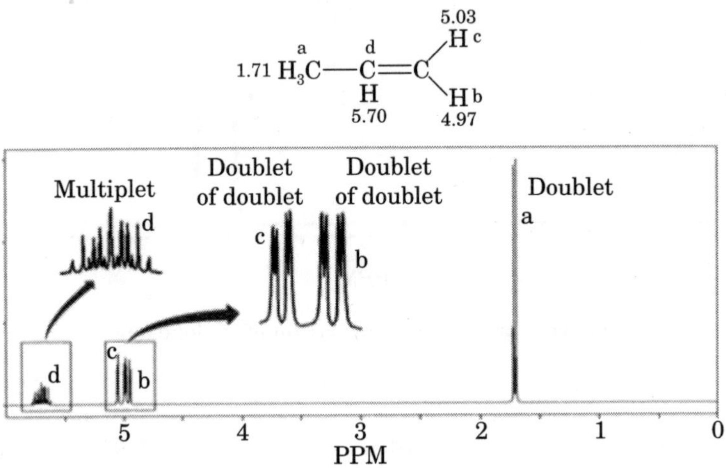

Figure 4.27. ^{1}H-NMR spectrum of prop-1-ene.

Interpretation

^{1}H-NMR δ (ppm): 1.71 (doublet, 3H of CH_3–a), 4.97 (doublet of doublet, 1H of CH–b), 5.03 (doublet of doublet, 1H of CH–c), and 5.7 (multiplet, 1H of CH–d).

4.5.3 Alkynes (Acetylenes)

^{1}H-NMR Spectral Characteristics of Alkynes

In the ^{1}H-NMR spectrum of alkynes, acetylenic hydrogens and hydrogens on carbons adjacent to acetylene group shows upfield and downfield respectively of NMR signals on a δ value scale. Some characteristic chemical shifts and *J* values of ^{1}H-NMR values observed for alkynes are as follows:

- **C ≡ CH**—Acetylenic hydrogen (*i.e.*, hydrogen on a triple bond) exhibits peak between 1.7–2.7 ppm. These hydrogens experiences shielding due to anisotropy of acetylenic bond.
- **C ≡ C–CH**—Allylic Hydrogens on carbon adjacent to the triple bond exhibits a peak between 1.6–2.5 (^4J = 2–3 Hz for H–C ≡ C–C–H) ppm. These hydrogens also experiences shielding due to anisotropy of acetylenic bond.
- **The order of shielding of hydrogens in methyl, methylene and methine groups in acetylenic compounds relative to TMS is C ≡ C–CH_3 > C ≡ C–CH_2 > C ≡ C–CH > C ≡ CH > C ≡ C–CH_2–C ≡ C.**
- **Alkynes commonly exhibits ^4J couplings.**

Example: *Hexa-1, 4-diyne* (^{1}H-NMR spectrum given in Figure 4.28).

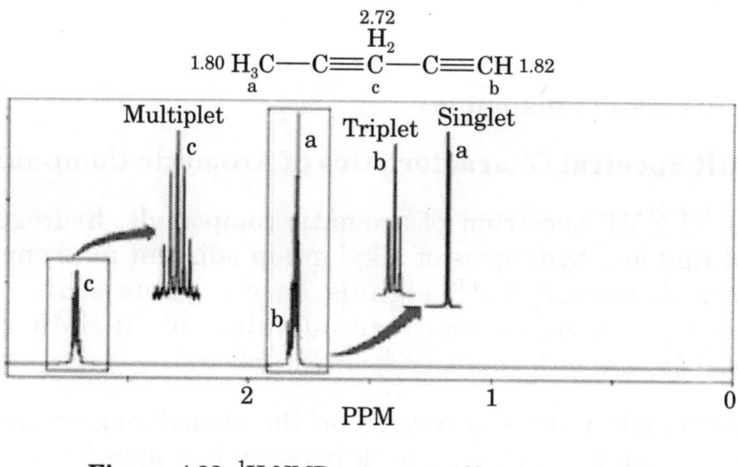

Figure 4.28: ^{1}H-NMR spectrum of hexa-1, 4-diyne.

Interpretation

^{1}H-NMR δ (ppm): 1.80 (Singlet, for 3H of CH_3–a), 1.82 (triplet due to long range coupling of $^4J_{H-H}$, for 1H of CH–b), 2.72 (multiplet, for 2H of CH_2–c).

Example: *Pent-1-yne* (^{1}H-NMR spectrum given in Figure 4.29).

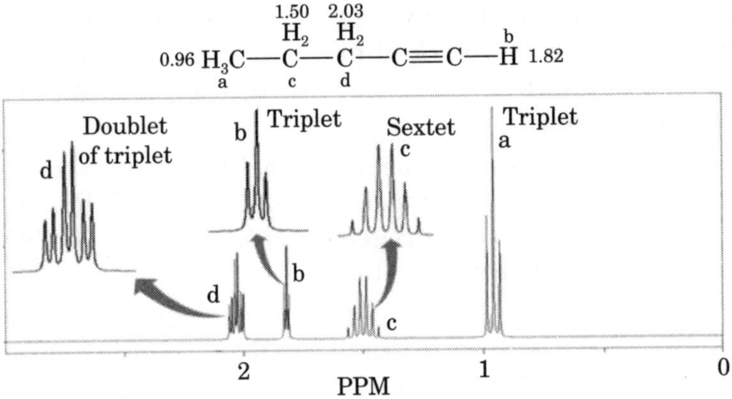

Figure 4.29: ^{1}H-NMR spectrum of pent-1-yne.

Interpretation

^{1}H-NMR δ (ppm): 0.96 (triplet, 1H of CH_3–a), 1.5 (Sextet, 2H of CH_2–c), 1.82 (triplet, 1H of CH–b), and 2.03 (doublet of triplet, 2H of CH_2–d).

4.5.4 Aromatic Compounds

^{1}H-NMR Spectral Characteristics of Aromatic Compounds

In the ^{1}H-NMR spectrum of aromatic compounds, hydrogens of phenyl ring and hydrogens of alkyl group adjacent to phenyl ring exhibits downfield NMR signals on a δ value scale. Some characteristic chemical shifts and *J* values of ^{1}H-NMR values observed for aromatic compounds are as follows:

- **Aromatic ring**—Hydrogens on the phenyl ring or aromatic ring exhibits *hydrogens* peak between 6–8 ppm. This greater deshielding is due to anisotropy of aromatic ring π electrons.
- **Ar–CH**—Hydrogens on the carbon adjacent to the aromatic ring exhibits a peak between 2.3–2.7 ppm. This deshielding is due to nearness to aromatic ring π electrons.
- **Ring hydrogens commonly exhibits ortho couplings (^3J between 7–10 Hz), meta couplings (^4J between 2–3 Hz), and para couplings (^5J between 0–1 Hz).**

- **The order of shielding of hydrogens in methyl, methylene and methine groups in phenyl compounds relative to TMS is Ar–CH_3 > Ar–CH_2 > Ar–CH.**

Example: *2-Ethyl-1-isopropyl-4-methylbenzene* (^{1}H-NMR spectrum given in Figure 4.30).

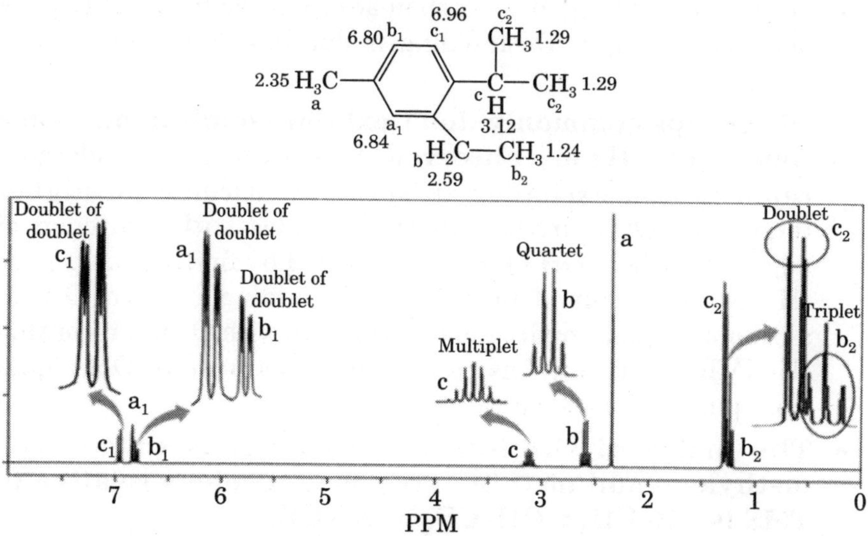

Figure 4.30: ^{1}H-NMR spectrum of 2-ethyl-1-isopropyl-4-methylbenzene.

Interpretation

^{1}H-NMR δ (ppm): 1.24 (triplet, 3H of CH_3–b_2), 1.29 (doublet, 6H of two CH_3–c_2), 2.35 (singlet, 3H of CH_3–a), 2.59 (quartet, 2H of CH_2–b), 3.12 (multiplet, 1H of CH–c), 6.8 (^3J & ^5J aroused doublet of doublet, 1H, Ar–H–b_1), 6.84 (^4J & ^5J aroused doublet of doublet, 1H, of Ar–H–a_1), and 6.96 (^3J & ^5J aroused doublet of doublet, 1H, of Ar–H–c_1).

4.5.5 Alcohols

^{1}H-NMR Spectral Characteristics of Alcohols

In the ^{1}H-NMR spectrum of alcohols, hydrogens of –OH group exhibits downfield NMR signals on a δ value scale. Some characteristic chemical shifts and *J* values of ^{1}H-NMR values observed for alcohols are as follows:

- **OH**—Hydrogen of OH exhibits, broad and highly variable signal between 0.5–5 ppm. This signal variability and broadening depends upon solvent, temperature and concentration.
- **H–O–CH**—Hydrogens of carbon adjacent to OH exhibits peak between 3–4 ppm. Deshielding is due to electronegativity of oxygen.
- **OH groups commonly don't exhibit coupling but some time ^{3}J of 5 Hz are observed.** It is because OH undergoes rapid proton exchange in solution. Alcohol on higher temperature/impure/moisturized couples with hydrogen of adjacent carbon. OH group is identified by firstly taking pure ^{1}H-NMR spectrum of alcohol. Nextly, adding a drop of D_2O in alcohol to replace hydrogen of OH group with D, and recording ^{1}H-NMR spectrum, this causes disappearance of OH signal from pure alcohol spectra.
- **The order of shielding of hydrogens in methyl, methylene and methine groups in alcohols relative to TMS is OH–CH_3 > OH–CH_2 > OH–CH.**

Example: *2-Methylpropan-1-ol* (^{1}H-NMR spectrum given in Figure 4.31).

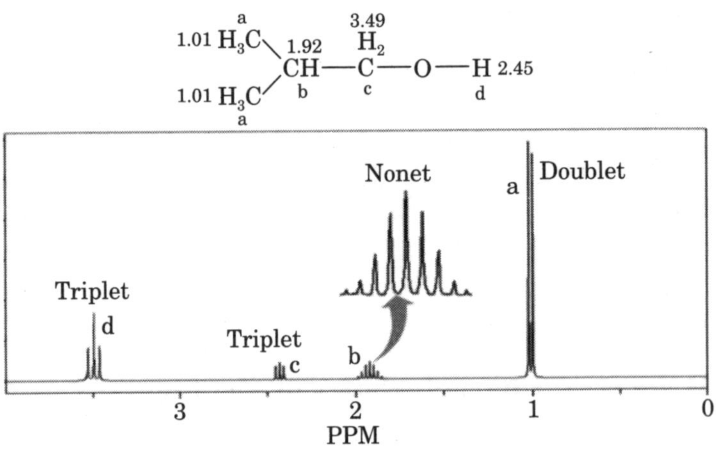

Figure 4.31: ^{1}H-NMR spectrum of 2-methylpropan-1-ol.

Interpretation

¹H-NMR δ (ppm): 1.01 (doublet, 6H of CH_3), 1.92 (nonet, 1H of CH), 2.45 (triplet, 1H of OH confirmed by D_2O treatment), 3.49 (triplet, 2H of CH_2).

4.5.6 Ethers

¹H-NMR Spectral Characteristics of Ethers

In the ¹H -NMR spectrum of ethers, the hydrogens on carbons adjacent to O atom exhibits a downfield NMR signal on a δ value scale. Some characteristic chemical shifts and *J* values of ¹H -NMR values observed for ethers are as follows:

- **O–CH**–Hydrogens of carbons adjacent to O atom exhibits peak between 3.2–3.8 ppm. This deshielding is due to electronegativity of oxygen.
- **The order of shielding of hydrogens in methyl, methylene and methine groups in ethers relative to TMS is O–CH_3 > O–CH_2 > O–CH.**

Example: *Methoxy ethane* (¹H-NMR spectrum given in Figure 4.32).

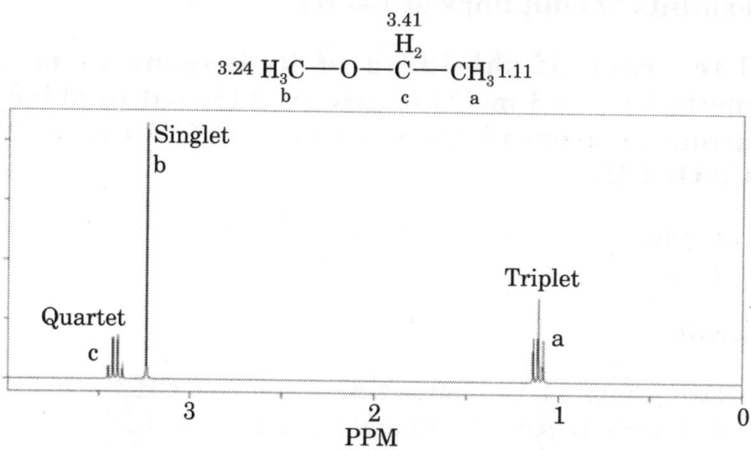

Figure 4.32: ¹H-NMR spectrum of methoxy ethane.

Interpretation

^{1}H-NMR δ (ppm): 1.11 (triplet, 3H of OCH_3), 3.24 (singlet, 3H of CH_3), and 3.41 (quartet, 2H of CH_2).

4.5.7 Aldehydes

^{1}H -NMR Spectral Characteristics of Aldehydes

In the ^{1}H -NMR spectrum of aldehydes, the hydrogen of aldehyde group and hydrogens of α-carbon adjacent to C = O group exhibits a downfield NMR signal on a δ value scale. Some characteristic chemical shifts and *J* values of ^{1}H-NMR values observed for aldehydes are as follows:

- **R–C(=O)–H**–Hydrogen of aldehyde group exhibits peak between 9–10 ppm. This greater deshielding is due to anisotropy of C = O group.
- **CH–C(=O)–H**–Hydrogens of α-carbon adjacent to C = O group exhibits peak between 2–2.5 ppm. Smaller deshielding is due to nearness to C = O group.
- **Aldehydic hydrogen and adjacent α-carbon commonly exhibits ^3J couplings of 1–3 Hz.**
- **The order of shielding of hydrogens in methyl, methylene and methine groups adjacent to aldehydes group relative to TMS is C(=O)–CH_3 > C(=O)–CH_2 > C(=O)–CH.**

Example: *Isobutyraldehyde* (^{1}H-NMR spectrum given in Figure 4.33).

Interpretation

^{1}H-NMR δ (ppm): 1.12 (doublet, 6H of geminal CH_3), 2.70 (doublet of septet, 1H of CH), and 9.72 [doublet, 1H of C(=O)H].

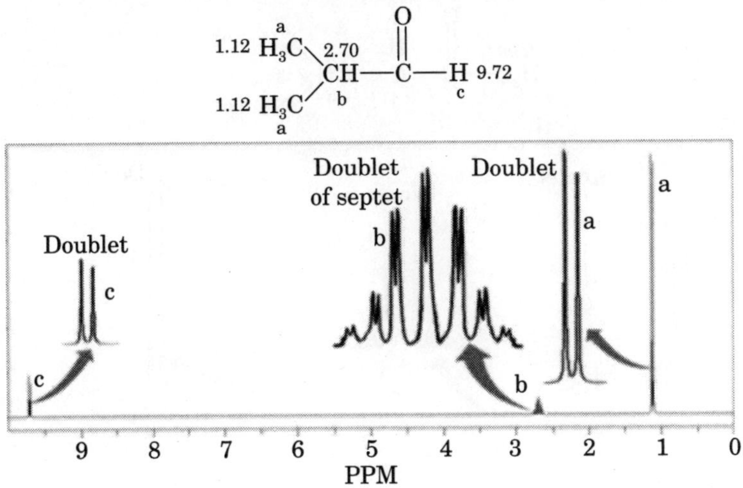

Figure 4.33: ¹H-NMR spectrum of isobutyraldehyde.

4.5.8 Ketones

¹H -NMR Spectral Characteristics of Ketones

In the ¹H -NMR spectrum of ketones, hydrogens of α-carbon adjacent to C = O group exhibits a downfield NMR signal on a δ value scale. Some characteristic chemical shifts and *J* values of ¹H -NMR values observed for ketones are as follows:

- **–C(=O)–CH**–Hydrogens of α-carbon adjacent to C = O group exhibits peak between 2–2.5 ppm. This deshielding is due to anisotropy of C = O group.
- **The order of shielding of hydrogens in methyl, methylene and methine groups in ketone relative to TMS is C(=O)–CH_3 > C(=O)–CH_2 > C(=O)–CH.**

Example: *4-Methylpentan-2-One* (¹H-NMR spectrum given in Figure 4.34).

Interpretation

¹H-NMR δ (ppm): 1.01 (doublet, 6H of geminal CH_3), 2.06 (nonet, 1H of CH), 2.09 (singlet, 3H of CH_3), 2.41 (doublet, 2H of CH_2).

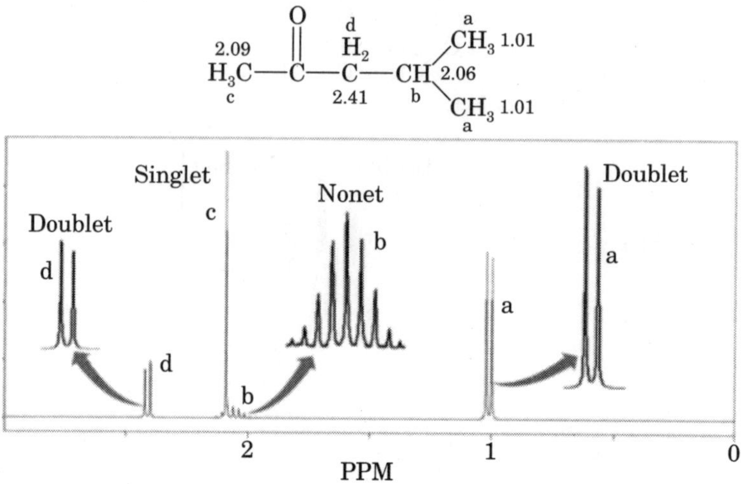

Figure 4.34: ^{1}H-NMR spectrum of 4-methylpentan-2-one.

4.5.9 Carboxylic Acids

^{1}H -NMR Spectral Characteristics of Carboxylic Acids

In the ^{1}H -NMR spectrum of carboxylic acids, hydrogens of α-carbon adjacent to C = O group exhibits a downfield NMR signal on a δ value scale. Some characteristic chemical shifts and *J* values of ^{1}H - NMR values observed for carboxylic acids are as follows:

- **C(=O)–OH**—Hydrogens of hydroxyl of carboxylic acid group exhibits peak between 11–12 ppm. This greater deshielding is due to electro negativity and anisotropy of oxygen and C = O groups.
- **HOOC–CH**—Hydrogens of α-carbon adjacent to COOH group exhibits deshielded peak between 2–2.5 ppm. is due to nearness to COOH group.
- **The order of shielding of hydrogens in methyl, methylene and methine groups in carboxylic acids relative to TMS is HOOC–CH_3 > HOOC–CH_2 > HOOC–CH.**

Example: *Propionic Acid* (^{1}H-NMR spectrum given in Figure 4.35).

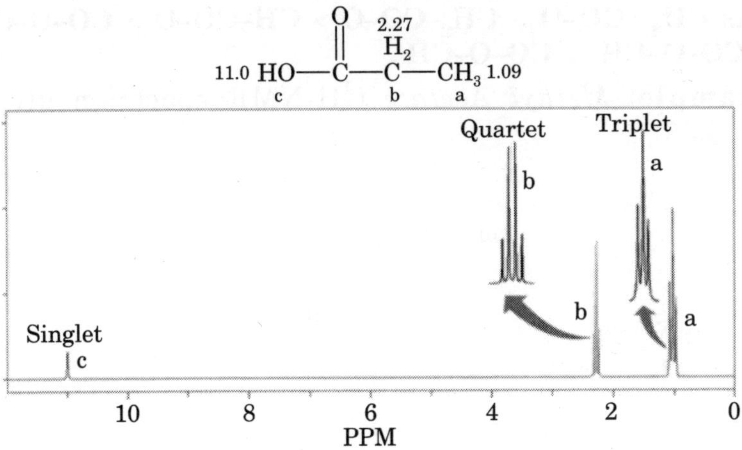

Figure 4.35: ¹H-NMR spectrum of Propionic acid.

Interpretation

¹H-NMR δ (ppm): 1.09 (triplet, 3H of CH_3), 2.27 (quartet, 2H of CH_2), and 11.0 (singlet, 1H of OH).

4.5.10 Esters

¹H -NMR Spectral Characteristics of Esters

In the ¹H -NMR spectrum of esters, hydrogens of α-carbon adjacent to C = O group and O atom exhibits a smaller and larger deshielded NMR signal respectively on a δ value scale. Some characteristic chemical shifts and *J* values of ¹H -NMR values observed for esters are as follows:

- **–CO–O–CH**—Hydrogens of α-carbon adjacent to oxygen atom of ester group exhibits peak between 3.5–5 ppm. This greater deshielding is a attributed to electronegative effect of adjacent oxygen atom.
- **CH–CO–O**—Hydrogens of α-carbon adjacent to carbonyl group of ester exhibits peak between 2–2.5 ppm. This deshielding is attributed to anisotropic effect of C = O group.
- **The order of shielding of hydrogens in methyl, methylene and methine groups in esters relative to TMS**

is CH_3 –CO–O > CH_2–CO–O > CH–CO–O > CO–O–CH_3 > CO–O–CH_2 > CO–O–CH .

Example: *Methyl Acetate* (^{1}H-NMR spectrum given in Figure 4.36).

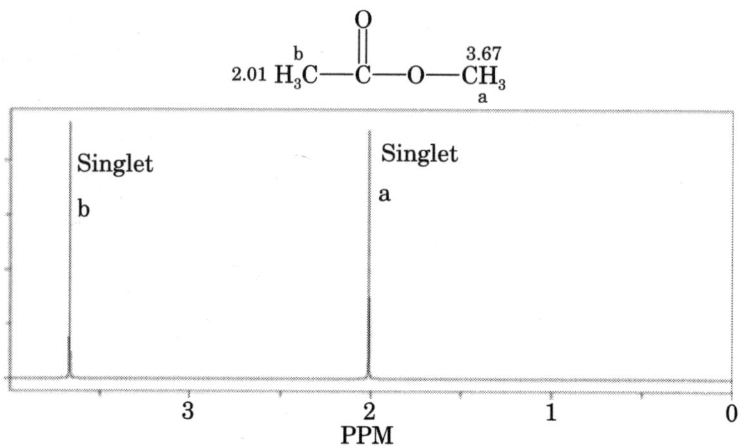

Figure 4.36: ^{1}H-NMR spectrum of methyl acetate.

Interpretation

^{1}H-NMR δ (ppm): 2.01 (singlet, 3H of CH_3CO), and 3.67 (singlet, 3H of OCH_3).

4.5.11 Amides

^{1}H -NMR Spectral Characteristics of Amides

In the ^{1}H-NMR spectrum of amides, hydrogens of α-carbon adjacent to C = O group of amide, NH group of amide and α-carbon adjacent to NH group of amide, exhibits a smaller and larger deshielded NMR signal respectively on a δ value scale. Some characteristic chemical shifts and *J* values of ^{1}H-NMR values observed for amides are as follows:

- **–CO–NH–CH**—Hydrogens of α-carbon adjacent to amino group of amides exhibits peak between 2–3 ppm. This greater deshielding is attributed to electronegative effect of adjacent nitrogen atom.

- **CH–CO–N—**Hydrogens of α-carbon adjacent to carbonyl group of amides exhibits peak between 2–2.5 ppm. This deshielding is attributed to anisotropic effect of C = O group.
- **-CO–NH**—Hydrogens of amino group of amides exhibits highly variable signal between 5–9 ppm. This signal variability depends upon solvent, temperature and concentration. In case of NH_2 two peaks appear.
- **The couplings that are commonly observed in amides are ^1J of 50 Hz with N–H, ^3J of 0–7 Hz with -NH–CH–, and ^2J of 0 Hz through N–CH– groups.**
- **The order of shielding of hydrogens in methyl, methylene and methine groups in amides relative to TMS is CH_3–CO–N > CH_2–CO–N > CH–CO–N > CO–N–CH_3 > CO–N–CH_2 > CO–N–CH.**
- Hydrogen of OH exhibits, broad and highly variable signal between 0.5–5 ppm. This signal variability and broadening depends upon solvent, temperature and concentration.

Example: *Propionamide* (^{1}H-NMR spectrum given in Figure 4.37).

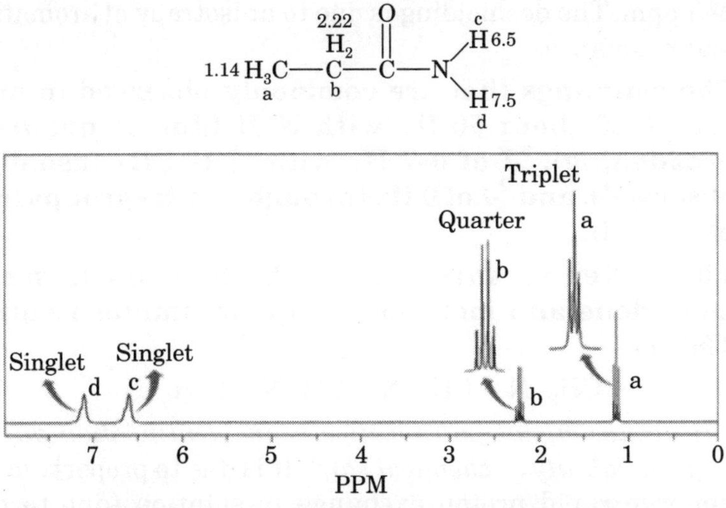

Figure 4.37: ^{1}H-NMR spectrum of propionamide.

Interpretation

^{1}H-NMR δ (ppm): 1.14 (triplet, 3H of CH_3–a), 2.22 (quartet, 2H of CH_2–b), 6.5 (singlet, 1H of NH–c) and 7.5 (singlet, 1H of NH–d).

4.5.12 Amines

^{1}H-NMR Spectral Characteristics of Amines

In ^{1}H-NMR spectrum of amines, hydrogens of amino group and α-carbon adjacent to nitrogen atom of amine exhibits a smaller and larger deshielded NMR signal respectively on a δ value scale. Some characteristic chemical shifts and *J* values of ^{1}H-NMR values observed for amines are as follows:

- **-N–CH**—Hydrogens of α-carbon adjacent to amino group exhibits peak between 2–3 ppm. Deshielding is due to electronegative nitrogen.
- **-C–NH**—Hydrogens of amino group exhibits variable signal between 0.5–5 ppm. This signal variability depends upon solvent, temperature, acidity and concentration. In case of NH_2 two peaks appear.
- **Ar–NH**—Hydrogens of amino group exhibits signal between 3–5 ppm. The deshielding is due to anisotropy of aromatic ring and resonance.
- **The couplings that are commonly observed in amines are ^1J of about 50 Hz with N–H (due to quadrupole broadening), ^3J of 0–7 Hz with -NH–CH- (usually not observed), and ^2J of 0 Hz through N–CH- groups (rarely observed).**
- **The order of shielding of hydrogens in methyl, methylene and methine groups in amines relative to TMS is**

$$CH_3–N > CH_2–N > CH–N > C_6H_5–N$$

- *NH groups signal poses difficulty in identification due to its high variability in chemical shift.* It is due to property of NH to undergo rapid proton exchange in solution (due to higher temperature, acidity, hydrogen bonding and solvent) and quadrupole broadening. Same as OH group, NH group identification is also done using D_2O to make the NH group to disappear thereby confirm NH group presence.

Example: *Ethylamine* (^{1}H-NMR spectrum given in Figure 4.38).

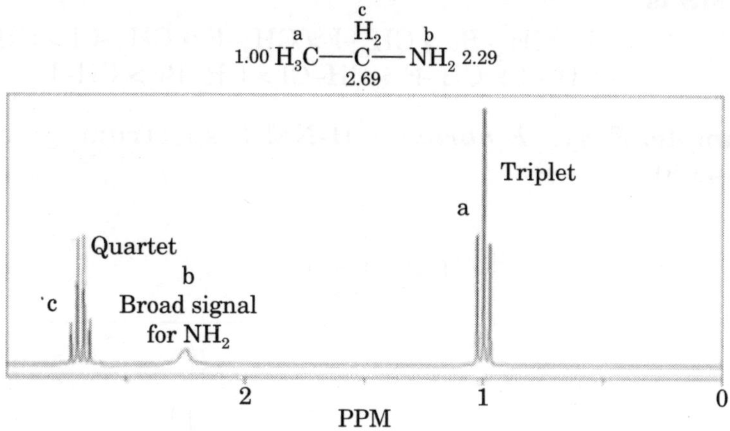

Figure 4.38: ^{1}H-NMR spectrum of ethylamine.

Interpretation

^{1}H-NMR δ (ppm): 1.0 (triplet, 3H of CH_3), 2.69 (quartet, 2H of CH_2), and 2.0 (broad signal, 2H of NH).

4.5.13 Alkyl Halides

^{1}H-NMR Spectral Characteristics of Alkyl Halides

In the ^{1}H-NMR spectrum of alkyl halides, hydrogens of same carbon of halogen atom of alkyl halide exhibits deshielded NMR signal on a δ value scale. Some characteristic chemical shifts and J values of ^{1}H-NMR values observed for alkyl halides are as follows:

- **X–CH**—Hydrogens on same carbon of halogen such as I, Br, Cl, and F, exhibits signals at 2–4, 2.7–4.1, 3–4.1, and 4.2–5 ppm respectively. Deshielding is due to electronegativity of halogen atom.
- **Generally in alkyl halides I, Cl, and Br do not exhibit coupling. But CH–F and CH–CF exhibits ^2J of about 50**

Hz on same carbon and ^3J of 20 Hz on different carbons respectively.

- **The order of shielding of hydrogens in methyl, methylene and methine groups in halides relative to TMS is**

$$CH_3\text{–}F > CH_3\text{–}Cl > CH_3\text{–}Br > CH_3\text{–}I > CH_2\text{–}F > CH_2\text{–}Cl > CH_2\text{–}Br > CH_2\text{–}I > CH\text{–}F > CH\text{–}Cl > CH\text{–}Br > CH\text{–}I$$

Example: *Ethyl Fluoride* (^{1}H-NMR spectrum given in Figure 4.39).

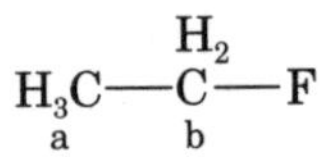

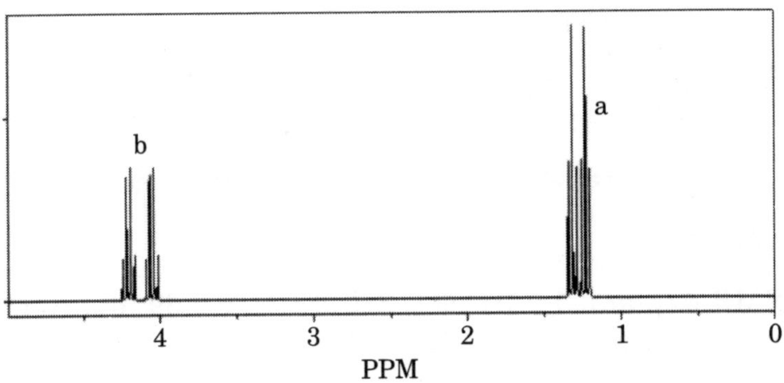

Figure 4.39: ^{1}H-NMR spectrum of ethyl fluoride.

Interpretation

^{1}H-NMR δ (ppm): 1.27 (multiplet, 3H of CH_3), 4.13 (multiplet, 2H of CH_2).

Example: *Ethyl Chloride* (^{1}H-NMR spectrum given in Figure 4.40).

Interpretation

^{1}H-NMR δ (ppm): 1.49 (triplet, 3H of CH_3), 3.42 (Quartet, 2H of CH_2).

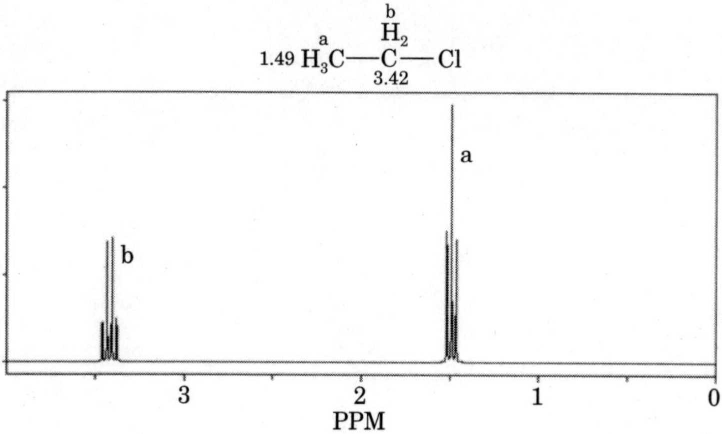

Figure 4.40: ^{1}H-NMR spectrum of ethyl chloride.

Example: *Ethyl Bromide* (^{1}H-NMR spectrum given in Figure 4.41).

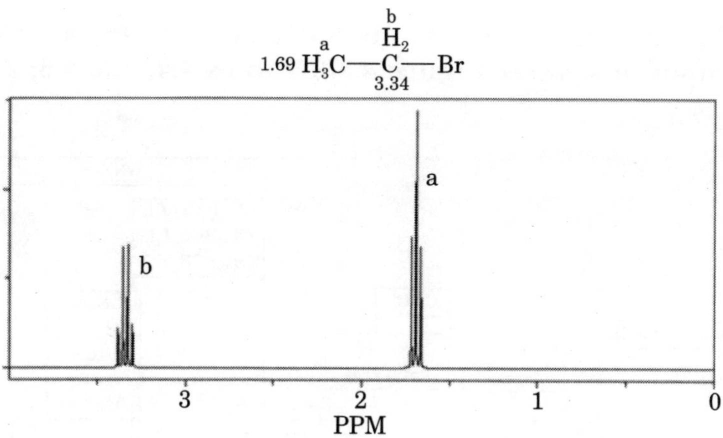

Figure 4.41: ^{1}H-NMR spectrum of ethyl bromide.

Interpretation

^{1}H-NMR δ (ppm): 1.69 (triplet, 3H of CH_3), 3.34 (Quartet, 2H of CH_2).

Example: *Ethyl Iodide* (^{1}H-NMR spectrum given in Figure 4.42).

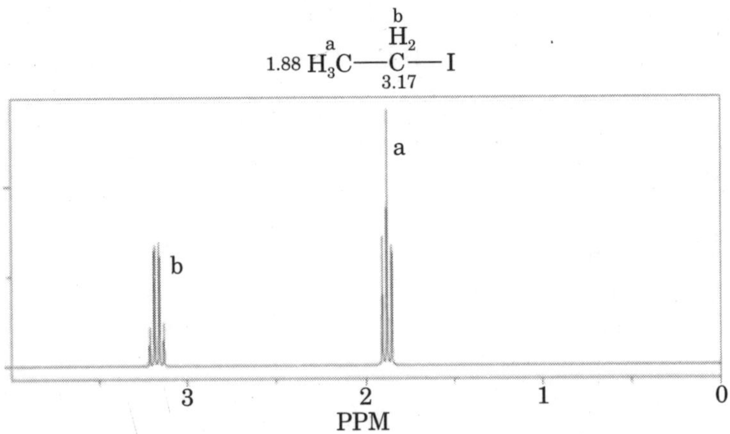

Figure 4.42: ^{1}H-NMR spectrum of ethyl iodide.

Interpretation

^{1}H-NMR δ (ppm): 1.88 (triplet, 3H of CH_3), 3.17 (Quartet, 2H of CH_2).

For Further Assistance in ^{1}H-NMR Spectra Interpretation of Compounds Refer Figure 4.43, Tables 4.4, 4.5, 4.6, 4.7, 4.8, and 4.9.

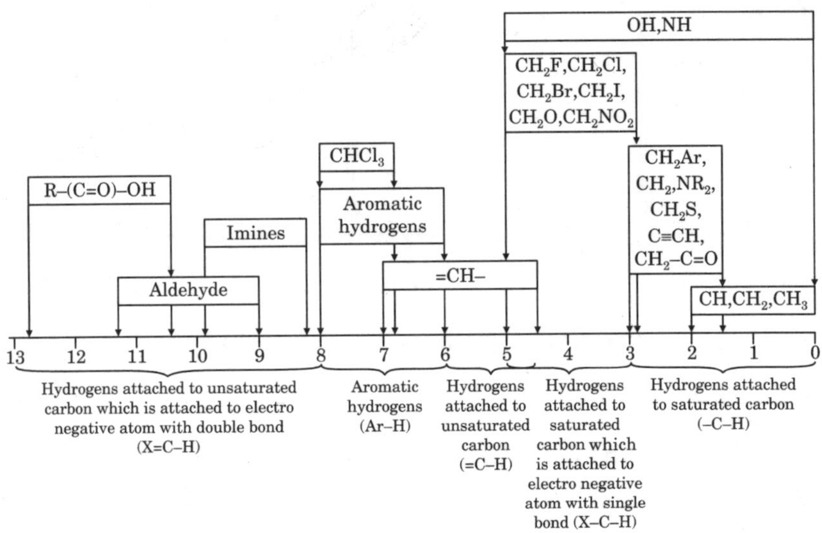

Figure 4.43: Chart representing ^{1}H chemical shift due to different chemical environments.

Table 4.4: Chemical shift of protons on alpha carbon atom to functional group.

	***Chemical shift* (δ-*value*)**		
Compounds	***αC = CH₃***	***αC = CH₂***	***αC = CH***
αC–CH_2–R	0.85	1.2	1.55
αC–C=C	1.6	2.05	2.6
αC–C	1.7	2.2	2.8
αC–Ph	2.25	2.6	2.85
αC–(C=O)–H	2.2	2.4	2.5
αC–(C=O)–R	2.1	2.3	2.65
αC–(C=O)–Ph	2.4	2.75	3.4
αC–(C=O)–OH	2.1	2.3	2.55
αC–(C=O)–OR	2.0	2.25	2.5
αC–(C=O)–N–	2.0	2.25	2.4
αC–CN	2.15	2.45	2.85
αC–NR_2	2.05	2.4	2.85
αC–NH_2	2.5	2.7	3.05
αC–N–Ph–R	2.6	3.05	3.6
αC–NR_3	2.95	3.05	3.6
αC–NH–(C=O)–R	2.85	3.35	3.8
αC–O–R	3.2	3.4	3.6
αC–O–H	3.2	3.4	3.8
αC–O–Ph	3.85	4.05	4.55
αC–O–(C=O)–R	3.65	4.1	4.95
αC–O–(C=O)–Ph	3.85	4.2	5.05
αC–I	2.15	3.15	4.25
αC–Br	2.7	3.4	4.1
αC–Cl	3.05	3.45	4.05
αC–F	4.25	4.45	4.8
αC–NO_2	4.05	4.15	4.4
αC–N=C	2.9	3.3	4.9
αC–SH	2.1	2.6	3.1
αC–SR	2.1	2.6	3.1
αC–SO–R	2.6	3.1	–
αC–SO_2–R	–	3.1	–
αC–SO_3–R	3	–	–

Table 4.5: Chemical shift of protons on beta carbon atom to functional groups.

	Chemical shift **(δ-value)**		
Compounds	$\beta C = CH_3$	$\beta C = CH_2$	$\beta C = CH$
βC–C–CH_2	0.85	1.2	1.55
βC–C–C=C	1	1.35	1.75
βC–C–C	1.2	1.5	1.8
βC–C–Ph	1.15	1.55	1.8
βC–C–(C=O)–R	1.05	1.55	1.95
βC–C–(C=O)–Ph	1.05	1.55	1.9
βC–C–(C=O)–OH	1.1	1.65	–
βC–C–(C=O)–OR	1.05	1.7	1.9
βC–C–(C=O)–N–	1.1	1.5	1.8
βC–C–CN	1.25	1.65	2
βC–C–NR_2	1.05	1.45	1.7
βC–C–N–Ph–R	1.1	1.5	1.8
βC–C–NR_3	1.4	1.75	2.05
βC–C–NH–(C=O)–R	1.1	1.5	1.9
βC–C–O–R	1.2	1.5	1.75
βC–C–OH	1.2	1.5	1.75
βC–C–O–Ph	1.3	1.55	2
βC–O–(C=O)–R	1.3	1.6	1.8
βC–O–(C=O)–Ph	1.65	1.75	1.95
βC–C–I	1.75	1.8	2.1
βC–C–Br	1.8	1.85	1.9
βC–C–Cl	1.55	1.5	1.95
βC–C–F	1.55	1.85	2.15
βC–C–NO_2	1.6	2.05	2.5
βC–C–SH	1.3	1.6	1.65
βC–C–SR	1.25	1.6	1.9

Table 4.6: Chemical shift of protons on mono substituted benzene rings.

	Chemical shift **(δ-value)**		
Compounds	***ortho***	***meta***	***para***
Benzene	7.25	7.25	7.25
CH_3	7.15	7.15	7.15
CH_3–CH_2	7.15	7.15	7.15
$(CH_3)_2CH$	7.15	7.15	7.15

Table 4.6: (*Contd...*)

Table 4.6: *(Contd...)*

Compounds	Chemical shift (δ-value)		
	ortho	***meta***	***para***
$(CH_3)_3C$	7.05	7.15	7.3
$CH_2=C$	7.25	7.25	7.25
CHC	7.25	7.45	7.45
CH_2OH	7.15	7.15	7.15
CH_2OR	7.25	7.25	7.25
$CH_2O(C=O)CH_3$	7.25	7.25	7.25
Phenyl	7.15	7.25	7.45
CF_3	7.5	7.5	7.5
CH_2–Cl	7.25	7.25	7.25
$CHCl_2$	7.35	7.35	7.35
CCl_3	7.5	8.1	8.1
F	7.25	6.95	7.05
Cl	7.25	7.25	7.25
Br	7.25	7.5	7.5
I	7.05	7.25	7.7
OH	7.05	6.75	6.9
OR	7.35	7	7.35
(C=O)H	7.5	7.95	7.65
$(C=O)CH_3$	7.55	7.9	7.9
(C=O)OH	7.35	7.9	7.45
(C=O)OR	7.45	8.15	7.55
CN	7.55	7.55	7.55
NH_2	7.15	6.5	6.85
$N(CH_3)_2$	7.15	6.7	7.15
NO_2	7.45	8.25	7.5
SR	7.25	7.25	7.25

Table 4.7: Chemical shift of protons on heteroatoms.

Protons	***Compounds***	***Chemical shift (δ-value)***
OH	Alcohols	0.5–4 (in DMSO 4.25–6.25)
	Phenols	4.5–7.5
	Phenols (intra molecular)	5.5–12.5
	Enols (Cyclic α-diketones)	6–7
	Enols (α-diketones)	14.5–16.5

Table 4.7: *(Contd...)*

Table 4.6: *(Contd...)*

Protons	*Compounds*	*Chemical shift* (δ-*value*)
	Carboxylic acids	10–13.5
	Sulphonic acids	10–12
	Water	4.5–5 (in DMSO 3.25–3.5 and in acetone 2.5–3)
NH_2	Alkyl and cyclic amines	0.5–3
and	Aryl Amines	3–5
NHR	Amides	5–8.75
	Urethanes	4.5–7.5
SH	Aliphatic	1.25–1.75
	Thiophenols	2.75–3.75

Table 4.8: Chemical shift of residual protons in commercially available deuterated solvents.

Solvent	*Chemical Shift*	*Multiplicity*
d_4-Acetic acid, (CD_3–COOD)	11.53 & 2.03	Singlet & Quintet
d_6-Acetone, (CD_3–CO–CD_3)	2.04	Quintet
d_3-Acetonitrile, (CD_3–CN)	1.93	Quintet
d_6-Benzene, (C_6D_6)	7.15	Broad
d_1-Chloroform, ($CDCl_3$)	7.26	Singlet
d_{12}-Cyclohexane, (C_6D_{12})	1.38	Broad
d_2-Deuterium Oxide, (D_2O)	4.63 & 4.67	Broad
d_{10}-Diethyl Ether, (C_2D_5–O–C_2D_5)	3.34 & 1.07	Multiplet
d_7-Dimethyl Formamide, $(CD_3)_2N(C{=}O)D$	8.01, 2.91 & 2.74	Broad, Quintet & Quintet
d_6-Dimethyl Sulfoxide, $(CD_3)_2S{=}O$	2.49	Quintet
d_8-Dioxane, ($C_4D_8O_2$)	3.53	Multiplet
d_6-Ethanol, (C_2D_5OD)	5.19, 3.55 & 1.11	Singlet, Broad & Multiplet
d_4-Methanol, (CD_3OD)	4.78 & 3.30	Singlet & Quintet
d_5-Pyridine, (C_5D_5N)	8.71 & 7.55	Broad

Table 4.9: Coupling constant value of *x* and *y* protons in different environments.

Proton Environments	***Coupling Constant (J_{xy})***
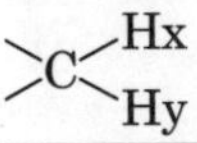	0–30 Hz (typical 12–15 Hz)
CH_x–CH_y	6–8 Hz (typical 7 Hz)
CH_x–C–CH_y	0–1 Hz (typical 0 Hz)
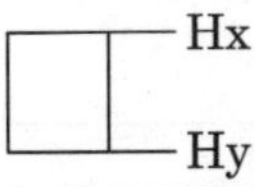	*cis* 4–12 *trans* 2–10 Hz
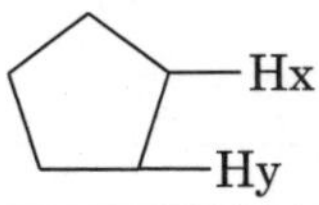	*cis* 5–10 *trans* 5–10 Hz
	Axial-Axial 6–14 Hz (typical 8–10 Hz) Axial-Equatorial 0–5 Hz (typical 2–3 Hz) Equatorial-Equatorial 0–5 Hz (typical 2–3 Hz)
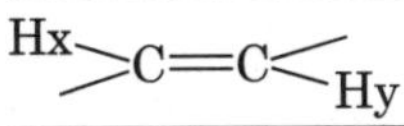	12–18 Hz (typical 17 Hz)
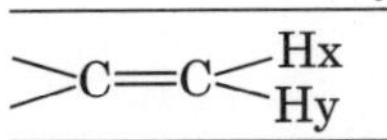	0–3 Hz (typical 0–2 Hz)
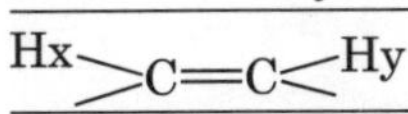	6–12 Hz (typical 10 Hz)
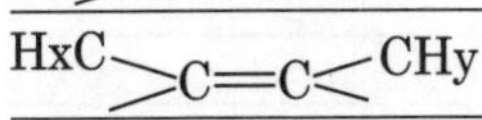	0–3 Hz (typical 1–2 Hz)
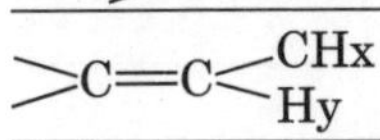	4–10 Hz (typical 7 Hz)
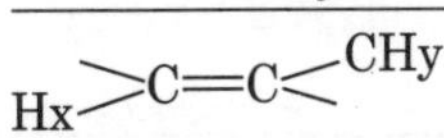	0–3 Hz (typical 1.5 Hz)
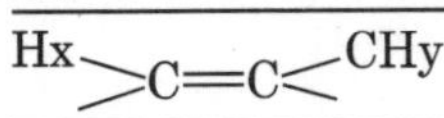	0–3 Hz (typical 2 Hz)
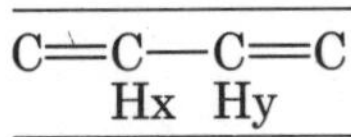	9–14 Hz (typical 10 Hz)
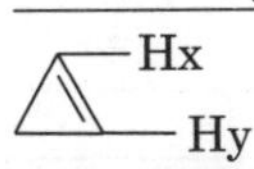	0.5–2 Hz
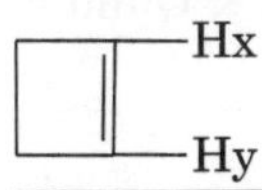	2.5–4 Hz

Table 4.9: (*Contd...*)

Table 4.9: *(Contd...)*

Proton Environments	***Coupling Constant*** **(J_{xy})**
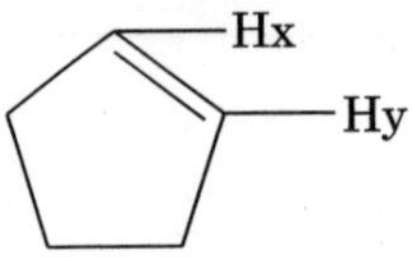	5–7 Hz
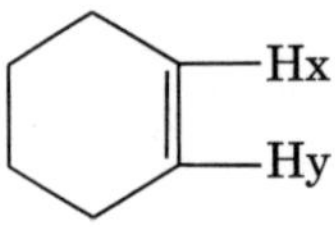	8.5–12 Hz
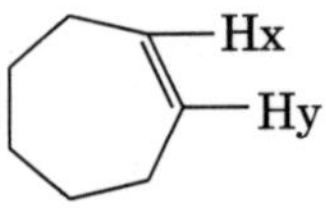	9–14 Hz
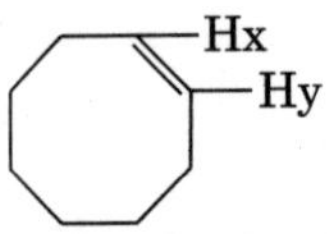	10–14 Hz
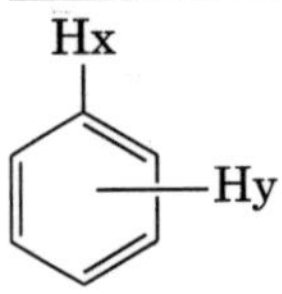	J_{ortho} 6–10 Hz (typical 9 Hz) J_{meta} 1–3 Hz (typical 3 Hz) J_{meta} 0–1 Hz (typical 0.5 Hz)
HxC—C ≡ CHy	2–3 Hz
HxC—C ≡ C—CHy	2–3 Hz
HxC—O—Hy	4–10 Hz (typical 5 Hz)
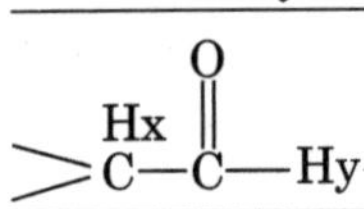	1–3 Hz (typical 2–3 Hz)
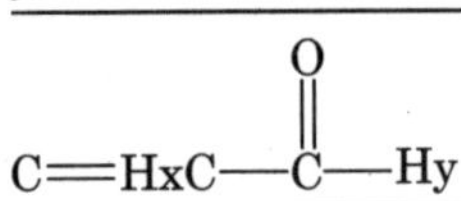	5–8 Hz (typical 6 Hz)
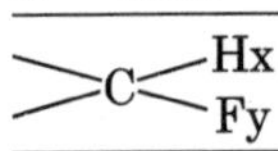	45–80 Hz
HxC—C—Fy	3–25 Hz
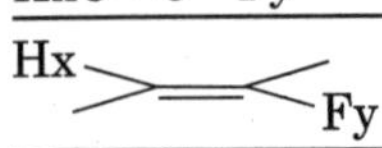	12–40 Hz
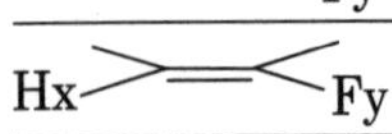	1–8 Hz

4.6 PROBLEMS BASED LEARNING

After reading the ^{1}H-NMR spectroscopy chapter, the readers are suggested/advised to test their learning by solving some below mentioned problems.

A. SUBJECTIVE PROBLEMS

1. Discuss various factors affecting chemical shift.
2. Give the principle of ^{1}H-NMR Spectroscopy. How spectroscopy helps in the structure elucidation of compounds.
3. Explain the principle of quadrature detector for FTNMR.
4. Describe decoupling techniques of ^{1}H-NMR.
5. Define and explain spin-spin coupling giving examples. Give applications of coupling constant in interpretation of spectra.
6. Discuss shielding and deshielding in NMR spectroscopy.
7. Write the significance of TMS as standard in NMR spectroscopy.

B. OBJECTIVE PROBLEMS

1. State True or False

(*i*) Angular momentum is described as spin quantum number (I).

(*ii*) Operating frequency of 300 MHz generates 48 excess nuclei.

(*iii*) Electron with drawing atom causes downfield effect.

(*iv*) Sterical hindrance causes Vander Waal's deshielding.

(*v*) Hydrogens bonding causes deshielding.

(*vi*) In ethyl chloride molecule, CH_2 group shows a Quartet in ^{1}H-NMR spectrum.

(*vii*) Electromagnets are used up to 100 MHz.

(*viii*) Superconducting magnets can be used above 100 MHz.

(*ix*) Old NMR spectrometers contains separate transmitter and receiver coils and modern spectrometer contain a single coil probe.

(*x*) Height of integral gives a relative number of hydrogens present in a molecule.

Ans.: 1. (*i*) True (*ii*) True (*iii*) True (*iv*) True (*v*) True (*vi*) True (*vii*) True (*viii*) True (*ix*) True (*x*) True.

2. Fill up the blanks

(*i*) Excess nuclei in lower spin state are called as .... population.

(*ii*) Angular momentum vector precesses about external field axis with an angular frequency is .... frequency.

(*iii*) .... is defined as phenomenon by which a proton absorbs radiofrequency and promotes to less favourable high energy state.

(*iv*) Alkyne hydrogens are more shielded than alkene due to ....

(*v*) .... magnets are temperature sensitive and require extensive cooling.

(*vi*) In .... NMR uses either frequency or field sweep method.

(*vii*) In .... NMR, all the nuclei are excited simultaneously.

Ans.: 2. (*i*) Spin state (*ii*) Larmor (*iii*) Resonance (*iv*) Anisotropy (*v*) Permanent (*vi*) Continuous wave (*vii*) Fourier Transformed.

3. Match the following

(*i*)	T_1 relaxation	(*a*)	Energy transfer to surrounding protons
(*ii*)	T_2 relaxation	(*b*)	Energy transfer among two precessing protons
(*iii*)	Geminal coupling	(*c*)	–25 cps at 105° and positive at 125° bond angle
(*iv*)	Vicinal coupling	(*d*)	Positive at 0 or 180° and negative at 90° angle

Ans.: 3. (*i*) – (*a*), (*ii*) – (*b*), (*iii*) – (*c*), (*iv*) – (*d*).

C. DATA INTEPRETATION PROBLEMS

1. Predict the class of compound which fits best for ^{1}H-NMR spectrum A.

   (*a*) Aliphatic ether (*b*) Aromatic ether

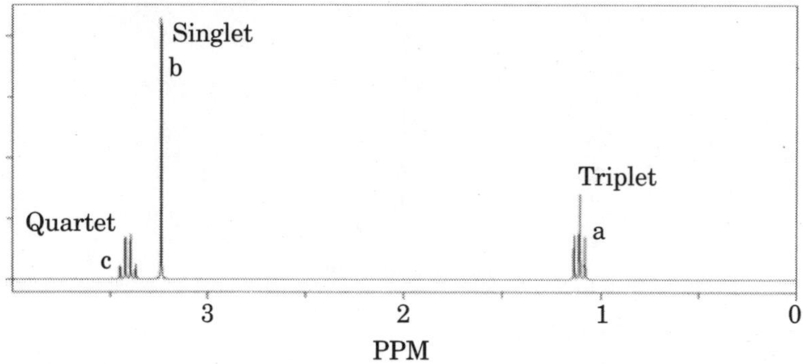

^{1}H-NMR spectrum—A

Ans.: *Aliphatic ether.*

2. Predict the class of compound which fits best for ^{1}H-NMR spectrum B.

   (*a*) Ethane (*b*) Ethene

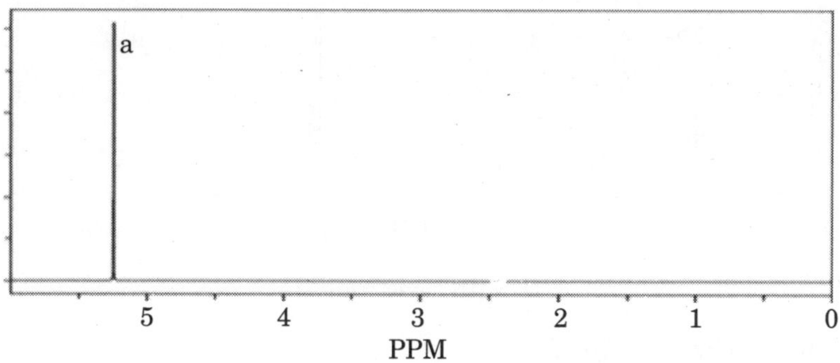

^{1}H-NMR spectrum—B

Ans.: *Ethene.*

3. Predict the class of compound which fits best for ^{1}H-NMR spectrum C.

(*a*) Methyl Acetate (*b*) Ethyl Acetate

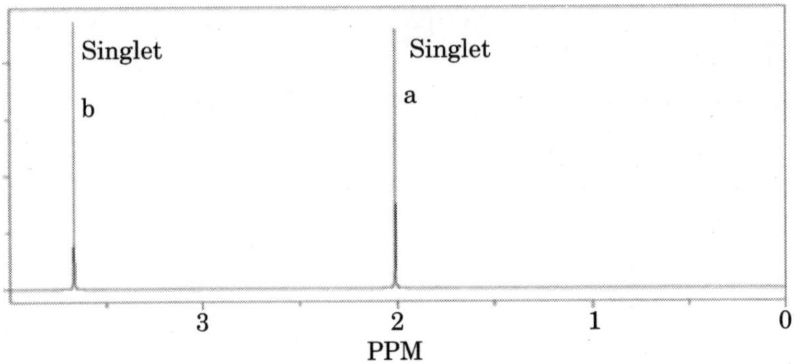

^{1}H-NMR spectrum—C

Ans.: *Methyl Acetate.*

4. Predict the class of compound which fits best for ^{1}H-NMR spectrum D.

(*a*) Ethyl Chloride (*b*) Ethyl Fluoride

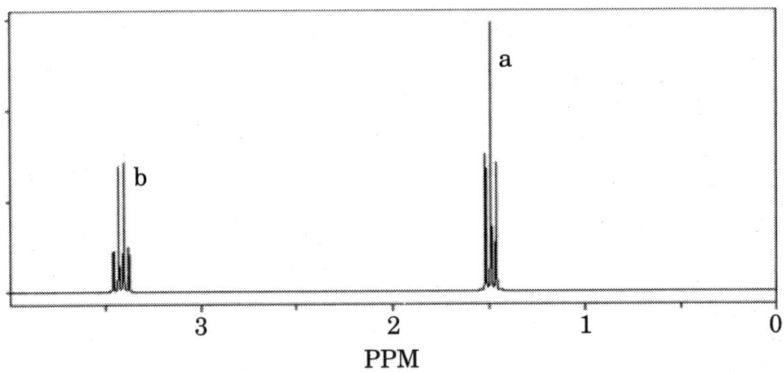

^{1}H-NMR spectrum—D

Ans.: *Ethyl chloride.*

4.7 SUGGESTED READINGS

[1] Keeler J., *Understanding NMR Spectroscopy,* 2nd Edition, Wiley, 2011.

[2] Pavia D. L, Lampman G. M., Kriz G. S., Vyvyan J. R., *Introduction to Spectroscopy*, 4th Edition, Brooks/Cole, 2009.

[3] Abraham R., Mobli M., *Modelling ^{1}H-NMR Spectra of Organic Compounds: Theory, Application, and NMR Prediction software*, Wiley & Sons, 2008.

[4] Skoog D.A., James Holler F., Crouch S.R., *Principles of Instrumental Analysis*, 6th Edition, Brooks Cole, 2007.

[5] Jacobsen N.E., *NMR Spectroscopy Explained: Simplified Theory, Applications and Example for Organic Chemistry and Structural Biology*, Wiley, N.J., 2007.

[6] Kalsi P.S., *Spectroscopy of organic compounds*, 6th Edition, New Age International, 2007.

[7] Silverstein R.M., Webster F.X. and David J. Kiemle; *Spectrometric Identification of Organic Compounds*, 7th Edition, John Wiley, 2005.

[8] Yadav L.D.S., *Organic Spectroscopy*, Kluwer academics publisher, Dordrecht, Netherland, 2005.

[9] Mitchell T.N., Costisella B., *NMR-From Spectra to Structures: An Experimental Approach*, Springer Verlag Berlin Heidelberg, N.Y., 2004.

[10] Schorn C., Taylor B.F., *NMR Spectroscopy: Data Acquisition*, 2nd Edition, Wiley-VCH, 2004.

[11] Nyquist R.A., *Interpreting Infrared, Raman and Nuclear Magnetic Resonance Spectra*, Volume-I, Academic press, Elsevier, USA, 2001.

[12] Dyer J.R., *Applications of Absorption Spectroscopy of Organic Compounds,* Prentice-Hall, India, 1965.

Chapter 5

^{13}C-Nuclear Magnetic Resonance (NMR) Spectroscopy

Objective

To make readers learn

- Basic Concepts, Theory of ^{13}C-NMR Spectroscopy.
- Instrumentation of ^{13}C-NMR Spectroscopy.
- ^{13}C-NMR Spectra/Data Interpretation for Structural Determination of Compounds.

5.1 INTRODUCTION

To determine the basic skeleton of organic compounds the carbon nuclei are need to be identified. In organic compounds two isotopes of carbon are present *i.e.*, ^{12}C and ^{13}C. Among ^{12}C and ^{13}C isotopes, ^{12}C are magnetically inactive (zero spin number) and highly abundant. Whereas ^{13}C isotopes are magnetically active (half spin number) but very less abundant *i.e.*, 1% of ^{12}C. Hence, carbon NMR experiment is done on ^{13}C nuclei but is less sensitive. As done in ^{1}H-NMR, **in ^{13}CNMR spectroscopy also ^{13}C nuclei are placed in an external magnetic field, and induced to absorb energy of radiofrequency (RF) range to undergo spin transitions with respect to applied magnetic field. The absorption frequencies peaks are plotted against peak intensities which generates a ^{13}C-NMR spectrum.** The ^{13}C-NMR spectroscopy suffers with two major problems. One is the low abundance of ^{13}C nuclei, that makes ^{13}C-NMR less sensitive. And second is the complexity of ^{13}C-NMR spectrum due to coupling of ^{13}C nuclei with ^{1}H nuclei. The first problem of low activity of ^{13}C-NMR is generally overcome by enhancing the ^{13}C-NMR instrument activity by replacement of continuous wave scanning with pulsed fourier transformed scanning (*i.e.*, simultaneous irradiation of all nuclei ^{13}C). Whereas the second problem of spectrum complexity is resolved by application of broad band decoupling of protons of hydrogens or other nuclei. To use ^{13}C-NMR as an analytical tool, it is necessary to understand the basic concepts of NMR spectroscopy.

5.2 BASIC CONCEPTS

5.2.1 Peak Splitting (Spin-spin Coupling) of ^{13}C Signal

In ^{13}C NMR homonuclear coupling between two carbons is very rarely observed, as probability of finding two ^{13}C nuclei to interact is very less. Whereas the heteronuclear coupling between ^{13}C and directly attached hydrogens is very common.

Therefore in ^{13}C-NMR one bond couplings ($^1J_{C-H}$) are generally studied. Same as ^{1}H-NMR experiment the splitting pattern in ^{13}C-NMR also follows n+1 rule.

Generally, the J value for ^{13}C-NMR experiment is represented with $^1J_{C-H}$ to represent coupling between C–H (*i.e.*, coupling between carbon and directly attached hydrogens). Usually, in ^{13}C-NMR experiment the $^1J_{C-H}$ value ranges from 100–250 Hz. **The spectrum that is generated after coupling of ^{13}C nuclei with ^{1}H nuclei is called as ^{13}C proton coupled spectrum** (or non decoupled spectrum). Table 5.1, shows the peak splitting pattern of ^{13}C nuclei using n + 1 rule in ^{13}C-NMR experiment.

Table 5.1: Peak splitting pattern of ^{13}C nuclei using n+1 rule in ^{13}C-NMR experiment.

Compound Structure and Name	*n = No. of attached hydrogens*	*n + 1 = Signal split pattern*
$-CH_3$ (Methyl carbon)	3	3 + 1 = 4 (quartet)
$-CH_2-$ (Methylene carbon)	2	2 + 1 = 3 (triplet)
–CH< (Methine carbon)	1	1 + 1 = 2 (doublet)
>C< (Carbon without hydrogen)	0	0 + 1 = 1 (singlet)

For example the proton coupled ^{13}C-NMR spectrum of ethyl methyl ether given in Figure 5.1a, shows a quartet for 'a' carbon at δ value of 14.5 ppm (J =127 Hz), quartet for 'b' carbon at 59 ppm (J =127 Hz), and triplet for 'c' carbon at 69.8 ppm.

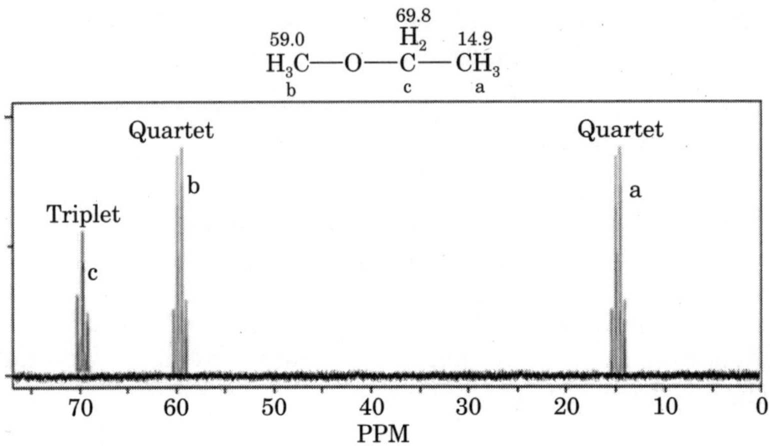

Figure 5.1a: Proton coupled ^{13}C-NMR spectrum of ethyl methyl ether.

5.2.2 Spin Decoupling (Double Resonance or Double Irradiation)

In spin decoupling a sample to be analyzed is irradiated with a broad spectrum of frequency and a second tunable radiofrequency such that the peak is decoupled or the observed spin-spin splitting is reduced to such a small value that they are no longer resolved.

Key Features

- Spin decoupling involves addition of an extra probe of radiation on a particular set of proton (*that may undergo coupling with other protons due to* ν_1 *pulse*) with a strong continuous wave RF (ν_2). This causes a saturation and very fast transition between two energy levels of irradiated proton. Hence through bond coupling is removed and proton is decoupled.
- In decoupling, two RF sources are used simultaneously for double irradiation.

- It is of great importance in determination of complex molecules by simplification of their spectra.
- This increases the sensitivity of instrument because it accumulates the intensities of all multiplet in a single peak.
- **The generated spectra of ^{13}C-NMR experiment, due to decoupling of ^{13}C nuclei with ^{1}H nuclei signals are often called as proton decoupled ^{13}C spectra.**
- For example the proton decoupled ^{13}C-NMR spectrum of ethyl methyl ether given in Figure 5.1b, shows all three carbons 'a', 'b' and 'c' carbon appears as singlet at 14.5 ppm, 59 ppm and 69.8 ppm.

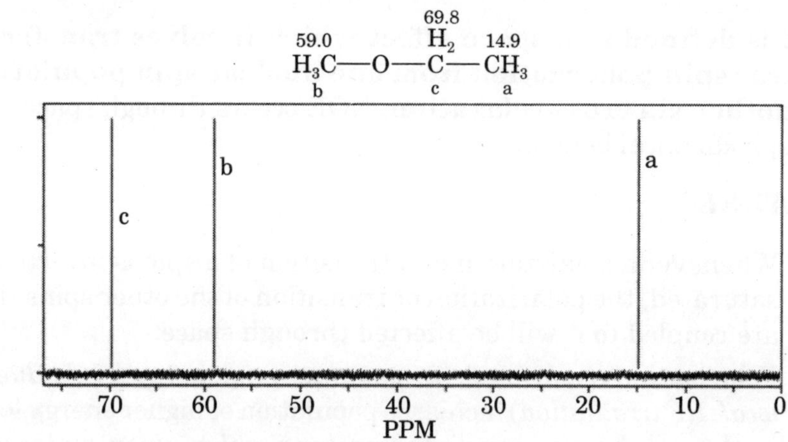

Figure 5.1b. Proton decoupled ^{13}C-NMR spectrum of ethyl methyl ether.

Decoupling is of three types *viz.*,

a. Homonuclear decoupling
b. Heteronuclear decoupling
c. Off resonance decoupling

8. **Homonuclear Decoupling**: The decoupling technique in which the nuclei being radio frequency (RF) irradiated are of the same isotope as the nuclei being observed (analyzed) in the spectrum is called as homonuclear decoupling.
2. **Heteronuclear Decoupling:** Technique of Decoupling in which when the nuclei being (RF) irradiated are of a different

isotope than the nuclei being observed in the spectrum is known as heteronuclear decoupling.

3. **Off Resonance Decoupling**: *Compromising decoupling technique in which sample is irradiated with a second RF either slightly upfield or downfield of normal proton resonances (i.e., off resonance) to preserves the one bond couplings and eliminate more than one bond couplings referred to off resonance decoupling*. This makes the spectra very simple and informative because retains the essential proton environment information also.

5.2.3 NOE Effect (Nuclear Over Hauser Effect) and Intensity

NOE is defined as a space effect which involves transfer of nuclear spin polarization from one nuclear spin population to another via cross-relaxation. NOE occurs through space, not through chemical bonds.

FEATURES

- Whenever a polarization or a transition of a spin is inverted or saturated, the polarization or transition of the other spins that are coupled to it will be affected through space.
- Polarization (*i.e., change in population of energy levels due to weak RF irradiation*) increases population of higher energy level in the nearby non irradiated protons which when undergoes T1 relaxation to lower energy level results in enhancement of nearby proton signal intensity.
- In NOE experiment the second tuner frequency RF (ν_2) is usually much weaker than in case of spin decoupling experiment.
- NOE effect can be homo nuclear or hetero nuclear.
- Atoms that are in close proximity to each other can give a NOE, whereas spin coupling is observed only when the atoms are connected by 2–3 chemical bonds.
- For example, imagine a molecule which contains H_a and H_b protons in a close proximity. But the two protons are separated by a very large number of bonds. *If we double irradiate a proton*

H_b. It causes stimulation the H_b proton, and this stimulation is transferred to the proton H_a, through relaxation mechanism. Due to increase in spin lattice relaxation of H_a, the signal intensity will increase by 50%. Therefore when H_b, is double irradiated this causes increment in intensity of H_a. This effect is called as NOE effect.

- This is why the proton decoupled spectrum of ^{13}C nuclei shows enhanced signal intensity for those carbons which are directly attached to hydrogen.

5.2.4 Cross Polarization

Theory of Cross polarization can be well understood as follows:

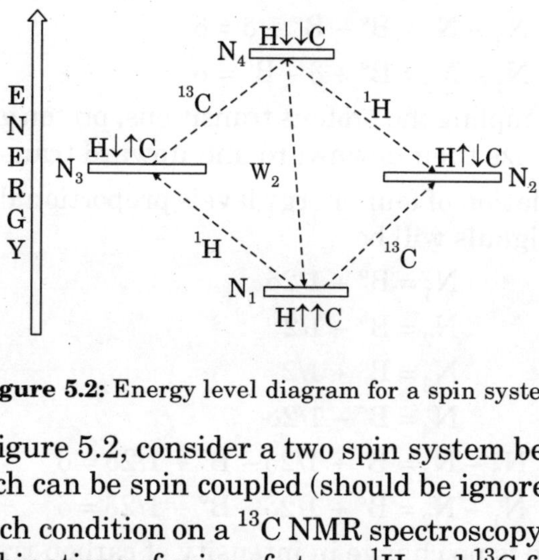

Figure 5.2: Energy level diagram for a spin system.

- As per Figure 5.2, consider a two spin system between ^{1}H and ^{13}C, which can be spin coupled (should be ignored).
- Apply such condition on a ^{13}C NMR spectroscopy. There can be four combinations of spins of atoms ^{1}H and ^{13}C for each energy level as N_1 (↑↑), N_2 (↑↓), N_3 (↓↑), and N_4 (↓↓).
- According to quantum mechanics *transitions which involve change of only one spin at a time are called as single quantum transition.*
- As spin coupling is ignored so both ^{1}H and ^{13}C transitions possess same energy.
- As four spin states possess different energies, so they possess different populations also.

- The populations of spin states N_3 and N_2 are almost equal as they possess similar energies.
- The populations of spin state N_1 will be more in δ value than and N_4.
- The spin states population of four energy levels proportional to intensities of NMR signals are represented as

$$N_1 = B° + \delta$$
$$N_2 = B°$$
$$N_3 = B°$$
$$N_4 = B° - \delta$$

- When spin states population of all four energy levels will be at equilibrium the intensities of carbon signals will be

$$N_3 - N_4 = B° - B° + \delta = \delta$$
$$N_1 - N_2 = B° + \delta - B° = \delta$$

- While decoupling the protons transitions, proton gets saturated, that equalize their downward and upward transitions.
- The population of four energy levels proportional to intensities of NMR signals will be

$$N_1 = B° + 1/2\delta$$
$$N_2 = B° - 1/2$$
$$N_3 = B° + 1/2$$
$$N_4 = B° - 1/2\delta$$

So, $$N_3 - N_4 = B° + 1/2\delta - B° + 1/2\delta = \delta$$
$$N_1 - N_2 = B° + 1/2\delta - B° + 1/2\delta = \delta$$

- Still there is no change in intensity of carbon transitions.
- Irradiation of proton signal causes non equilibrium condition, the population of spin states gets disturbed.
- Further due to relaxation process equilibrium condition of population is restored.
- While relaxation follows *double quantum transition in which both spin changes simultaneously*. This relaxation pathway is labeled as W2, which restores equilibrium from N_4 to N_1. Number of spins relaxed by such pathway is denoted with symbol d.

- Hence spin states population is

$$N_1 = B° + 1/2\delta + d$$
$$N_2 = B° - 1/2\delta$$
$$N_3 = B° + 1/2\delta$$
$$N_4 = B° - 1/2\delta - d$$

- So, the intensities of carbon signal are represented as

$$N_3 - N_4 = B° + 1/2\delta - B° + 1/2\delta + d = \delta + d$$
$$N_1 - N_2 = B° + 1/2\delta + d - B° + 1/2\delta = \delta + d$$

- Hence, intensity of ^{13}C increases with d (*i.e.*, number of spins relaxed of around value of 2.998).
- Commonly, the value of NOE enhancement is less than 2.998.
- The N_3 to N_2 relaxation are ignored because it does not involve net change in total no. of spins (*so called zero quantum transition which decreases the NOE*).
- In case of small molecules W2 pathway is more important than second relaxation as generates strong NOE.

5.2.5 Pulse, Pulse Sequence and Decoupling

A. ***Pulse*:** short burst of energy having wide range of frequencies which is generated by a source by machining it off and on very quickly is called as pulse. The timing and duration of excitation pulse is programmed with the help of computer.

B. ***Pulse Sequence*:** As shown in Figure 5.3, a pulse sequence includes by three basic features *i.e.*,

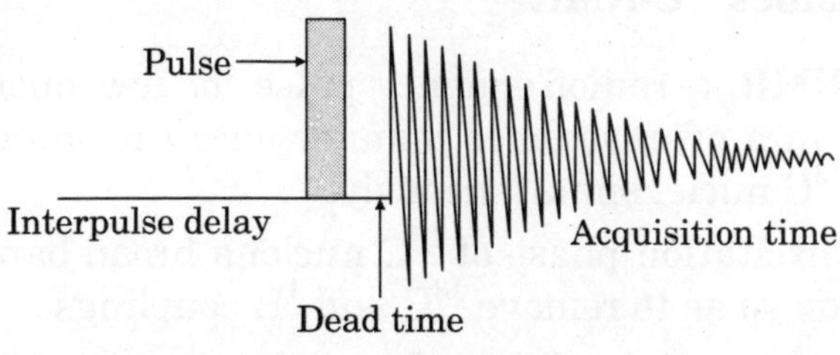

Figure 5.3: Schematic diagram of pulse sequence.

- **Excitation pulse** (Initial Pulse).

- **Acquisition time** (time during which free induction decay pattern is collected).
- **Relaxation delay** (time during which nuclei are allowed to relax so as to re-establish the equilibrium populations of two spin states).
- **Second pulse** (after relaxation time makes the cycle of pulse sequence).

C. ***Decoupling in Pulse Sequence*****:** During pulse sequence for decoupling the timings of another tunable RF are made very precisely. Usually, in the pulse sequences two techniques of decoupling are used such as:

(*a*) **Gated Decoupling**—Gated decoupling is commonly referred to NOE enhanced proton coupled technique. In this technique the decoupler is kept 'ON' during the phase or time before the pulse of ^{13}C nuclei is introduced. At the time of excitation pulse transmission the decoupler is made 'OFF'. During the relaxation delay again decoupler is made 'ON' until next excitation pulse. Such pulse sequence gives ^{13}C and ^{1}H coupled spectrum with NOE enhancement.

(*b*) **Inverse Gated Decoupling**—In this technique before the excitation pulse is transmitted decoupler is made OFF. During excitation pulse and acquisition time decoupler is kept 'ON'. During the relaxation delay again decoupler is made 'OFF' until next excitation pulse. Such pulse sequence gives ^{13}C and ^{1}H decoupled spectrum without NOE enhancement.

5.2.6 Mechanics ^{13}C-NMR

- In ^{13}C-NMR, a radiofrequency pulse for few microseconds is applied in a off resonance manner onto the sample to excite all the ^{13}C nuclei simultaneously.
- During excitation phase of ^{13}C nuclei a broad band decoupler is kept on so as to remove ^{13}C and ^{1}H couplings.
- All the ^{13}C nuclei exhibits free induction decay, (*where free induction decay is defined as exponentially decaying sine wave possessing a frequency equal to difference between applied frequency and resonance frequency of a nuclei*) because they

are influenced by a pulse sequence that is supplied at slightly off resonance.

- When a compound contains more than one ^{13}C nuclei, it commonly exhibits two characteristic components that is firstly a sine wave and secondly interference called beat.
- The obtained FID data are digitized and stored in a computer interface.
- Similarly, a series of pulses are applied, and respectively their signal acquisition time and time difference between pulses are recorded in the computer, which builds the signal when plotted on a time domain.
- This generates a time domain curve.
- Later on, with the help of a computer, the obtained time domain data/plot is further Fourier transformed and converted into a frequency domain curve.
- The Figure 5.4a & 5.4b, shows FID (time domain) curve and a fourier transformed (frequency domain) spectrum of methanol.

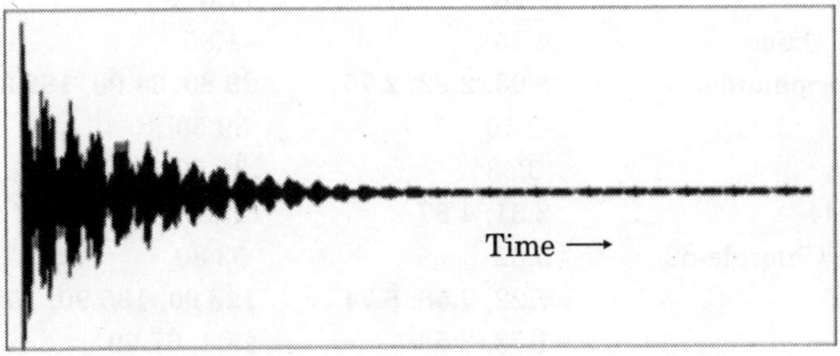

Figure 5.4a: Time domain curve of methanol.

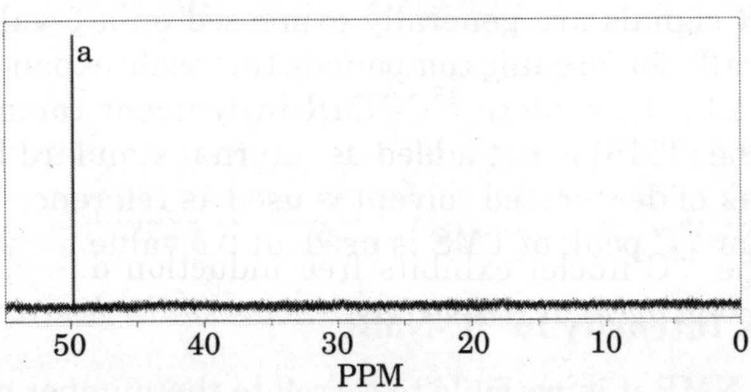

Figure 5.4b: Frequency domain curve of methanol.

5.2.7 Sampling, Scale, and Internal Standard of ^{13}C-NMR

In practice for a ^{13}C-NMR experiment at 75 MHz the sample requirement is about 10 mg. Sample is dissolved in about 0.4 ml of solvent in a 5 mm o.d. (outside diameter) NMR tube, made up of borosilicate glass.

In ^{13}C-NMR samples are usually dissolved in $CDCl_3$ or some other deuterated solvent. Usually, the solvents that are used for ^{1}H-NMR experiments are used in ^{13}C-NMR experiment also.

A list of commonly used solvents for ^{1}H-NMR and ^{13}C-NMR are given in Table 5.2.

Table 5.2: List of commonly used solvents in ^{1}H-NMR and ^{13}C-NMR.

Solvent	*^{1}H*	*^{13}C Signal at δ Value*
Acetic acid-d4	2.04, 11.65	20.0, 179.00
Acetone-d6	2.05	29.90, 206.70
Acetonitrile-d3	1.94	1.39, 118.70
Benzene-d6	7.16	128.0
Chloroform-d	7.26	77.0
Deuterium Oxide	4.75	–1.85
Dimethylformamide-d7	8.03, 2.92, 2.75	29.80, 34.90, 162.30
DMSO-d6	2.50	39.50
Dioxane-d8	3.53	39.50
Methanol-d4	3.31, 4.87	49.10
Methylene Chloride-d2	5.32	53.80
Pyridine-d5	7.22, 7.58, 8.74	123.90, 135.90, 150.40
THF-d8	1.73, 3.58	25.4, 67.60
Trifluoroacetic acid-d	11.50	116.6, 164.2

^{13}C-NMR signals are generally expressed on a δ value scale in ppm. Generally for organic compounds this scale expands from 0 to more than 240. In modern ^{13}C-NMR instrument commonly tetra methyl silane (TMS) is not added as internal standard but instead of it ^{13}C peak of deuterated solvent is used as reference. But still in the spectrum ^{13}C peak of TMS is used at 0 δ value.

5.2.8 Peak Intensity in ^{13}C-NMR

As with ^{1}H-NMR it is possible to correlate the number of hydrogen nuclei with peak intensity. Whereas **in a ^{13}C-NMR spectrum one**

cannot correlate the peak intensity of any signal with number of nuclei. This is because in general ^{13}C-NMR experiment the relaxation time (T1) of ^{13}C nuclei varies such that it does not relax equally between pulses. Commonly, in a ^{13}C-NMR spectrum, the non protonated carbons undergoes longer T1 relaxation and a negligible NOE effect so exhibits smaller peaks. Whereas the protonated carbon undergoes spin lattice relaxation mechanism. For example nitriles, carbonyls, non protonated alkenes and alkynes exhibits low intensity signals. To have better results a longer delays between pulses can be used but with utmost care such that a sufficient number of pulses are allowed with a long enough interval between pulses so that the weak signal may not loss with baseline noise.

5.3 INTERPRETATION OF ^{13}C -NMR SPECTRA

This section describes the general ^{13}C-NMR absorptions of various classes of compounds, using some relevant examples.

5.3.1 Alkanes

Example: *Ethane* (^{13}C-NMR spectrum given in Figure 5.5).

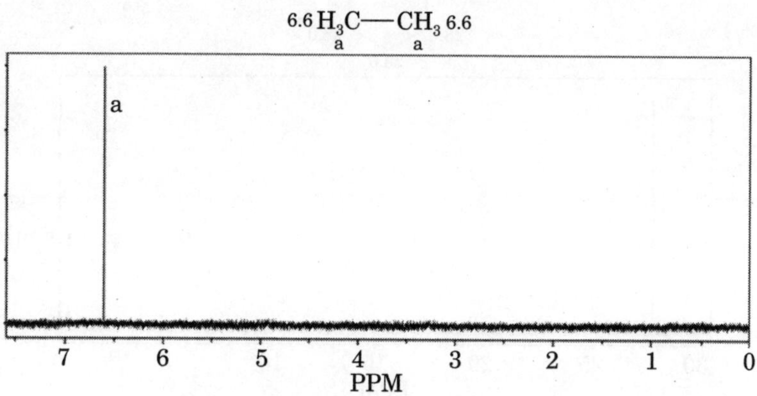

Figure 5.5: ^{13}C-NMR spectrum of ethane.

Interpretation

^{13}C-NMR δ (ppm): 6.6 (2C of CH_3).

Example: *Butane* (^{13}C-NMR spectrum given in Figure 5.6).

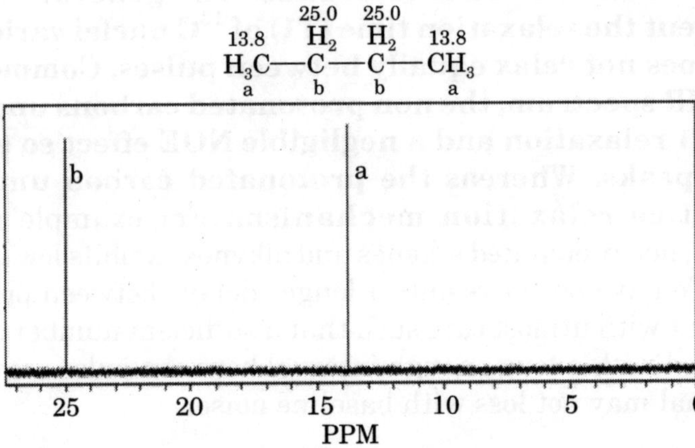

Figure 5.6: ^{13}C-NMR spectrum of butane.

Interpretation

^{13}C-NMR δ (ppm): 13.8 (2C of two CH_3) and 25 (2C of two CH_2).

Example: *Cyclohexane* (^{13}C-NMR spectrum given in Figure 5.7)

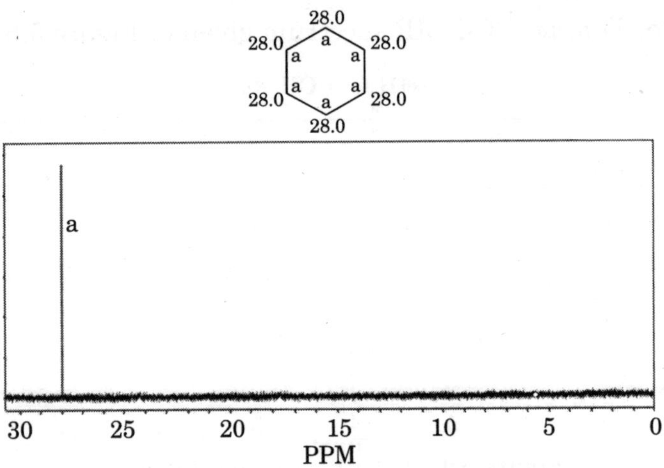

Figure 5.7: ^{13}C-NMR spectrum of cyclohexane.

Interpretation

^{13}C-NMR δ (ppm): 28 (6C of CH_2).

5.3.2 Alkenes

Example: *Ethene* (^{13}C-NMR spectrum given in Figure 5.8).

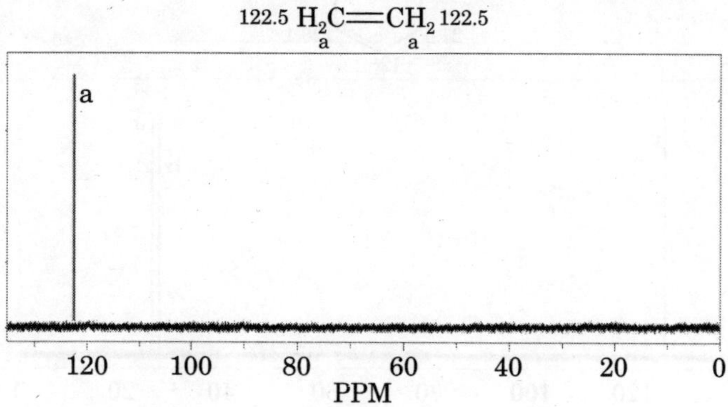

Figure 5.8: ^{13}C-NMR spectrum of ethene.

Interpretation

^{13}C -NMR δ (ppm): 122.5 (2C of $=CH_2$).

Example: *2-propene* (^{13}C-NMR spectrum given in Figure 5.9).

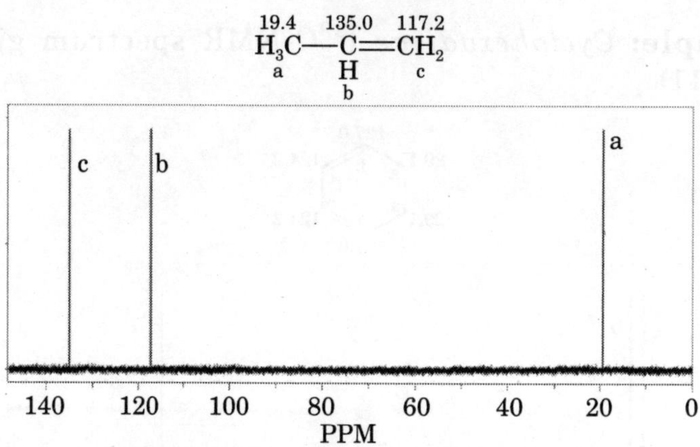

Figure 5.9: ^{13}C-NMR spectrum of propene.

Interpretation

^{13}C -NMR δ (ppm): 19.4 (1C of CH_3), 117.2 (1C of CH_2), 135 (1C of CH).

Example: *Cyclohexene* (^{13}C-NMR spectrum given in Figure 5.10).

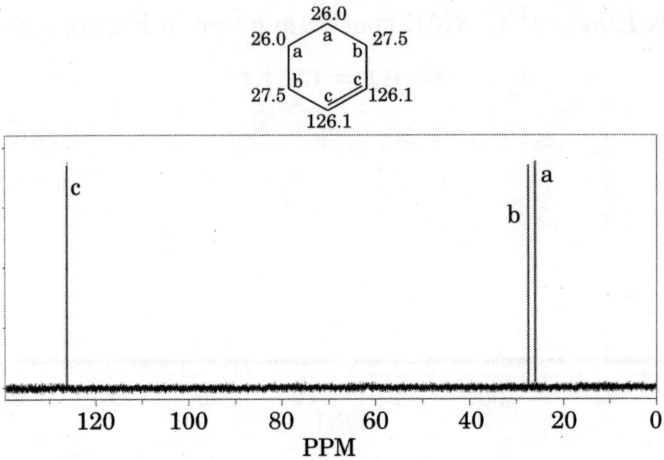

Figure 5.10: ^{13}C-NMR spectrum of cyclohexene.

Interpretation

^{13}C -NMR δ (ppm): 26 (2C of two beta CH_2), 27.5 (2C of two alpha CH_2), 126.1 (2C of =CH).

Example: *Cyclohexadiene* (^{13}C-NMR spectrum given in Figure 5.11).

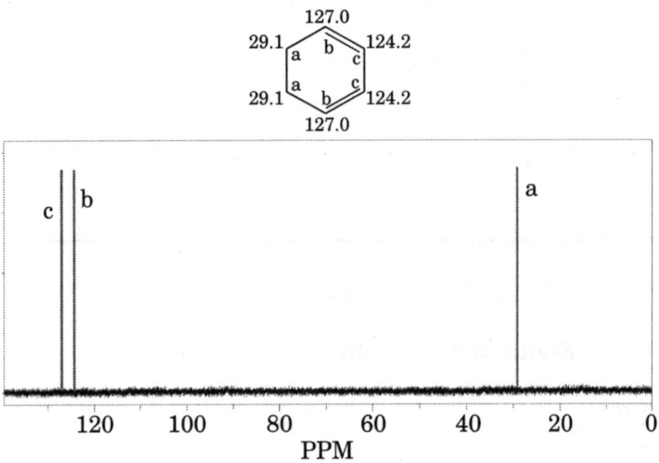

Figure 5.11: ^{13}C-NMR spectrum of cyclohexadiene.

Interpretation

^{13}C-NMR δ (ppm): 29.1 (2C of CH_2), 124.2 (2C of CH_{2b} & CH_{2c}), 127 (2C of CH_{2a} & CH_{2d}).

5.3.3 Alkenes with Heteroatom

Example: *Schiff base or Imine (N-ethylidenemethanamine)* (^{13}C-NMR spectrum given in Figure 5.12).

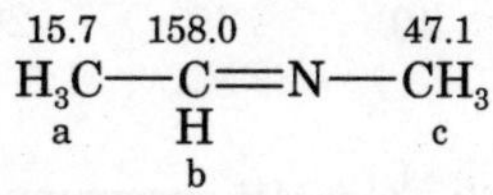

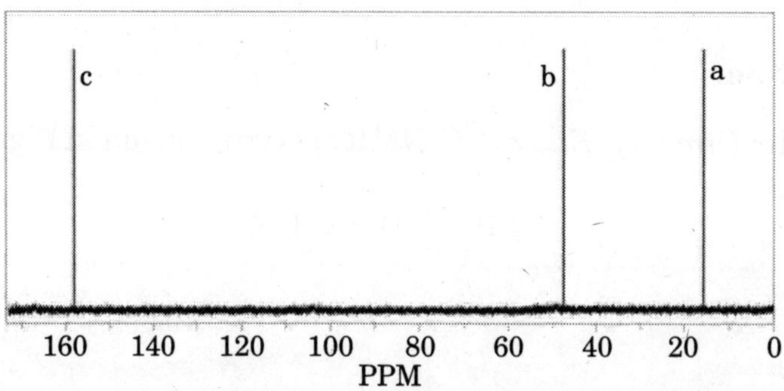

Figure 5.12: ^{13}C-NMR spectrum of N-ethylidenemethanamine.

Interpretation

^{13}C -NMR δ (ppm): 15.7 (1C of CH_3), 47.1 (1C of CH), and 158 (1C of N–CH_3).

5.3.4 Alcohols

Example: *Ethanol* (^{13}C-NMR spectrum given in Figure 5.13).

Interpretation

^{13}C-NMR δ (ppm): 17.4 (1C of CH_3), 57.9 (1C of CH_2).

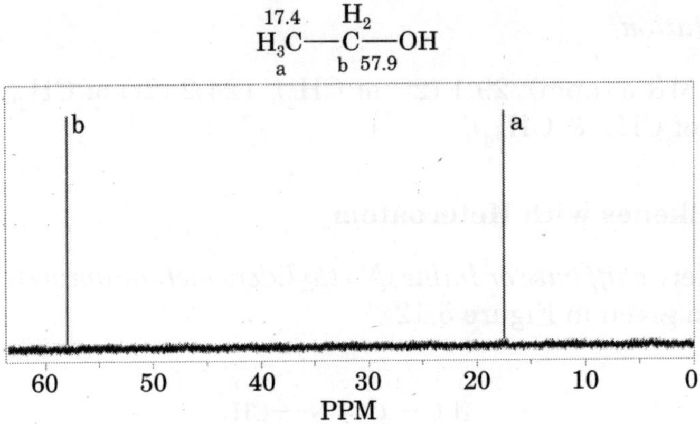

Figure 5.13: ^{13}C-NMR spectrum of Ethanol.

5.3.5 Ether

Example: *Dimethyl Ether* (^{13}C-NMR spectrum given in Figure 5.14).

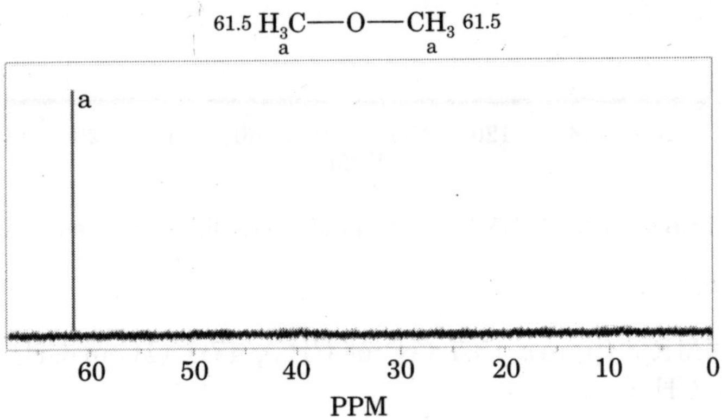

Figure 5.14: ^{13}C-NMR spectrum of Dimethyl ether.

Interpretation

^{13}C-NMR δ (ppm): 61.5 (2C of two CH_3).

Example: *Ethyl Methyl Ether* (^{13}C-NMR spectrum given in Figure 5.15).

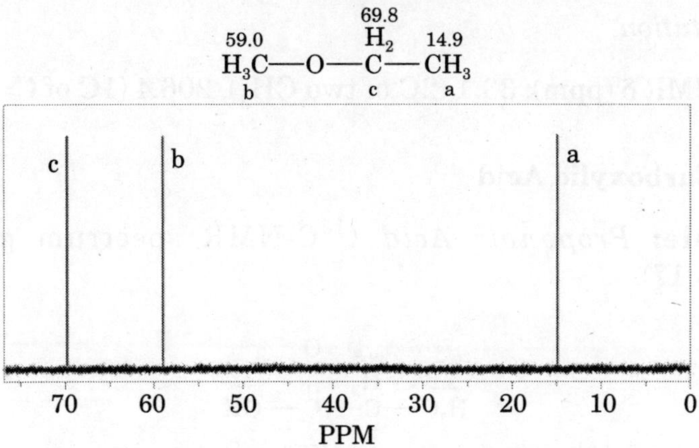

Figure 5.15: ^{13}C-NMR spectrum of ethyl methyl ether.

Interpretation

^{13}C-NMR δ (ppm): 14.9 (1C of CH_3), 59 (1C of O–CH_3), 69.8 (1C of O–CH_2).

5.3.6 Ketones

Example: *Acetone* (^{13}C-NMR spectrum given in Figure 5.16).

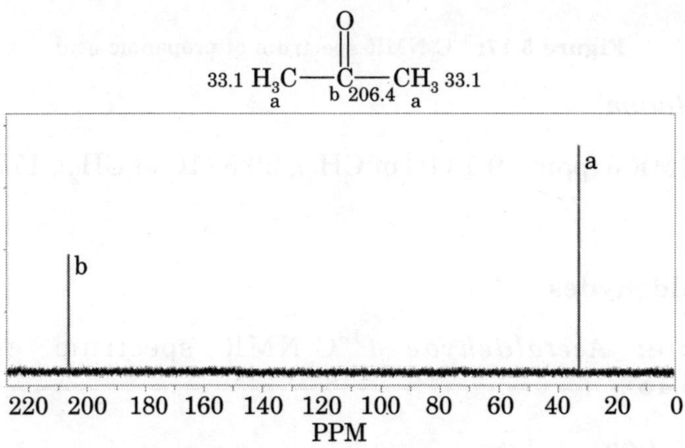

Figure 5.16: ^{13}C-NMR spectrum of acetone.

Interpretation

^{13}C-NMR δ (ppm): 33.1 (2C of two CH_3), 206.4 (1C of C=O).

5.3.7 Carboxylic Acid

Example: *Propanoic Acid* (^{13}C-NMR spectrum given in Figure 5.17).

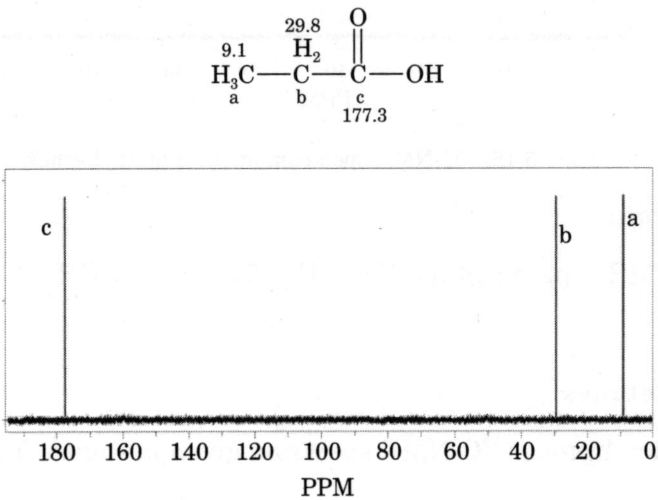

Figure 5.17: ^{13}C-NMR spectrum of propanoic acid.

Interpretation

^{13}C-NMR δ (ppm): 9.1 (1C of CH_3), 29.8 (1C of CH_2), 177.3 (1C of C=O).

5.3.8 Aldehydes

Example: *Acetaldehyde* (^{13}C-NMR spectrum given in Figure 5.18).

Interpretation

^{13}C-NMR δ (ppm): 30.7 (1C of CH_3), 199.8 (1C of CHO).

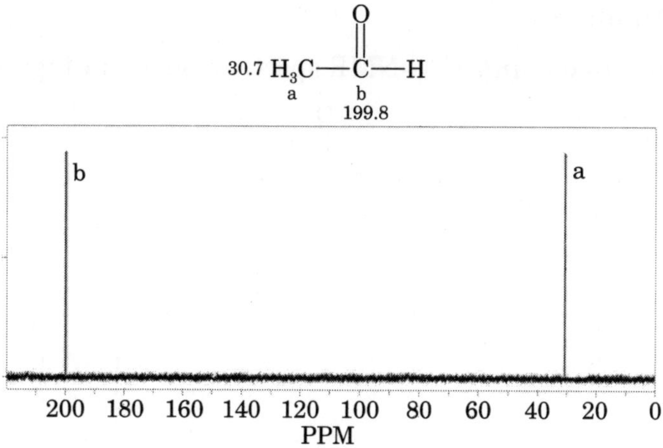

Figure 5.18: ^{13}C-NMR spectrum of acetaldehyde.

5.3.9 Ester

Example: *Methyl Propionate* (^{13}C-NMR spectrum given in Figure 5.19).

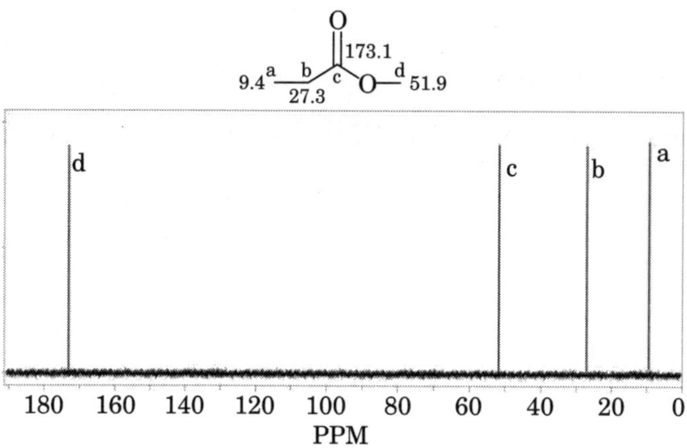

Figure 5.19: ^{13}C-NMR spectrum of methyl propionate.

Interpretation

^{13}C-NMR δ (ppm): 9.4 (1C of CH_3), 27.3 (1C of CH_2), 51.9 (1C of O–CH_3), 173.1 (1C of C=O).

5.3.10 Amide

Example: *Proanamide* (^{13}C-NMR spectrum given in Figure 5.20).

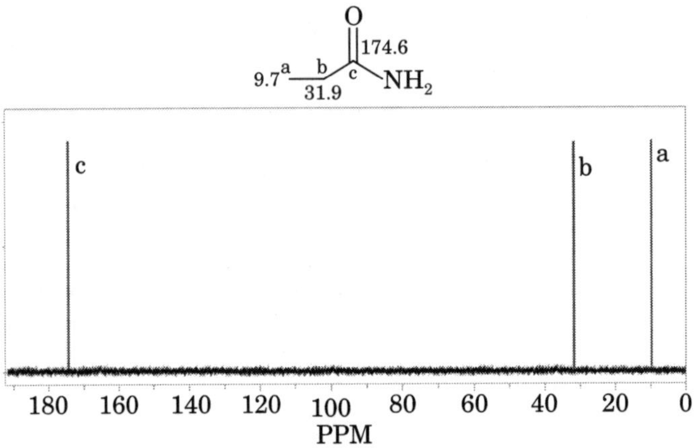

Figure 5.20: ^{13}C-NMR spectrum of propanamide.

Interpretation

^{13}C-NMR δ (ppm): 9.7 (1C of CH_3), 31.9 (1C of CH_2), and 174.6 (1C of C=O).

5.3.11 Alkyl Halides

Example: *Ethyl Chloride* (^{13}C-NMR spectrum given in Figure 5.21).

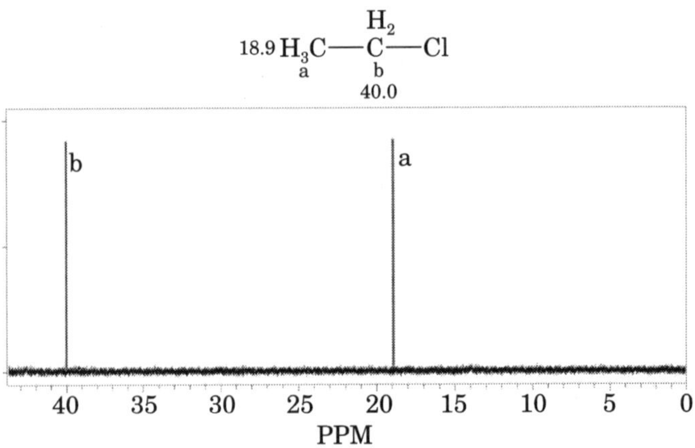

Figure 5.21: ^{13}C-NMR spectrum of ethyl chloride.

Interpretation

^{13}C-NMR δ (ppm): 18.9 (1C of CH_3), 40 (1C of CH_2).

Example: *Ethyl Chloride* (^{13}C-NMR spectrum given in Figure 5.22).

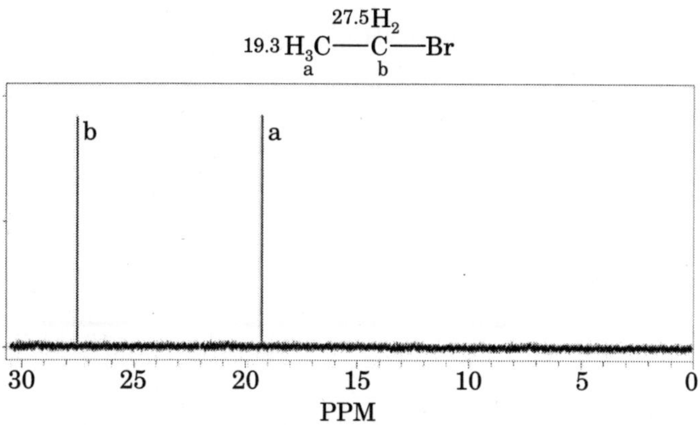

Figure 5.22: ^{13}C-NMR spectrum of ethyl bromide.

Interpretation

^{13}C-NMR δ (ppm): 19.3 (1C of CH_3), 27.5 (1C of CH_2).

5.3.12 Amines

Example: *Ethyl Amine* (^{13}C-NMR spectrum given in Figure 5.23).

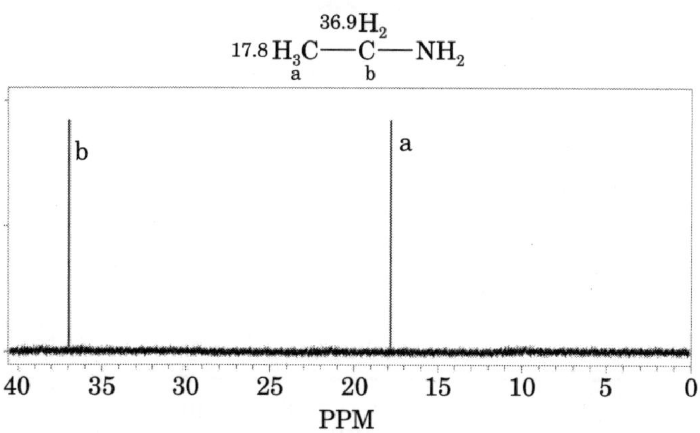

Figure 5.23: ^{13}C-NMR spectrum of ethyl amine.

Interpretation

^{13}C-NMR δ (ppm): 17.8 (1C of CH_3), 36.9 (1C of CH_2).

Example: *N-methylethanamine* (^{13}C-NMR spectrum given in Figure 5.24).

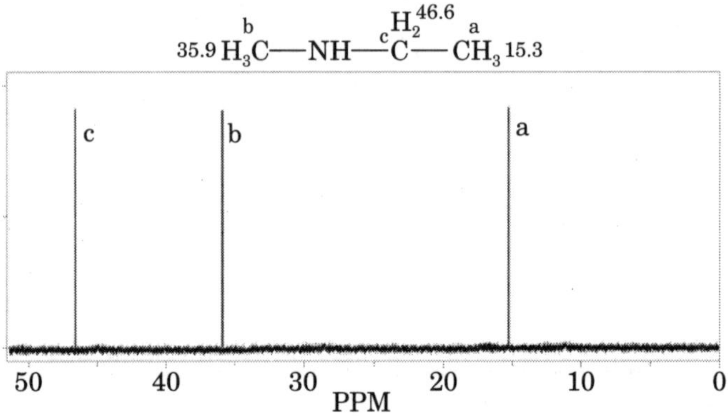

Figure 5.24: ^{13}C-NMR spectrum of butane-2-amine.

Interpretation

^{13}C-NMR δ (ppm): 15.3 (1C of CH_3), 35.9 (1C of CH_3), 46.6 (1C of CH_2) .

5.3.13 Aromatic Compounds

Example: *Benzene* (^{13}C-NMR spectrum given in Figure 5.25).

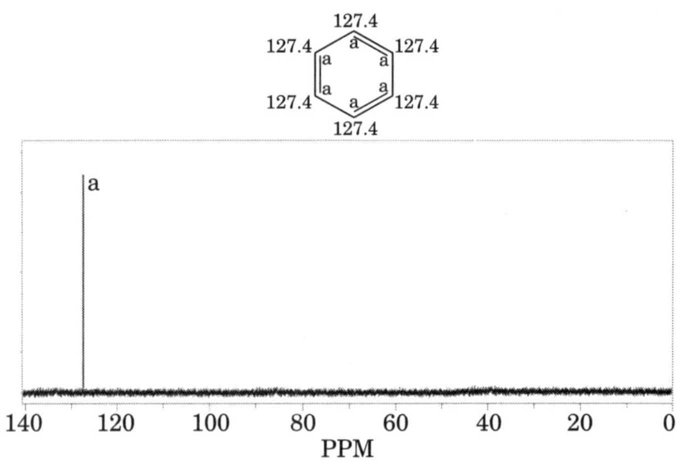

Figure 5.25: ^{13}C-NMR spectrum of benzene.

Interpretation

^{13}C-NMR δ (ppm): 127.4 (6C of Aromatic C–a).

Example: *Ethyl Benzene* (^{13}C-NMR spectrum given in Figure 5.26).

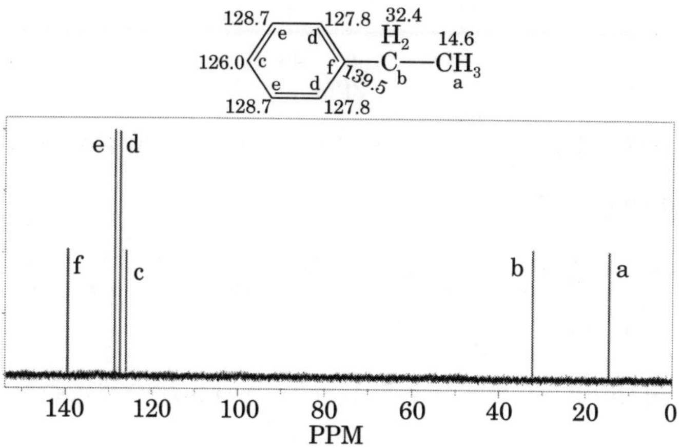

Figure 5.26: ^{13}C-NMR spectrum of ethyl benzene.

Interpretation

^{13}C-NMR δ (ppm): 14.6 (C–a), 32.4 (C–b), 126 (C–c), 127.8 (C–d), 128.7 (C–e), 139.5 (C–f).

5.3.14 Complex Compounds

Example: *2-(4-Chloro-3-Methylphenoxy) Acetate* (^{13}C-NMR spectrum given in Figure 5.27).

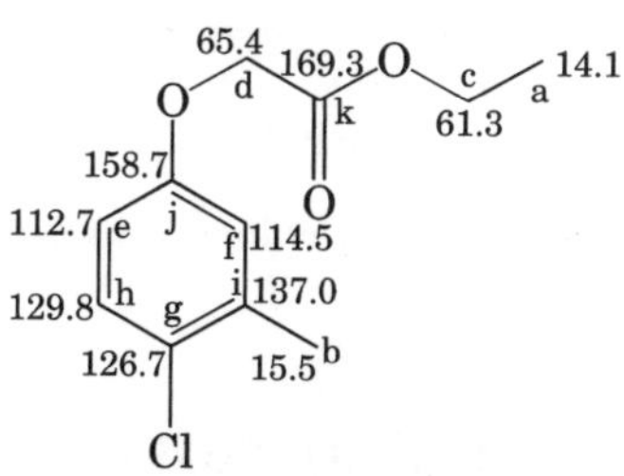

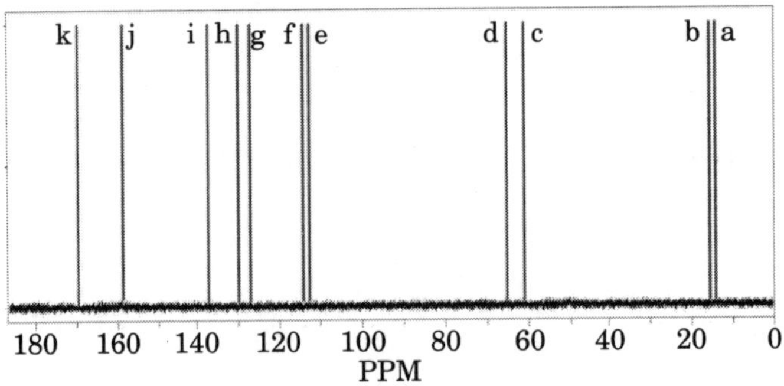

Figure 5.27: ^{13}C-NMR spectrum of 2-(4-chloro-3-methylphenoxy) acetate.

Interpretation

^{13}C-NMR δ (ppm): 14.1 (C–a), 15.5 (C–b), 61.3 (C–c), 65.4 (C–d), 112.7(C–e), 114.5 (C–f), 126.7 (C–g), 129.8 (C–h), 137.0 (C–i), 158.7 (C–j), 169.3 (C–k).

Example: *2-(4-Chloro-3-Methylphenoxy) Acetohydrazide* (^{13}C-NMR spectrum given in Figure 5.28).

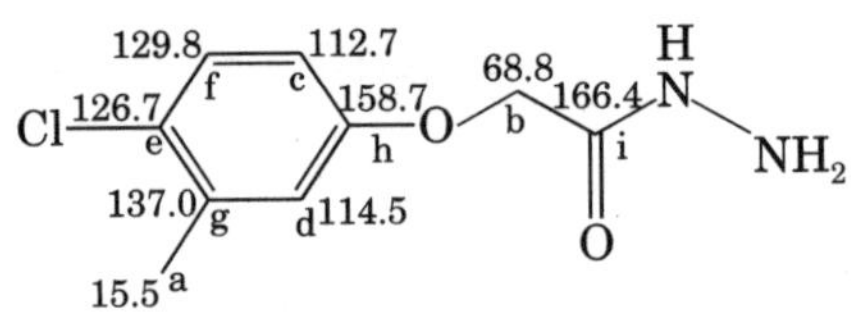

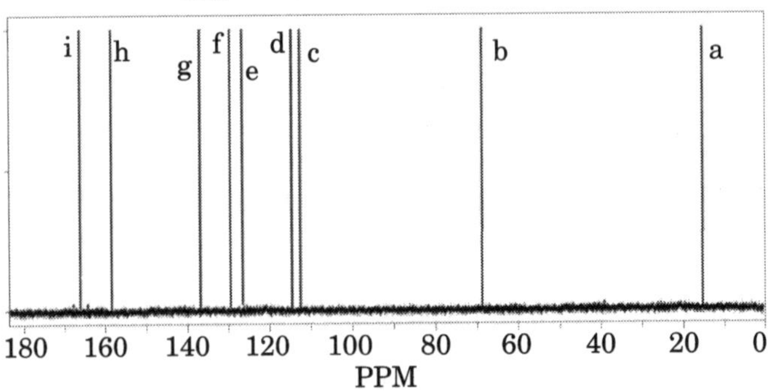

Figure 5.28: ^{13}C-NMR spectrum of 2-(4-chloro-3-methylphenoxy) acetohydrazide.

Interpretation

^{13}C-NMR δ (ppm): 15.5 (C–a), 68.8 (C–b), 112.7 (C–c), 114.5 (C–d), 126.7 (C–e), 129.8 (C–f), 137.0 (C–g), 158.7 (C–h), 166.4 (C–i).

Example: *N-{4-(Dimethylamino) Benzylidene}-2-(4-Chloro-3-Methylphenoxy) Aceto Hydrazide* (^{13}C-NMR spectrum given in Figure 5.29).

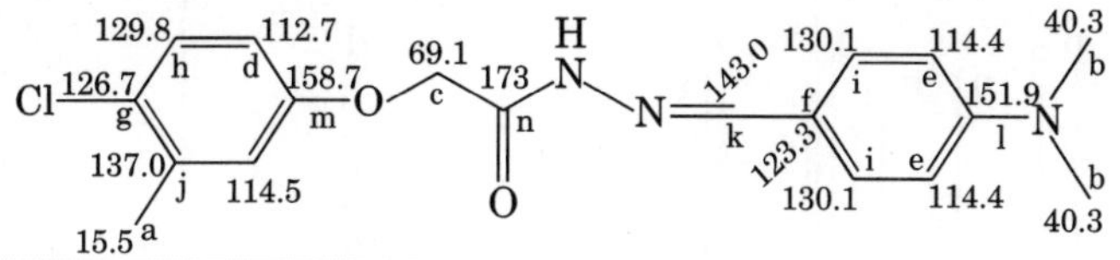

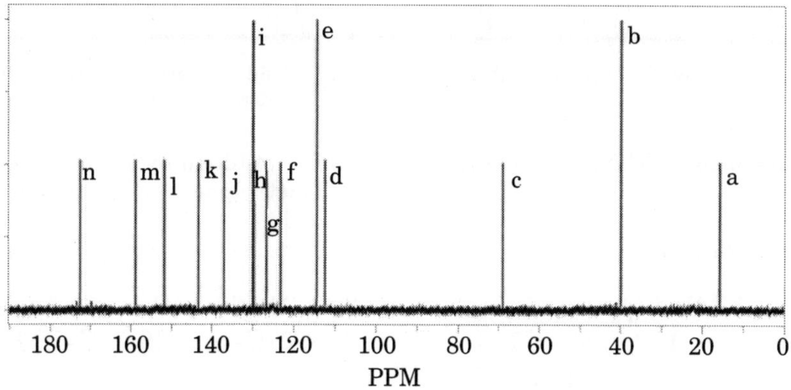

Figure 5.29: ^{13}C-NMR spectrum of n-{4-(dimethylamino)benzylidene}-2-(4-chloro-3-methylphenoxy) aceto hydrazide.

Interpretation

^{13}C-NMR δ (ppm): 15.5 (C–a), 40.3 (C–b), 69.1 (C–c), 112.7 (C–d), 114.4 (C–e), 123.3 (C–f), 126.7 (C–g), 129.8 (C–h), 130.1 (C–i), 137 (C–j), 143 (C–k), 151.9 (C–l), 158.7 (C–m), 173 (C–n).

Example: *N-[3-Chloro-2-{4-(Dimethylamino) Phenyl}-4-Oxo Azetidin-1-Yl]-2-(4-Chloro-3-Methyl Phenoxy) Acetamide* (^{13}C-NMR spectrum given in Figure 5.30).

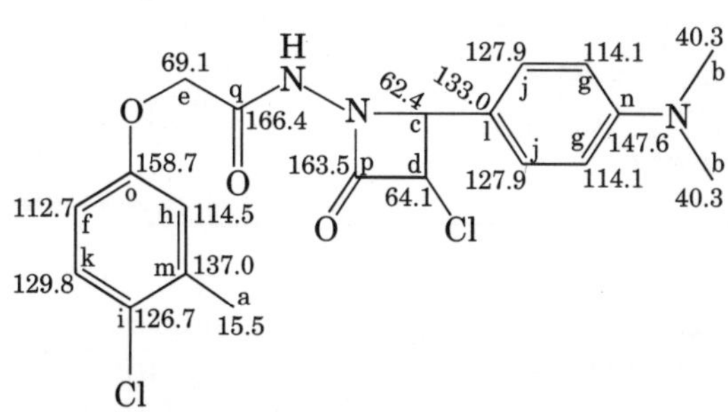

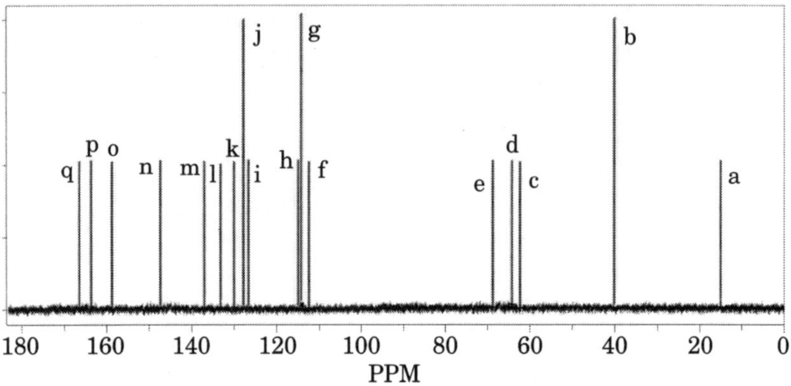

Figure 5.30: ^{13}C-NMR spectrum of (3-chloro-2-(4-(dimethylamino) phenyl)-4-oxoazetidin-1-yl)-2-(4-chloro-3-methylphenoxy) acetamide.

Interpretation

^{13}C-NMR δ (ppm): 15.5 (C–a), 40.3 (C–b), 62.4 (C–c), 64.1 (C–d), 69.1 (C–e), 112.7 (C–f), 114.1 (C–g), 158.7 (C–h), 126.7 (C–i), 127.9 (C–j), 129.8 (C–k), 133 (C–l), 137 (C–m), 147.6 (C–n), 158.7 (C–o), 163.5 (C–p), 166.4 (C–q).

For Further Assistance in ^{13}C-NMR Spectra Interpretation of Compounds Refer Figure 5.31.

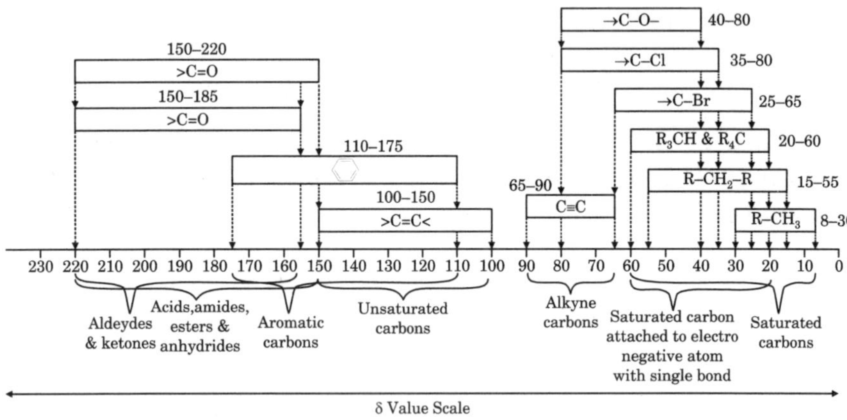

Figure 5.31: Chart representing ^{13}C chemical shift due to different chemical environments.

5.4 PROBLEMS BASED LEARNING

After reading the ^{13}C-NMR spectroscopy chapter, the readers are suggested/advised to test their learning by solving some below mentioned problems.

A. SUBJECTIVE PROBLEMS

1. Give the principle of ^{13}C-NMR Spectroscopy.
2. Explain the peak splitting in ^{13}C-NMR .
3. Describe double resonance or double irradiation.
4. Write a note on NOE Effect in ^{13}C-NMR.
5. Explain cross polarization.
6. Describe decoupling in ^{13}C-NMR.

B. OBJECTIVE PROBLEMS

1. State True or False

(*i*) Sample irradiation with a broad spectrum of frequency and a second tunable radiofrequency causes decoupling of peak.

(*ii*) Transitions which involve change of only one spin at a time so called single quantum transition.

(*iii*) Transitions which involve change of only one spin at a time so called single quantum transition.

(*iv*) In double quantum transition both spin changes simultaneously.

(*v*) Zero quantum transition does not involve net change in total no. of spins.

(*vi*) During acquisition time free induction decay pattern is collected.

(*vii*) During relaxation delay time nuclei are allowed to relax so as to re-establish the equilibrium populations of two spin states.

Ans.: 1. (*i*) True (*ii*) True (*iii*) True (*iv*) True
(*v*) True (*vi*) True (*vii*) True.

2. Fill up the blanks

(*i*) In .... decoupling nuclei being (RF) irradiated are of a different isotope than the nuclei being observed in the spectrum.

(*ii*) In .... decoupling nuclei being (RF) irradiated are of a same isotope than the nuclei being observed in the spectrum.

(*iii*) In .... decoupling sample is irradiated with a second RF either slightly upfield or downfield of normal proton resonances.

(*iv*) Transfer of nuclear spin polarization from one nuclear spin population to another via cross-relaxation is called as .... effect.

(*v*) Short burst of energy having wide range of frequencies generated by a source by making it off and on very quickly is called as pulse.

(*vi*) Exponentially decaying sine wave with frequency equal to difference between applied frequency and resonance frequency of a nuclei is called as ....

Ans.: 2. (*i*) Heteronuclear (*ii*) Homonuclear (*iii*) Off Resonance (*iv*) NOE (*v*) Pulse (*vi*) FID.

C. DATA INTERPRETATION PROBLEMS

1. Predict the class of compound which fits best for ^{13}C-NMR spectrum A.

(*a*) Butane (*b*) Butene

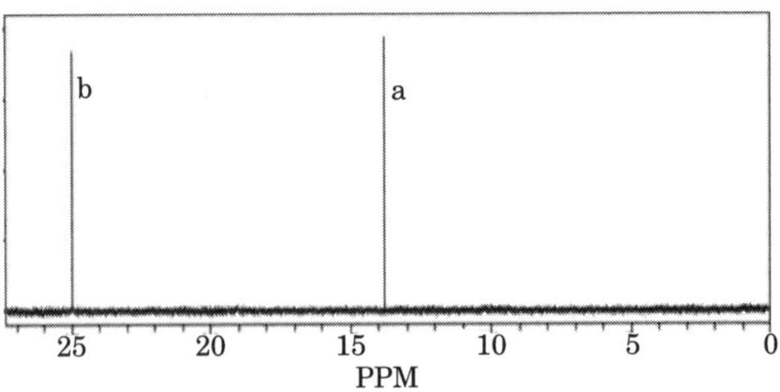

^{13}C-NMR spectrum—A

Ans.: *Butane.*

2. Predict the class of compound which fits best for ^{13}C-NMR spectrum B.

(*a*) Aliphatic ether (*b*) Aromatic ether

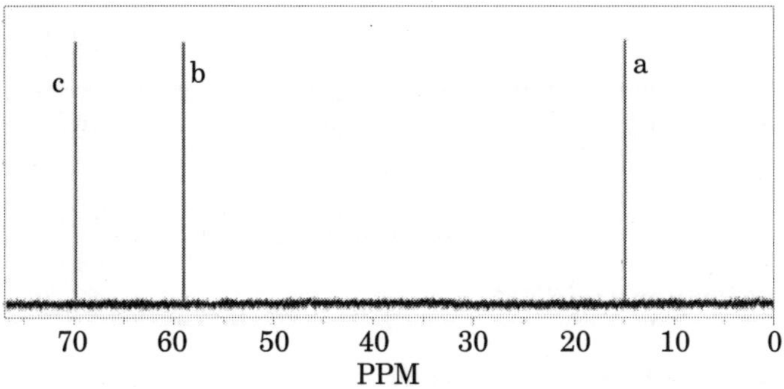

^{13}C-NMR spectrum—B

Ans.: *Aliphatic ether.*

3. Predict the class of compound which fits best for ^{13}C-NMR spectrum C.

(*a*) Ketone (*b*) Carboxylic acid

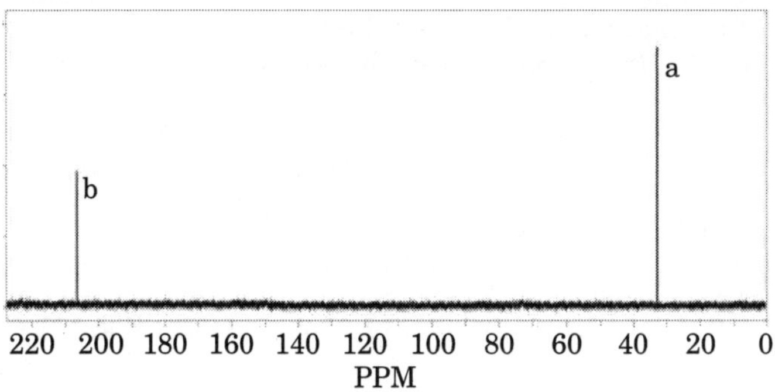

^{13}C-NMR spectrum—C

Ans.: *Ketone.*

4. Predict the class of compound which fits best for ^{13}C-NMR spectrum D.

(*a*) Propanamide (*b*) Propyl amine

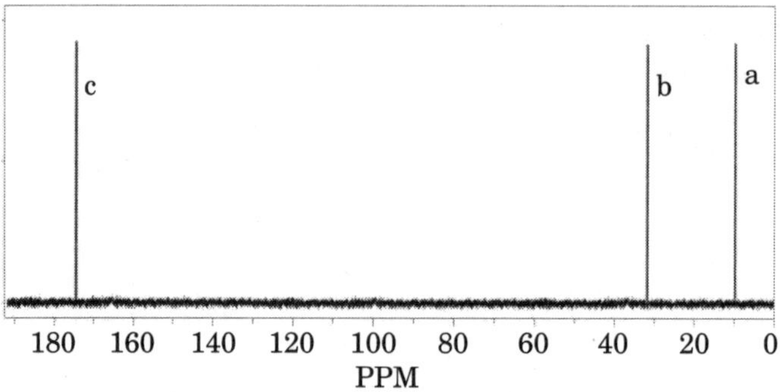

^{13}C-NMR spectrum—D

Ans.: *Propanamide.*

5.5 SUGGESTED READINGS

[1] James Keeler, *Understanding NMR Spectroscopy*, 2nd Edition, Wiley, 2011.

[2] Pavia D.L., Lampman G.M., Kriz G.S., Vyvyan J.R., *Introduction to Spectroscopy*, 4th Edition, Brooks/Cole, 2009.

[3] Jacobsen N.E., *NMR Spectroscopy Explained: Simplified Theory, Applications and example for organic chemistry and structural biology*, Wiley, N.J, 2007.

[4] Skoog D.A., James Holler F., Crouch S.R., *Principles of Instrumental Analysis*, 6th Edition, Brooks Cole, 2007.

[5] Silverstein R.M., Webster F.X., Kiemle D.J., *Spectrometric Identification of Organic Compounds*, 7th Edition, John Wiley, 2005.

[6] Yadav L.D.S., *Organic Spectroscopy*, Kluwer Academics Publisher, Dordrecht, Netherland, 2005.

[7] Mitchell T.N., Costisella B., *NMR-from Spectra to Structures: An Experimental Approach*, Springer Verlag Berlin Heidelberg, N.Y., 2004.

[8] Dyer J.R., *Applications of Absorption Spectroscopy of Organic Compounds*, Prentice-Hall, India, 1965.

[9] Breitmaier E., Voelter W., *13C NMR Spectroscopy: Methods and Applications*, Verlag Chemie, 1974.

[10] Kalsi P.S., *Spectroscopy of organic compounds,* 6th Edition, New Age International, 2007.

Chapter 6

Other Nuclei NMR

Objective

To make readers learn

- Fundamentals and Interpretation of ^{15}N-NMR, ^{19}F-NMR, ^{31}P-NMR, ^{29}Si-NMR, and ^{17}O-NMR Spectroscopy.

6.1 ^{15}N- NMR (NUCLEAR MAGNETIC RESONANCE)

6.1.1 Introduction

Nitrogen possess two isotopes (^{15}N and ^{14}N) which are magnetically active. Among two isotopes of nitrogen, isotope nuclei ^{14}N bears around 99% of natural abundance but possesses a spin of 1, which produces electric quadrupole moment (see Table 6.1). So, due to electric quadrupole relaxation and low sensitivity ^{14}N nuclei are not considered for conducting NMR experiments.

Table 6.1: Data on natural abundance, sensitivity, and net spin of some nuclei.

Nuclei	*Natural Abundance*	*Sensitivity*	*Net Spin*
^{1}H	99.9	1	1/2
^{29}Si	4.7	3.69×10^{-4}	1/2
^{31}P	100	6.63×10^{-2}	1/2
^{23}Na	100	9.25×10^{-2}	3/2
^{15}N	0.37	3.85×10^{-6}	1/2
^{14}N	99.6	1.01×10^{-3}	1
^{13}C	1.1	1.76×10^{-4}	1/2
^{19}F	100	0.83	1/2
^{17}O	3.7×10^{-2}	3.85×10^{-6}	5/2

^{15}N nuclei also possess very low sensitivity which is attributed to their very low abundance. But as ^{15}N nuclei possesses a spin number of ½ and also with advance instruments of NMR this problem of sensitivity is resolved, hence these ^{15}N isotopes of nitrogen are commonly used for NMR experiment.

6.1.2 Factors Affecting ^{15}N-NMR Signals

A ^{15}N-NMR signal are affected by several factors such as,

A. T_1 Phase Timings: ^{15}N nuclei relaxes in a very slowly manner. In case of T_1 phases duration is longer than 80 seconds. So either long pulse delay has to be used in pulse sequence or a different spin relaxation route should be used. To overcome such problem commonly a paramagnetic substance such as chromium (III) acetylacetonate is used as a catalyst. The lone pair of electron present on this paramagnetic substance results in stimulation of transfer of spin. In case the T_1 phase's timings are unknown, pulse delays and pulse angles should be delivered with utmost care so that the signals from one or more ^{15}N should not be as slow as to miss their signals.

B. Nuclear Over Hauser Effect: As by NOE protons attached to any nuclei can be decoupled due to which intensity of signals of the proton attached nuclei can be enhanced. This change in signal intensity of proton attached nuclei is attributed to polarization of spin population away from the predicted Boltzmann distribution. Generally the magnitude of enhancement is attributed to two factors such as

- Firstly, half of the magnetogyric ratio of nuclei *i.e.*, $\gamma_H/2\gamma_x$ (where x can be any nuclei attached to protons). This magnetogyric ratio contributes maximum to the signal enhancement of nuclei. For example the signal enhancement in case of ^{13}C nuclei by decoupling the attached protons should be around 1.98 *i.e.*, threefold increase in sensitivity, this is because $\gamma_C = 6728$ and $\gamma_H = 26753$.
- Secondly, the extent of dipolar relaxation contributes to the actual enhancement. The actual enhancement can be at any value between 0–1.98 attributed to mechanism of relaxation.

As compare to ^{13}C nuclei the ^{15}N nuclei exhibits negative signal enhancement such as equal to –4.9. Magnetogyric ratio of ^{15}N is very small and negative. Hence it can be generalized that nuclei of ½ spin if possesses a positive magnetogyric ratio will give positive signal enhancement. Whereas nuclei of ½ spin if possesses a negative magnetogyric ratio will give negative signal enhancement. The case where ^{15}N–^{1}H dipolar spin lattice relaxation occurs together, the signal becomes inverted and intensity becomes four times to that obtained in absence of ^{1}H irradiation. ^{15}N proton decoupling leads NOE effect to occur from 0 to –4.93 or 1 to –3.93.

6.1.3 Internal Standard for ^{15}N-NMR Experiment

Generally, liquid ammonia is used as a universal standard for reference in ^{15}N NMR spectrum and set to zero in ppm. Earlier many compounds such as ammonium nitrate, nitric acid, and nitro methane etc. were also been used as standard.

Commencement of ^{15}N NMR experiment by using liquid ammonia as a standard is a tedious job. So, despite of the fact that nitro methane generates a negative chemical shift, it is common practice to use nitro methane as internal standard and add 380 ppm to the shift that is obtained in reference to nitro methane in a ^{15}N NMR experiment.

6.1.4 Chemical Shift

Nitrogen nuclei experiences the similar electronic effects as with ^{13}C nuclei. Therefore the chemical shifts of nitrogenous organic compounds are almost similar to chemical shifts of carbon. For nitrogenous organic compounds the chemical shifts range is upto 500 ppm. Due to large chemical shift range in ^{15}N NMR spectrum, usually the chances of overlapping of nitrogen nuclei signal are even lesser than that in spectrum of ^{13}C nuclei.

Lone pair of electrons over the nitrogen nuclei is one of the major factor that affects the chemical shift of ^{15}N nuclei. This is why because lone pair of electrons on nitrogen undergoes protonation (in solvent), thereby influences or causes the ^{15}N signal to shift on a δ value scale (*i.e.*, either right or left). This magnitude of shifting depends upon the type of nitrogen available.

In ^{15}N NMR proton coupled spectrum $^{1}J_{N-H}$ (with 75–135 Hz), $^{2}J_{N-H}$ (with 0–20 Hz) and $^{3}J_{N-H}$ (with 0–10 Hz) couplings are very common.

6.1.5 Interpretation

Example A: *^{15}N-NMR Proton Decoupled Spectrum of Formamide* (Figure 6.1).

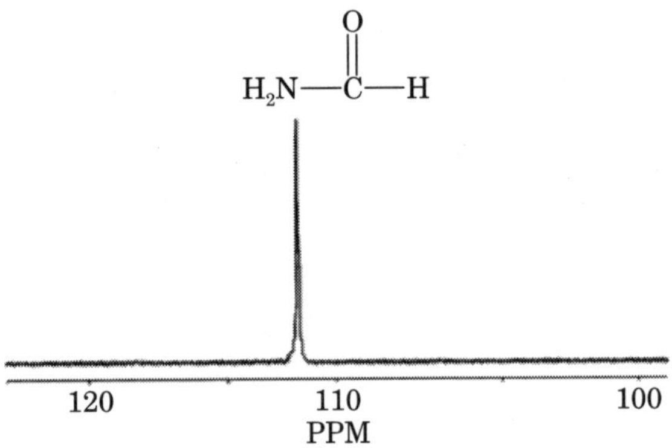

Figure 6.1: ^{15}N-NMR proton decoupled spectra of formamide.

Interpretation

- Formamide shows negative NOE enhancement with proton decoupling.
- A peak of nitrogen appears at 112.5 as a singlet on a δ value scale.

Example B: *^{15}N-NMR proton decoupled and coupled spectra of Ethylenediamine* (Figure 6.2).

Interpretation

- A peak of nitrogen appears at 12.2 on a δ value scale in the form of a quintet with a coupling constant of 2–3 Hz.

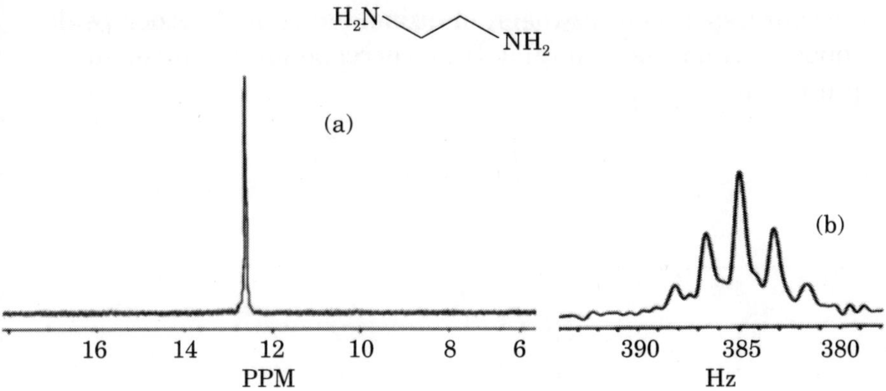

Figure 6.2: ^{15}N-NMR proton decoupled (a) and ^{15}N-NMR proton coupled (b) spectra of ethylenediamine.

Example C: *^{15}N-NMR proton decoupled Spectrum of Pyridine* (Figure 6.3).

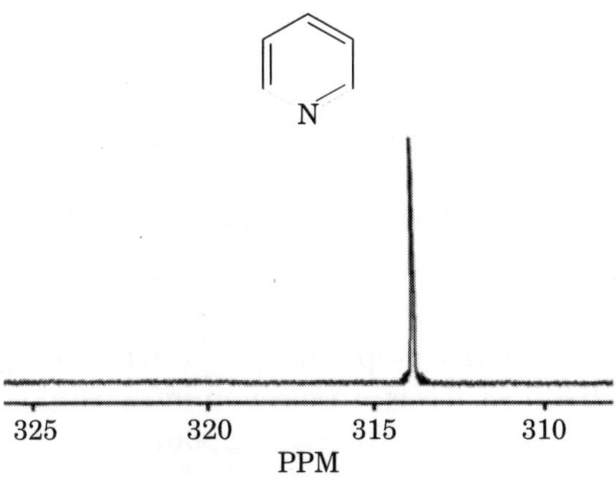

Figure 6.3: ^{15}N-NMR proton decoupled spectrum of pyridine.

Interpretation

- The ^{15}N-NMR spectrum of pyridine (given in Figure 6.3) shows a ^{15}N peak at 314 on a δ value scale as a singlet.

Example D: *^{15}N-NMR Spectrum of Quinine* (Figure 6.4).

Interpretation

The ^{15}N-NMR spectrum of quinine shows two ^{15}N peaks. One peak appears at 295 on a δ value scale as a singlet that corresponds to nitrogen of quinolone ring. Another peak appears at 29 as a singlet that corresponds to nitrogen of quinuclidinyl ring.

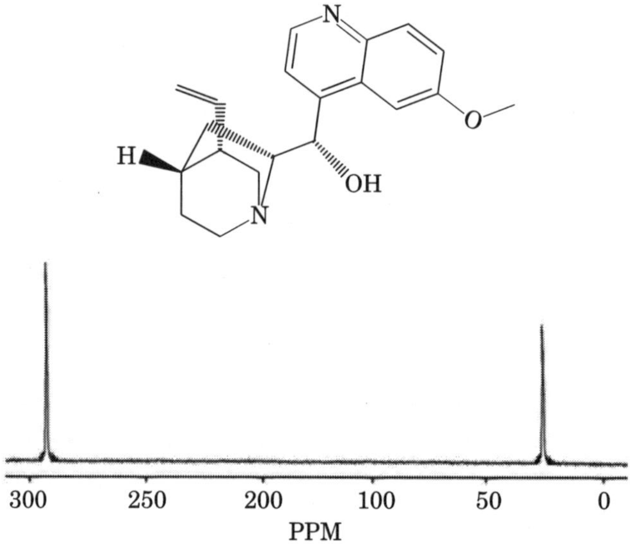

Figure 6.4: ^{15}N-NMR proton decoupled spectrum of quinine.

6.2 ^{19}F-NMR (NUCLEAR MAGNETIC RESONANCE)

6.2.1 Introduction

Fluorine possesses only one isotope that is ^{19}F with an appreciable sensitivity that is 0.83. As ^{19}F is uncommon in natural compounds and most commonly exists in synthetic compounds, hence ^{19}F NMR focus mainly on the synthetic compounds. In fact proton decoupling of ^{19}F NMR spectrum does not causes NOE enhancement of ^{19}F signals. More over just like ^{1}H NMR, ^{19}F NMR is also recorded in a continuous wave instrument.

6.2.2 Internal Standard for ^{19}F-NMR Experiment

Currently, $CFCl_3$ (Trichloro Fluoro Methane) is used as a standard in ^{19}F NMR. The reason behind the choice of $CFCl_3$ as a standard is its volatility, inertness, also its single ^{19}F resonance.

6.2.3 Interpretation

Example A: *^{1}H-NMR coupled, ^{19}F-NMR proton decoupled and ^{19}F-NMR proton coupled spectra of Fluoro Acetone* (Figure 6.5).

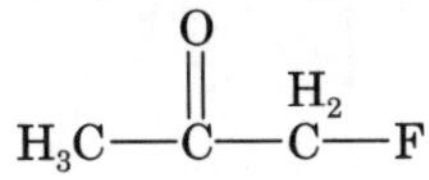

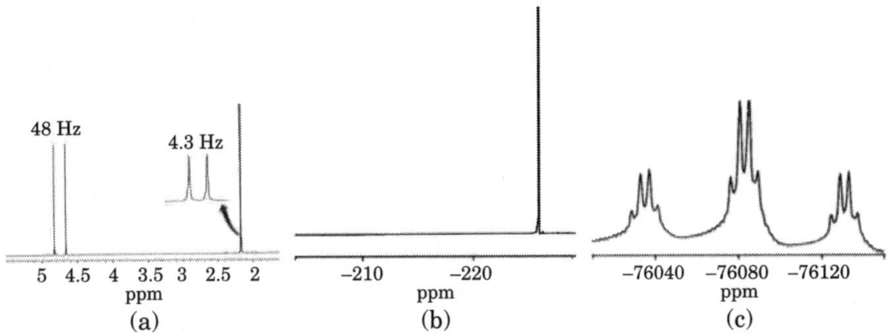

Figure 6.5: ^{1}H-NMR coupled spectrum (a), ^{19}F-NMR proton decoupled spectrum (c) & ^{19}F-NMR proton coupled spectrum (a) of fluoroacetone.

Interpretation

- In case of ^{1}H-NMR coupled spectrum (a) of fluoroacetone, the methyl and methylene groups are splitted into two doublets by ^{19}F at 4.65 and 4.75 respectively.
- In ^{19}F NMR proton decoupled spectrum (b) of fluoroacetone, only a singlet appears for fluorine nuclei at –226.
- Whereas, in case of ^{19}F NMR proton coupled spectrum (c) of fluoroacetone, firstly the fluorine nuclei produces a triplet peak on coupling with hydrogens of CH_2 group. Secondly, the same triplet peak is further splitted into quartet of triplet on further coupling with hydrogens of CH_3 group.

Example B: *^{19}F-NMR proton decoupled and proton coupled spectra of Para Fluoro Acetophenone* (Figure 6.6).

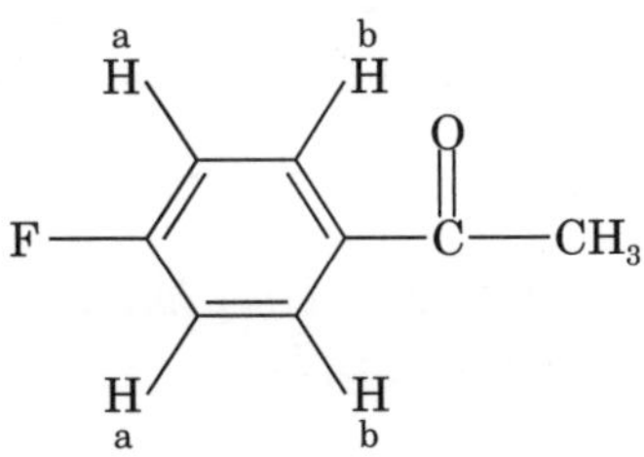

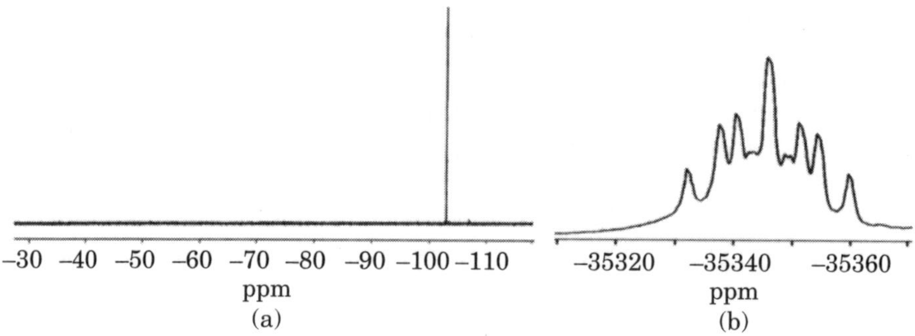

Figure 6.6: (a) ^{19}F-NMR proton decoupled and (b) ^{19}F-NMR proton coupled spectra of para fluoro acetophenone.

Interpretation

- In the ^{19}F NMR proton decoupled spectrum of para fluoro acetophenone, only a singlet appears for fluorine nuclei.
- Whereas in case of ^{19}F NMR proton coupled spectrum of para fluoro acetophenone, firstly the peak of fluorine nuclei is splitted into a triplet on coupling with 'a' aromatic hydrogens of phenyl group. Secondly, the same triplet peak is further splitted into triplet of triplet on further coupling with 'b' aromatic hydrogens of phenyl group.

6.3 ^{29}SI-NMR (NUCLEAR MAGNETIC RESONANCE)

6.3.1 Introduction

^{29}Si is an isotope of silicon that possesses 4.7% of natural abundance and a magnetic moment. Silicon is another nuclei that does not occur in the natural organic compounds. The sensitivity of ^{29}Si nucleus is twice of the ^{13}C nuclei. It is of high importance for synthetic chemists in organics and polymers.

In ^{29}Si-NMR tetra methyl silane is commonly employed as an internal standard.

Compare to ^{13}C, the chemical shift of ^{29}Si is very small.

6.3.2 Interpretation

Example A: *^{29}Si NMR proton decoupled and proton coupled spectrum of Triethyl Silane* (Figure 6.7).

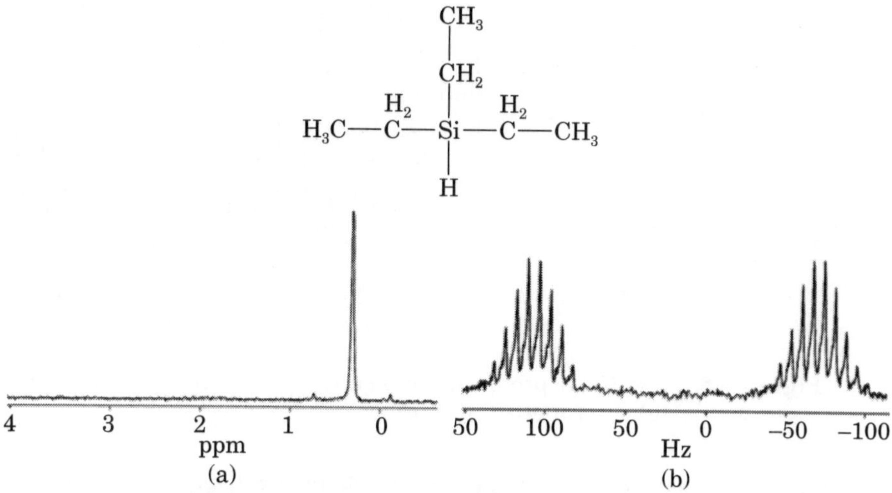

Figure 6.7: ^{29}Si-NMR proton decoupled (a), and ^{29}Si -NMR proton coupled (b) spectra of triethyl silane.

Interpretation

- In the ^{29}Si NMR proton coupled spectrum of triethyl silane molecule, the ^{29}Si nuclei exhibits a coupling constant of 175 Hz due to $^1J_{Si-H}$.
- In the proton decoupled spectrum of triethyl silane a singlet peak is observed that is on a slightly downfield region than standard tetramethyl disiloxane.

Example B: *^{29}Si-NMR proton decoupled and proton coupled spectrum of tetra ethyl disiloxane* (Figure 6.8).

Interpretation

- The ^{29}Si NMR proton coupled spectrum tetraethyl disiloxane shows a larger one bond coupling constant ($^1J_{Si-H}$) *i.e.*, 215 Hz. Whereas decoupled spectrum shows a singlet peak on a much more upfield region as compare to tetra methyl silane.

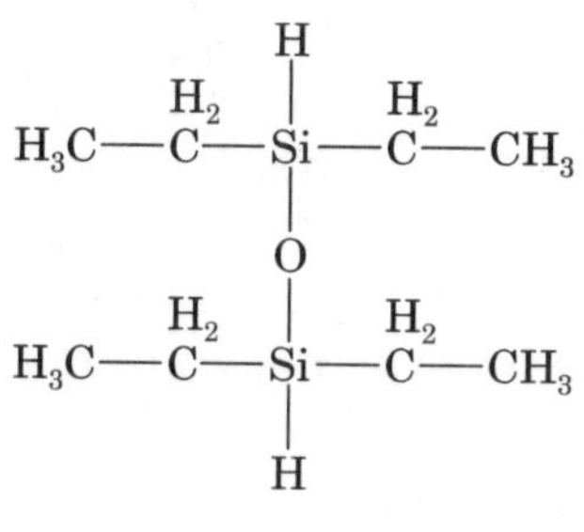

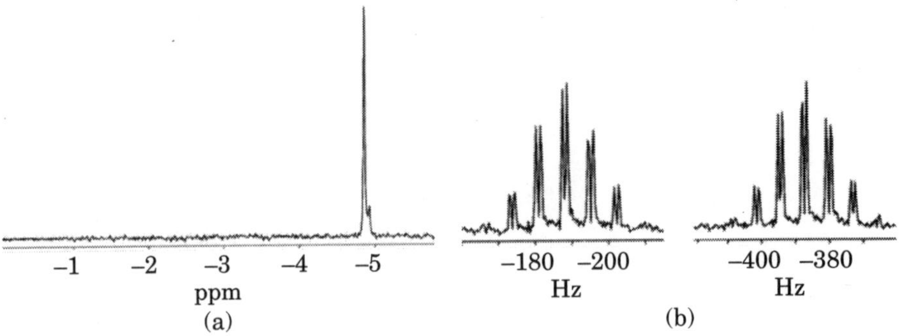

Figure 6.8: ^{29}Si-NMR proton decoupled (a), and proton coupled spectrum (b) of tetra ethyl disiloxane.

6.4 ^{31}P-NMR (NUCLEAR MAGNETIC RESONANCE)

6.4.1 Introduction

^{31}P is an isotope of phosphorus that is commonly available in inorganic and organic of natural as well as synthetic class of compounds. The sensitivity of ^{31}P nucleus is 6.63×10^{-2} and it possess a net half spin. It is highly abundant and very useful for organic chemists as well biochemists due to its presence in nucleic acids and other bioorganic molecules.

In ^{31}P-NMR 85% of H_3PO_4 is the only internal standard that is used in ^{31}P-NMR .

6.4.2 Interpretation

Example: *^{31}P-NMR Proton Decoupled and Proton Coupled Spectra of Diethyl Chloro Phosphate* (Figure 6.9).

Interpretation

In case of ^{31}P-NMR proton coupled spectrum of di-ethyl chloro phosphate, the peak of ^{31}P nuclei is splitted into a quintet due to coupling with hydrogens of methylene group. As there is no further splitting of ^{31}P nuclei quintet peaks it shows that ^{31}P nuclei peak does not undergoes any further coupling with hydrogens of methyl group.

In case of proton decoupled spectrum of ^{31}P-NMR of di-ethyl chloro phosphate, only a singlet peak is observed at 5 ppm.

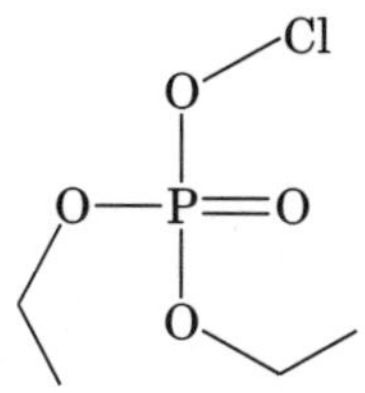

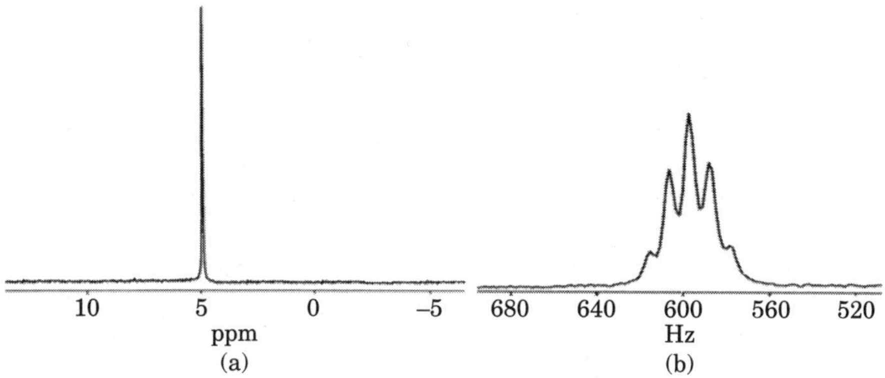

Figure 6.9: ^{31}P-NMR proton decoupled (*a*), and ^{31}P-NMR proton coupled (*b*) spectrum of ethyl chloro phosphate.

6.5 ^{17}O- NMR (NUCLEAR MAGNETIC RESONANCE)

6.5.1 Introduction

^{17}O is an isotope of oxygen that is available in very less abundance *i.e.*, 0.037% in both inorganic and organics that too in natural as well as synthetic class of compounds. Generally, ^{17}O nucleus possess net 5/2 of spin. It is highly abundant. Very useful for organic chemists as well biochemists due to its presence in many organic and bioorganic molecules.

In case of ^{17}O-NMR, water or dioxane are commonly used as internal standards. In case of ^{17}O-NMR among proton coupled and decoupled spectra, commonly the proton coupled spectra are mostly preferred. This is because coupled spectra are encompasses loss of sensitivity and broadness of signals.

6.5.2 Interpretation

Example: *^{17}O-NMR Spectra of Primary, Secondary, and Tertiary Alcohols* (Figure 6.10).

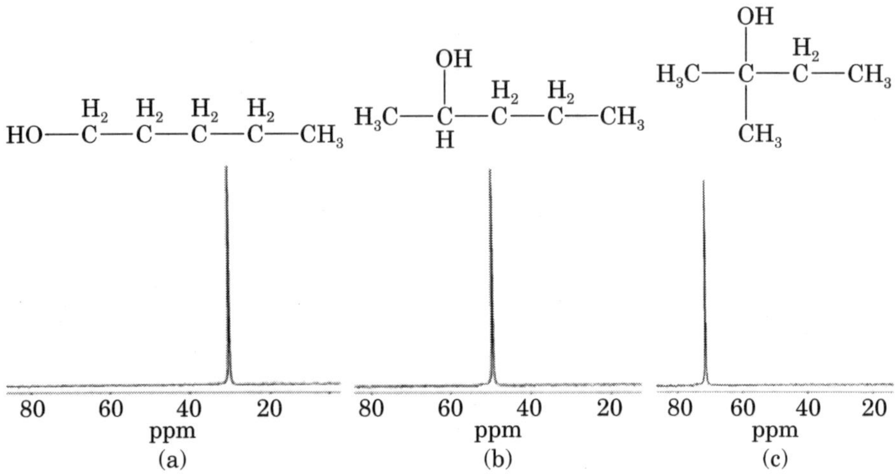

Figure 6.10: ^{17}O-NMR spectrum of primary (a), secondary (b), and tertiary (c), forms of alcohols.

Interpretation

The primary, secondary and tertiary forms of pentanol decoupled ^{17}O-NMR spectrum shows different signals at 30, 50 and 70 ppm.

6.6 PROBLEMS BASED LEARNING

After reading the other nuclei NMR spectroscopy chapter, the readers are suggested/advised to test their learning by solving some below mentioned problems.

A. SUBJECTIVE PROBLEMS

1. Discuss ^{15}N-NMR spectroscopy, along with the factor affecting their signals.
2. Describe ^{19}F-NMR spectroscopy and the associated internal standard.

3. Write a note on ^{29}Si-NMR.
4. Explain ^{31}P-NMR.
5. Briefly write about the aspects of ^{17}O-NMR.

B. OBJECTIVE PROBLEMS

1. State True or False

(*i*) ^{17}O is available in very less abundance *i.e.*, 0.037%.

(*ii*) Sensitivity of ^{31}P nucleus is 6.63×10^{-2}.

(*iii*) ^{19}F possess an appreciable sensitivity of 0.83.

(*iv*) ^{14}N possess 99% of natural abundance with spin of one, Whereas ^{15}N possess 0.37% of natural abundance with spin of half.

(*v*) Chances of overlapping of nitrogen nuclei signal are negligible.

(*vi*) In ^{15}N NMR proton coupled spectrum $^{1}J_{N-H}$ (75–135 Hz), $^{2}J_{N-H}$ (0–20 Hz) and $^{3}J_{N-H}$ (with 0–10 Hz) couplings are very common.

(*vii*) Decoupling in ^{19}F NMR does not causes NOE enhancement of ^{19}F signals.

(*viii*) ^{19}F-NMR is recorded in a continuous wave instrument.

(*ix*) Sensitivity of ^{29}Si nucleus is twice of the ^{13}C nuclei.

Ans.: 1. (*i*) True (*ii*) True (*iii*) True (*iv*) True
(*v*) True (*vi*) True (*vii*) True (*viii*) True
(*ix*) True.

2. Fill up the blanks

(*i*) In .... NMR chromium (III) acetylacetonate is used as catalyst.

(*ii*) .... nucleus possess net 5/2 of spin.

(*iii*) .... nuclei possess 4.7% of natural abundance.

(*iv*) In case of ^{15}N-NMR spectrum of formamide nitrogen signal appears at ....

(*v*) In case of ^{15}N-NMR spectrum of pyridine nitrogen signal appears at ....

(*vi*) In ^{19}F-NMR spectrum of fluoroacetone fluorine signal appears at ....

Ans.: 2. (*i*) ^{15}N (*ii*) ^{17}O (*iii*) ^{29}Si
(*iv*) 112.5 (*v*) 314 (*vi*) –226.

3. Match the following

(*i*) ^{15}N-NMR	(*a*) Liquid ammonia
(*ii*) ^{19}F-NMR	(*b*) $CFCl_3$
(*iii*) ^{29}Si-NMR	(*c*) TMS
(*iv*) ^{31}P-NMR	(*d*) H_3PO_4
(*v*) ^{17}O-NMR	(*e*) Dioxane

Ans.: 3. (*i*) – (*a*), (*ii*) – (*b*), (*iii*) – (*c*), (*iv*) – (*d*), (*v*) – (*e*).

6.7 SUGGESTED READINGS

[1] James Keeler, *Understanding NMR Spectroscopy*, Second Edition, Wiley, 2011.

[2] Pavia D.L., Lampman G.M., Kriz G.S., Vyvyan J.R., Introduction to Spectroscopy, 4th Edition, Brooks/Cole, 2009.

[3] Neil E. Jacobsen, *NMR Spectroscopy Explained: Simplified Theory, Applications and example for organic chemistry and structural biology*, Wiley, N.J., 2007.

[4] Douglas A. Skoog, F. James Holler and Stanley R. Crouch, *Principles of Instrumental Analysis*, Brooks Cole; 6 edition, 2007.

[5] Kalsi P.S., *Spectroscopy of organic compounds*, 6th Edition, New Age International, 2007.

[6] Silverstein R.M., Webster F.X. and Kiemle D.J.; *Spectrometric Identification of Organic Compounds*, 7th edition, John Wiley, 2005.

[7] Yadav L.D.S., *Organic Spectroscopy*, Kluwer Academics Publisher, Dordrecht, Netherland, 2005.

[8] Terence N. Mitchell, Burkhard Costisella, *NMR-from Spectra to Structures: An Experimental Approach*, Springer Verlag Berlin Heidelberg, N.Y., 2004.

[9] Dyer J.R., *Applications of Absorption Spectroscopy of Organic Compounds*, Prentice-Hall, India, 1965.

[10] Breitmaier E., Voelter W., *13C NMR Spectroscopy: Methods and Applications*, Verlag Chemie, 1974.

Chapter 7

Advance NMR Spectroscopy

Objective

To make readers learn

- Basic Concepts, Features and Interpretation of Spectra/Data of Advance NMR Experiments such as DEPT, Two Dimensional NMR, Three Dimensional NMR, Multiple Dimension NMR.
- Concepts and Theory of Solid State NMR.

7.1 INTRODUCTION

Conventional NMR spectroscopy of single nuclei (such as ^{1}H, ^{13}C, Si, P, N, O and S) is not sufficient for gathering complete structural information of an organic compound. So, with a view to generate more information about organic compound structure, a wide number of advance/modern techniques are developed such as

- DEPT.
- 2D-NMR (COSY, HETCOR, HMBC, INADEQUATE, TOCSY, HOHAHA Technique, NOESY, ROESY, HOESY).
- 3D-NMR.
- Multiple Dimension NMR.

7.2 DEPT (DISTORTION LESS ENHANCEMENT BY POLARIZATION TRANSFER)

DEPT is an advance technique which is used to determine the number of hydrogens attached to a given carbon atom. DEPT employs introduction of a complex sequence of pulses

in both ^{1}H and ^{13}C channel. As a result different phases of ^{13}C carbon atoms are generated, that in turn depends upon the number of hydrogens attached to each carbon. DEPT records different experiments in ^{1}H and ^{13}C channel by using different pulses. In DEPT experiment pulses are selected on the basis of their flip angle such as $(\pi/2)_z$ with flip angle of 45°, $(\pi/2)_y$ with 90°, $(\pi/2)_x$ with 135° and also their delay time. After introduction of complex pulses, carbon atoms contained in CH_3, CH_2 and CH groups shows different phases and recorded. The phase of carbons depends upon duration of delays in the pulse sequence. DEPT is much complicated than off resonance decoupling. One of the pulse sequence for DEPT is shown in Figure 7.1.

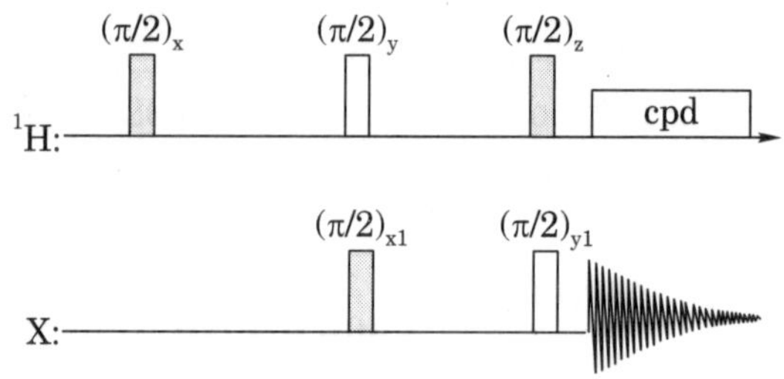

Figure 7.1: Pulse sequence for DEPT.

Key Concepts

- In case of DEPT-45 experiment, carbon that possesses hydrogen (number 1 or 2 or 3) shows peak. Where as carbon that does not possess hydrogens does not show any peak.
- In DEPT-90 experiment a higher duration of delay is applied than DEPT-45. As a result carbon which possesses only one hydrogen, exhibits peak in the spectrum. On the other hand carbon which possesses two or three hydrogens does not show any peak.
- DEPT-135, makes use of further higher duration of delay, than DEPT-90. Net result of such experiment is that carbon which possesses only one and three hydrogens shows positive peaks (*peak appears above the axis or upside*). Whereas carbon that possess two hydrogens shows a negative peak (*peak appears below the axis or downside*).

- Hence, carbons having different number of hydrogens vary in behavior with different DEPT experiments. This variation indicates the presence of number of hydrogens on each carbon.
- DEPT requires a Fourier transformed pulse spectrometer with a computer interface.

Example: *DEPT Spectrum of Ethyl-2-Butenoate* (spectrum given in Figure 7.2).

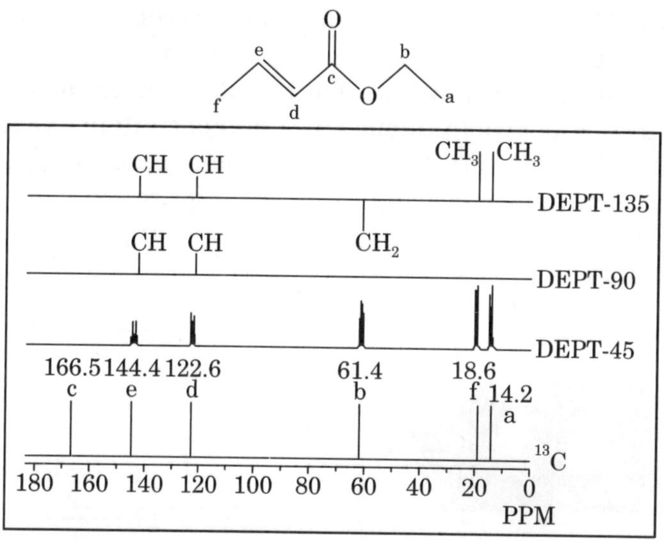

Figure 7.2: DEPT spectrum for ethyl-2-butenoate.

Interpretation

In DEPT spectrum of ethyl-2-Butenoate due to different pulses four plots, are observed.

- The down 4th line ^{13}C spectrum shows ^{13}C signal at 14.2, 18.6, 61.4, 122.6, 144.4, and 1.66.5 for a, f, b, d, e, and c carbons respectively.
- The 3rd line spectrum of DEPT 45, exhibits signals for those carbons that possess hydrogen (*one*, *two* or *three*).
- 2nd Line spectrum of DEPT 90, generates signals only for those carbons, which possess only one hydrogen (*i.e., carbons with two or three hydrogens does not show any peak).*

- The first line spectrum of DEPT-135, shows positive peak only for carbons which possess only one and three hydrogen shows (*i.e., peak above the axis or upside*). Whereas those carbons that possess two hydrogens shows a negative peak (*i.e., peak below the axis or downside*).

7.3 2D (TWO DIMENSIONAL) NMR

Simple ^{1}H NMR or ^{13}C or other nuclei spectroscopy possess one frequency axis and one intensity axis. So, conventionally called as 1D-NMR due to one frequency axis. On the other hand correlation spectroscopy possesses two frequency axis and one intensity axis so termed as 2D-NMR. The correlation of two nuclei generates a stack plot or contour plot comprising contour peaks. Figure 7.3, shows the stack and contour plots of peaks or signals that appears in a 2D NMR experiments.

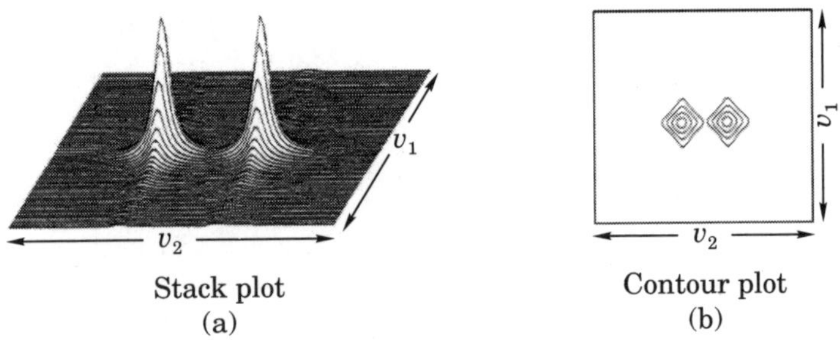

Figure 7.3: Stack plot (a) and contour plot (b) for peaks or signals that appears in a 2D NMR experiement.

2D-NMR involves different experiments a few of which are as follows

- COSY (SQF, DQF, TQF & MQF)
- HETCOR (HSQC, HMQC & HMBC)
- INADEQUATE
- TOCSY
- NOESY

- HOESY
- ROESY

7.3.1 COSY (Correlation Spectroscopy)

COSY is defined as a homonuclear 2D-NMR correlation spectroscopy. COSY correlates the chemical shift of two hydrogen nuclei located on two different carbons that are separated by a single bond via *J* coupling. Thereby detects the chemical shift for hydrogen's on both F1 and F2 axis.

Types of COSY

Based on type of pulse sequence COSY experiment is categorized as follows:

- Simple COSY
- DQF COSY
- TQF COSY
- MQF COSY

7.3.1.1 *Simple COSY or SQF (Single Quantum Filtered) COSY*

Simple COSY technique incorporates a simple pulse sequence in which firstly a $(\pi/2)_x$ pulse is introduced in ^1H channel to create an evolution phase. There after some time a second $(\pi/2)_y$ pulse is introduced to create an acquisition phase. The schematic diagram for pulse sequence of simple COSY is given in Figure 7.4.

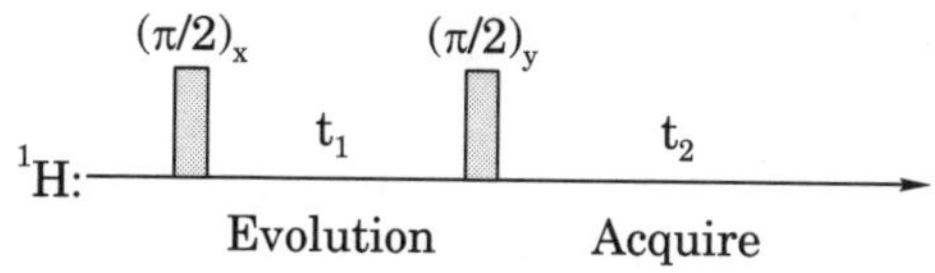

Figure 7.4: Schematic diagram for pulse sequence of simple COSY.

Key Concepts of simple COSY

- In ^1H–^1H COSY pulse sequence contains variable relaxation delay time (t_1) and acquisition time (t_2).

- For identification of coupling among protons in a molecule, the coupling interaction is allowed to commence during t_1. In this period due to spin-spin coupling the individual nuclear magnetization spreads and modifies the signal obtained during t_2.
- Experiment is repeated with different values of t_1 and t_2. So that the value of t_1 is increased at regular intervals, to generate a series of different FID data during t_2.
- COSY offers three bond coupling ($^3J_{H-H}$).

Theory

Let us assume a spin system of two nuclei that undergoes coupling. In this system, two protons are coupled to each other.

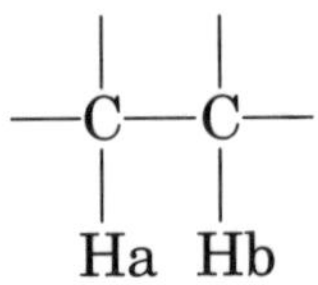

The pulse and initial t_1 phase produces a spin system by rotating magnetization vector of nuclei by 90°. In t_1 phase nuclei undergoes precession under influence of chemical shift and mutual spin coupling, so that each nuclei contains spin of only one of two protons. This precession modifies signal obtained during t_2. Where mutual coupling of spin causes a single spin nuclei to possess magnetic component of both nuclei.

After t_1 phase, another 90° pulse so called mixing pulse is applied. This introduces a mixing period where it distributes magnetization into various spin states of coupled nuclei.

The distributed magnetization containing chemical shift component of t_1 phase, when detected at another chemical shift during t_2 phase generates two types of peaks in 2D plot enumerated as:

- **Diagonal peaks, attributed to two nuclei having same precessional frequency *i.e.*, $v_1 = v_2$.** These are of less importance because less informative.

- **Off diagonal peak or cross peak, that are attributed to two nuclei having different precessional frequency *i.e.*,** $v_1 \neq v_2$. These peaks are very informative because these are result of magnetic combination of chemical shift of t_1 and after mixing period of one nuclei with t_2 precession of another nuclei. These peaks are useful to determine correlation of coupled nuclei and further confirms the coupling between two protons, because if there is no coupling-no diagonal peaks appears.

Lastly, the obtained data are Fourier transformed, which generates chemical shift of first proton on F_1 axis and second proton on F_2 axis.

Example: *^{1}H–^{1}H COSY Spectrum of Ethyl-2-Butenoate* (spectrum given in Figure 7.5).

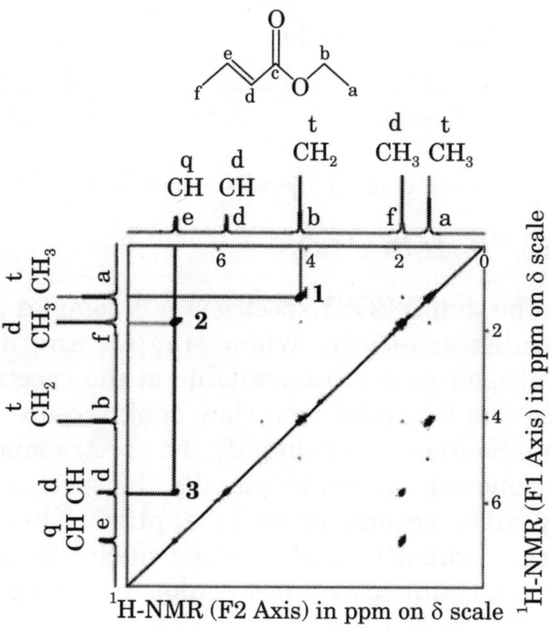

Where d = doublet, t = triplet, and q = quartet

Figure 7.5: ^{1}H–^{1}H simple COSY spectrum for ethyl-2-butenoate.

Interpretation

- The off diagonal peaks at point 1, 2, and 3 represents coupling of hydrogen protons of a with b, e with f, and e with d protons.

There is no cross peak for c because it does not possess any hydrogens.

7.3.1.2 *^{1}H–^{1}H DQF (Double Quantum Filtered) COSY*

^{1}H–^{1}H DQF COSY is a technique of COSY that incorporates a typical pulse sequence, in which firstly a $(\pi/2)_x$ pulse is introduced in ^{1}H channel. Nextly, after first pulse a second $(\pi/2)_y$ pulse with just immediate third $(\pi/2)_z$ pulse is introduced to eliminate singlet peaks. This simplifies the complex COSY spectrum. The schematic diagram for pulse sequence of ^{1}H–^{1}H DQF COSY experiment is shown in Figure 7.6.

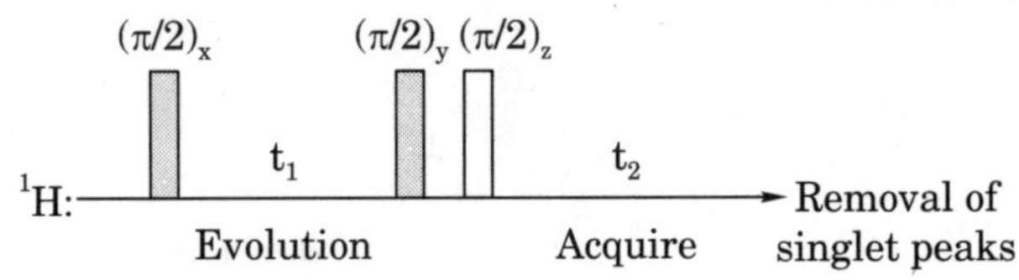

Figure 7.6: Schematic diagram for pulse sequence of ^{1}H-^{1}H DQF COSY.

Key Concepts of ^{1}H-^{1}H DQF COSY

Generally, the simple COSY spectra is very complex to understand in big molecules especially when singlet, and doublets and triplets or multiplet peaks are available in the spectrum. Isolated nuclei offers a singlet peaks, and they don't require to be studied for correlation. So in order to simplify the spectra single peaks can therefore be removed. Hence, to simplify the spectra a third pulse immediately after second pulse is applied. This removes all single quantum transitions of isolated nuclei from the spectra and preserves all multiplet cross peaks. Hence spectra become simplified.

Example: *^{1}H-^{1}H DQF COSY Spectrum of Ethyl Acetate* (Spectrum given in Figure 7.7).

The significance of ^{1}H-^{1}H DQF COSY experiment can be very well understood from ^{1}H-^{1}H simple COSY and ^{1}H-^{1}H DQF COSY spectra of ethyl acetate given in Figure 7.7.

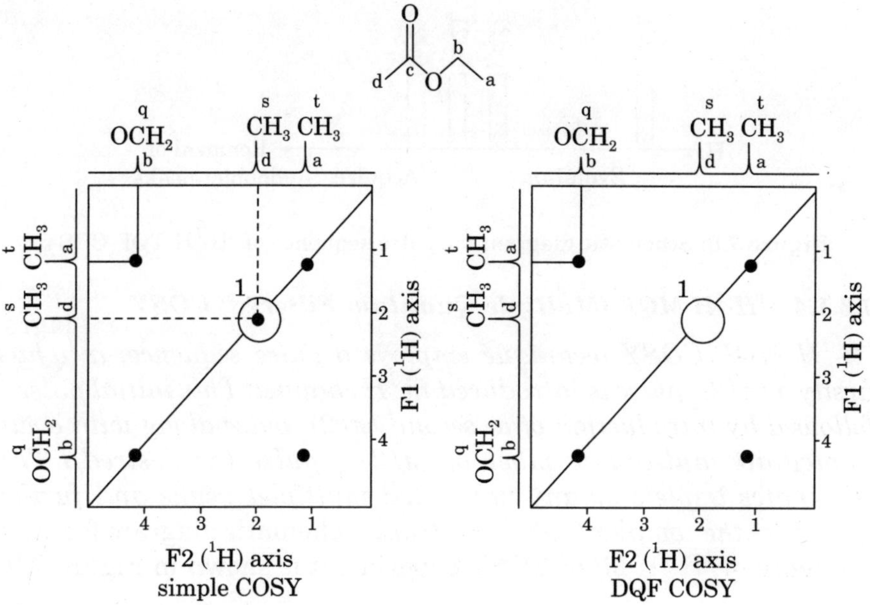

Figure 7.7: ^{1}H-^{1}H simple COSY and ^{1}H-^{1}H DQF COSY spectra of ethyl acetate.

Interpretation

It can be clearly observed that in simple COSY spectrum of ethyl acetate the isolated 'd' hydrogen produces a singlet cross peak at spot '1'. But in case of ^{1}H-^{1}H DQF COSY spectrum of ethyl acetate spot '1' is removed and the spectra is simplified.

7.3.1.3 *^{1}H-^{1}H TQF (Triple Quantum Filtered) COSY*

^{1}H-^{1}H TQF COSY employs further typical pulse sequence, in which initially a $(\pi/2)_x$ pulse is introduced in ^{1}H channel. This first pulse is followed by introduction of a second $(\pi/2)_y$ pulse along with just immediate third $(\pi/2)_z$ and fourth $(\pi/2)_{z1}$pulse. This eliminates singlet and doublet peaks and simplifies the complex COSY spectra. Schematic diagram for pulse sequence of TQF COSY is shown in Figure 7.8.

Key Concepts of ^{1}H-^{1}H TQF COSY

When spectra is too complex to be resolved even after removal of singlet peaks, than after second pulse immediately third and fourth pulse are applied to remove doublet peaks also.

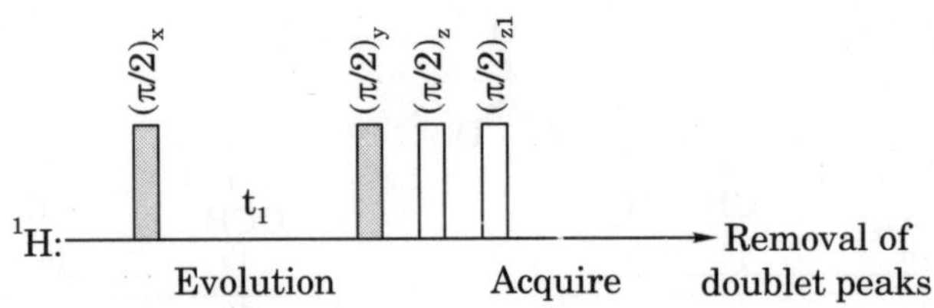

Figure 7.8: Schematic diagram for pulse sequence of ^1H-^1H TQF COSY.

7.3.1.4 *^1H-^1H MQF (Multiple Quantum Filtered) COSY*

^1H-^1H MQF COSY technique employs a pulse sequence, in which firstly a $(\pi/2)_x$ pulse is introduced in ^1H channel. This initial pulse is followed by introduction of a second $(\pi/2)_y$ pulse along with a just immediate multiple number of $(\pi/2)_{zm}$ pulse (as desired). This eliminates triplets or any unwanted multiplet peaks and further simplifies the complex COSY spectrum. Schematic diagram for pulse sequence of ^1H-^1H MQF COSY experiment is shown in Figure 7.9.

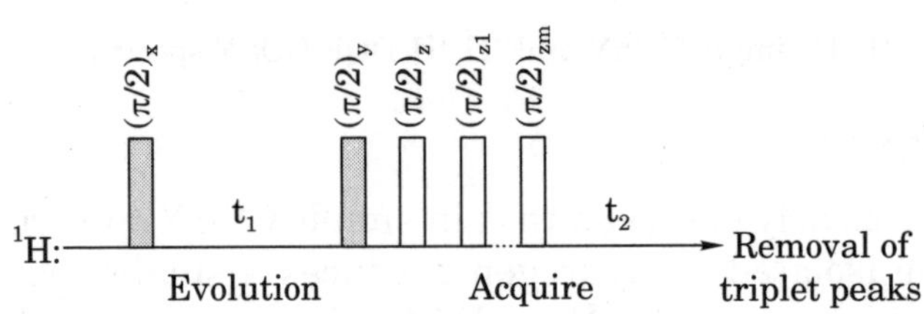

Figure 7.9: Schematic diagram for pulse sequence of ^1H-^1H MQF COSY.

Key Concepts of ^1H-^1H MQF COSY

When spectra is too complex to be resolved even after removal of triplet peaks, than after second pulse immediately 'm' number of peaks (that is more than three pulses as per the type of multiplet) can be applied so as to remove any unwanted multiplet peak.

7.3.2 Hetero Nuclear Correlation Spectroscopy (HETCOR)

HETCOR is a heteronuclear 2D-NMR correlation spectroscopy, which correlates the chemical shift coupling of hetero nucleus *e.g.*, ^{13}C or ^{15}N with hydrogens that are separated by a single bond, that detects chemical shift for hydrogen's

on F_1 axis and hetero nucleus on F_2 axis. HETCOR follows one bond couplings. For example in case of correlation of ^{13}C with ^{1}H $^{1}J_{C\text{-}H}$ is followed. The schematic diagram for pulse sequence of HETCOR experiment is shown in Figure 7.10.

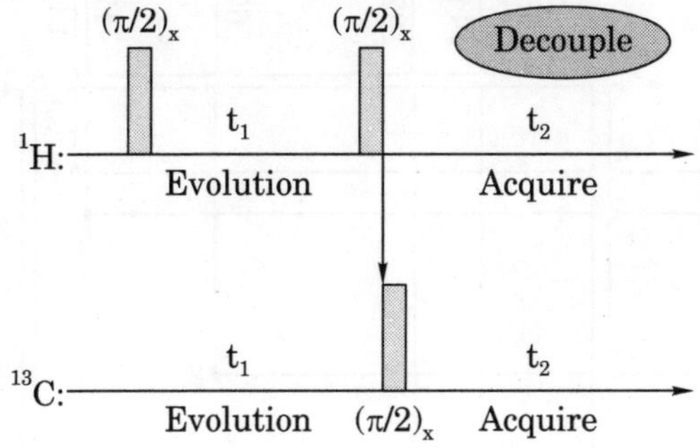

Figure 7.10: Schematic diagram for pulse sequence of HETCOR.

Key Concepts of HETCOR

- In HETCOR $\pi/2$ read pulse is applied in ^{13}C channel.
- As a result FID occurs in t_2. Therefore for ^{13}C (v_2) scale occurs on F_2 axis and ^{1}H (v_1) scale occurs on F1 axis due to incremental delay.
- Decoupling is applied in ^{1}H channel (acquisition phase), due to which ^{13}C signal appears in singlet forms such that one bond coupling ($^{1}J_{C\text{-}H}$) are observed.
- As in 2D experiment correlation occurs during t_1 so during ^{1}H evolution phase decoupler is made off.
- Interpretation is very simple *i.e.*, a line is dropped vertically from ^{13}C to encounter a proton peak.
- If no cross peak appears, it signifies that no hydrogen is present over a particular carbon.
- If one cross peak appears, it indicates that there can be one, two or three hydrogens.
- If two cross peaks appears it indicates the presence of a diastereotype carbon (hydrogen).

Example: *HETCOR Spectrum of Ethyl-2-Butenoate* (spectrum given in Figure 7.11).

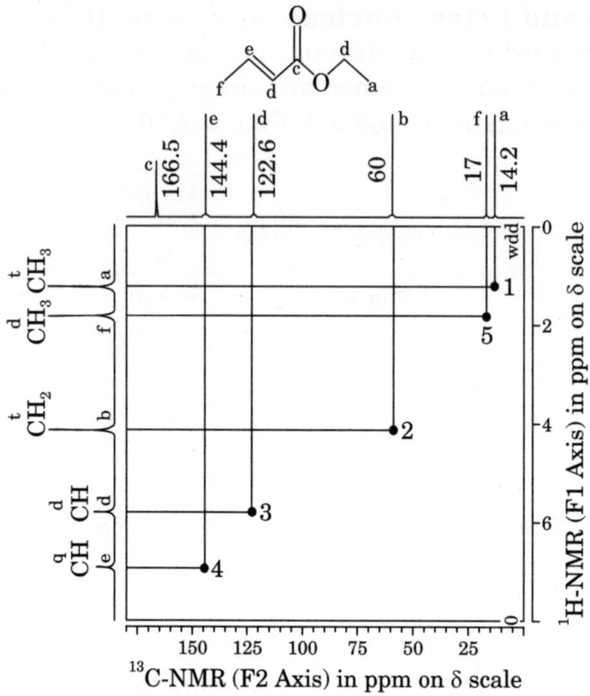

Figure 7.11: HETCOR spectrum of ethyl-2-butenoate.

Interpretation

- In HETCOR spectrum of ethyl-2-Butenoate the contour (cross peaks) at point **1**, **2**, **3**, **4** & **5** accounts for **a**, **b**, **d**, **e** & **f** carbons and attached hydrogens with them.
- In the spectrum. No cross peak is observed for carbon 'c' and its directly attached protons of hydrogens, because carbon 'c' does not contains any hydrogen so no coupling. Hence, with HETCOR, one can easily determine the single bonded hetero nuclear correlation.

7.3.3 Reverse Phase HETCOR Spectroscopy

As earlier HETCOR $^1J_{C\text{-}H}$ presented many experimental difficulties, so inverse detection technique is applied over earlier HETCOR experiments. **Reverse Phase HETCOR is defined as a heteronuclear 2D-NMR correlation spectroscopy, which correlates the chemical shift of hetero nucleus *e.g.*, ^{13}C or**

^{15}N via coupling with hydrogens separated by a single bond. The reverse phase HETCOR detects chemical shift for hydrogen's on F2 axis and hetero nucleus on F1 axis. In reverse HETCOR, results are same as HETCOR but here ^{13}C is F1 axis and ^{1}H is F2 axis *i.e.*, reverse phase. F2 axis is ^{1}H detected because correlation occurs more in ^{1}H channel. This is donebecause ^{13}C requires less resolution derived from t_1 and this is important for cases where ^{13}C have close resonances *i.e.*, superimposing peaks. Most of new instruments are able to run both HETCOR and reverse phase. Reverse Phase HETCOR is of two types.

7.3.3.1 *Heteronuclear Multiple Quantum Coherence* (*HMQC*)

In HMQC experiment, during evolution time both ^{1}H and X magnetization (e.g., X = ^{13}C) are allowed to evolve. It is affected by homonuclear ^{1}H coupling during evolution period. The homonuclear ^{1}H coupling manifests itself as broadening of X dimension.

7.3.3.2 *Heteronuclear Single Quantum Coherence* (*HSQC*)

In HSQC technique only X magnetization is allowed to evolve. It is unaffected by homonuclear ^{1}H coupling during evolution period because no ^{1}H-magnetization occurs during evolution time.

Example: *HMQC and HSQC Spectra of Menthol* (Spectra given in Figure 7.12a and 7.12b).

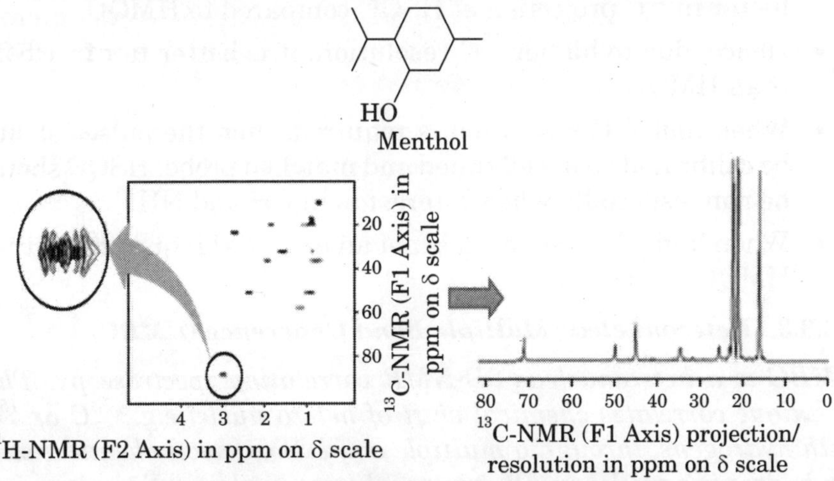

Figure 7.12a: HMQC spectrum of menthol.

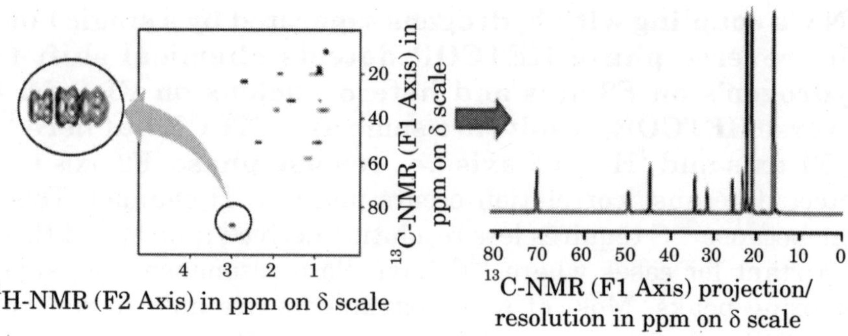

Figure 7.12b: HSQC spectrum of menthol.

Interpretation

- The HMQC spectrum [given in Figure 7.12(a)] at 2.99 shows an expanded $^{1}H/^{13}C$ cross peak (*i.e.*, more expanded resonance and broad ^{13}C dimension) in comparison to HSQC spectrum [given in Figure 7.12(b)]. The cross peak at 2.99 in both HMQC and HSQC is shown in a circle magnified outside and nonmagnified inside the spectrum.
- The right hand side of two spectra shows corresponding ^{13}C projections of HSQC and HMQC spectra.
- It can be very well seen in the ^{13}C resolution/projections of two HMQC and HSQC spectra of menthol, that resolution is much better in ^{13}C projection of HSQC compared to HMQC.
- Hence, due to higher ^{13}C resolution, it is better to run HSQC than HMQC.
- When high ^{13}C resolution is required, then the pulses should be calibrated on a well tuned and matched probe. HSQC should be run, especially when interest is in CH and NH.
- When high ^{13}C resolution is not an essential requirement then HMQC can be run.

7.3.3.3 *Heteronuclear Multiple Bond Coherence* (*HMBC*)

HMBC is a heteronuclear 2D-NMR correlation spectroscopy. This technique correlates chemical shift of hetero nuclei e.g., ^{13}C or ^{15}N with hydrogens, through a multiple bond. So, detects chemical shift for hydrogen's nuclei on F2 axis and hetero nuclei on F1 axis.

Key Concepts of HMBC

- This technique provides indirect C–C correlations.
- It is an aid to trace skeleton through indirect 'C' connectivity's because quaternary 'C' can be correlated with nearby protons (which is not possible with ordinary HETCOR).
- HMBC is nonflexible as it yield unexpected results *i.e.*, $^2J_{C-H}$ are always found and $^3J_{C-H}$ are occasionally absent, because of variation in magnitude of $^2J_{C-H}$ and $^3J_{C-H}$.
- Over two or three bonds (J = 2–15 Hz) are utilized.
- Tedious technique as correlates $^2J_{C-H}$, $^3J_{C-H}$ between ^1H & ^{13}C that are 2–3 bond away.
- Direct one-bond $^1J_{C-H}$ are removed and provides more correlation.
- Widely used because based on proton-detection, offers high sensitivity compared to carbon-detected COLOC (correlation spectroscopy for long range coupling). In addition, long-range proton-carbon coupling constants can be measured from resulting spectra.

Example: *HMBC of Ethyl Benzene* (spectrum given in Figure 7.13).

Interpretation

- When a vertical line is drawn parallel to ^1H axis and vertical to ^{13}C axis, from 16.2 ppm corresponding to C–2′, it intersects three cross peaks of which one corresponds to H–1′, which shows correlation of C–2′ carbon with two hydrogens attached on C–1′. Whereas other cross peaks do not show any correlation.
- Vertical line when drawn for C–1′ at 29 ppm, intersects one cross peak, which shows correlation of C–1′ to hydrogens present on C–2′ carbon.
- Similarly, correlation exists for C–1 with hydrogens attached to C–1′, C–2′ and C–2; correlation exists for C–2 with hydrogens attached to C–3, C–4, and C–1′; correlation exists for C–4 with hydrogens attached to C–3; correlation exists for C–3 with hydrogens attached to C–2 and C–4; also correlation exists for C–4 with hydrogens attached to C–3 and C–2.

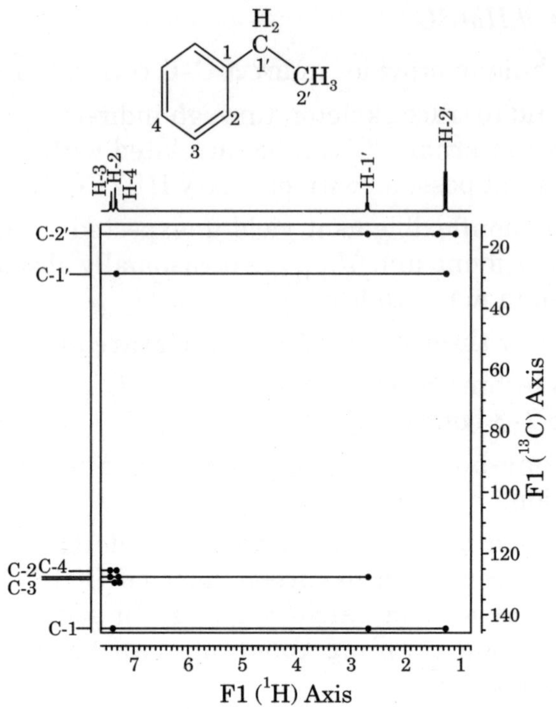

Figure 7.13: HMBC spectrum of ethylbenzene.

Example: *HMBC Spectrum of Ipsenol* (Spectrum given in Figure 7.14).

Interpretation

- When a vertical line is drawn parallel to ^{1}H axis and vertical to ^{13}C axis, from 68 ppm corresponding to C4, it intersects five cross peaks. Among these five cross peaks, four peaks corresponds to two pairs of diastereotypic methylene groups (2.48, 2.22, 1.45 and 1.28) and these represents $^2J_{C-H}$ couplings. And none of five peaks corresponds to $^1J_{C-H}$ at 3.8 ppm. The fifth cross peak corresponds to $^3J_{C-H}$ coupling with isopropyl methine at 1.82. As the other carbon atom in beta position to C4, do not possess any protons so there is no correlation found. As no cross peak is observed at 3.4 it show there is no $^1J_{C-H}$ interaction.
- When another line is drawn for C–5 at 41 ppm than correlations to attached protons 2.48 and 2.22 are absent. There is only one alpha carbon. One corresponding peak occurs for C–4 carbinol

methine at 3.83. There are 3 beta carbons with protons. C–3 methylene carbon shows indirect correlations through both of its diastereotype protons at 1.45 and 1.28. C–7 olefinic methine protons gives a cross peak at 6.39. The same is protons of methylene group attached at C–6 at 5.16 and 5.09.

- Similarly, correlation exists for C–1 & C–1′ with hydrogens attached to C–2 and C–3; correlation exists for C–2 with hydrogens attached to C–1 & C–1′, C–3, and C–4; correlation exists for C–3 with hydrogens attached to C–1 & C–1′, C–2, C–4 and C–5; correlation exists for C–6 with hydrogens attached to C–4, C–5, C–7, C–8 and C–9; correlation exists for C–7 with hydrogens attached to C–5, C–8, and C–9; correlation exists for C–8 with hydrogens attached to C–7; also correlation exists for C–9 with hydrogens attached to C–5 and C–7.

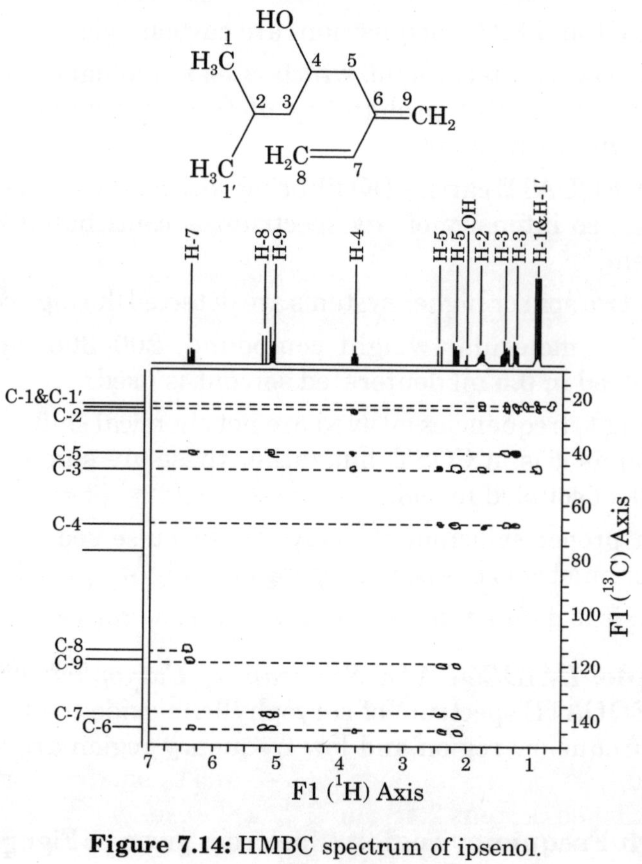

Figure 7.14: HMBC spectrum of ipsenol.

7.3.4 INADEQUATE (Incredible Natural Abundance Double Quantum Transfer Experiment) Spectroscopy

HMBC traces 'C' skeleton of molecules via indirect C–C connectivities, which is tedious, as we do not know whether correlation is due to $^{2}J_{C-H}$ or $^{3}J_{C-H}$. Hence, a new technique is introduced called INADEQUATE. **INDEQUATE is a homonuclear 2D NMR correlation spectroscopy, which correlates the chemical shift of C–C ($^{1}J_{C-C}$ coupling) separated by a single bond, that detects chemical shift for carbon on both F1 and F2 axis.** This follows one bond couplings.

Key Concepts of INADEQUATE

- INADEQUATE correlates between directly attached carbons (C–C).
- Both F1 and F2 (t_2 acquisition) are carbon axis.
- Most powerful technique, which is least ambiguous & easy to interpretate. But probability of finding two adjacent ^{13}C together is 1 in 10000.
- INADEQUATE carries DQ filtering which removes isolated ^{13}C atoms, so intensity of its spectrum is contributed by 2 spin system.
- Only two spin or higher systems are detected during acquisition.
- For low molecular weight compound, 200–300 mg sample dissolved in 0.5 ml deuterated solvent is used.
- During t_1 frequencies evolved are not chemical shifts of coupled nuclei as it is in COSY, but rather these are sum of chemical shifts of coupled nuclei.
- With proper selection of delays (τ) in pulse sequence selects larger number of one bond coupling *i.e.*, only $^{1}J_{C-C}$ can be selected.
- F1 axis is defined in Hz and it is twice the range of F2 axis.

Example: *INADEQUATE Spectrum of Caryophyllene Oxide*— In INADEQUATE spectrum of caryophyllene oxide the two regions *i.e.*, high frequency region and low frequency region are explained separately.

A. High Frequency region (Spectrum given in Figure 7.15a).

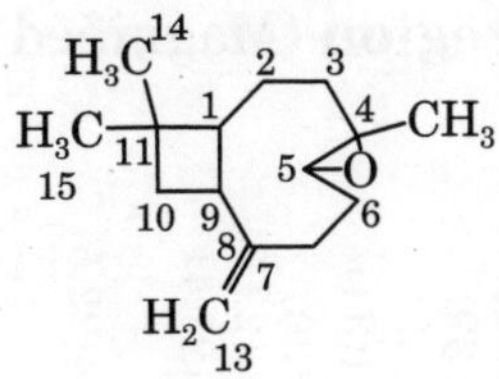

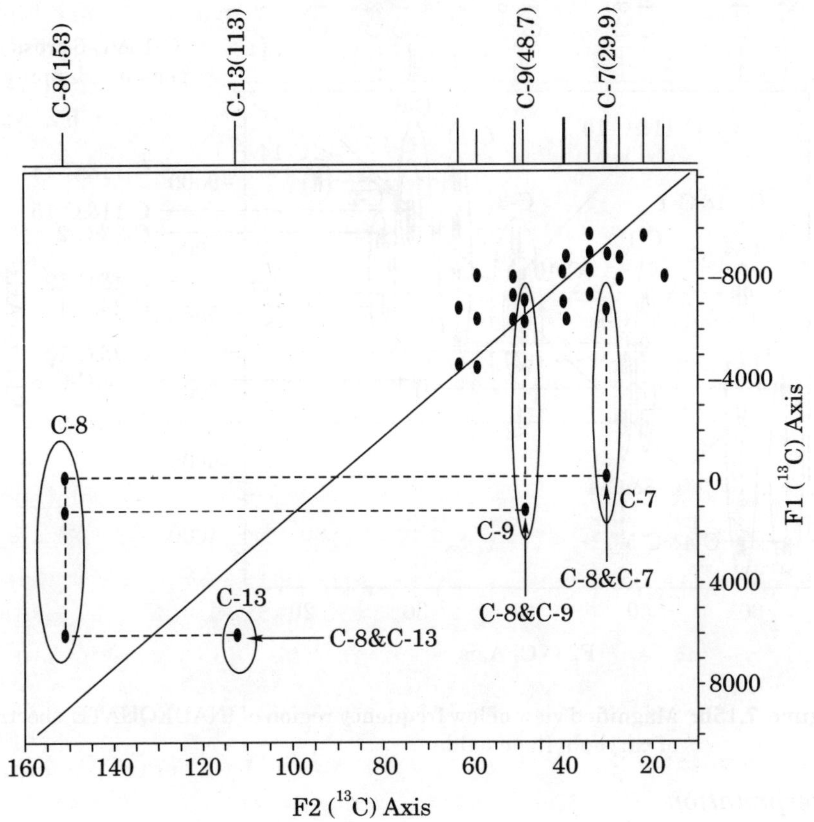

Figure 7.15a: High frequency region of INADEQUATE spectrum of caryophyllene oxide.

Interpretation

- When a line from C–8 at 153 is dropped vertically on F2, it intersects three cross peaks. Which further connects horizontally with C–7 at 29.9, C–9 at 48.7, and C–13 at 113. This show the connectivity of C–8 with C–7, C–9 and C–13.
- C–13 has only one cross peak which is connected to C–8 at 153.

B. Low Frequency region (Magnified spectrum view given in Figure 7.15b).

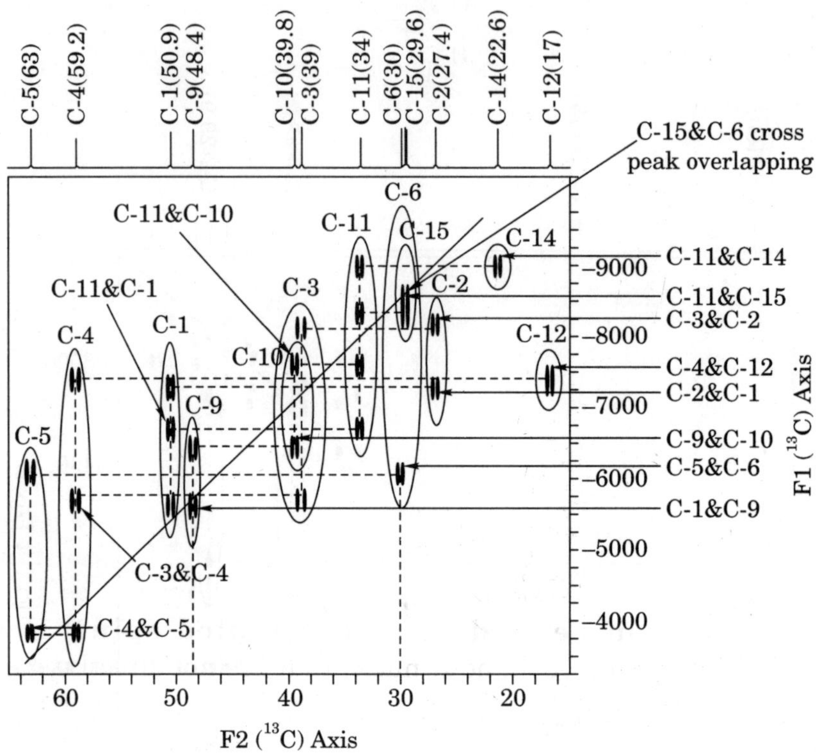

Figure 7.15b: Magnified view of low frequency region of INADEQUATE spectrum of caryophyllene oxide.

Interpretation

- When a vertical line is drawn at 34 vertical to F2 axis for C–11 (quaternary carbon), it intersects four cross peaks which further connects horizontally to cross peaks at 22.6, 30, 39.8 and 50.9. This shows the connectivities of C–11 to C–14 at 22.6, C–15 at 30, C–10 at 39.8 and C–1 at 50.9.
- Vertical line when drawn at 59.2 vertical to F2 axis for C–4, it intersects three cross peaks which further connects horizontally to cross peaks at 17, 39 and 63. This shows the connectivities of C–4 to C–12 at 17, C–3 at 39, and C–5 at 63.

- Vertical line when drawn at 63 vertical to F2 axis for C–5, it intersects two cross peaks which further connects horizontally to cross peaks at 30 and 59.2. This shows the connectivities of C–5 to C–4 at 30, and C–6 at 59.2.
- Vertical line when drawn at 50.9 vertical to F2 axis for C–1, it intersects three cross peaks which further connects horizontally to cross peaks at 27.4, 34 and 48.4. This shows the connectivities of C–1 to C–2 at 27.4, C–11 at 34 and C–9 at 48.4.
- Similarly, C–9 shows connectivities with C–10 and C–1; C–10 shows connectivities with C–11 and C–9; C–3 shows connectivities with C–4 and C–2; C–2 shows connectivities with C–1 and C–3.
- C–6 and C–7 undergoes overlapping in the spectrum.

7.3.5 TOCSY (Total Correlation Spectroscopy)

TOCSY is a homonuclear 2D NMR correlation spectroscopy, which correlates the chemical shift of ^{1}H-^{1}H ($^mJ_{H\text{-}H}$ coupling) nucleus separated by multiple bonds by transfering the magnetization beyond directly coupled spins. Hence correlations can be seen among nuclei that are not directly coupled but are within the same spin system. In TOCSY instead of second π/2 pulse a mixing period is inserted during which magnetization spin is locked on Y axis. The schematic diagram for pulse sequence of TOCSY is given in Figure 7.16.

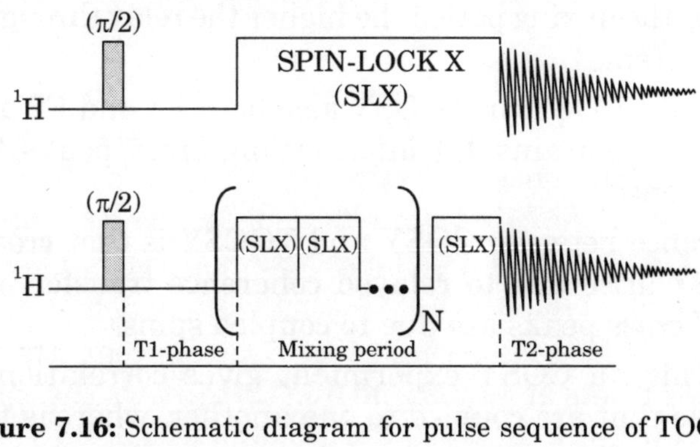

Figure 7.16: Schematic diagram for pulse sequence of TOCSY.

TOCSY experiment is conducted by two ways such as:

- 2D TOCSY
- 1D TOCSY

7.3.5.1 *2D-TOCSY*

The 2D TOCSY (Total Correlation Spectroscopy) sequence is a homonuclear experiment which produces a COSY-like plot. It possess two dimensions or axis of ^{1}H-NMR. A 2D-TOCSY, gives correlations between all protons in a given spin system. The mixing time in 2D TOCSY is so long that magnetic coherence is transferred almost throughout each spin system of molecule. Here resolution is little poor and still overlapping occurs but most of individual proton resonances are made clear.

Key Concepts of 2D TOCSY

- Commonly in correlation spins are allowed to evolve during t_1 under the influence of directly coupled nuclear spins.
- In this technique objective is to transfer the magnetization beyond directly coupled spins, hence correlations can be seen among nuclei that are not directly coupled but are within the same spin system.
- In TOCSY instead of second $\pi/2$ pulse a mixing period is inserted during which magnetization is spin locked on Y axis.
- During this mixing period magnetization is relayed from one spin to the neighbor, and then to the next neighbour and so on. Longer the mixing period the higher the relay throughout entire spin system.
- Same like COSY, in TOCSY also both F1 and F2 axis are ^{1}H; Diagonal contains 1D information; cross peaks have same appearance.
- Difference between COSY and TOCSY is that cross peaks in TOCSY arise due to relayed coherence transfer, whereas in COSY cross peaks are due to coupled spins.
- Generally, a COSY experiment gives correlations between protons that are coupled to one another, whereas the TOCSY

experiment gives correlations between all protons in a given spin system.

- On long mixing time in TOCSY all spins within a spin system appears to be coupled.

Example: *2D-TOCSY and COSY Spectrum of 3-Heptanone* (Spectra given in Figures 7.17a and 7.17b).

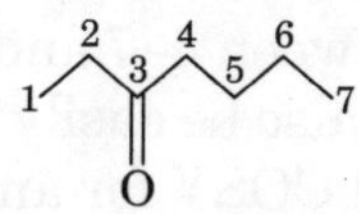

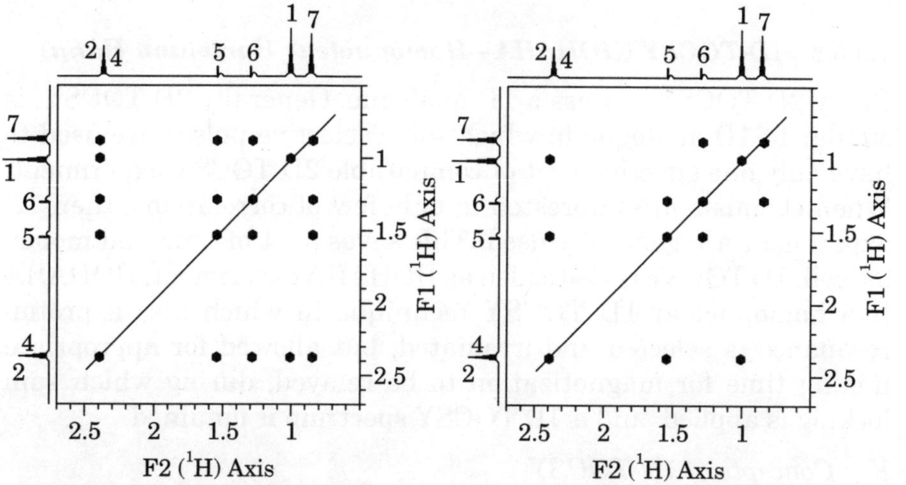

(a) TOCSY spectrum (b) COSY spectrum

Figure 7.17: 2D-TOCSY and COSY spectrum of 3-heptanone.

Interpretation

- The TOCSY plot of 3-heptanone (7.17a) is just like a COSY plot (Figure 7.17b).
- In the TOCSY spectrum of 3-heptanone, there occurs two ^{1}H spin systems *i.e.*, first is the protons on carbons 1 and 2 and the second is the protons on carbons 4, 5, 6 and 7.
- The first spin system with protons on carbons 1 and 2 generates a TOCSY spectrum which is similar to COSY.
- The second spin system of protons on carbons 4, 5, 6 and 7 generates a TOCSY spectrum which is different from COSY

spectrum. This is because in TOCSY spectrum all the protons are correlated to each other *i.e.*, it follows $^{n}J_{H-H}$ couplings. Whereas in COSY spectrum only $^{3}J_{H-H}$ couplings occurs.

- The TOCSY and COSY spectrum of 3-heptanone can be easily distinguished on the basis of correlation, such as COSY correlations occurs between 4–5, 5–6 and 6–7 hydrogens, but not between 4–7 and 5–7 hydrogens. Whereas TOCSY correlations occurs between 4–7 and 5–7 hydrogens. Similarly, other correlations can also be easily distinguished between two spectra of TOCSY and COSY for any given compound.

7.3.5.2 *1D-TOCSY (HOHAHA—Homonuclear Hartmann Hann)*

Every 2D TOCSY possess a 1D analogue. Generally 2D TOCSY, is similar to 1D analogue in which some selective pulses are used to have only one effective part of comparable 2D TOCSY experiment. When chemists are interested in only few of correlations, then 1D experiment are generally used. This saves a lot of time and money as well. 1D TOCSY is also called as HOHAHA experiment. HOHAHA is a homonuclear 1D TOCSY technique in which also a proton resonance is selected and irradiated; but allowed for appropriate mixing time for magnetization to be relayed, during which spin locking is applied and a 1D TOCSY spectrum is obtained.

Key Concepts of 1D-TOCSY

- In this 1D TOCSY spectrum only those signals are selected on whom magnetization is transferred. Whereas other signals which are outside the spin system do not appear.
- Like wise a series of 1D TOCSY are run, during which a systematic increase in mixing time is done while proton irradiated is kept constant.

Example: *Beta Lactose–with two separate spin system* (Spectral features given in Figure 7.18).

- Irradiate the anomeric protons of alpha anomer of glucose ring in lactose at 5.21 and run a series of experiments with mixing time ranging from 19–232 micro seconds (ms).
- At a mixing time of 19 ms only one H2 resonance is found which is seen as dd; after 33 ms mixing time H3 appears as dd; at

62 ms H4 is appears and H5 emerges; at 90, H5 appears and H6 emerges; finally at 232 coherence is completely transferred from anomeric proton to H–6 protons.

- Long mixing time causes loss of both signal and resolution *i.e.*, how H2 signal broadens from 19–232 ms. So, FID are used.
- For the remaining sugar residues a series of experiments are used using other anomeric protons as irradiation resonances.

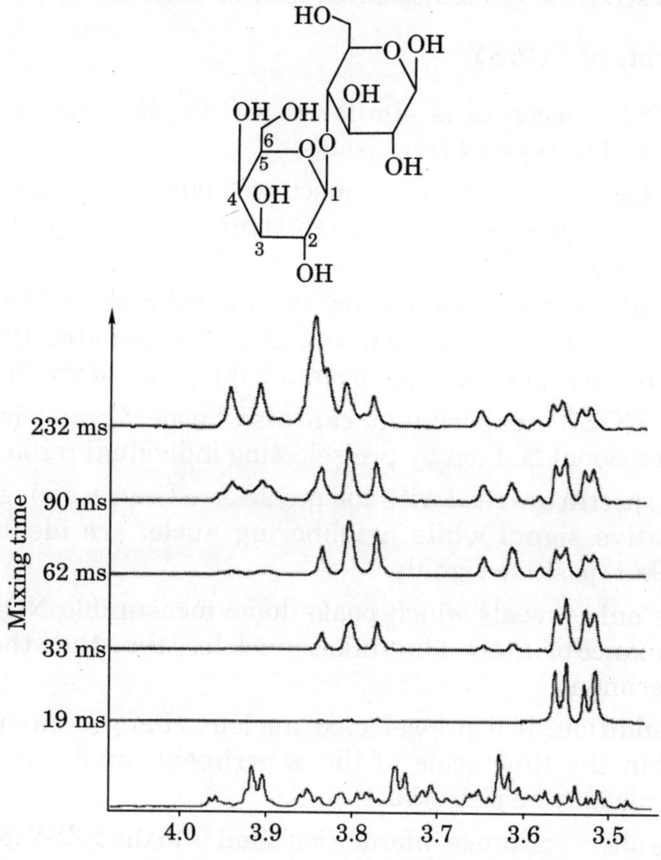

Figure 7.18: 1D-TOCSY experiment series on variable mixing time.

7.3.6 NOESY (Nuclear Overhauser Effect Spectroscopy)

NOESY is a homonuclear correlation spectroscopy that correlates between the nuclei which are physically close to

each other regardless of whether there is a bond between them exists or not. NOE is a consequence of modulation of the dipole-dipole coupling by motion of the molecule in solution due to which nearby atoms (within about 5 Å) undergo cross relaxation by a mechanism related to spin–lattice relaxation. In NOESY, the Nuclear Overhauser cross relaxation between nuclear spins during the mixing period is used to establish the correlations. NOESY is applied to study large biomolecules such as protein.

Key Concepts of NOESY

- NOESY spectrum is similar to COSY, because it possess diagonal peaks and cross peaks.
- In NOESY cross peaks connect resonances from nuclei that are spatially close rather than those that are through-bond coupled to each other.
- NOESY spectra also contain extra *axial peaks* which do not provide extra information and can be eliminated through a difference experiment by reversing the phase of the first pulse.
- The NOESY experiment can also be performed in a one-dimensional fashion by pre-selecting individual resonances.
- The spectra are read with the pre-selected nuclei giving a large, negative signal while neighboring nuclei are identified by weaker, positive signals.
- This only reveals which peaks have measurable NOE to the resonance of interest but takes much less time than the full 2D experiment.
- In addition, if a pre-selected nucleus changes environment within the time scale of the experiment, multiple negative signals may be observed.
- This offers exchange information similar to the EXSY (exchange spectroscopy) NMR method.
- For example, presence of a cross peak (a correlation off the diagonal) on a COSY dataset is a result of nuclei coupling through a bond whereas a NOESY dataset measures NOE's (Nuclear Overhauser Effect) through space regardless of the number of bonds separating the nuclei.

- A NOE is typically observed for nuclei that are separated not more than 5 A° apart.

Example: *NOESY Spectrum of Octahydro Quinolone Enantiomer* (Spectrum given in Figure 7.19a).

Significance of NOESY experiment can be well understood from NOESY and simplified COSY spectra of octahydroquinolone enantiomer given in Figures 7.19a and 7.19b.

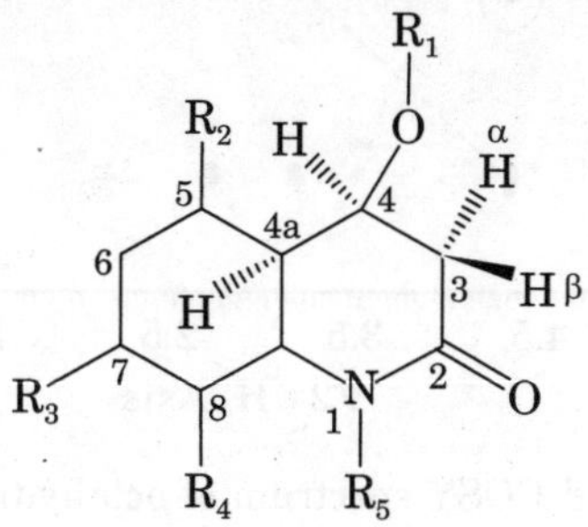

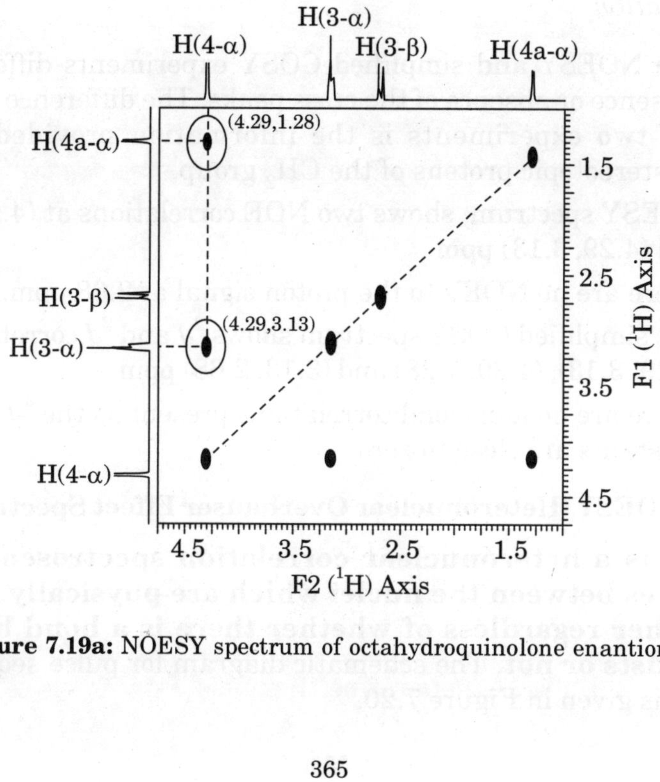

Figure 7.19a: NOESY spectrum of octahydroquinolone enantiomer.

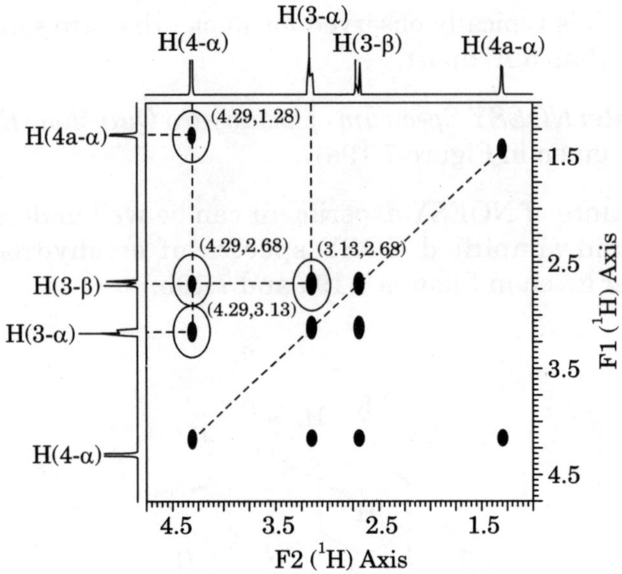

Figure 7.19b: Simplified COSY spectrum of octahydroquinolone enantiomer.

Interpretation

- The NOESY and simplified COSY experiments differ in the presence or absence of the cross peaks. The difference between the two experiments is the information provided on the diastereotopic protons of the CH_2 group.
- NOESY spectrum, shows two NOE correlations at (4.29, 1.28) and (4.29, 3.13) ppm.
- There are no NOE's to the proton signal at 2.68 ppm.
- The simplified COSY spectrum shows ^2J and ^3J correlations at (4.29, 3.13), (4.29, 1.28) and (3.13, 2.68) ppm.
- There are no four-bond correlations present as the ^4J coupling constants are close to zero.

7.3.7 HOESY (Heteronuclear Overhauser Effect Spectroscopy)

HOESY is a heteronuclear correlation spectroscopy that correlates between the nuclei which are physically close to each other regardless of whether there is a bond between them exists or not. The schematic diagram for pulse sequence of HOESY is given in Figure 7.20.

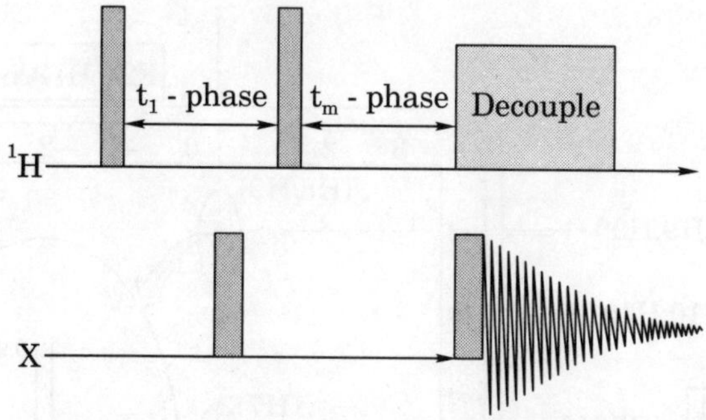

Figure 7.20: Schematic diagram for pulse sequence of HOESY.

Key Concepts of HOESY

- In the pulse sequence of HOESY, firstly a 90° pulse is applied which creates a transverse H magnetization.
- During the evolution time t_1 H chemical shift is characterized (coupling with X nuclei is removed with 180° (X) applied at mid evolution).
- The second 90° pulse on H returns magnetization to Z axis where it is left to exchange with NOE during the mixing period.
- At the end of the mixing, a 90° pulse on X-nuclei creates NOE induced X-magnetization which is then detected during t_2.

Example: *HOESY Spectrum of Complex Organic Compound of Phosphate Sodium* (Spectrum given in Figure 7.21).

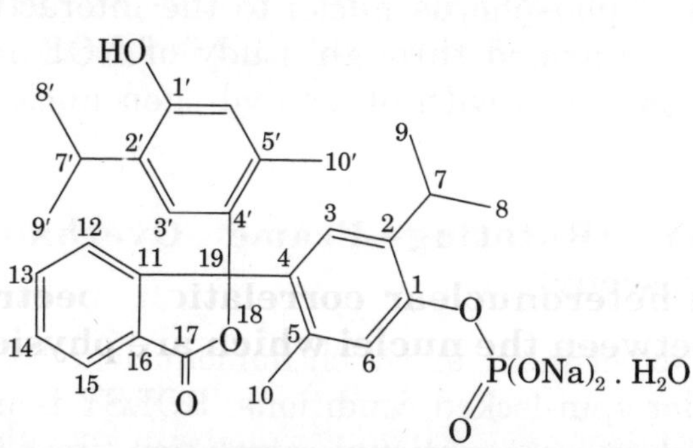

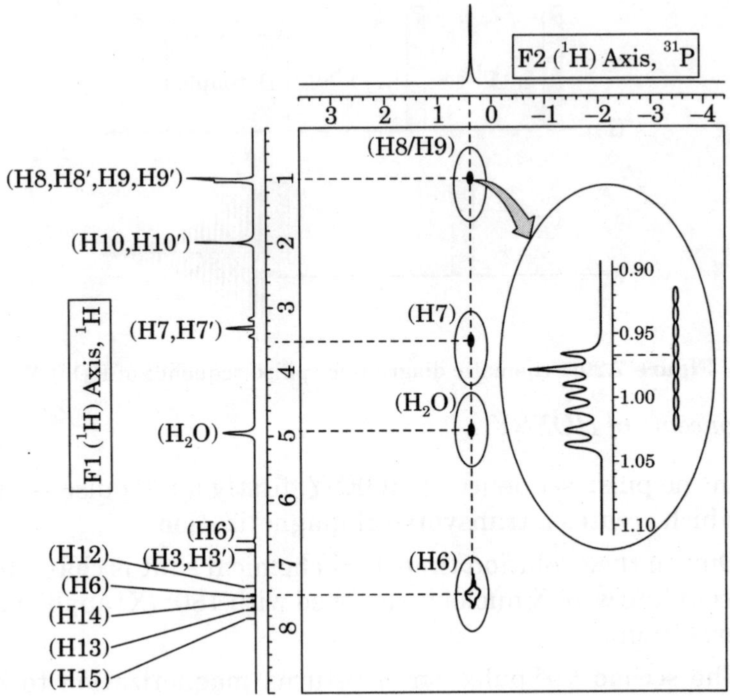

Figure 7.21: HOESY spectrum of complex organic compund of phosphate sodium.

Interpretation

- From HOESY spectrum of the phosphate sodium it can be clearly interpretated that phosphorus nuclei interacts with H8, H9, H7, H6 and H_2O through NOE. Therfore connectvity and closeness of phosphorus nuclei to the interacting hydrogen nuclei is confirmed through study of NOE intercation of phosphorus nuclei with other hydrogen nuclei on different positions.

7.3.8 ROESY (Rotating—Frame Overhauser Effect Spectroscopy)

ROESY is an experiment in which homonuclear NOE effects are measured under spin-locked conditions. ROESY is used for those molecules which possess motional correlation times (τc) such that ωτc ~1 (where ω is the angular frequency ω = γB). In such cases the laboratory-frame NOE is nearly zero, but the rotating-frame NOE (or ROE) is always positive and increases with increase in values of τc. In ROESY the mixing time is the spin-lock period. During this time spin exchange occurs among spin-locked magnetization components of different nuclei. Different spectral density functions makes relevance for ROESY rather than for NOESY. Hence ROE is positive for all values of τc. The ROESY pulse sequence is shown in Figure 7.22. In ROE first pulse must be set to appropriate 90° time and pulse fifteenth is a spinlock pulse, during which ROE buildup occurs.

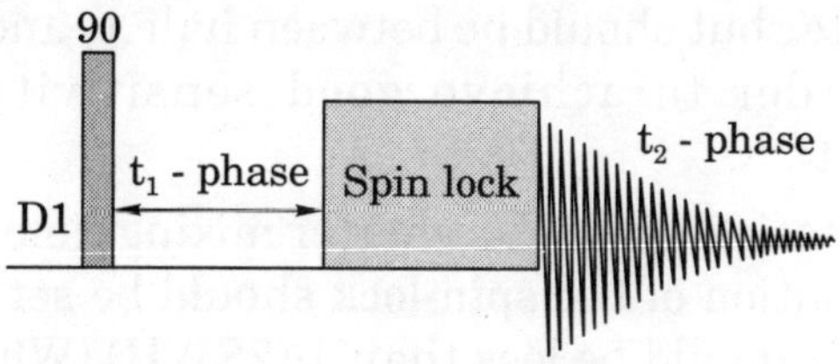

Figure 7.22: Schematic diagram for pulse sequence of ROESY.

Key Concepts of ROESY

- ROESY is similar to NOESY, except that the initial state is different.
- Instead of observing cross relaxation from an initial state of *z*-magnetization, the equilibrium magnetization is rotated onto the *x* axis and then spin-locked by an external magnetic field so that it cannot precess.
- ROESY is useful for molecules whose rotational correlation time falls in a range where the Nuclear Overhauser effect is too weak to be detectable, usually molecules with a molecular weight around 1000 daltons, because ROESY has a different dependence between the correlation time and the cross-relaxation rate constant.
- In NOESY cross-relaxation rate constant goes from positive to negative as the correlation time increases, giving a range where it is near zero, whereas in ROESY the cross-relaxation rate constant is always positive.

- ROESY is sometimes called "cross relaxation appropriate for minimolecules emulated by locked spins" (CAMELSPIN).
- This technique is useful in determining the signals arise from protons that are close to each other in space even if they are not bonded.
- ROESY detects chemical and conformational exchange. Also, ROESY spectrum yields through space correlations *via* spin-spin relaxation.
- The conventional ROESY pulse sequence is commonly used in NMR experiment. The mixing time (t_m) should never be longer than 300 ms. This is done to prevent overheating of sample or spectrometer but should be between half t_1 and t_2 (t_2 is not too long in order to achieve good sensitivity for spectral assignment).
- For quantitative ROESY, a shorter mixing time should be used. The attenuation of the spin-lock should be set so that the 90° pulse width should be less than 1/(2SWH) (Where SWH is the spectral width in Hz).
- An attenuation of 22 dB with a 50 W amplifier yielding a 90° pulse width of 110 ms is typical. However, for quantitative ROESY that is used for distance and exchange rate measurement, short mixing times and long relaxation delays are required.
- The strength of the ROE signal is proportional to the inverse sixth power of the distance between the atoms, $I\mu \propto 1/r^6$.
- Comparison of the cross-peak integrals in a quantitative ROESY is used as a measure of the distance between the protons.
- Sensitivity of ROESY is increased by increasing the transverse relaxation time either by choosing a low-viscosity solvent such as acetone-d_6 and/or removing dissolved oxygen from the sample.

Example: *ROESY Spectrum of 12,14-Di Butylbenzo Tertiary Chrysene*—Through-space correlations in ROESY determines the connectivity of 12, 14-di tertiary butylbenzo chrysene so characterize full spectra.

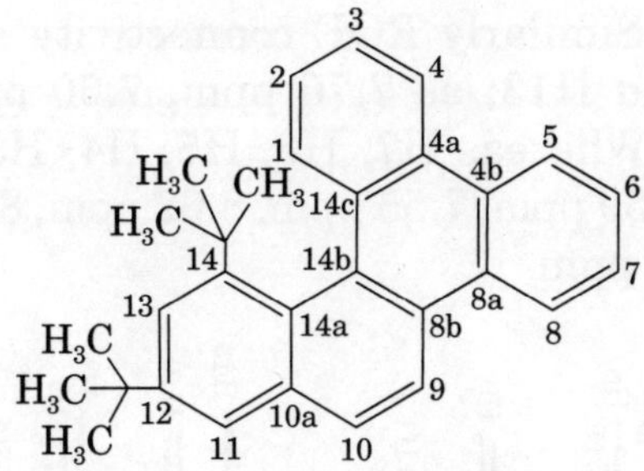

12, 14-di butylbenzo tertiary chrysene

- The ROESY spectrum given in Figure 7.23a, represents the correlations of butyl groups and ring systems of 12, 14-di butylbenzo tertiary chrysene.

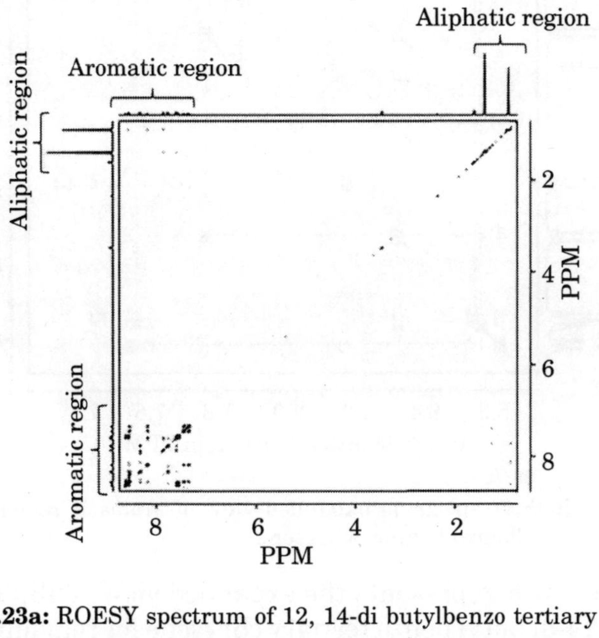

Figure 7.23a: ROESY spectrum of 12, 14-di butylbenzo tertiary chrysene.

- An expanded view of ROESY spectrum of 12, 14-di butyl benzo tertiary chrysene is given in Figure 7.23b. The spectrum represents aromatic region and shows three and four bond correlations. This determines neighboring protons, for example proton H8 and 9 which at 8.56 and 8.32 ppm shows ROE connectivity. Similarly ROE connectivity asists in assigning H10, H11 and H13; as 7.76 ppm, 7.60 ppm and 7.86 ppm respectively. Whereas H7, H6, H5, H4, H3, H2 and H1; are assigned at 7.59 ppm, 7.55 ppm, 8.62 ppm, 8.54, 7.44 ppm, 7.34 ppm and 8.17 ppm.

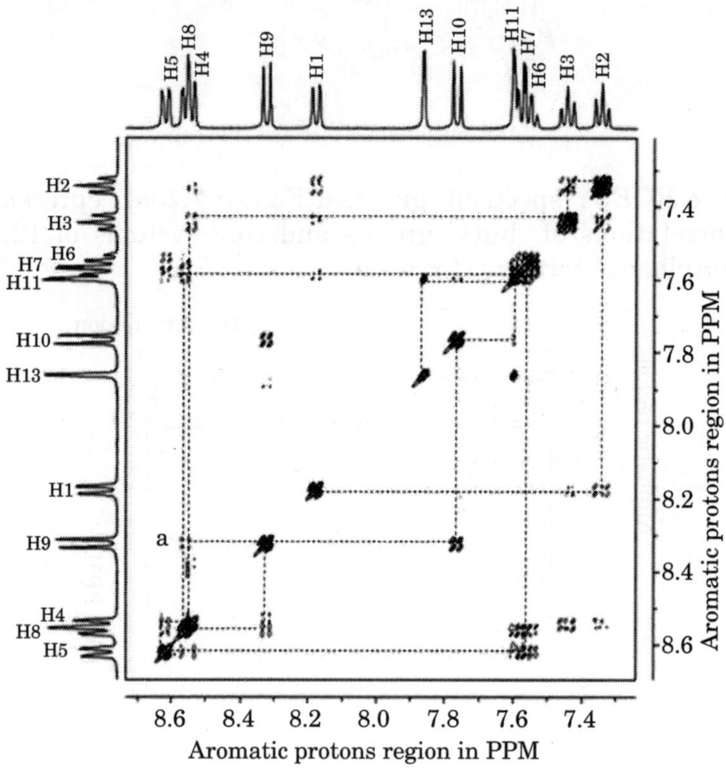

Figure 7.23b: ROESY spectrum expanded view of aromatic region of 12, 14-di butylbenzo tertiary chrysene.

- Figure 7.23c, represents the expanded view of aliphatic region of 12, 14-di butyl benzo tertiary chrysene for remaining tertiary butyl groups. The protons of tertiary butyl group at 14 position shows a correlation with H1 and H13 at 0.92 ppm. Whereas protons of tertiary butyl group at C–12 shows a correlation at 1.40 ppm with H11 and 13 and confirms the ^{1}H assignment through ROESY.

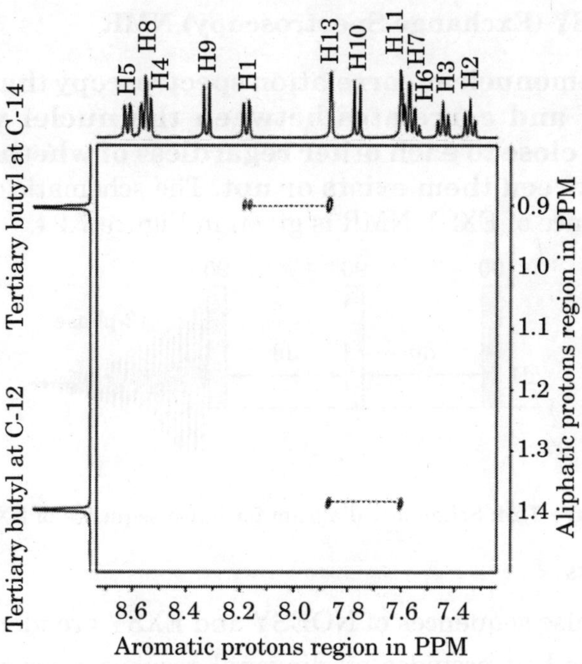

Figure 7.23c. ROESY spectrum expanded view of aliphatic region of 12, 14-di butylbenzo tertiary chrysene.

7.3.9 TRNOE (Transfered Nuclear Overhauser Effect) NMR

- NOE is a consequence of modulation of the Dipole-Dipole coupling by motion of the molecule in solution.
- Irradiation of free-ligand resonance A result in NOE's slowly building up on HB.
- Transfer of spin saturation information via chemical exchange to bound state results in a much more rapid build-up of NOE's on protons H_C.
- This NOE is transferred back to free resonance via a chemical exchange, where it decays relatively slowly.
- The net result is that changes in intensity in free-ligand signals reflect predominantly bound ligand NOE's, provided certain kinetic conditions are met.

7.3.10 EXSY (Exchange Spectroscopy) NMR

This is a homonuclear correlation spectroscopy that is similar to NOESY and correlates between the nuclei which are physically close to each other regardless of whether there is a bond between them exists or not. The schematic diagram for pulse sequence of EXSY NMR is given in Figure 7.24.

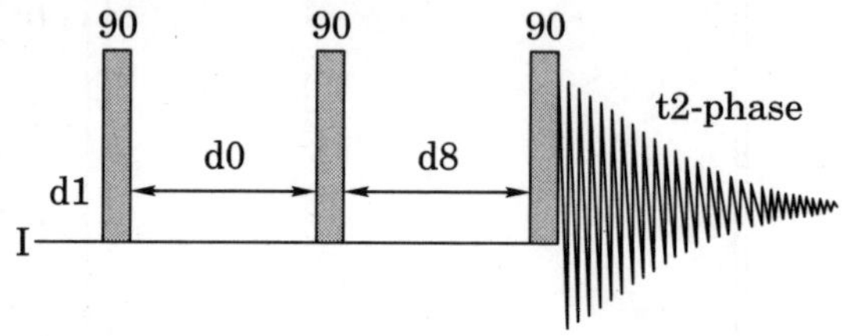

Figure 7.24: Schematic diagram for pulse sequence of EXSY.

Key Concepts

- The pulse sequences of NOESY and EXSY are identical.
- The method provides off-diagonal responses for spins which exchange slowly with one another (either conformationally or chemically) and also between spins with NOE's.
- The EXSY method is useful for showing exchange when the rate of the exchange is greater than or of the same order as the T1 relaxation rate (1/T1) but less than the frequency difference between the two spins (in the absence of exchange).
- Depending on the experimental conditions, the responses due to exchange are often much more intense than those due to NOE's.

Example: *EXSY Spectrum of N, N-Dimethyl Acetamide* (given in Figure 7.25).

- The Figure 7.25, shows the 300 MHz 2D ^{1}H-EXSY spectrum of N,N-dimethyl acetamide at room temperature. At this temperature, the molecule exhibits slow rotation about the $(CH_3)_2N$–C bond such that both methyl groups exchange with one another rotationally yet are distinct in the spectrum. This is evident by the cross peaks in the spectrum between the two methyl groups on the nitrogen.

- It is important to note that off diagonal peaks are of same phase as diagonal peaks.

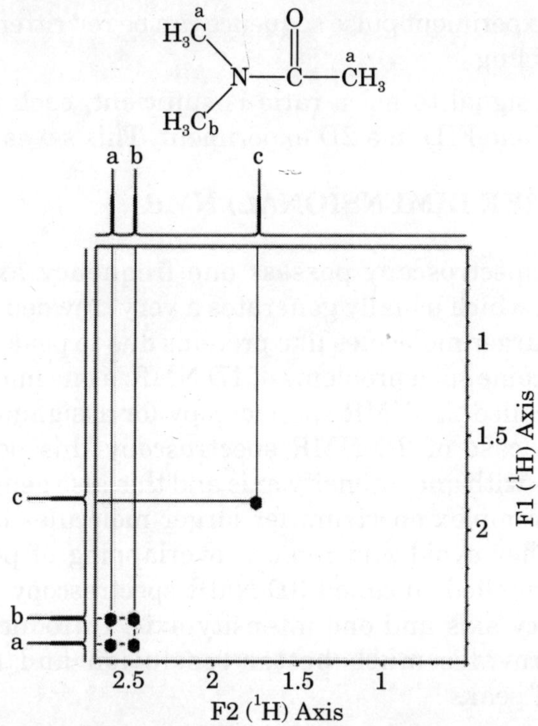

Figure 7.25: EXSY spectrum of complex organic compund of phosphate sodium.

7.3.11 PFG NMR

- Earlier efforts were made to avoid gradients that is to make field homogeneous as much possible.
- Today most of modern high field NMR instruments are well equipped with hardware coils that rapidly ramps field along one of three mutual orthogonal axis.
- These magnetic field gradients are incorporated into pulse sequence in a large range of applications.
- In correlation experiments anywhere phase cycling is done from 4–64 to produce one FID.
- If signal to noise is poor then identical cycles are repeated till sufficient signal is acquired.

- These phase cycles wastes time that is why any 2D experiment takes so long time.
- In PFG experiment pulse sequence can be rewritten to eliminate phase cycling.
- Hence if signal to noise ratio is sufficient, each pulse can be saved as one FID in a 2D experiment. This saves time.

7.4 3D (THREE DIMENSIONAL) NMR

As 1D-NMR spectroscopy possess one frequency axis with one intensity axis, which usually generates a very crowded and complex spectrum for larger molecules like proteins due to peak overlapping. Hence to overcome such problems of 1D NMR, a one more dimension is applied so called 2D-NMR spectroscopy for assignment of peaks. If we see the case of 2D-NMR spectroscopy this possesses two frequency axis with one intensity axis and this also generates a very crowded and complex spectrum for larger molecules like proteins. Hence to further avoid and remove overlapping of peaks a third dimension is applied so called 3D-NMR spectroscopy that possess three frequency axis and one intensity axis. Introduction of third dimension provides much better resolution and reduces the overlapping of peaks.

For example combination of 2D-NOESY and 2D-TOCSY offers a new 3D-NMR *i.e.*, NOESY-TOCSY spectroscopy.

- In case of the protein structure given in Figure 7.26, the proton of NH group of amino acid unit 'B' can be correlated with protons of NH group of amino acid unit 'C' and CHα of amino acid unit 'A'.

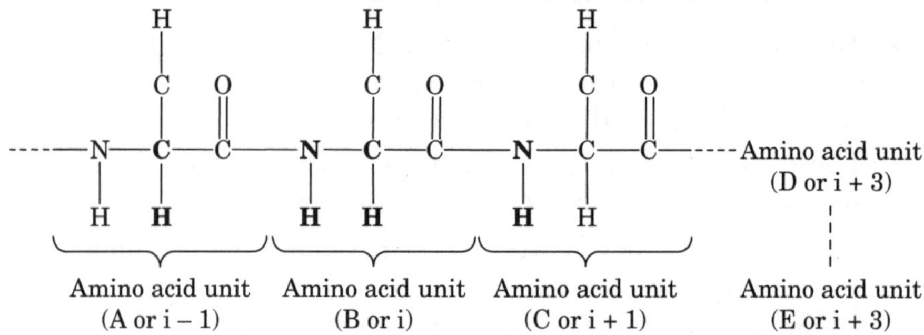

Figure 7.26: Amino acid sequence of protein.

- Figure 7.27, shows the pulse sequence and spectrum of 2D-NOESY, which provides information about neighbouring nuclei arrangement in space.

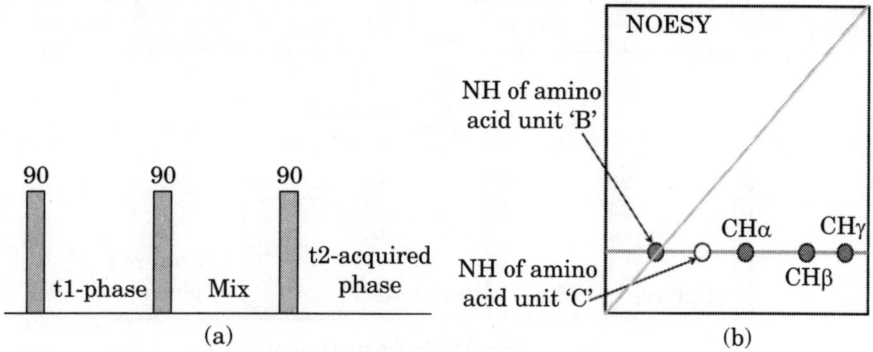

Figure 7.27: Pulse sequence (a) and 2D-NOESY spectrum (b).

- Figure 7.28, shows the pulse sequence and spectrum of 2D-TOCSY, that provides information about links of various protons through chemical bond. As in 2D-TOCSY there is no proton coupling between amino acid units that are separated by C=O group, hence this technique is useful for the assignment of protons within a single amino acid unit.

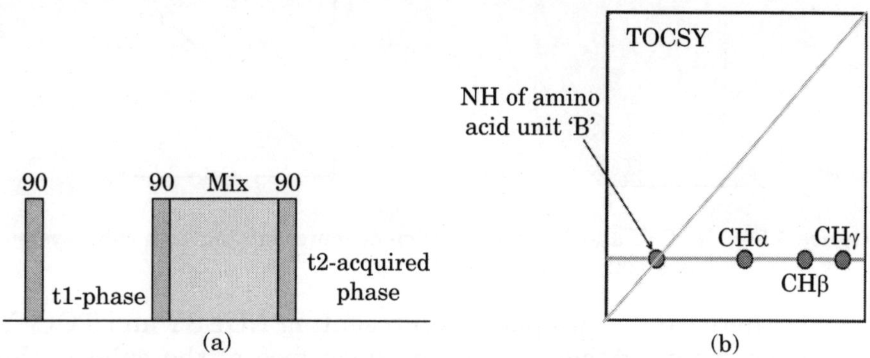

Figure 7.28: Pulse sequence (a) and 2D-TOCSY spectrum (b).

- Hence in order to simplify the overcrowded spectra of proteins, as per Figure 7.29, the two techniques of NOESY and TOCSY are combined together in one experiment to offer a homonuclear 3D spectrum.

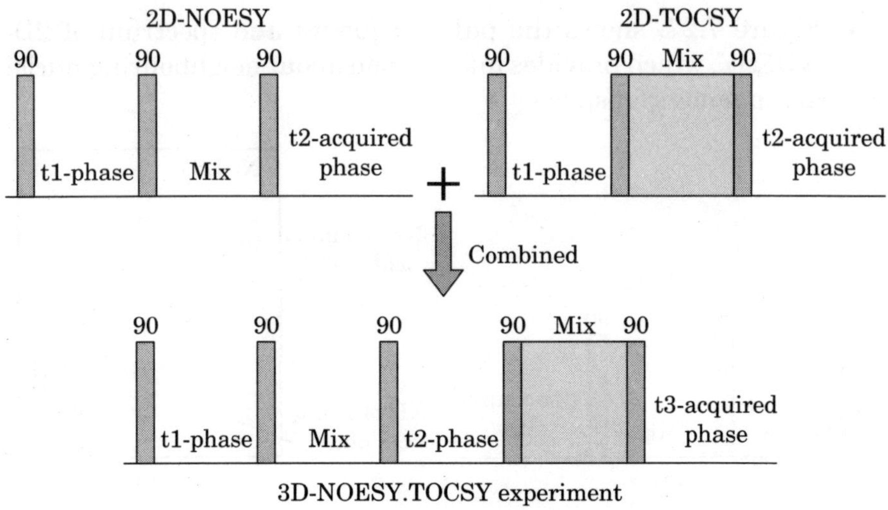

Figure 7.29: Combined pulse sequence of NOESY and 2D-TOCSY spectrum.

- A 3D-NMR of combined NOESY and TOCSY experiment can be best understood when represented through a cubic system shown in Figure 7.30, that possess three axis F1, F2 and F3.

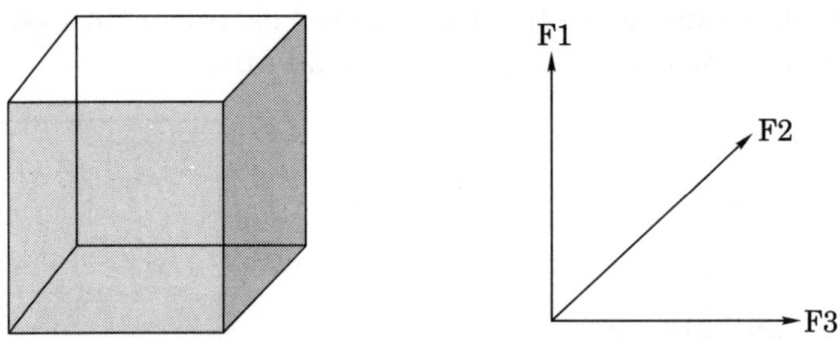

Figure 7.30: NOESY and TOCSY experiment representation in a cubic system possessing three axis.

- As there can be two planes representing NOESY and TOCSY experiments. Assume that the dark face of the cube is the NOESY plane and light face is the TOCSY plane (Figure 7.31). The cube contains F1, F2 and F3 as the three axes.
- In the cubic system the peaks appearing on the body diagonal (line that joins one corner of the cube to the other corner) possess identical coordinates along three frequency axes. Also peak

which appears on this diagonal doesn't experience any magnetization exchange. This 3D-NMR diagonal is similar to the diagonal of 2D-NMR.

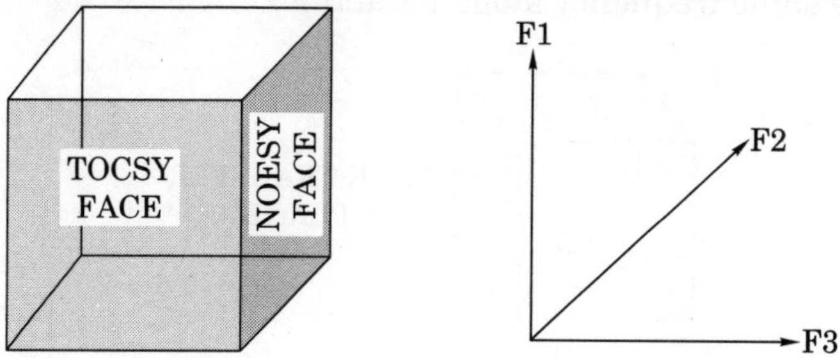

Figure 7.31: Two planes representing NOESY and TOCSY face.

- As per Figure 7.32, it is obvious that for the dark faced NOE plane diagonal will run through the TOCSY face. The peak that will appear along NOE plane will have F2 = F3, and will only account for NOE exchange along NOE experiment (*i.e.*, no TOCSY peaks).

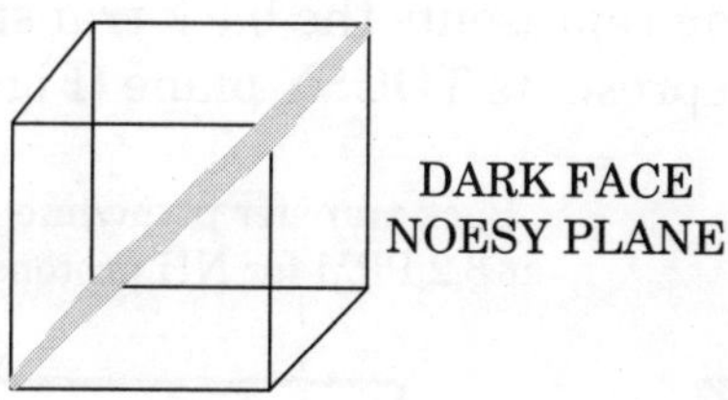

Figure 7.32: Dark faced NOESY plane diagonal passing through TOCSY face.

- As per Figure 7.33, in case of light face TOCSY plane diagonal will move through NOE face. The peak that will appear along TOCSY plane will have F1 = F2, and will only account for TOCSY experiment peaks (*i.e.*, no NOE peaks).

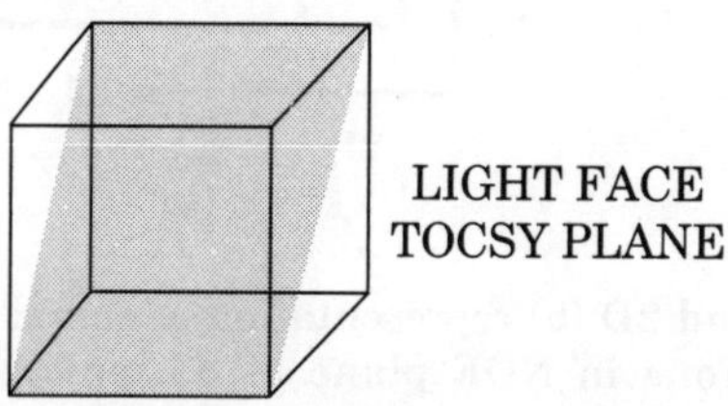

Figure 7.33: Dark faced TOCSY plane diagonal passing through TOCSY face.

- As per Figure 7.34, there is one more plane called back transfer plane shown with the dotted face. This plane will have the diagonal along the cube. The peak appearing in this plane will have same frequency along F1 and F3.

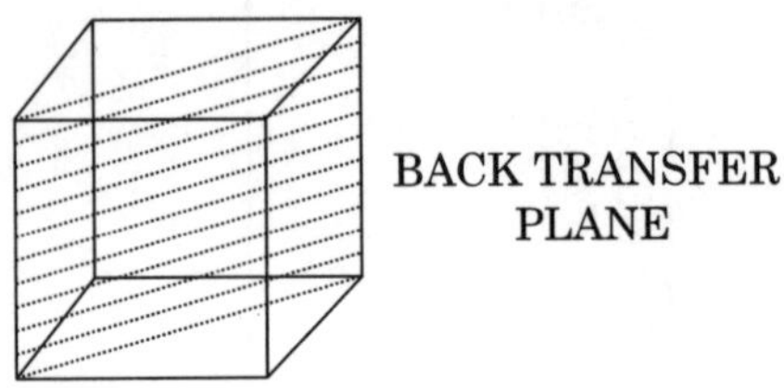

Figure 7.34: Dotted plane or back transfer plane diagonal passing along cube.

- Therefore during NOESY experiment nucleus-1 transfer's magnetization to nucleus-2 and during TOCSY experiment magnetization is transferred back to nucleus-1.
- The 3D-NMR analysis progresses from 2D face *i.e.*, first case along NOE face. In this horizontal line represents the NOE line, vertical line represents the back transfer plane, where as diagonal line represents TOCSY plane (Figure 7.35).

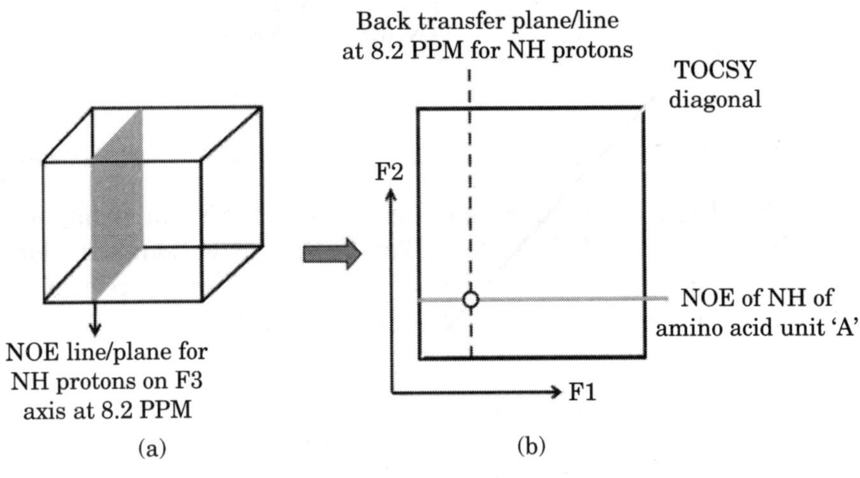

Figure 7.35: 3D (a) and 2D (b) representation of signal for amino acid unit 'b' NH protons in NOE plane at 8.2 ppm on F3 axis on TOCSY diagonal.

- In the NOE plane case the proton of NH group of amino acid unit 'B' of a protein shows a peak at 8.2 ppm, on horizontal NOE, Back transfer, and Diagonal line. So NOE plane of NH group of amino acid unit 'B' shows a signal in F3 axis at 8.2 ppm. In NOESY dimensional plane along diagonal *i.e.*, TOCSY plane an interaction between proton of NH group of amino acid unit 'B' and proton of CHα group of amino acid unit 'B' was observed (TOCSY interaction). This proton of CHα group was found to undergo interaction (through space *i.e.*, NOE) with proton of NH group of amino acid unit 'C' residue.
- In the TOCSY face the the NOE interaction of proton of CHα group of amino acid unit 'A' residue can be observed. This proton of CHα group of amino acid unit 'A' residue is further found to be connected through bonds with other protons of same amino acid unit.
- During protein structural assignment, sequential assignment is the first step which involves assignment of chemical shift of various amino acid units.
    - For this firstly proton of amide group is determined that is nonoverlapping to other amide group and gives a signal at downfield region *e.g.*, NH at 8.2 (representation shown in Figure 7.36).

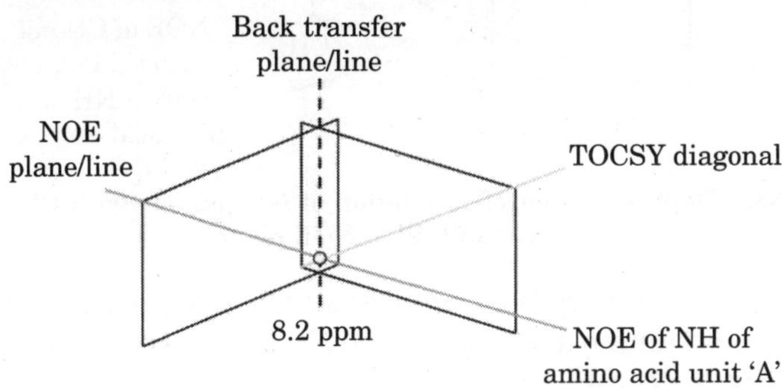

Figure 7.36: Representation of signal for amino acid unit 'b' NH protons in NOE plane at 8.2 ppm on F3 axis on TOCSY diagonal.

- The protons of such NH group of the amino acid unit 'B' is correlated with CHα group through TOCSY diagonal in same amino acid unit 'B' (correlation is shown in Figure 7.37).

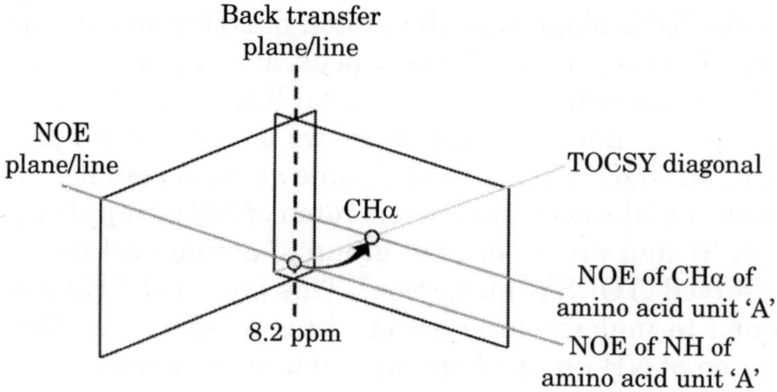

Figure 7.37: Representation of correlation of NH protons with CHα group in same amino acid unit 'B'.

- The proton of CHα group of amino acid unit 'B' further exerts NOE interaction with the protons of NH group of next amino acid unit 'C' (correlation is shown in Figure 7.38).

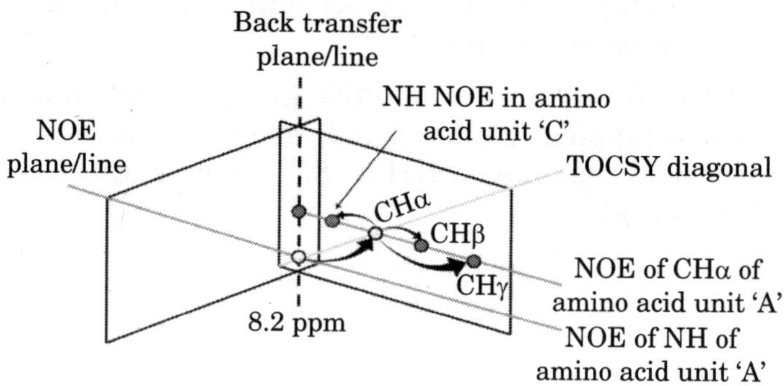

Figure 7.38: Representation of correlation of NH protons with CHα group in same amino acid unit 'B' in NOE plane.

- By determining the chemical shift of the proton of NH group of next amino acid unit 'C' the NOE plane can be determined (Figure 7.39).
- In the same manner proteins sequential assignement is done in the NOE plane.
- The proton of NH of the amino acid unit 'B' is further correlated with CHα (through TOCSY) in the same amino acid unit 'B'. Like wise this CHα of amino acid unit 'B' is

correlated through NOE to proton of NH group of further next amino acid unit 'C' similarly sequential assignment in amino acid unit 'C'.

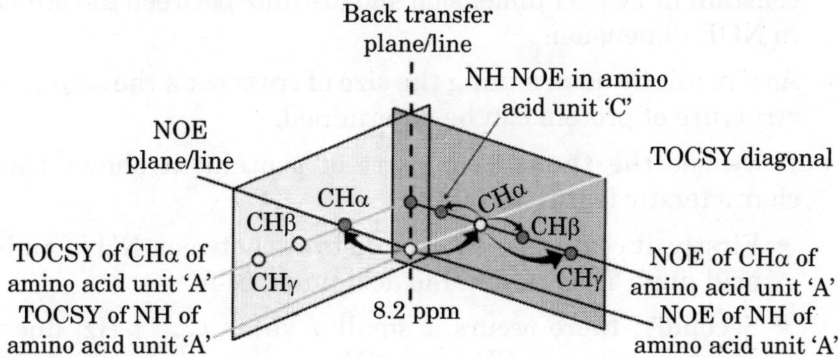

Figure 7.39: Representation of correlation of NH protons at 8.6 ppm with CHα group in next amino acid unit 'C' in NOE and TOCSY plane.

- Similarly the NH of amino acid 'C' is correlated with CHα of amino acid unit 'C' which further inturn is correlated through space (NOE) to amino acid 'D', 'E', 'F', ... and so on (Figure 7.40).

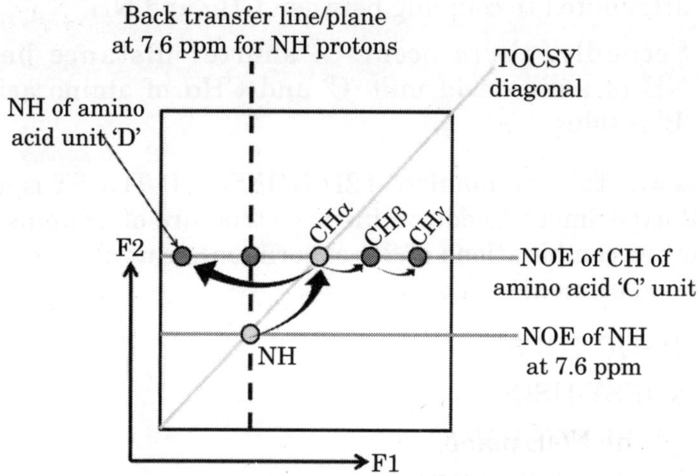

Figure 7.40: Representation of correlation of NH protons at 7.6 with CHα group in same amino acid unit 'D' in NOE and TOCSY plane.

  - Nextly, this correlation NMR analysis is progressed into the next NH plane.
- Hence in 3D-NMR the peak intensity depends upon the coupling constant in TOCSY dimension and distance between the protons in NOE dimension.
- As a result by determining the size of cross peak the secondary structure of protein can be determined.
- If we see the the α helix part of protein, it shows three characterstic features such as
  - Firstly, it comprises shorter distance between NH of amino acid units 'B–C' andamino acid units 'B–A'.
  - Secondly, there occurs a small *J* value *i.e.*, 4 Hz, due to coupling between CHα and NH.
  - Thirdly, the distance is shorter when between NH and CHα but longer when between NH and CHα of same amino acid unit 'B' residue.
- In β helix part of protein, generally two characterstic features are commonly observed such as
  - Firstly, there occurs a large *J* value *i.e.*, 8–10 Hz, which is attriibuted to coupling between CHα and NH.
  - Secondly, there occurs a shorter distance between NH of amino acid unit 'C' and CHα of amino acid unit 'B' residue.

In this way this combinaion of 2D-NOESY-2D-TOCSY is used for 3D-NMR experiment to determine the structure of proteins. Apart from it other combinations of 2D experiments can also be used for 3D-NMR experiments such as

- 2D HSQC-TOCSY
- 2DNOESY-HSQC
- ^{1}H-^{15}N ROESY-HMQC
- ^{I}H-^{13}C ROESY-HMQC
- ^{1}H-^{15}N HMQC-NOESY-HMQC etc.

7.5 MD (MULTIDIMENSIONAL) NMR

7.5.1 Introduction

Over the last few decades tremendous progress have occurred in NMR spectroscopy for identification of proteins with utmost accuracy. Introduction of multiple dimensions into NMR have identified many proteins such as C terminal domain of cellobiohydrolase, antennapedia homeodomain, interleukin-8, a zinc finger domain from a human enhancer binding protein and human thioredoxin etc.

The structural elucidation of larger proteins of 150–250 residues by 1D or 2D-NMR, pose two major problems to investigators such as

- Chemical shift overlapping or degeneracy.
- Peak widths.

The probable solution for these problems is to increase the resolution of the spectra using two different approaches such as

- *Firstly, Conventional Approach of Improvising 2D-NMR Spectra* by increasing the digital resolution or using strong resolution enhancement digital filtering.
- *Secondly, Alternative Approach of Increasing Dimensions of Spectrum* by separating individual proton resonances according to chemical shifts of directly bonded hetero nuclei and simultaneously yield additional information about the system (*i.e.*, ^{15}N and ^{13}C chemical shifts).

In 3D-NMR and 4D-NMR, spectra digitization lowers the resolution. Therefore spectra with equivalent resolution are recorded on magnetic field lower than 600 MHz. For example multi-dimensional spectra at 500 MHz and 600 MHz offers similar resolution but the signal-to-noise ratio is better at 600 MHz than at 500 MHz.

7.5.2 Theory and Concept of Multi-dimensional NMR

A 2D experiment is commonly comprised of a preparation pulse, an evolution period (t_1*-during which evolution period spins are labelled according to their chemical shifts in indirectly detected dimension*), a mixing period (*M-during which spins are correlated with each other*), and a detection period (t_2). This experiment is repeated several times with successive incremented values of evolution period t_1 which generates a data matrix s(tx, t_2). Fourier transformation in t_2 dimension yields a set of n number of 1D spectra (*in which intensities of resonances are modulated sinusoidally as function of t_1 phase*). As a result subsequent Fourier transformation in the t_1 dimension yields desired 2D spectrum $S(\omega_1, \omega_2)$.

A 3D pulse sequence is constructed by combining two 2D sequences, and leaving out the detection period of the first experiment and the preparation pulse of the second. Therefore, in 3D experiment, two evolution periods are incremented independently that yields a data matrix $s(t_1, t_2, t_3)$.

7.5.2.1 *4D-NMR*

As it is with 2D and 3D-NMR, in 4D-NMR experiment a pulse sequence is made up of three 2D sequences. Also in 4D experiments the detection periods of the first and second experiments, and the preparation pulse of the second and third experiments are left out. Hence, in 4D experiment, there are three independently incremented evolution periods that yields data matrix (t_1, t_2, t_3, t_4).

Just like normal NMR experiments, the 3D and 4D-NMR experiments are also categorized into two types of experiments *i.e.*,

A. Homonuclear Experiment—The earlier 3D-NMR on proteins were use to be homonuclear experiments, in which a HOHAHA sequence was combined with a NOESY, using selective or non-selective pulses. For example in case of the 3D HOHAHA-NOESY spectrum of any protein. The homonuclear experiment employs a selective pulse centred around the NH and $C^{\alpha}H/C^{\beta}H$ regions in the F1 and F2 dimensions, respectively, and cross peaks arise from the pathway.

$$\mathrm{NH} \xrightarrow{\text{J+Relayed}} \mathrm{C^{\alpha}H/C^{\beta}H} \xrightarrow{\text{NOE}} \mathrm{NH/Aliphatic}$$

Although these homonuclear 3D or 4D experiments are conceptually acceptable, but their applicability to larger proteins is very low due to some problems such as:

- The efficiency of scalar correlation severely reduces on increase in the line widths. As a result the sensitivity for experiments of 3D or 4D homonuclear experiments decreases on increase in proteins molecular weight above 10 kDa.
- The 3D or 4D homonuclear experiments are difficult to analyze as the number of peaks present in the spectrum are much larger than in the corresponding 2D experiments. This is due to the presence of single (A→A→B and A→B→B) and back transfer (A→B→A) peaks which contain no new information, in addition to genuine 3D (A→B→C) peaks.
- Apart from this other homonuclear experiments such as ^{13}C/ ^{13}C-edited NOESY can also be used.

B. Heteronuclear Experiment (for example ^{13}C/^{15}N-edited NOESY experiment)—The deficiencies of homonuclear experiments are commonly overcome by heteronuclear methods using uniformly (>95%) labelled ^{15}N and/or ^{13}C proteins. Heteronuclear 4D-NMR experiments creates high interest due to their two very important experimental characteristics such as

- Efficient transfer of magnetization through bonds during Heteronuclear 4D-NMR experiment.
- Increased spectral resolution on increase in number of dimensions, without an increase in spectral complexity. So, that the number of cross peaks present in entire spectrum remains unchanged, irrespective of the dimensionality of the spectrum.

A 4D-NMR can be best understood with spectra given in Figure 7.41, which illustrates the progression and relationship between ^{15}N/^{13}C edited 2D, 3D and 4D-NOESY experiments.

Key Concepts of 4D NMR (Figure 7.41)

- Circular dots in the block diagram represents NOE cross peaks between NH and aliphatic protons.
- 3D experiments records a data set of more than 30 complex points in heteronuclear dimensions, more than 125 complex points in the indirectly detected ^{1}H dimension and from 510–1025 real points in ^{1}H acquisition dimension.

- 4D experiments records 8–16 complex points in the heteronuclear dimensions, 64 complex points in the indirectly detected ^{1}H dimension and 256–512 real points in the ^{1}H acquisition dimension. The data are then processed with one zero-filling in all dimensions, and, when appropriate, the time domain data in the severely truncated dimensions can be extended by linear prediction.
- Commonly, it takes 2–3 days to record heteronuclear 3D experiments, whereas 3–4 days for 4D.

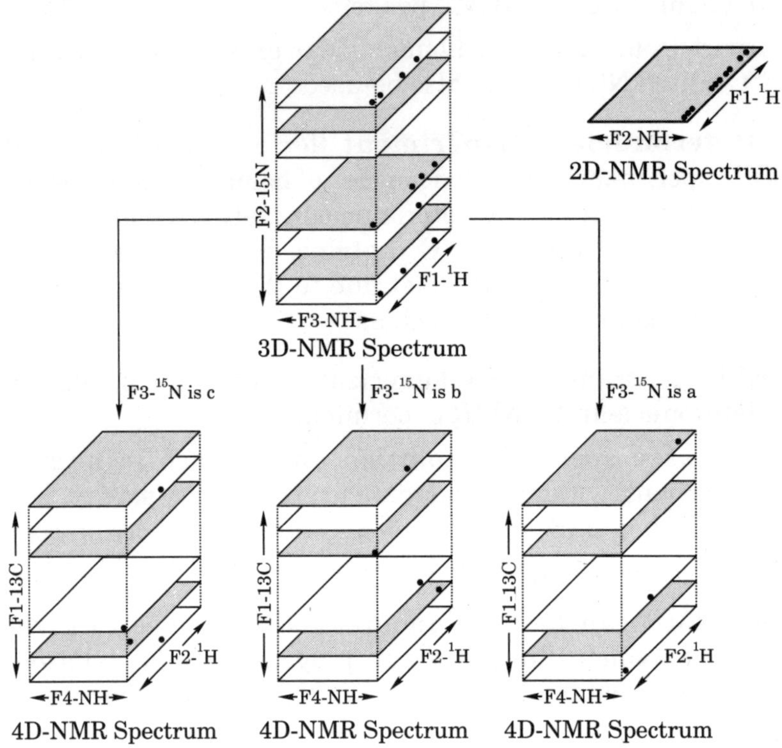

Figure 7.41: Schematic representation of correlation and relationship between heteronuclear $^{13}C/^{15}N$-edited 2D, 3D and 4D-NOESY spectra.

Peak Assignment

From schematic Figure 7.41, a few outcomes can be made such as

1. **During 2D-NMR experiment**, the NH protons of F2 dimension and aliphatic protons of F1 dimension exhibits NOE

to occur in a single plane. As represented in the figure there are cross peaks from 11 aliphatic protons to three NH protons at a single NH chemical shift.

2. **During the 3D NMR experiment**, NOE spreads to 3D cube between planes of F3(NH)-to F1(^{1}H) as per chemical shift of the directly attached ^{15}N atoms in the F2 dimension.

   The peaks are located in three planes, which indicates that 11 NOE involves three different NH protons, thus resolves confusion related to NH chemical shift degeneracy.

   The aliphatic protons involved in NOE effects, can only be identified by their ^{1}H chemical shifts. As chemical shift overlap in aliphatic region is more extensive than NH region of spectrum, therefore it becomes difficult to assign originating aliphatic protons.

   These difficulties can be resolved by extending or introducing one more dimension (of chemical shift of the directly attached ^{13}C nuclei for these aliphatic protons) in the 3D experiment *i.e.*, 4D.

3. **Hence, each plane of specific N chemical shift in the 3D spectrum constitutes a cube in 4D spectrum.** Therefore each cube is further subdivided into a series of planes based on ^{13}C chemical shift of the ^{13}C atoms directly attached to the aliphatic protons of FI.

   Thereby in 4D experiment, the identity of the originating aliphatic protons can be established with great ease on the basis of xH and associated xaC chemical shifts. Also, the identification of NH protons can be done by XH and associated ^{15}N chemical shifts.

Assignment Procedure

Assignment of peaks can be done by two ways such as

- By identification of direct and relayed through-bond connectivities through COSY and HOHAHA the conventional homonuclear ^{1}H-^{1}H 2D-NMR experiments.
- By through-space connectivities by NOESY experiments to obtain sequential type C^αH/NH/C^βH(i)–NH(i+1), C^αH(i)–NH(i+2, 3, 4) and C^αH(i)–C^βH(i+3) along the polypeptide chain.

- To remove the overlapping problems, two of the 2D experiments of ^{1}H-^{15}N relayed HMQC-NOESY and HMQC-HOHAHA spectroscopy can be used. Where the NH chemical shifts are replaced by the corresponding ^{15}N chemical shifts.
- Similarly in 3D heteronuclear NMR experiment also such strategy can be used. Despite of it a number of triple resonance experiments involving correlations of ^{1}H, ^{13}C and ^{15}N chemical shifts can also be used to determine sequential connectivities by means of heteronuclear J couplings, without NOE.

Experiments of 3D and 4D resolves the problems of resonance overlapping & large line widths, and also opened the new doors for investigations. The new 3D and 4D-NMR experiment developed a new era for protein structure determination and opened the doors for studies of large number of medium sized proteins of 15–40 kDa.

7.5.2.2 *5D and 6D NMR*

Intrinsically disordered proteins (globular proteins) are highly difficult to be characterized due to their property of rapid inter conversion rate into various conformations. As a result even small disordered protein fragments of IDP, causes characterization obstacles by averaging the chemical shifts, poor peak separation, and also difficulty in resonance assignment. All such obstacles can be handled by multidimensional NMR methods *i.e.*, by utilizing non-uniform sampling of indirectly detected dimensions.

The conventional backbone resonance assignment of globular proteins is comprised of group of triple-resonance experiments to determines the sequential connectivities using carbonyl ^{13}C, aliphatic ^{13}C and aliphatic ^{1}H chemical shifts. This approach very often fails in the case of IDP because of poor dispersion of side-chain chemical shifts.

IDP—*Intrinsically disordered proteins (globular proteins) or macromolecules with peculiar structure, dynamics and interactions due to which they plays a numerous biological functions.*

Non-uniform Sampling of Indirectly Detected Dimensions—*This sampling facilitates acquisition of high resolution and high-dimensional spectra.*

For effective backbone resonance assignment of IDPs, among different available strategies 5D and 6D experiments were used with sparsed sampling.

In 5D NMR commonly three types of NMR pulse sequences like HACA(N)CONH, HNCOCACB, (HACA)CON(CA)CONH, (H)NCO (NCA)CONH) can be used. Whereas for 6D-NMR one pulse sequences for example (H)NCO(N)CACONH NMR can be used. These experiments employs non-uniform sampling which makes to achieve high resolution in indirectly detected dimensions. These experiments facilitates in resonance assignment of intrinsically disordered proteins. For example these can be used for resonance assignment of δ subunit (20 kDa) of Bacillus subtilis RNA polymerase possessing 81-amino acid disordered part containing several repetitive sequences.

Usually, it takes several minutes to measure resonance in 2D-NMR, several hours in 3D-NMR, several days in 4D-NMR, and several weeks in 5D and 6D-NMR with conventional FTNMR methods. This is because of number of free inductions decays, that is number of data points sampled in the indirect dimensions, increases exponentially with increase in the number of dimensions. Such that to scan 16 complex points in each indirect dimension when it takes one second per FID, then it takes 30 minutes for 3D-NMR, and 12 days for 5D-NMR of measurement time. The time required for data collection is a rate limiting step for structural characterization or resonance measurements.

So, inorder to make the NMR measurements fast following strategies can be used such as

(a) **Ultrafast (single-scan) NMR**—Slice-selective excitation.

(b) **Non-linear sampling**—Restricted time-domain sampling in the indirect dimension by omission of increments randomly or in a selective manner followed by spectral reconstruction

(*i*) Maximum Entropy Reconstruction

(*ii*) Frequency Diagonalization Method

(*iii*) Three-way decomposition (Multiway decomposition)

(*c*) **Time-shared approach (Simultaneous Data Acquisition)** —Parallel execution oftwo or more magnetization transfer pathways in the RF pulse sequence.

(*d*) **Projection NMR spectroscopy**—Phase sensitive or non-phase sensitive joint sampling of two or more chemical shifts in the indirect dimension(s)

- (*i*) RD-NMR
- (*ii*) GFT NMR
- (*iii*) GFT-based data collection followed by spectral re-construction

(*e*) **Hadamard NMR spectroscopy**—Selective excitation and encoding of phases of signals using Hadamard matrix.

(*f*) **Fast pulsing methods**—Reduction of relaxation delay period between scans

- (*i*) L-optimization
- (*ii*) SO-FAST NMR
- (*iii*) Variable relaxation delay

7.6 SS (SOLID STATE) NMR

7.6.1 Introduction

Solid-state NMR (SSNMR) spectroscopy is an advance nuclear magnetic resonance (NMR) spectroscopy, which is associated with anisotropy or directionally dependent interactions. As a spin is usually associated with a magnetic or an electric field, and therefore the spatial proximity among two atoms may lead to nuclear interactions. Commonly, such types of interactions depends upon orientation of molecules. That is why, less or non mobile system of crystals, powders, vesicles, molecular aggregates etc. shows maximum anisotropic interactions which affects their nuclear spins. In liquid-state NMR experiments the Brownian motion causes averaging of anisotropic interactions. Therefore in liquid-state NMR such interactions are neglected. Solid-state NMR spectroscopy act as an important tool of analysis in organic chemistry, inorganic chemistry also physics. SSNMR is also a valuable tool to study local dynamics, kinetics, and thermodynamics of a variety of systems. For example SSNMR is applied in the study of substances like membrane proteins and amyloid fibrils.

7.6.2 Theory/Principle of SSNMR

In solid-state NMR experiments, the two direction interactions are very common such as

- Quadrupolar Interactions,
- Chemical Shift Anisotropy (CSA) and
- Inter nuclear dipolar coupling.

Anisotropic interactions causes modification of nuclear spin energy levels in a molecule which leads to line broadening in NMR spectrum. Despite of the fact of line broadening still there are instances when anisotropic interactions becomes highly important because these determines the structural parameters such as orientation of molecules.

In solid state NMR a high-resolution can be achieved by

- Magic angle spinning.
- Macroscopic sample orientation.
- Combinations of both techniques.
- Enhancement of mobility by highly viscous sample conditions.
- Different radiofrequency (RF) irradiation patterns.

7.6.2.1 *Magic Angle Spinning (MAS)*

Sample rotation is a technique to reduce or remove the chemical shift anisotropy (dipolar coupling) and most commonly called as magic angle spinning. The name "magic-angle spinning" was coined by Cornelis J. Gorter, in 1960. Magic angle spinning can be best understood from figure 7.42. According to this figure a sample rotates with high frequency inside the main magnetic field (B_0). If the axis of rotation of sample is tilted with a the magic angle θ_m, with respect to the direction of B_0. This spinning of sample at magic angle θ with respect to direction of magnetic field, leads to narrowing of broad lines (*broad lines are the result of dipolar, chemical shift anisotropy, quadrupolar spin interactions*). Which leads to increase in resolution of spectrum. The MAS causes narrowing of broad lines by averaging the dipole-dipole interaction to zero, chemical shift anisotropy to non zero, and quadrupolar spin interactions to partial average.

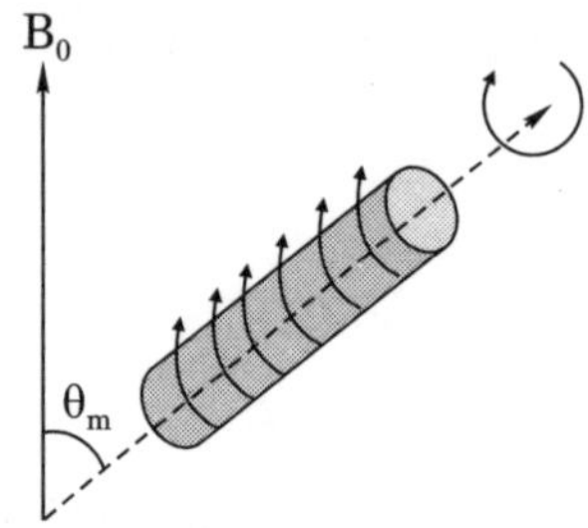

Figure 7.42: Schematic diagram for representation of magical angle spinning.

The spinning of the sample can be done by using a rotor or cylindrical air turbine (Figure 7.42). These turbines spun by air or nitrogen gas. The rotors are made up of different materials such as ceramics *e.g.*, ziconia, silicon or polymers. Polymers such as polyoxy methylene (POM) or polymethyl metha acrylate (PMMA) are most commonly used in SSNMR. The samples are packed into the rotors and then sealed with a single or double end cap. These caps are made up of different materials *e.g.*, zirconia or boron nitride, vespel, depending upon application.

7.6.2.2 *Peak Widths in Solid-State NMR*

As the line-broadening is attributed to microscopic inhomogeneities and can be reduced by using some suitable methods of sample preparation. Commonly, at MAS rate of 5–15 kHz, the line width of ^{13}C nuclei ranges from from 0.5–2 ppm. Whereas MAS rates of 20 kHz or more, causes only partial suppression of non linear groups of same nuclei via homonuclear dipolar interactions, which causes 0.5 ppm or above of line widths (that is more than solution state NMR). Some factors that affects peak widths in SSNMR are as follows:

(*a*) ***Dipolar interaction Truncation*:** As dipolar coupling between two nuclei is inversely proportional to cube of distance between them, as a result the polarization transfer mediated by dipolar interaction can be cut off by presence of a third nucleus (*all nuclei of same kind such as ^{13}C*) close to any one nuclei. Such an effect is called as dipolar truncation. This is another major factor that affects determination of inter nuclear distances, that are crucial in the structural analysis of bio molecules. However with the help of labeling schemes or pulse sequences, such problems can be overcome.

(b) ***Anisotropic Interactions*:** In solution-state NMR, the motional averaging of dipolar interactions is required. The magnitude of residual dipolar couplings (RDC) are of a few Hz, and are not sufficient to destroy the high-resolution conditions. Therefore provides much information about molecular domain orientations.

(c) ***Quadrupolar Interaction*:** Depending upon the strength of interaction, a quadrupolar interaction leads to increase in line widths up to thousands of ppm. The first order quadrupolar broadening can be suppressed by fast MAS, but the second-order quadrupolar broadening possesses a different angular dependence and cannot be removed by spinning at one angle alone.

Hence, isotropic line shapes for quadrupolar nuclei can be refined by spinning the sample at two angles simultaneously, which can be achieved by

- DOR—Double Angle Rotation
- Sequential DAS—Double Angle Spinning or
- Refocusing the second-order quadrupolar interaction with 2D experiment such as—Multiple Quantum Magic Angle Spinning (MQMAS) or Satellite Transition Magic Angle Spinning (STMAS).

7.6.2.3 *Spin Interactions in SSNMR*

The nuclear spin interactions that one commonly come across in SSNMR experiments are as follows:

(a) ***Quadrupolar Interaction*:** Those nuclei which possesses a spin greater than one-half and bears a non spherical charge distribution are called as quadrupolar nuclei. The non spherical charge distribution of quadrupole nuclei can interact with an electric field gradient produced as result of non symmetry thereby causes to change the energy level. This interaction is the largest interaction of NMR *i.e.*, more than Zeeman interaction. As it is too large, so it is expected to be in two forms *i.e.*,

- *First order quadrupolar interaction* that depends upon magnetic field of 3 $\cos^2 \theta - 1$ which is P2 Legendre polynomial. As a result spinning of the sample at $\theta = \arctan \sqrt{2}$, averages out first order interaction over one rotor period.
- *Second order interaction* that depends upon magnetic field of P4 Legendre polynomial. angular dependency has zero points at 30.6° and 70.1°. This can be made advantageous using techniques like DOR, DAS, MQMAS, STMAS.

(*b*) ***Chemical Shielding*:** Chemical shielding is defined as magnetic property of a nuclei that depends on the external magnetic field. In the experiment the external magnetic field induces currents of the electrons in molecular orbital's and thereby the induced currents generates a local magnetic fields that varies across molecular structure in such a way that nuclei in molecular experiences unique local filed. Fast magic angle spinning, or in solution-state NMR, removes directionally dependent character of chemical shielding.

(*c*) ***J-Coupling*:** The *J*-coupling is also called as scalar coupling or indirect spin-spin coupling describes the interaction of nuclear spins via chemical bonds.

(*d*) ***Dipolar Coupling*:** Interaction of dipole moment of one nuclei with dipole moment of other nuclei is called as dipolar coupling (Figure 7.43). The magnitude of dipolar interaction depends upon internuclear distance, spin species and orientation of the vector connecting the two nuclear spins with respect to the external magnetic field *B*.

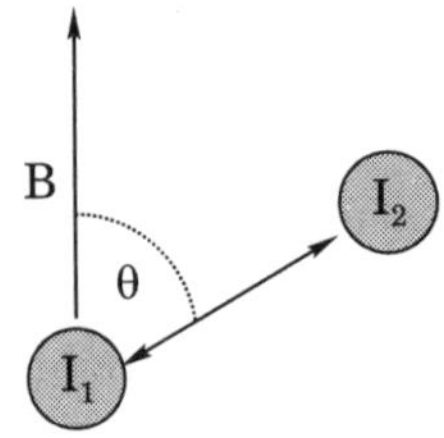

Figure 7.43: Schematic representation of dipolar coupling between two spinning nuclei in magnetic field.

Dipolar coupling constant *d*, is expressed as

$$d = \frac{h.\mu_0}{4\pi}.\frac{\gamma_1.\gamma_2}{r^3} \quad ...(7.1)$$

where r = distance between the nuclei,

γ_1 & γ_2 = gyromagnetic ratios of the nuclei.

In strong magnetic field, the dipolar coupling depends on the orientation of the inter nuclear vector with external magnetic field and expressed as follows:

$$D \propto 3.\cos^2\theta - 1 \quad ...(7.2)$$

Hence if two nuclei having a dipolar coupling vector at an angle of θ_m=54.7° are placed in a strong external magnetic field, causes D to become zero, possesses zero dipolar coupling.

7.6.3 Methods and Techniques in SSNMR

To achieve improved sensitivity, better resolution and more structural information in SSNMR following are the three common processes that are commonly employed.

A. Cross Polarization

Commonly, SSNMR is initiated by cross-polarization (CP) that is a result of fundamental RF pulse sequence and acts as a building block (Figure 7.44). CP is used for two basic reasons as follows:

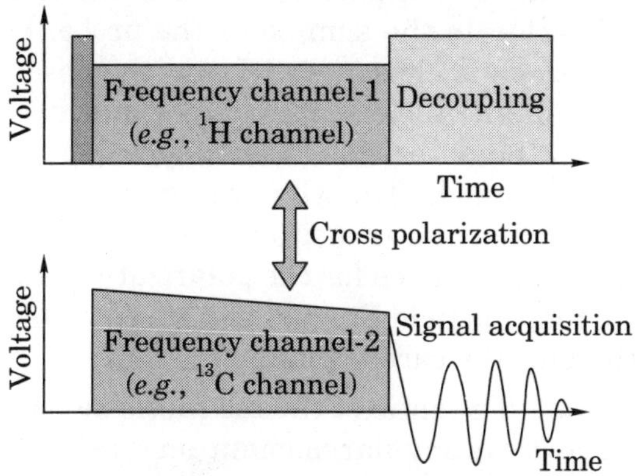

Figure 7.44: CP pulse sequence.

1. For enhancement of signal of nuclei with a low gyromagnetic ratio such as ^{13}C and ^{15}N nuclei by magnetization transfer from nuclei with a high gyromagnetic ratio such as ^{1}H nuclei.
2. For spectral editing, such as directing CP from ^{15}N to ^{13}C in protein NMR.

Magnetization transfer is establish by applying RF pulses in two frequency channels that satisfies conditions of Hartmann-Hahn experiment. At the same time application of magical angle spinning, creates a relationship between rate of sample rotation and voltage through the RF coil. So, routinely such experimental programming are commonly performed during SSNMR study. CP-MAS acts a building block for most of the pulse sequences of SSNMR. Application of pulse sequence for direct excitation of ^{1}H spin polarization, followed by CP transfer to perform signal detection of nuclei such as ^{13}C, ^{15}N or similar nuclei is a common process of CP experiment. When CP is applied with MAS, such an experiment is called as CP-MAS-SSNMR, which is the beginning of SSNMR investigation.

Note:

Cross Polarisation—Cross Polarisation (CP) is a double resonance experiment that produces a much stronger signal than direct excitation. CP allows faster repetition rate which depends upon the T_1 of the proton, instead of nuclei which is of lower gyromagnetic ratio. CP application requires high power irradiation, that may deteriorate the sample or the probe. For example the abundant nucleus is chosen as ^{1}H and the observed nucleus as ^{13}C. The abundant nucleus is excited, and its energy is then transferred to the observed nucleus by using a long low power pulse on both channels. The RF power ratio of these pulses needs to be tuned so that the transition energy for both nuclei is the same. So that for instance for the polarization transfer from a ^{1}H to a ^{13}C, the rotating field B_1 must be 4 times weaker for the proton channel than for the carbon channel.

Gyromagnetic Ratio—The ratio of the magnetic moment of a spinning charged nuclei to its angular momentum is called as gyromagnetic ratio or g-factor.

B. Decoupling

Decoupling is another technique to increase the resolution of SSNMR spectra. Decoupling causes removal of nuclear spin interactions and thereby isolates.

Commonly, in SSNMR experiment there are two ways for decoupling such as

- *Homonuclear decoupling*—experiments which involves decoupling of spin interactions of similar nuclei that are the same as those being detected.
- *Heteronuclear decoupling*—experiments which involves decoupling of spin interactions of different nuclei.

C. Recoupling

Coupling is a useful tool to gather information about structure of atoms in a molecule, such as inter atomic distance. This is because dipolar couplings are distance dependent, so particularly in isotopically labelled molecules dipolar couplings are used to calculate inter atomic distances. Also, because sample spinning phases out many of the dipolar interactions, so, it becomes necessary to conduct recoupling experiments to re produce desired dipolar couplings. For example in NMR crystallographic studies of amorphous solids Rotational Echo Double Resonance (REDOR) experiment is used recoupling technique.

7.7 PROBLEMS BASED LEARNING

After reading the advance-NMR spectroscopy chapter, the readers are suggested/advised to test their learning by solving some below mentioned problems.

A. SUBJECTIVE PROBLEMS

1. Define the terms one dimension, two dimension, three dimension and multiple dimension.
2. Explain DEPT experiment with one example.

3. Define COSY and discuss various COSY experiment with at least one example.
4. Explain HETCOR and Reverse phase HETCOR.
5. Differentiate between HSQC and HMQC.
6. Write a note on HMBC.
7. Describe INADEQUATE and TOCSY.
8. Differentiate between NOESY, and ROESY.
9. Define and Explain HOESY experiment.
10. Define Solid state NMR and discuss various techniques used in it.

B. OBJECTIVE PROBLEMS

1. State True or False

(*i*) HETCOR correlates the chemical shift of hetero nucleus with hydrogen.

(*ii*) In HMQC during evolution time both ^1H and X magnetization (*e.g.*, X = ^{13}C) are allowed to evolve.

(*iii*) In EXSY responses due to exchange are often much more intense than those due to NOE's.

(*iv*) In 3D-NMR experiment a pulse sequence is made up of two 2D sequences.

(*v*) In 4D-NMR experiment a pulse sequence is made up of three 2D sequences.

(*vi*) Quadrupolar interactions, chemical shift anisotropy and inter nuclear dipolar coupling are the two directional interactions.

(*vii*) Quadrupolar interaction can lead to line widths of thousands of ppm.

(*viii*) Dipolar coupling $d = (h \cdot \mu_0/4\pi) \times (\gamma_1 \cdot \gamma_2/r^3)$.

Ans.: 1. (*i*) True (*ii*) True (*iii*) True (*iv*) True
(*v*) True (*vi*) True (*vii*) True (*viii*) True.

2. Fill up the blanks

(*i*) HETCOR that correlates chemical shift of hetero nucleus *e.g.*, ^{13}C via ^{2-3}J coupling with hydrogens through a multiple bond is called as ....

(*ii*) Homonuclear 2D correlation that causes $^{1}J_{C\text{-}C}$ coupling is called as ....

(*iii*) Technique that involves $^{m}J_{H\text{-}H}$ coupling is called as ....

(*iv*) 2D spectroscopy that correlates between nuclei which are physically close, regardless of bond between them is called as ....

(*v*) .... is a heteronuclear correlation spectroscopy that correlates between physically close nuclei regardless of their distance.

(*vi*) Experiment in which homonuclear NOE effects are measured under spin-locked conditions is called as ....

(*vii*) .... NMR is a spectroscopy, that is associated with anisotropy or directionally dependent interactions.

(*viii*) .... spinning is also called as sample rotation.

Ans.: 2. (*i*) HMBC (*ii*) Inadequate (*iii*) TOCSY
(*iv*) NOESY (*v*) HOESY (*vi*) ROESY
(*vii*) Solid state (*viii*) Magic angle.

3. Match the following

(*i*) DEPT-45	(*a*) Signals only for carbon having 1 or 2 or 3 hydrogen
(*ii*) DEPT-90	(*b*) Signals only for carbon having one hydrogen
(*iii*) DEPT-135	(*c*) Signals only for carbon having 1 and 2 hydrogens

Ans.: 3. (*i*) – (*a*), (*ii*) – (*b*), (*iii*) – (*c*).

4. Match the following

(*i*) SQF COSY	(*a*) (π/2)a & (π/2)b pulse
(*ii*) DQF COSY	(*b*) (π/2)a, (π/2)b and (π/2)c pulse
(*iii*) TQF COSY	(*c*) (π/2)a, (π/2)b, (π/2)c and (π/2)d pulse
(*iv*) MQF COSY	(*d*) (π/2)a ... (π/2)m pulse

Ans.: 4. (*i*) – (*a*), (*ii*) – (*b*), (*iii*) – (*c*), (*iv*) – (*d*).

7.8 SUGGESTED READINGS

[1] *Organic structure determination using 2D-NMR spectroscopy*, 2nd Edition, Academic Press, 2012.

[2] *Annual Reports on NMR Spectroscopy*, Volume 76, Academic Press, 2012.

[3] Lian L., Roberts G., *Protein NMR Spectroscopy: Principal techniques and Applications*, 1st Edition, John Wiley & Sons, 2011.

[4] Pavia D.L., Lampman G.M., Kriz G.S., Vyvyan J.R., *Introduction to Spectroscopy*, 4th Edition, Brooks/Cole, 2009.

[5] Holzgrabe U., Wawer I, Diehl B., *NMR Spectroscopy in Pharmaceutical Analysis*, 1st Edition, Elsevier, 2008.

[6] Kalsi P.S., *Spectroscopy of Organic Compounds*, 6th Edition, New Age International, 2007.

[7] Silverstein R.M., Webster F.X. and Kiemle D.J., *Spectrometric identification of Organic Compounds*, 7th Edition, John Wiley, 2005.

[8] Melinda J. Duer, M.J., *Introduction to Solid-State NMR Spectroscopy*, Wiley, 2005.

[9] Berger S., Braun S., *200 and More NMR experiment—A Practical Course*, Wiley-VCH, 2004.

[10] Sarkar, S.K., *NMR Spectroscopy and its Application to Biomedical Research*, Elsevier, 1996.

[11] Lambert J.B., Rittner R., *Recent Advances in Organic NMR Spectroscopy*, Norell Press, 1987.

Chapter 8

UV-Visible Spectroscopy

Objective

To make readers learn

- Basic Concepts, Theory and Underlying Principles of UV-visible Spectroscopy.
- Instrumentation of UV-visible Spectroscopy.
- Application of Woodward-Fieser and Fieser-Kuhn Rules for Calculation of Absorption Maxima for Various Compounds.

8.1 INTRODUCTION

UV-Visible spectroscopy is a form of absorption spectroscopy which involves treatment and absorption of ultra violet & visible radiations onto the organic compounds. This causes molecules to undergo electronic excitations from one energy level to another energy level. Molecules containing σ-electrons, π-electrons or non-bonding electrons (*n-electrons*) absorbs energy in the form of ultraviolet or visible light to excite their electrons to higher anti-bonding molecular orbitals. The more easily the excitation of electron (*i.e., lower energy gap between the HOMO and LUMO*) occurs, the longer the wavelength of light is absorbed. The absorption of radiation is characteristic and depends upon the nature of electrons in a molecule. Compounds that are colourless absorbs radiations in the ultraviolet region (200–400 mμ). Whereas compounds that are coloured absorbs radiations in the visible region (400–800 mμ). In case of both UV and visible spectroscopy it is only valence electrons of atoms of a molecule which are responsible for absorption of energy. To use UV-visible spectroscopy as an analytical tool, it is necessary to understand the theory and basic principles of UV-visible spectroscopy.

8.2 THEORY/BASIC PRINCIPLES

8.2.1 UV-visible Radiation Absorption Laws

The absorption of UV-visible radiations is governed by two laws *viz.*,

a. Beer's law
b. Lambert's law

A. Beer's Law

According to Beer's law, '*when a beam of monochromatic radiation passes through a solution of absorbing medium, then rate of decrease of intensity of radiation is proportional to the intensity of incident light and concentration of solution.*

Accordingly, decrease in intensity of incident light (dI) with concentration (c) is directly proportional to the intensity of incident light (I); represented as

$$-\frac{dI}{dc} \propto I \qquad ...(8.1)$$

Introducing the constant of proportionality gives

$$-\frac{dI}{dc} = k.I \qquad ...(8.2)$$

Rearrangement of the data gives,

$$-\frac{dI}{I} = k.d.c \qquad ...(8.3)$$

Integrating the equation gives,

$$-\ln I = k.c + b \qquad ...(8.4)$$

Where, b is constant of integration.

At zero concentration absorbance is zero, therefore I = Io. So, substituting the value of I in equation 8.4 gives,

$$-\ln I = k.0 + b \qquad ...(8.5)$$

$$-\ln I = b \qquad ...(8.6)$$

Substituting the value of 'b' in equation 8.4 gives,

$$-\ln . I = k.c - \ln . I_0 \qquad ...(8.7)$$

$$\ln . I_0 = \ln . I = k.c \qquad ...(8.8)$$

As, log A – log B = log A/B, therefore applying same approach in Eqn. 8.8 gives,

$$\ln . \frac{I_0}{I} = k.c \qquad ...(8.9)$$

Removing the natural logarithm from Eqn. 8.9 gives,

$$\frac{I_0}{I} = e^{k.c} \qquad ...(8.10)$$

Putting inverse on both sides of Eqn. 8.10 gives,

$$\frac{I}{I_0} = e^{-k.c} \qquad ...(8.11)$$

Hence, Beer's law expression is given as:

$$I = I_0.e^{-k.c} \qquad ...(8.12)$$

B. Lambert's Law

According to Lambert's law, '*when a beam of monochromatic radiation passes through a homogenous absorbing medium, then rate of decrease of intensity of radiation with thickness of medium is proportional to intensity of incident light*'.

Accordingly, decrease in the intensity of incident light (dI) with thickness of media is directly proportional to the intensity of incident light (I); represented as

$$-\frac{dI}{dt} \propto I \qquad ...(8.13)$$

The above equation 8.13 is simplified as equation 8.12 and modified by replacing the term 'c' with 't' to give

$$I = I_0.e^{-k.t} \qquad ...(8.14)$$

Beer-Lambert's Law Combination

Beer's law equation 8.12 and Lamberts law equation 8.14 are combined together to formulate a new expression for combined Beer-Lamberts law. The expression can be derived as follows:

$$I = I_0.e^{-k.c.t} \quad \text{...(8.15)}$$

Applying the natural logarithm in equation 8.15 gives,

$$I = I_0.10^{-k.c.t} \quad \text{...(8.16)}$$

Rearranging the terms in equation 8.16 gives,

$$\frac{I}{I_0} = 10^{-k.c.t} \quad \text{...(8.17)}$$

Putting inverse on both sides of equation 8.17 gives,

$$\frac{I_0}{I} = 10^{k.c.t} \quad \text{...(8.18)}$$

Putting log on both sides of equation 8.18 gives,

$$\log\frac{I_0}{I} = k.c.t \quad \text{...(8.19)}$$

As, Transmittance (T) $= \frac{I}{I_0}$ and Absorbance (A) $= \frac{\log 1}{\text{T}}$

Hence, Absorbance

$$(\text{A}) = \frac{\log 1}{\frac{I_0}{I}}$$

Therefore,

$$\text{A} = \log\frac{I_0}{I} \quad \text{...(8.20)}$$

As from equation 8.19 and 8.20, it is known that

$$\text{A} = \log.\frac{I_0}{I} = k.c.t$$

So, for combined Beer-Lambert's law, the mathematical relation can be given as:

$$A = k.c.t \text{ or } A = \varepsilon.c.t \quad ...(8.21)$$

Where, A = Absorbance or extinction coefficient

ε = Molecular extinction coefficient

c = Sample concentration mol/lt

t = Path length

C. Deviations From Beer's Law

In usual procedure different absorbances are measured on a UV spectrometer against different solution concentrations of standard compound (*in microgram*). There after a plot is drawn for absorbance against standard and concentration of unknown sample can be determined. In this procedure two conditions may occur *viz.*,

- There can be appearance of linearity or straight line in the plot (concentration versus absorbance) of analyte under investigation. In such a condition the analyte system is said to be in agreement or obeys Beer's law. The straight line is obtained by joining the maximum number of points. Further, a regression line can be used to determine the concentration of analyte solution. Figure 8.1, shows a plot of concentration versus absorbance of analyte system obeying Beer's law.

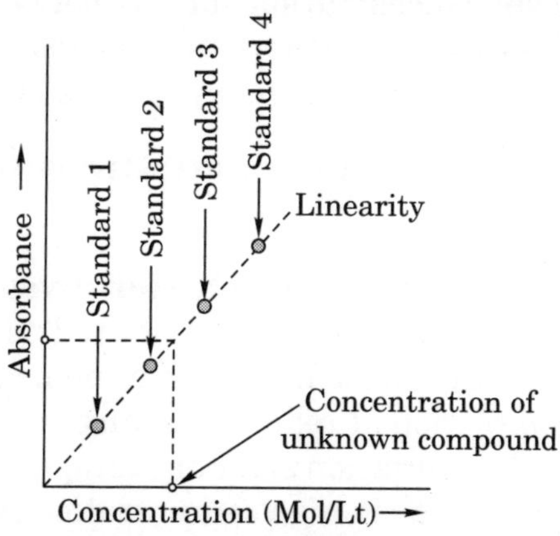

Figure 8.1: Plot of concentration versus absorbance of analyte system obeying Beer's law.

- A straight line may not be obtained or a non-linear curve can be obtained, in such a condition the analyte system is deviates from Beer's law. These deviations can be either positive or negative. Figure 8.2 a & b, shows plots of concentration versus absorbance of analyte system deviating from Beer's law.

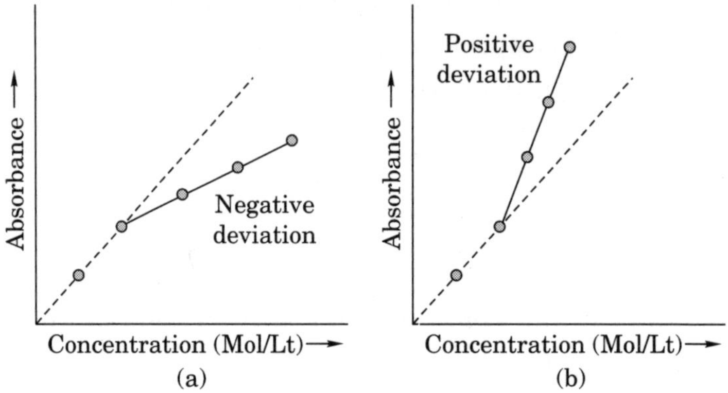

Figure 8.2: Plots of system deviating from Beer's law a) negative and b) positive.

In such deviating plots, positive deviation occurs due to small change in concentration occurs which causes greater change in absorbance. Whereas negative deviations occurs due to a large change in concentration occurs that causes smaller change in absorbance. Commonly many of the analyte systems offers linearity or agreement to Beer's law at lower concentration such as between 10–50 μg/ml, but on further higher concentrations such as above 50 μg/ml shows deviations from linearity or do not obeys Beer's law.

These deviations from Beer's law are attributed to three major reasons.

1. **Deviations Attributed to Solution System**—Frequently deviations from Beers law is attributed to physiochemical properties of solution system, such as unhomogeneity of analyte solution, incompletion of reaction, change in pH or ionisation, association and dissociation etc. For example the benzyl alcohol in carbon tetrachloride in high concentration exists in polymeric form, but commonly undergoes polymeric dissociation on dilution which causes absorbance.

Analyte in the form of suspensions also do not follows Beer's law, but can be estimated with different known concentration reference curve.

2. **Deviations Attributed to Instrument**—Another major cause for Beers law deviation is due to faults or shortfalls at the instrument level such as noise, fluctuations in the radiation source, use of stray light (or radiations), slit dimension (improper slit width), imperfect or contaminated cell windows, temperature variations and some monochromator defects (such that unable to monochromatize the radiations).

- Insufficiency of time such as rapid absorbance scanning also leads to deviation from beer's law.
- It is rare that radiations after monochromator are completely unpolarized. So, polarization of the incident radiation and polarization dependent response of photo detector is also one cause to deviation from Beer's law.
- The best results (minimum error) can be expected when absorbance is in the range 0.2 to 0.8. But when absorbance is below 0.1 and above 1.0, then considerable errors can be expected. Furthermore the error in results depends upon performance characteristics of the individual spectrophotometer also.

3. **Deviations Attributed to Sample**—Sample instability, scattering error (error due to presence of particulate matter, emulsions or micelles in sample), sample errors (due to photochemical reaction, formation of aggregates or adsorption on the cell wall), fluorescence error (sample excitement to fluoresce by the incident radiation beam leading to erroneous reading) and also presence of impurities in the analyte are few sample characteristics that commonly results in deviation from beers law. Apart from solute solvents interactions are also the basic cause for Beer's law deviation.

8.2.2 Electronic Transitions

UV-Visible absorption spectra arise as a result of transitions or electrons within a molecule or ion from a lower to a higher electronic energy level. Whereas emission spectra arises from the reverse type of transition. For radiation to cause electronic excitation, it should be in the region of the electromagnetic spectrum.

In a molecule there occurs three types of electrons such as:

1. **Sigma (σ) electrons:** *Electrons that are associated with saturated bonds are called as sigma electrons.* These electrons are absorbed at a very low wavelength region (below 200 mμ) *i.e.*, in the vacuum region. Treatment of σ electrons with UV radiations excites them into anti bonding state σ* electrons and this transition is represented as σ → σ*.

   Example: 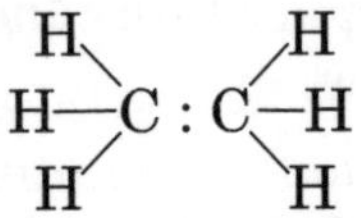

2. **Pi (π) electrons:** *Electrons that are associated with unsaturated bonds are termed as pi electrons.* The π electrons on double or triple bonds are absorbed on higher wavelength. Treatment of π electrons with UV radiations excites them into anti bonding state π* electrons and this transition is represented as π → π*.

   Example: 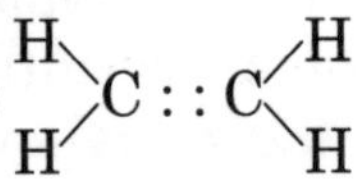

3. **Nonbonding (n) electrons:** *Electrons that are not associated with any bond are known as non bonding electrons.* For example lone pair electrons of S, O, N and X hetero atoms (attached with saturated and unsaturated carbons) treatment with UV radiations excites them into two anti bonding states *i.e.*, σ* and π* respectively. The transitions for the two cases are represented as n → σ* and n → π*.

   Example: 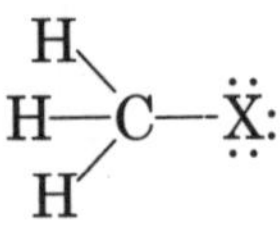

Based on the type of electrons, the electronic transitions are classified into four types

(a) σ → σ* **Transitions:** *Transitions which involves excitation of electron from bonding σ orbital to corresponding anti bonding σ* orbital are referred to σ → σ* transitions.* As sigma bonds are very strong hence these types of transitions are high energy process (requires high energy of shorter wavelength). These transitions occurs below 200 mμ. For study of such transitions generally path is made evacuated. Hence these transitions occurs in evacuated region. Absorption maxima due to $\sigma \rightarrow \sigma^*$ transitions are not seen in typical UV-Visisble spectra (200–700 mμ).

- Commonly, saturated hydrocarbons are accompanied by $\sigma \rightarrow \sigma^*$ transitions and are likely to be absorbed near 150 mμ.

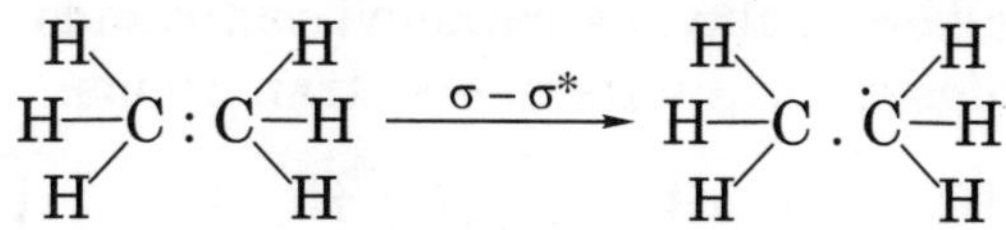

For example, methane (which has only C–H bonds, and can only undergo $\sigma \rightarrow \sigma^*$ transitions) shows an absorbance maximum at 125 mμ.

(b) **n → σ* Transitions:** *Transitions which involves excitation of lone pair of electrons (non-bonding electrons) of heteroatom to corresponding antibonding σ* orbital are termed as n → σ* transitions.* Saturated compounds containing atoms with lone pairs (non-bonding electrons) are capable of $n \rightarrow \sigma^*$ transitions. These transitions usually needs relatively lesser energy than $\sigma \rightarrow \sigma^*$ transitions, so absorbs on a higher wavelength. These transitions are initiated by light whose wavelength is in the range 150 – 250 mμ. The number of organic functional groups with $n \rightarrow \sigma^*$ peaks in the UV region are relatively small.

- Commonly, saturated hydrocarbons containing a heteroatom are accompanied by $n \rightarrow \sigma^*$ transitions and are likely to be absorbed between 150–250 mμ.

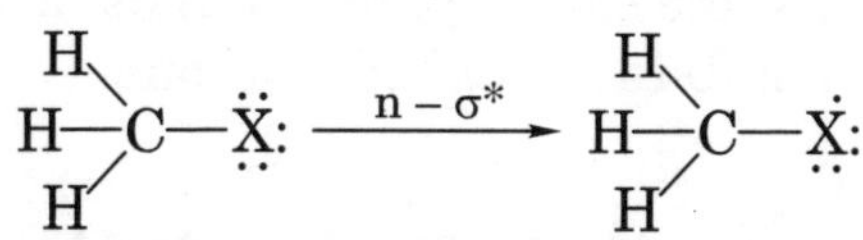

For example, methyl chloride absorbs at 175 mμ and methyl iodide absorbs at 258 mμ.

(c) **π → π* Transitions:** *Those Transitions which involves excitation of electron from π orbital to corresponding antibonding* *π* orbital are called as π → π* Transitions.* These transitions usually needs further lesser energy than n → σ* transitions, so absorbs on a further higher wavelength.

- The solvent in which absorbing species is dissolved also has an profound effect on the spectrum of the species. Such that, on increasing the polarity of solvent, the peaks resulting from π → π* transitions are shifted to longer wavelengths (called bathochromic or red shift).
- The unsaturated hydrocarbons containing double bond or triple bond like alkenes, alkynes, carbonyl compounds, azo compounds and cyanides etc. exhibits π → π* transitions.

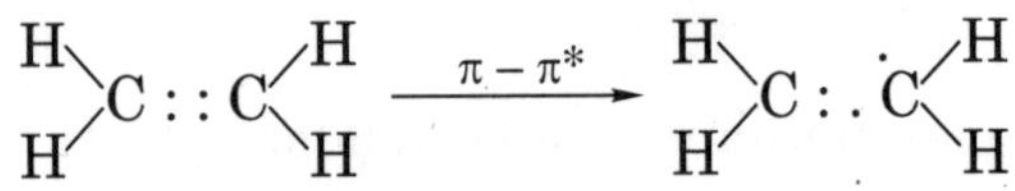

For example, unconjugated alkenes shows a band due to π → π* between 170–190 mμ.

(*d*) n → π* Transitions: *Those Transitions which involves excitation of lone pair (non-bonding electron) of electron of heteroatom to corresponding antibonding π* orbital are called as n → π* Transitions.* As compare to other transitions n → π* transitions requires least energy, hence exhibits absorption on highest wavelength than other transitions. n → π* transitions commonly shows molar absorptivities between 1000 and 10,000 mol^{-1} cm^{-1}.

- The increase in the polarity of solvent (in which species are dissolved), causes the peaks resulting from n → π* transitions to shift towards shorter wavelengths (called hypsochromic or blue shift).
- The compounds like carbonyl compounds, azo compounds and cyanides etc. exhibits n → π* transitions.

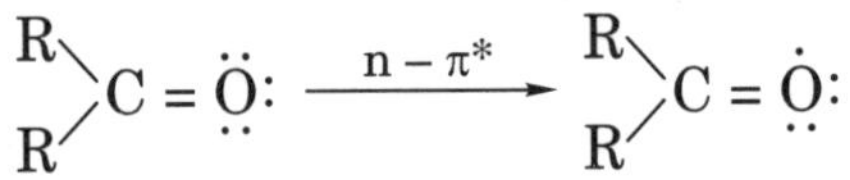

For example, saturated aldehyde shows an absorption at 290 mμ and 180 mμ due to n → π* and π → π* transitions respectively.

The energy requirement for transition of three types of electrons can be best understood from the Figure 8.3.

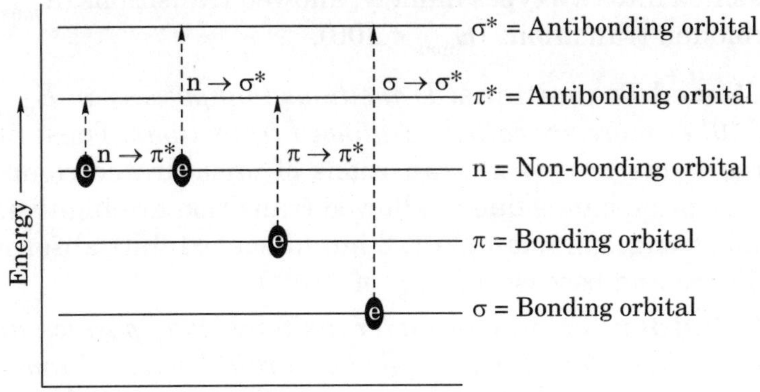

Figure 8.3: Energy diagram for various electronic transitions.

Example: *Transitions for carbonyl compounds*—In case of carbonyl compounds commonly two types of transitions occurs such as

1. High energy transitions (n → σ* and π → π*) and
2. Low energy transitions (n → π*).

8.2.3 Electronic Transition Probability

It is not essential that exposure of a compound to UV or visible light always cause electronic transitions. In fact the probability of a particular electronic transition has been found to depend upon the value of molar extinction coefficient and certain other factors. Also, it is not always necessary that these excitations of an electron takes place from a bonding orbital or lone pair to an anti bonding or nonbonding orbital when exposed to UV or visible light.

It can be shown that

$$E_{max} = 0.87 \times 10^{20}\,\text{p.a}$$

Where p – Transition probability (value 0–1)

a – Target area of absorbing system (Chromophore)

A. Types of Transitions

Depending upon symmetry and value of E_{max}, electronic transitions are classified into two types namely, allowed transitions ($E_{max} > 10^4$) and forbidden transitions ($E_{max} < 100$).

1. **Allowed Transitions:** *Transitions that possess an E_{max} value of 10^4 or more are called as allowed transitions.* These usually occurs due to $\pi \rightarrow \pi^*$ transitions of unsaturated compounds. Absorption bands due to allowed transition are highly intense and strong. For example 1, 3-butadiene exhibits absorption at 217 mμ and possess an E_{max} of 21000.
2. **Forbidden Transitions:** *Transitions that possess an E_{max} value of less than 100 are called as forbidden transitions.* These are the transitions that possess an E_{max} of generally less than 10^4. These usually occurs due to $n \rightarrow \pi^*$ transition of a saturated aldehyde or ketone showing an absorption near 290 mμ and having E_{max} less than 100. For Example benzophenone shows absorption at 325 mμ and E_{max} of 180 mμ.

- In order to confirm whether transition is allowed or forbidden for any molecule it is important to consider following factors:
    - The geometry of the molecular orbital in the ground state.
    - The geometry of the molecular orbital in the excited state.
    - Orientation of the electric dipole of the incident light.

B. Absorption Intensity Measurement

It may be noted that the intensity of absorption is directly proportional to the transition probability. An allowed transition always possess E_{max} value greater than 10000 and whereas low probability transitions possess E_{max} value less than 1000.

C. Electronic Transitions Versus Symmetry Restrictions

Existence of transitions whether allowed or forbidden depends upon two aspects.

(*a*) Firstly, geometries of the lower and the higher energy molecular orbital's.

(*b*) Secondly, the symmetry of the molecule as a whole.

Symmetrical molecules cause more restrictions on their electronic transitions than unsymmetrical molecules. For example, benzene is a highly symmetrical molecule. Hence many restrictions apply to the electronic transitions and thus its electronic absorption spectrum is simple.

In unsymmetrical molecules, no restrictions occur on electronic transitions, so their transitions are observed among all vacant molecular orbitals except filled orbitals. Such molecules results in a complex electronic spectrum. For example, cyclohexane compound results in a very complex spectrum.

8.2.4 Chromophore

An isolated covalently bonded group that shows a characteristic absorption in the UV or visible region is called as chromophore. For example nitro compounds exhibits yellow colour due to presence of nitro group as a chromophore. In a similar manner in azo dyes, the azo group acts as a chromophore responsible for providing colour.

Some other important examples for chromophores are ethylenic, acetylenic, carbonyls, acids, esters, nitro group etc.

There are three types of chromophores *viz.*,

1. **Chromophore bearing π electrons:** Chromophoric groups containing π electrons in their molecular structure undergoes $\pi \rightarrow \pi^*$ transitions. For example, ethylenes and acetylenes act as chromophores bearing π electrons.
2. **Chromophore bearing n and π electrons:** Chromophoric groups containing n and π electrons in their molecular structure undergoes two types of transitions *i.e.*, $\pi \rightarrow \pi^*$ and $n \rightarrow \pi^*$. For example, carbonyl group in aldehydes or ketones exhibits chromophoric property.
3. **Chromophore bearing n and σ electrons:** Chromophoric groups containing n and σ electrons in their molecular structure and therefore undergoes two types of transitions *i.e.*, $\sigma \rightarrow \sigma^*$ and $n \rightarrow \sigma^*$. For example a C–O group in ethers or alcohols commonly acts as a chromophoric group.

Chromophore Identification Tips

- Existence of a band near 300 mμ shows presence 2–3 conjugated units.
- Presence of a band near 270–350 mμ with very low intensity *i.e.*, E_{max} 10–100 are due to n → π* transitions of the carbonyl group.
- High values E_{max} from 10000 to 20000 indicates presence of simple conjugated chromophores such as dienes or α, β- unsaturated ketones.
- Characteristic absorption with E_{max} value from 1000 to 10000 indicates presence of an aromatic system.

8.2.5 Auxochrome

A group that itself does not act as a chromophore but whose presence causes shift of the absorption band towards the longer wavelength region of the UV visible is called as auxochrome. As presence of an auxochromic group causes colour enhancement, so auxochrome is also called as colour enhancers.

Auxochrome causes absorption enhancement by extending the conjugation of a chromophore by sharing non-bonding electrons. Thus results in new chromophore which possesses a different value of the absorption max as well as the extinction coefficient.

The maximum absorption enhancement due to auxochrome group is 200 mμ. Various groups like $-NH_2$, –NHR, $-NR_2$, –SH, OH, and –OR etc. are some common groups that exhibit auxochromic properties.

For example, benzene exhibits an absorption maximum at 255 mμ, whereas substitution of benzene with an $-NH_2$ group (*i.e.*, aniline) makes it to absorb at 280 mμ. Therefore in this case amino $-NH_2$ group acts as an auxochromic group.

8.2.6 Absorption (λ_{max}) Shifts

Beside the nature of chromophoric group present in the molecule or compounds under investigation, there are certain other factors or effects that cause shifts in absorption (λ_{max}).

Based on extent of shifts in absorption, the absorption shifts are categorised into two types namely,

1. Bathochromic shift and
2. Hypsochromic shift.

1. **Bathochromic Shift (Red Shift)**—*Bathochromic shift is defined as an effect which involves shifting of absorption maximum* (λ_{max}) *is towards longer wavelength or red end of the spectrum*. Bathochromic shift exhibits an increase in λ_{max} value from its normal value. The Red shift is generally attributed to change in polarity of solvent and presence of auxochrome.

   For example, $n \rightarrow \pi^*$ transitions of carbonyl compounds exhibit bathochromic shift on change in polarity of solvent.

2. **Hypsochromic Shift (Blue Shift)**—*Hypsochromic shift is an effect which causes shifting of absorption maximum* (λ_{max}), *which is shifted towards shorter wavelength or blue end in the spectrum*. Hypsochromic shift encompasses a decrease in absorption (λ_{max}) value from its normal value. Blue shift is generally attributed to removal of conjugation and change in polarity of solvent.

   For example, aniline shows an absorption maximum at 280 mμ in normal conditions. But experiences an hypsochromic shift in absorption maximum from to 280 mμ when converted into acidic solution form.

8.2.7 Intensity (E_{max}) Shifts

Similar to absorption shifts there are two types of intensity shifts *viz.*,

1. Hyperchromic shift and
2. Hypochromic shift.

1. **Hyperchromic Shift**—*Hyperchromic shift is defined as an effect in which the intensity of absorption maximum* (E_{max}) *increases*. Hyperchromic shift reflects an increase in E_{max} value from its normal value. Hyperchromic shift is generally attributed to introduction of some group that disturbs the geometry of molecule under investigation.

For example, normally pyridine shows 'B' band at 257 mμ with an E_{max} value (intensity) of 2750. But when pyridine is substituted with an auxochromic methyl group it shows a further increase in E_{max} value *i.e.*, 3560 and absorbs at 262 mμ.

2. **Hypochromic Shift**—*Hypochromic shift is an effect due to which the intensity of absorption maximum* (E_{max}) *decreases.* Hypochromic shift causes an increase in E_{max} value from its normal value. The hypochromic shift is generally attributed to presence of auxochrome.

   For example, biphenyl compounds shows absorption at 250 mμ and E_{max} value (intensity) of 19000. But when biphenyl compound is substituted with an auxochromic methyl group, it shows a further decrease in E_{max} value *i.e.*, 10250 with absorption at 237 mμ.

8.2.8 Absorption Bands

As excitation of molecules with UV-visible radiation causes different types of electronic transitions, so the spectrum is accompanied by different types of bands described as:

a. K-bands,
b. R-bands,
c. B-bands and
d. E-bands.

1. **K-bands (Konjugierte bands)**—*The high intensity fundamental bands which originates due to* $\pi \rightarrow \pi^*$ *transitions in a compound that contains a conjugated system are commonly termed as K-bands.* K-bands are also called as allowed bands. These K-bands are commonly generated by compounds that contain conjugated double bonds such as dienes, polyenes, and enones etc. Aromatic compounds that are substituted by chromophore also commonly generate K-bands. The intensity of K-bands is usually more than 10000. Table 8.1, represents some examples of compounds that exhibits K-bands.

Table 8.1: Some examples of compounds exhibiting K-bands.

Compounds	*Transitions*	E_{max}	λ_{max} *in mμ*
1, 3-butadiene	$\pi \rightarrow \pi^*$	21000	217
1, 3-pentadiene	$\pi \rightarrow \pi^*$	22600	223
Acetophenone	$\pi \rightarrow \pi^*$	13000	240
Acrolein	$\pi \rightarrow \pi^*$	11500	210

2. **R-bands (Radical bands)**—*The low intensity or forbidden bands which originates due to n → π* transitions of a single chromophoric group in a compound that contains at least one lone pair of electrons on the hetero atom are known as R-bands.* R-bands are also sometimes called as forbidden bands as the intensity of these bands (E_{max}) is below 100. Table 8.2, represents some examples of compounds that exhibits R-bands.

Table 8.2: Some examples of compounds exhibiting R-bands.

Compounds	*Transitions*	E_{max}	λ_{max}
Acetone	$n \rightarrow \pi^*$	15	270
Acetaldehyde	$n \rightarrow \pi^*$	12	293
Acetophenone	$n \rightarrow \pi^*$	50	319
Acrolein	$n \rightarrow \pi^*$	14	315

3. **B-bands (Benzenoid bands)**—*Bands which exhibits an intensity* (E_{max}) *below 10000 and above 100, also originates due to π → π* transitions in a aromatic or hetero aromatic compound, are commonly referred to B-bands.* These bands originates due to $\pi \rightarrow \pi^*$ transition in aromatic or hetero aromatic molecules. Usually, these bands are stronger in intensity than R-bands and lesser in intensity than K-bands. Table 8.3, represents some examples of compounds that exhibits B-bands.

Table 8.3: Some examples of compounds exhibiting B-bands.

Compounds	*Transitions*	E_{max}	λ_{max}
Benzene	$\pi \rightarrow \pi^*$	215	255
Phenol	$\pi \rightarrow \pi^*$	1450	270
Acetophenone	$\pi \rightarrow \pi^*$	1110	278
Quinoline	$\pi \rightarrow \pi^*$	2500	315

4. **E-bands (Ethylenic bands)**—*A couple of bands (one with larger E_{max} and another with lower E_{max}) which are a result of electronic transitions are commonly termed as E-bands*. As these bands exists in the form of a couple, so categorized into two types *i.e.*, E_1 and E_2.
   (*a*) **E_1 band:** These bands usually appears on a shorter wavelength but exhibits very high intensity such that some time it is more than the fundamental K-bands.
   (*b*) **E_2 band:** These bands commonly appears on a longer wavelength and exhibits lower intensity in comparison to E_1 bands. Actually, the intensity of E_2 band is also high but usually less than K-bands and more than R-band.

   For example: Benzene exhibits E_1 and E_2 bands with an intensity of 50000 and 7900 respectively E_1 band for the same compound which appears at longer λ. Table 8.4, represents some examples of compounds that exhibits E-bands.

Table 8.4: Some examples of compounds exhibiting E-bands.

Compounds	***E_1 Band***		***E_2 Band***	
	E_{max}	**λ_{max}**	***E_{max}***	**λ_{max}**
Benzene	50000	184	7900	204
Naphthalene	133000	221	9300	286
Anthracene	180000	256	9000	375
Quinoline	40000	288	3162	270

8.2.9 Woodward-Fieser Rules

Robert Burns Woodward in 1945 developed certain rules for calculation of λ_{max} to determine structure of molecules. Louis Frederick Fieser in 1959 modified those Woodward rules with more experimental data, thereby known as Woodward-feiser Rules. Woodward-fieser rules are the theoretical assistance for the practical values of λ_{max} that are generated on a UV-visible spectrometer for a given set of compounds. The benefit of Woodward-fieser rules is that with these rules without subjecting the sample to UV-visible spectrometer, the λ_{max} value of sample molecule can be determined in advance. The theoretical λ_{max} value generated through Woodward-fieser calculations are almost near to the λ_{max} value generated on a UV-visible spectrometer.

Terminologies Commonly Used in Double Bonds Containing System

1. *Open Chain/Alicyclic Conjugated Dienes*—compounds that contains a 'butadiene' as basic unit are called as open chain conjugated dienes.

   For example, following structure contains a basic unit of open chain conjugated diene.

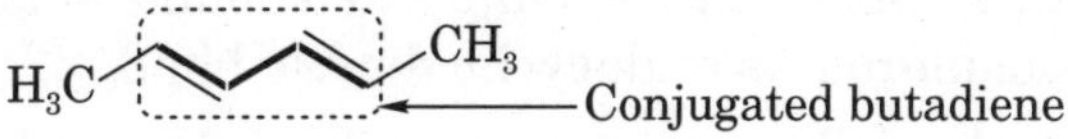

2. *Homoannular Conjugated Dienes*—compounds that contains a conjugated butadiene in a single ring are called as homoannular conjugated dienes.

   For example, following structure contains a basic unit of homoannular conjugated diene.

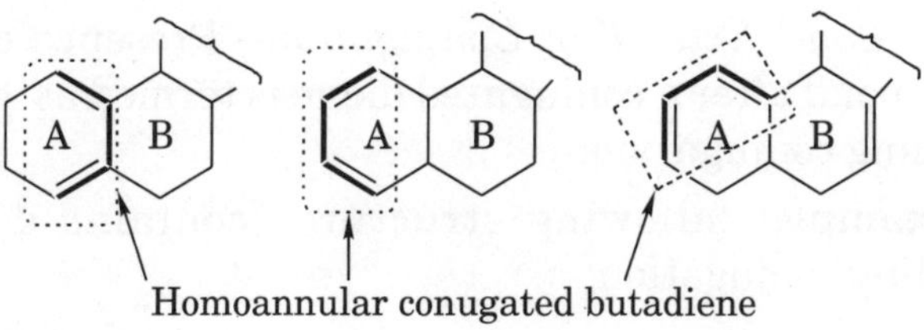

3. *Heteroannular Conjugated Dienes*—compounds that contains a conjugated diene in two rings are called as heteroannular conjugated dienes.

   For example, following structure contains a basic unit of homoannular conjugated diene.

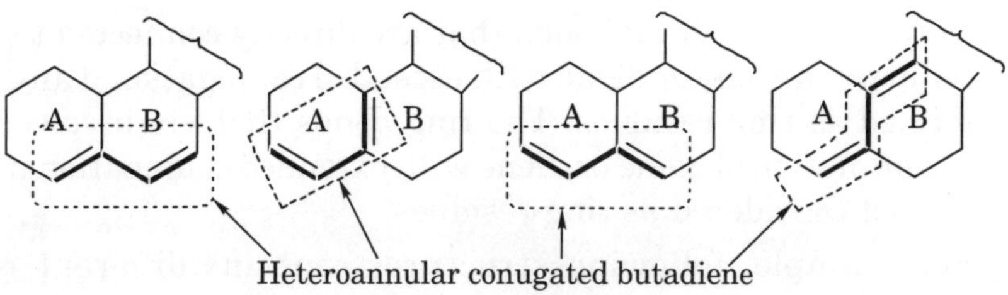

4. *Exocyclic Double Bond*—Conjugated double bond out to any ring is considered as exocyclic double bond.

   For example, following structures contains exocyclic double bond.

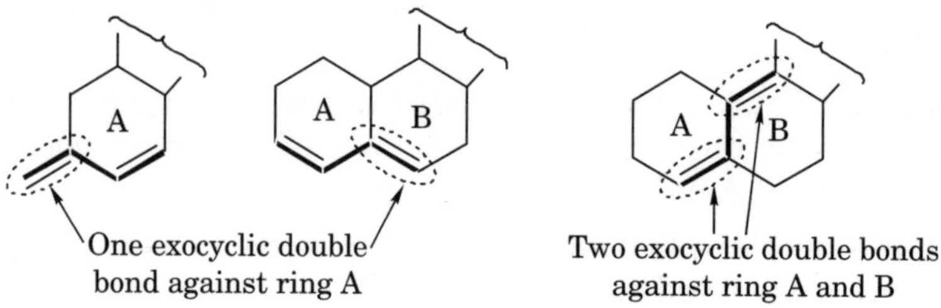

5. *Endocyclic Double Bond*—Conjugated double bond inside of any ring is considered as endocyclic double bond.

   For example, following structure contains endocyclic double bond.

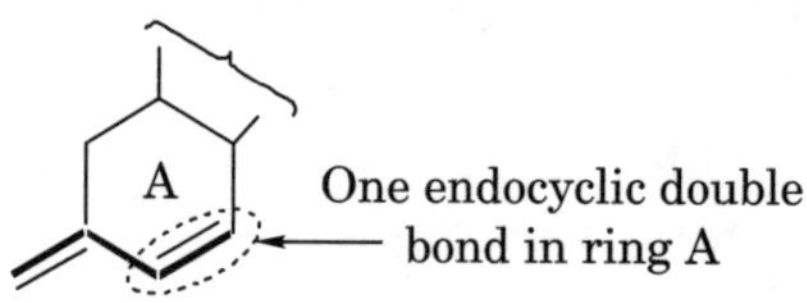

6. *Double Bond Extending Conjugation*—Presence of alternate double bond after a conjugated diene is termed as double bond extending conjugation.

   For example, following structures contains double bond extending conjugation.

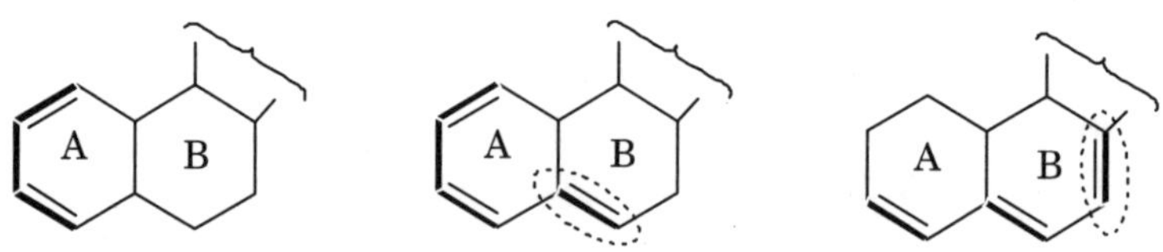

No double bond extending conjugation | One double bond extending conjugation | One double bond extending conjugation

7. *Ring Residues*—Ring bonds that are directly connected to the conjugated diene or diene with extended conjugation if any are termed as ring residues. The ring bonds that are involved in the formation of diene or diene with extended conjugation itself are not considered as ring residues.

   For example, following structures contains different ring residues.

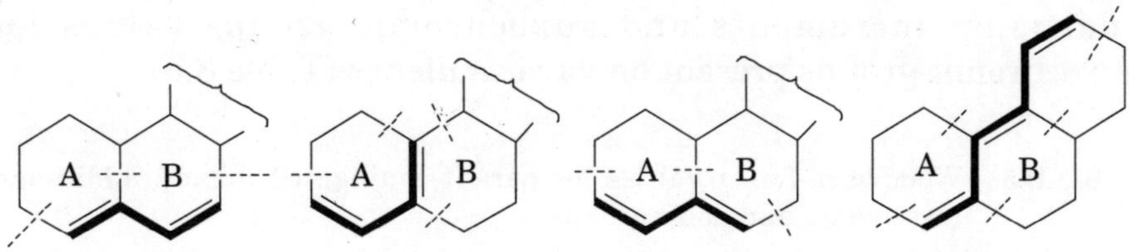

Three ring residues | Four ring residues | Three ring residues | Five ring residues

8. *Alkyl Residues*—Alkyl bonds that are directly connected to the conjugated diene or diene with extended conjugation if any are termed as alkyl residues. The alkyl bonds that are involved in the formation of diene or diene with extended conjugation itself are not considered as ring residues.

   For example, following structures contains different ring residues.

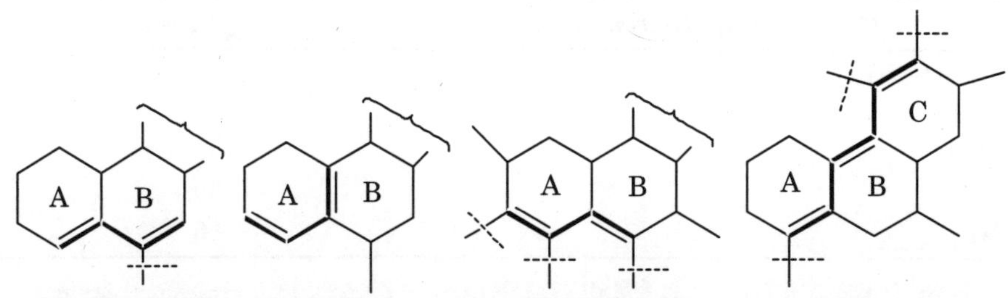

One alkyl residue on ring B | No alkyl residue | Three alkyl residue- two on ring A and one on ring B | Three alkyl residue- one on ring A and two on ring C

The class of compounds whose λ_{max} value can be calculated through Woodward-fieser rules are as follows:

- Dienes
- Polyenes
- α, β-Unsaturated Carbonyl Compounds
- Benzene Chromophores

8.2.9.1 *Woodward-fieser Rules for λ_{max} Calculation of Conjugated Dienes*

To calculate the λ_{max} value of conjugated dienes, Woodward-fieser suggested some parent absorption values for certain class of dienes, additional increments and auxochromic groups values for auxochromic groups present on various dienes (Table 8.5).

Table 8.5: Woodward-fieser values for parent conjugated dienes, additional increments and polar groups.

Parent Double Bond System	λ_{max} **in mμ**
Base value for open chain/salicyclic conjugated dienes	217
Base value for heteroannular conjugated dienes	214
Base value for homoannular conjugated dienes	253
Additional Increments	λ_{max} **in mμ**
Alkyl substituent or ring residues	5
Exocyclic double bond	5
Double bond extending conjugation	30
Polar Group Increments	λ_{max} **in mμ**
-O-CO-CH$_3$	0
-Cl/Br	5
-O-R	6
-S-R	30.
-NR$_2$	60

Following are the examples which illustrates the Woodward-fieser rules based calculation of λ_{max} for dienes.

Example 1.

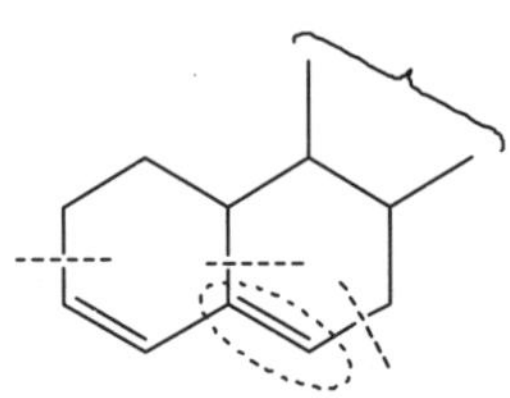

For heteroannular diene system	=	214 mμ
Alkyl substituent and ring residues	=	15 mμ
One exocyclic double bond	=	5 mμ
Calculated λ_{max} value	**=**	**234 mμ**
Observed value	=	235 mμ

Example 2.

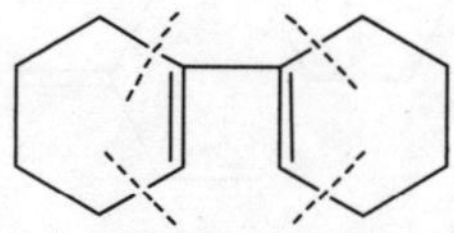

For heteroannular diene system	=	214 mμ
Alkyl substituent and ring residues	=	20 mμ
Calculated λ_{max} value	**=**	**234 mμ**
Observed value	=	234 mμ

Example 3.

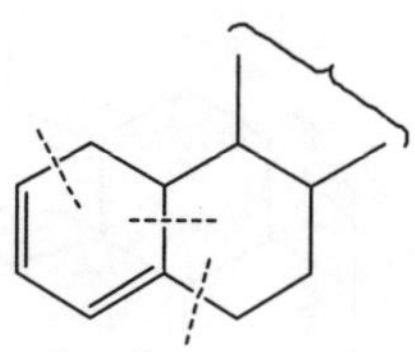

For homoannular diene system	=	253 mμ
Alkyl substituent and ring residues	=	15 mμ
One exocyclic double bond	=	5 mμ
Calculated λ_{max} value	**=**	**273 mμ**
Observed value	=	274 mμ

Example 4.

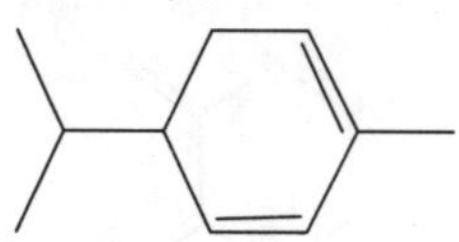

For homoannular diene system	=	253 mμ
Alkyl substituent and ring residues	=	15 mμ
Calculated λ_{max} value	**=**	**268 mμ**
Observed value	=	263 mμ

Example 5.

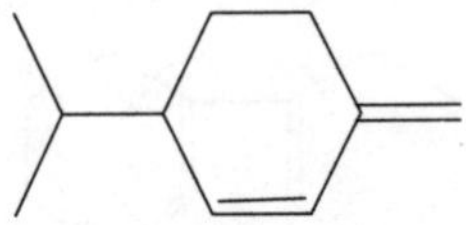

For heteroannular diene system	=	214 mμ
Alkyl substituent and ring residues	=	10 mμ
One exocyclic double bond	=	5 mμ
Calculated λ_{max} value	**=**	**229 mμ**
Observed value	=	232 mμ

Example 6.

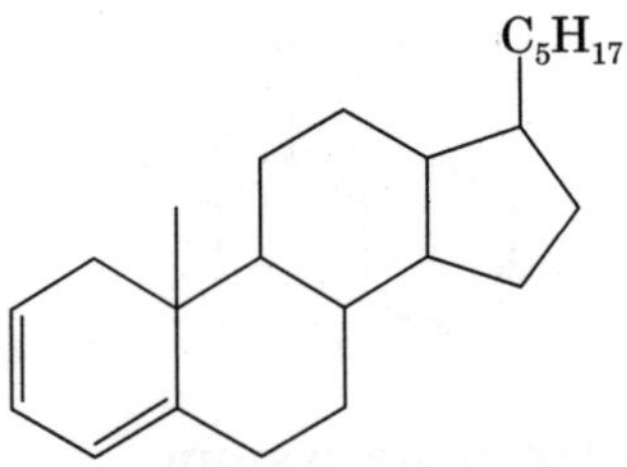

For homoannular diene system	=	253 mμ
Alkyl substituent and ring residues	=	15 mμ
One exocyclic double bond	=	5 mμ
Calculated λ_{max} value	**=**	**273 mμ**
Observed value	=	273 mμ

Example 7.

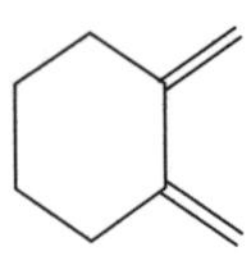

For open chain/alicyclic conjugated diene system	=	217 mμ
Alkyl substituent and ring residues	=	10 mμ
Two exocyclic double bond	=	10 mμ
Calculated λ_{max} value	**=**	**237 mμ**
Observed value	=	220 mμ

8.2.9.2 *Woodward-fieser Rules for λ_{max} Calculation of* α, β *Unsaturated Carbonyl Compounds*

To calculate the λ_{max} value of α, β unsaturated carbonyl compounds, Woodward-fieser suggested some parent absorption values for certain class of α, β unsaturated carbonyl compounds and different absorption increment values for polar groups present on various α, β unsaturated carbonyl compounds summarized in Table 8.6.

Table 8.6: Woodward-fieser values for parent α, β unsaturated carbonyl compounds, additional increments and polar groups.

Parent α, β Unsaturated Carbonyl System	λ_{max} **in mμ**
Base value for acyclic α, β unsaturated ketones	214
Base value for six membered α, β unsaturated ketones	215
Base value for five membered α, β unsaturated ketones	202
Base value for α, β unsaturated aldehydes	210
Base value for carboxylic acid and esters	195
Additional Increments	λ_{max} **in mμ**
Alkyl substituent or ring residues	
at α-position	10
at β-position	12
at γ-position	18

Polar Groups	***At α-Position***	***At β-Position***	***At γ-Position***	***At δ and higher***
–OH	35	30	0	50
$–OCO–CH_3$	6	6	6	6
$–O–CH_3$	35	30	17	31
–Cl	15	12	0	0
–Br	25	30	0	0
–SR	0	85	0	0
$–NR_2$	0	95	0	0

Following are the examples which illustrates the mentioned Woodward-fieser rules based calculation of λ_{max} for α, β Unsaturated carbonyl compounds.

Example 1.

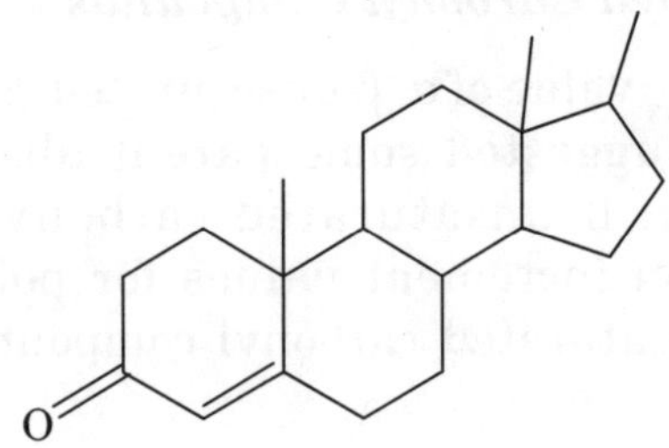

For six membered α, β Unsaturated ketone	=	215 mμ
2β ring residues	=	24 mμ
One exocyclic double bond	=	5 mμ
Calculated λ_{max} value	**=**	**244 mμ**
Observed value	=	241 mμ

Example 2.

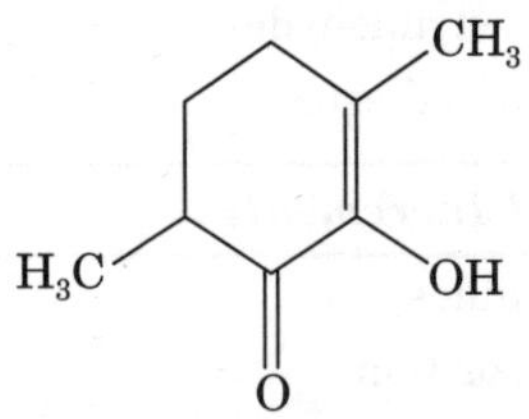

For six membered α, β Unsaturated ketone	=	215 mμ
1β ring residue	=	12 mμ
1β alkyl residue	=	12 mμ
–OH at α-position	=	35 mμ
Calculated λ_{max} value	**=**	**274 mμ**
Observed value	=	275 mμ

Example 3.

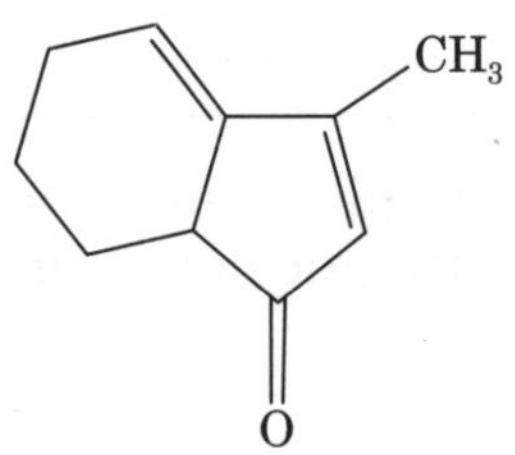

For five membered α, β Unsaturated ketone	=	202 mμ
1β alkyl residue	=	12 mμ
1γ ring residue	=	18 mμ
1δ ring residue	=	18 mμ
one exocylic double bond	=	5 mμ
one double bond extending conjugation	=	30 mμ
Calculated λ_{max} value	=	**285 mμ**
Observed value	=	287 mμ

Example 4.

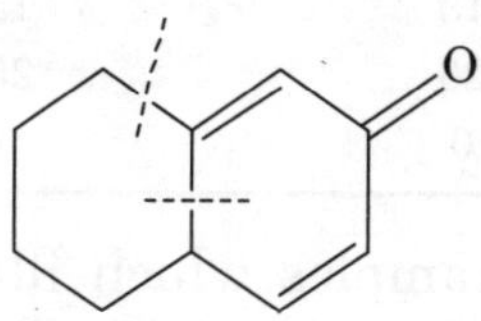

For six membered α, β Unsaturated ketone	=	215 mμ
2β ring residues	=	24 mμ
one exocylic double bond	=	5 mμ
Calculated λ_{max} value	=	**244 mμ**
Observed value	=	245 mμ

8.2.9.3 *Woodward-fieser Rule for λ_{max} Calculation of Aromatic Compounds*

To calculate λ_{max} value of aromatic compounds, Woodward-fieser suggested some parent absorption values for certain class of aromatic compounds and different absorption increment values for polar group present on various aromatic compounds summarized in Table 8.7.

Table 8.7. Woodward-fieser values for parent aromatic compounds, additional increments and polar groups.

Parent Aromatic System	**λ_{max} in mμ**
Base value for ArCOR	246
Base value for ArCHO	250
Base value for ArCOOH	230
Base value for ArCOOR	230

Additional Increments	λ_{max} **in mμ**
Alkyl substituent or ring residues	
at ortho and meta position	3
at para position	10

Polar Groups	***At ortho Position***	***At meta Position***	***At para Position***
–OH, $-OCH_3$, –OR	7	7	25
$-O^-$	11	20	78
–Cl	0	0	10
–Br	2	2	15
$-NH-CO-CH_3$	20	20	45
$-NH_2$	13	13	58
$-N(CH_3)_2$	20	20	85
$-NH-CH_3$	0	0	73

Following are the examples which illustrates the mentioned Woodward-fieser rules based calculation of λ_{max} for aromatic compounds.

Example 1.

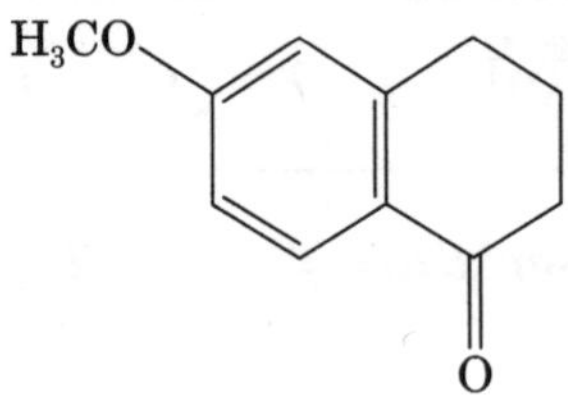

Base value for ArCOR	=	246 mμ
Ring residue in o-position	=	3 mμ
Polar group $-OCH_3$ in p-position	=	25 mμ
Calculated λ_{max} value	**=**	**274 mμ**
Observed value	=	276 mμ

Example 2.

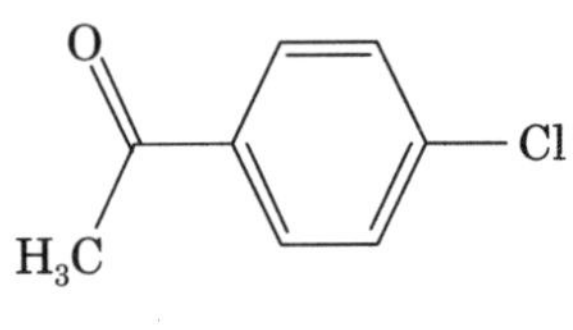

Base value for ArCOR	=	246 mμ
Ring residue in o-position	=	3 mμ
Polar group-Cl in p-position	=	10 mμ
Calculated λ_{max} value	**=**	**256 mμ**
Observed value	=	254 mμ

Example 3.

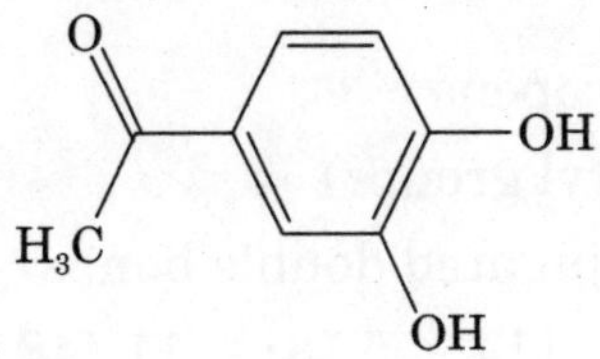

Base value for ArCOR	=	246 mμ
–OH at meta-position	=	7 mμ
–OH at meta-position	=	25 mμ
Calculated λ_{max} value	**=**	**278 mμ**
Observed value	=	281 mμ

8.2.10 Fieser-kuhn Rules for λ_{max} Calculation of Polyenes

Usually, Woodward-fieser rules works best only for conjugated polyenes having four double bonds or less. But in case of conjugated polyenes with more than four double bonds the Fieser-Kuhn rules are commonly used. The benefit of Fieser-Kuhn rule is that with this approach both λ_{max} and E_{max} as well can be related to number of conjugated double bonds and other structural units present in a polyene compound. The expression is given as follows:

$$\lambda_{max} = 114 + 5M + n\,(48 - 1.7n) - 16.5\,R_{endo} - 10\,R_{exo}$$

Also, $$E_{max} = (1.74 \times 10^4)\,.\,n$$

where,

n = Number of conjugated double bonds

M = Number of alkyl groups

R_{endo} = Number of endocyclic double bonds

R_{exo} = Number of exocyclic double bonds

Following are the examples of direct application of Fieser-Kuhn expression to determine the λ_{max} and E_{max} of polyenes.

Example 1. *Lycopene*

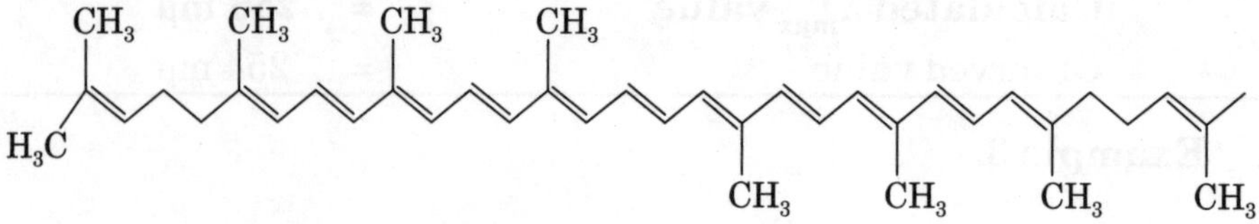

In a molecule of Lycopene

M (Number of alkyl groups) = 08

n (Number of conjugated double bonds) = 11

Therefore, λ_{max} = 114 + 5 (8) + 11 (48 – 1.7 × 11) – 16.5 (0) – 10 (0) = 476 mμ

and E_{max} = $(1.74 \times 10^4) . 11 = 19.14 \times 10^4$

Example 2. β-*Carotene*

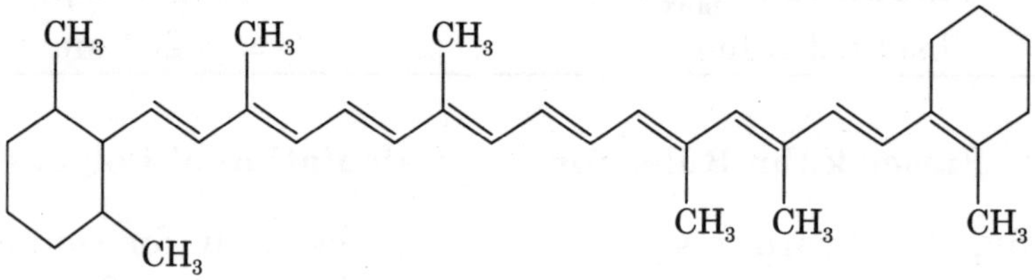

In β-Carotene molecule,

M (Number of alkyl groups) = 10

n (Number of conjugated double bonds) = 11

R_{endo} (Number of endocyclic double bonds) = 02

Therefore, λ_{max} = 114 + 5 (10) + 11 (48 – 1.7 × 11) – 16.5 (2) – 10 (0) = 463 mμ

and E_{max} = $(1.74 \times 10^4) . 11 = 19.14 \times 10^4$

8.3 INSTRUMENTATION

Commonly the U.V. and Visible spectrometers are comprised of following components.

- Sources (UV and visible)

- Wavelength selectors—Filters and Monochromators
- Sample Handling
- Detector
- Signal processor and readout

8.3.1 Radiation Sources

The radiation sources used in the UV-visible spectrometer generally involves electronic treatment in the vapour or gaseous phase. This leads to electronic, vibrational and rotational excitation in the gas molecules or the electron collisions with the gas molecule. When the pressure of the gas is low only line spectra are emitted. But if the pressure of gas is high, band spectra and continuous spectra will be obtained.

Ideal characteristics of an UV-visible radiation source are as follows:

- Should emit radiations on entire wavelength region.
- Should emit stable and continuous radiations.
- Must be of sufficient intensity for the transmitted energy to be detected.

A. Deuterium Lamps

Deuterium lamps generates radiations in a UV region of 190 – 370 nm. As Deuterium lamps operates on high temperatures, so these do not use normal glass casing because glass absorbs radiation of wavelengths less than 350 nm. Therefore, deuterium lamps are made up of quartz or magnesium fluoride envelope. Deuterium lamps are always used along with a tungsten halogen lamp to allow measurements to be performed in both the UV and visible regions. The electrical excitation of deuterium at low pressure causes production of continuous UV spectrum. The mechanism of production of such radiations involves formation of an excited molecular species, which further breaks up to produce two atomic species and an ultraviolet photon. This process can be shown as

$$D_2 + \text{electrical energy} \rightarrow D_2^* \rightarrow D' + D'' + hv$$

These lamps are most widely used in place of hydrogen lamp because the intensity of radiation emitted is usually 3–5 times more than the intensity of a hydrogen lamp. These are more expensive and used when high intensity is required. Deuterium lamps produces low noise which limits their performance. The intensity of deuterium lamp commonly decreases steadily over time such that a typical service life is 1,000 hours. Figure 8.4 shows an image of deuterium lamp.

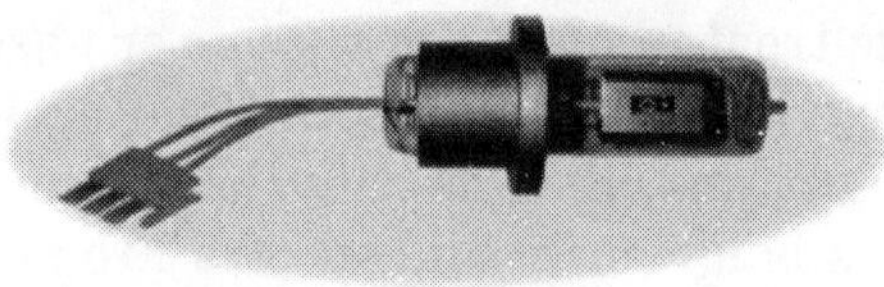

Figure 8.4: Image of deuterium lamp.

B. Hydrogen Discharge Lamp

Hydrogen discharge lamps are similar in construction to deuterium lamps except that here instead of deuterium, hydrogen gas is used. In hydrogen lamps the hydrogen gas is stored at high pressure. Whenever an electrical discharge is passed through the lamp, then same as in case of deuterium lamp the molecules of hydrogen are excited and ultimately produces UV radiations. The electrical excitation of hydrogen at low pressure produces a continuous UV spectrum.

Both deuterium and hydrogen lamps emits radiations in the range 160 – 375 nm. Quartz windows must be used in these lamps, and quartz cuvettes must be used.

C. Tungsten Filament Lamp

Tungsten (W) light sources are generally referred to incandescent lamps. These lamps consists of a thin, coiled tungsten wire that is sealed in an evacuated glass bulb. A tungsten filament lamp is commonly used as the visible light source—this is just like a normal light globe. When this filament is heated electrically it generates white heat. The intensity of generated radiations is low (below 350 nm) or at short wavelength. Figure 8.5, shows a diagram of a tungsten filament lamp.

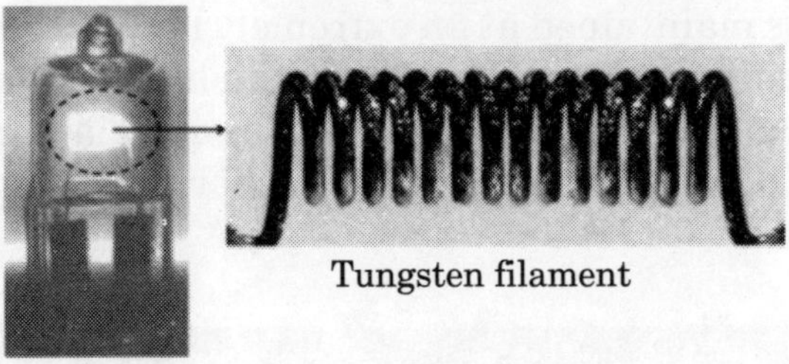

Figure 8.5: Schematic image of tungsten filament lamp.

The intensity of this lamp varies with wavelength. The tungsten filament lamp is commonly used as a source of visible light. This type of lamp is used in the wavelength range of 350 – 2500 nm. The radiation emission by a tungsten filament lamp is proportional to the fourth power of the operating voltage. To make these lamps stable electronic voltage regulators or constant-voltage transformers are used.

Now a day's tungsten halogens are commonly used. Tungsten/ halogen lamps contains a small amount of iodine in a quartz "envelope" which also contains the tungsten filament. The iodine reacts with gaseous tungsten, formed by sublimation, producing the volatile compound WI_2. When molecules of WI_2 hits the filament commonly undergoes decomposition, and further redeposits tungsten back on filament. The lifetime of a tungsten/halogen lamp is approximately double that of an ordinary tungsten filament lamp. These lamps produces very low noise and a long service life of typically 10,000 hours. The wavelength range for this type of lamp extends across the entire visible range and extends into the ultraviolet (300 nm to 3000 nm). Tungsten/halogen lamps are very efficient, and their output extends well into the ultra-violet. They are used in many modern spectrophotometers.

D. Xenon Lamp

Xenon lamps are made up of fused quartz that contains two tungsten electrodes separated by 8 mm. When a low voltage is applied it excites the xenon gas molecules to emit the radiations. Commonly the intensity of UV radiations produced in xenon lamp is much higher than hydrogen lamp. To achieve maximum efficiency, the xenon gas inside lamps is maintained at an extremely high pressure that is up to 30 atmospheric pressure. They require a high voltage pulse in the 20–50/Kv range to start the lamp, and require an extremely well regulated DC power source. Figure 8.6, shows the image of a xenon lamp.

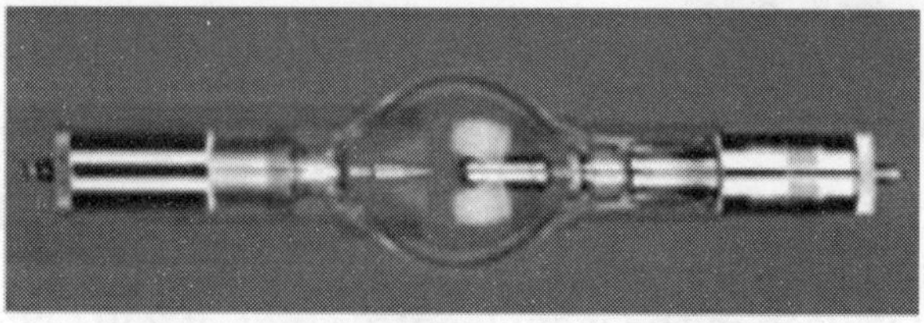

Figure 8.6: Image of xenon lamp.

E. Mercury Lamp

Mercury lamp source is composed of a quartz jacketed tube that contains mercury vapours. When electric current is passed through the tube, it tends mercury vapours to emit radiations. Based on application of pressure mercury lamps are classified into two types, namely, low pressure and high pressure mercury lamps. Low pressure mercury lamp is a highly efficient source of UV light for short wavelength which mainly emits radiation between 254 nm and 184 nm. Whereas high pressure mercury lamps irradiates light of varying wavelength that includes visible light also. Figure 8.7, shows the image of a mercury lamp.

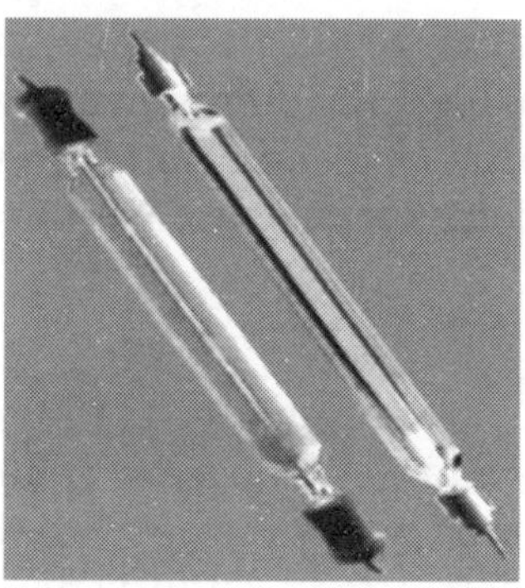

Figure 8.7: Image of mercury lamp.

F. Carbon Arcs

A carbon arc occurs when an electrical discharge occurs between two carbon electrodes. These sparks are incredibly bright reaching temperature as hot as 6000 K. The carbon sources emits radiations in entire range of visible region. Carbon arc gives high outputs. The Figure 2.6, shows the image of carbons arc source. These requires more maintenance because carbon electrodes are consumed in the arcing process, so it is necessary to continuously feed new rod forward to maintain the arc. Also carbon rods requires an inert atmosphere to avoid combustion of the carbon.

8.3.2 Wavelength Selectors

In spectrophotometry for measurements a narrow band of wavelengths of light is generally required. Commonly, less expensive instruments uses filters as wavelength selectors for isolation of radiation. There are many conditions when we are required to continuously vary the wavelength. In those cases monochromators are used. So, wavelength selectors are classified into two types *i.e.*, filters and monochromators.

A. Filters

Filters are the wavelength selector devices that converts polychromatic light into monochromatic light. On the basis of mechanics of wavelength selection, filters are generally divided into two types *i.e.*, absorption and interference filters. These are commonly used for visible radiations.

(*a*) Absorption Filters: *Filters that absorbs the unwanted radiations and transmits the required radiations for colorimetry are called absorption filters.* These filters are commonly made up of glass, coated with pigments or made up of dyed gelatin.

The selection of absorption filters according to colour of sample solution can be done by two ways

- Firstly, by using filter wheel circle given in Figure 8.8. In a colour wheel, the colours, which face one another, are said to be complementary to each other. Suppose the sample solution colour is red then according to filter wheel circle a green colour filter should be used because it is complementary. In the colour wheel terms P, S and T are used to indicate the primary, secondary and tertiary colours, respectively.

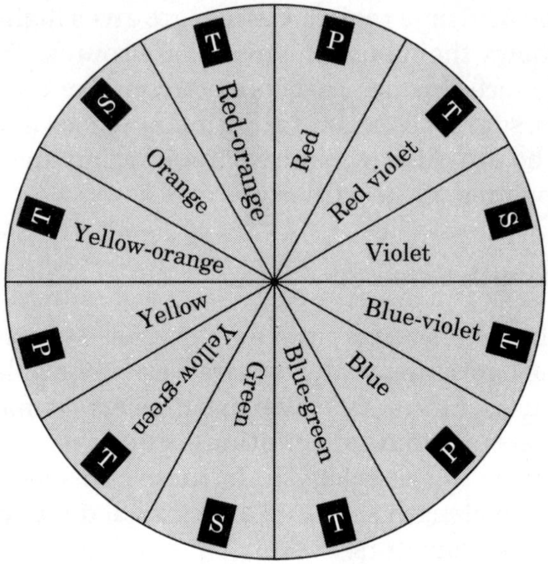

Figure 8.8: Filter wheel circle.

- Secondly, by using complementary colours chart given in Table 8.8.

Table 8.8: Complementary Colours Chart.

Wavelength region	*Colour absorption*	*Residual light complementary colour*
400 to 450	Violet	Yellow-green
450 to 480	Blue	Yellow
480 to 490	Green-blue	Orange
490 to 500	Blue-green	Red
500 to 560	Green	Purple
560 to 580	Yellow-green	Violet
580 to 600	Yellow	Blue
600 to 650	Orange	Green-blue
650 to 750	Red	Blue-green

Merits

- Filter selection is easy.
- Economical.
- Simple construction.

Demerits

- Less Accuracy.
- Radiation Intensity is Reduced.

(*b*) Interference Filters: *Interference filters works on the principle of constructive interference, such that the radiations reflected by second film and incoming radiations undergoes constructive interference and thereby monochromatic radiations are produced.* These filters are called as Febry-Perot filters also.

- These are composed of dielectric spacer film made up of CaF_2, MgF_2, or SiO_2, which is placed between two parallel reflecting silver films.
- If the thickness of dielectric spacer film is 1/2λ then it is termed as first order. Whereas if the thickness is 2 λ/2 then it is referred to second order. Also, if the thickness is of 3 λ/2 then it is called as 3rd order.

B. Monochromators

Monochromator is a device which converts polychromatic light into a monochromatic light. In practice monochromators are commonly used for UV-visible radiations. For UV-visible radiations commonly the prisms as well as gratings are used as monochromators. For UV radiations gratings made up of quartz are of much use. The mono-chromators are discussed in detail in section 2.3.1.2 of Chapter 2.

8.3.3 Sample Handling

In sample handling two key elements are of utmost importance *viz.*,

a. Sample Cell and
b. Solvents.

A. Sample Cell (Cuvettes)

The UV-VIS absorptions are commonly determined in vapour phase or in solution. *To perform the UV determination the analyte is taken in a cell known as cuvette*. Cuvette is made up of quartz is transparent to the wavelength of light passing through it. These cuvettes are of varying path lengths and are equipped with inlet and outlets. For solutions determinations in visible region the cuvettes made up of glass can be used. But, cuvettes made up of glass cannot be used in the UV region. Therefore, most spectrophotometers employ quartz cuvettes, as these can be used for both visible and UV region. Commercially, square cuvettes having internal path length 1.0 cm are commonly used. Although cuvettes of 0.1 mm or 0.05 mm path lengths are also available. Figure 8.9, shows the image of cuvette for liquid sampling in UV spetrometry.

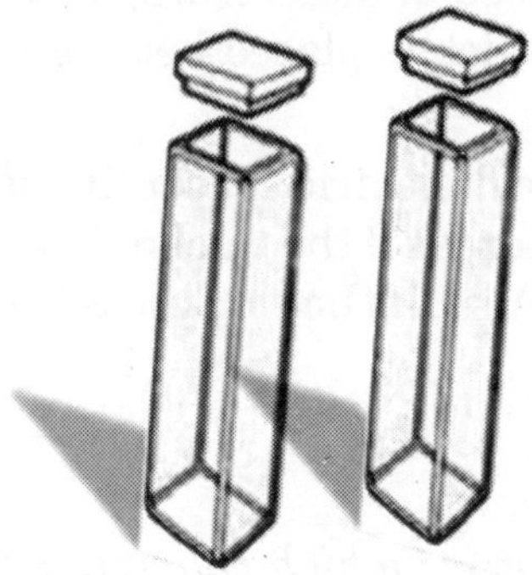

Figure 8.9: Image of cuvettes for liquid sampling in spectrometry.

Requirements for a Sample Cell

- Must be uniform in construction, the thickness must be constant and surfaces facing the incident light must be optically flat.
- Material of construction should be inert to solvents.
- Must transmit light of the wavelength used.

B. Solvents

The analyte is dissolved in a solvent which is transparent in the UV region. Hexane, 95% ethanol, methanol and 1, 4 dioxane are commonly employed as solvents. Now 'spectral grade' solvents are also available that gives high purity and have negligible absorption in the region of absorption by the chromophore.

8.3.4 Detectors

Detectors in UV spectrometer usually converts a light signal into electrical signal that can be measured and transformed into an output. In most of instruments detectors generates a signal, which is linear in transmittance *i.e.*, they respond linearly to radiant power falling on them. There are three types of detectors which are used in modern spectrophotometers.

A. Barrier layer (Photovoltaic) cells

Barrier layer cells are composed of a copper or iron plate on which a semiconducting layer of cuprous oxide or selenium is placed. This layer is covered by a light-transmitting layer of metal which acts as a collector electrode. As illumination passes through the electrode to the semiconducting layer, an electron flow is induced in the semiconducting layer, and this flow can be sensed by an ammeter. These detectors are rugged, relatively inexpensive, and sensitive from the UV region up to about 1000 nm. No external power is required, and the photocurrent produced is directly proportional to the radiant-energy intensity. Barrier layer cells exhibit the fatigue effect, which means that on illumination, the current rises above the apparent equilibrium value and then gradually decreases.

B. Phototube

A phototube consists of a photo emissive cathode (*in the form of a half-cylinder of metal with high sensitivity*) and an anode (*Present in tube fixed along the axis of the tube*) in an evacuated tube with a quartz window as shown in Figure 8.10. These two electrodes are subjected to high voltage (about 100 V) difference. When a photon enters the tube and strikes cathode, an electron is ejected out which is attracted by anode resulting in a flow of current that can be amplified and measured. Response of photo emissive material is wavelength dependent and different phototubes are available for different regions of spectrum. Using quartz or fused silica windows, range of photocells can be increased.

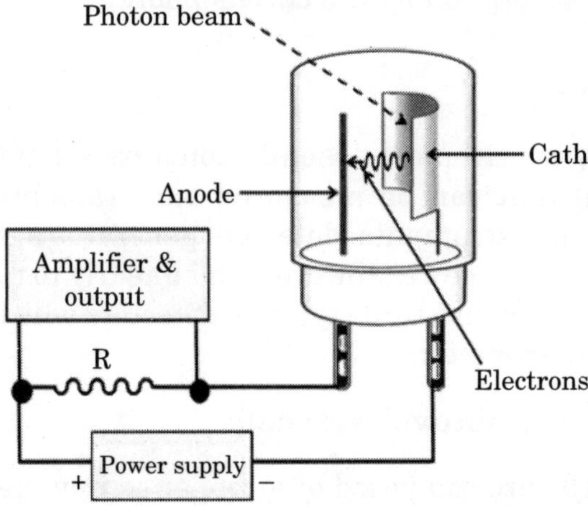

Figure 8.10: Schematic diagram of a photo tube.

C. Photomultiplier Tubes

Photomultiplier tube comprises a photo emissive cathode (*a cathode which emits electrons when struck by photons of radiation*), several dynodes (*which emit several electrons for each electron striking them*) and an anode. A photomultiplier tube can be well understood from Figure 8.11.

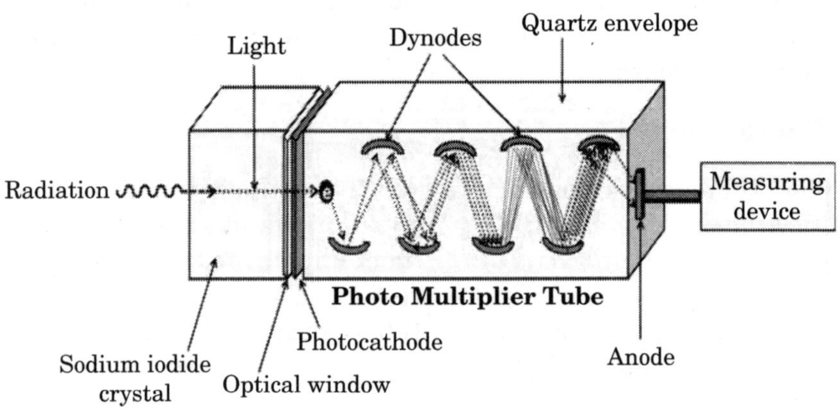

Figure 8.11: Schematic diagram of a photo multiplier tube.

A photomultiplier tube is composed of an electron tube that is capable of amplifying the current. When a photon of radiation entering the tube strikes the cathode (*made of a light-sensitive metal that can absorb radiant energy and emit electrons in proportion to the radiant energy that strikes the surface of the light-sensitive metal*), it causes emission of several electrons. These electrons are accelerated towards the first dynode (which is 90 V more positive than the cathode). The electrons strike the first dynode, causing the emission of several electrons for each incident electron. These electrons are then accelerated towards the second dynode, to produce more electrons which are accelerated towards dynode three and so on. As many as 15 stages (or dynodes) are present in today's photomultiplier tubes. Photomultiplier tubes have rapid response times, do not show as much fatigue as other detectors, and are very sensitive. Eventually, the electrons are collected at the anode. By this time, each original photon has produced $10^6 - 10^7$ electrons. The resulting current is amplified and measured.

Photomultipliers are very sensitive to UV and visible radiation. They have fast response times. Intense light damages photomultipliers; they are limited to measuring low power radiation.

D. Photodiodes

Photodiodes are semiconductors that change their charged voltage (usually 5 V) upon being struck by light. The change is converted to current and is measured. A diode array is a two-dimensional matrix composed of hundreds of thin semiconductors spaced very closely together. Light from the instrument is dispersed by either a grating or prism onto the photodiode array. Each position or diode on the array is calibrated to a specific wavelength. Each diode is scanned, and the resultant electronic change is calculated to be proportional to absorption. The entire spectrum is essentially recorded within milliseconds.

The linear photodiode array is an example of a *multichannel photon detector*. These detectors are capable of measuring all elements of a beam of dispersed radiation simultaneously.

A linear photodiode array comprises many small silicon photodiodes formed on a single silicon chip. There can be between 64 to 4096 sensor elements on a chip, the most common being 1024 photodiodes. For each diode, there is also a storage capacitor and a switch. The individual diode-capacitor circuits can be sequentially scanned.

In use, the photodiode array is positioned at the focal plane of the monochromator (*after the dispersing element*) such that the spectrum falls on the diode array. They are useful for recording UV-Vis. absorption spectra of samples that are rapidly passing through a sample flow cell, such as in an HPLC detector.

8.3.5 Signal Processer and Recorder

The electrical signal from the transducer is amplified and processed before it is sent to the recorder. The substraction of the solvent spectrum from that of the solution is done electronically. The output plot between the wavelength and the intensity of absorption is the resultant of the subtraction process and is characteristic of the absorbing species.

8.3.6 Types of UV-visible Spectrophotometers

A. Single Beam Spectrophotometers

These instruments contain a single beam of light. The same beam is used for reading the absorption of the sample as well as the reference. The schematic diagram of a typical single beam UV-visible spectrometer is given in Figure 8.12.

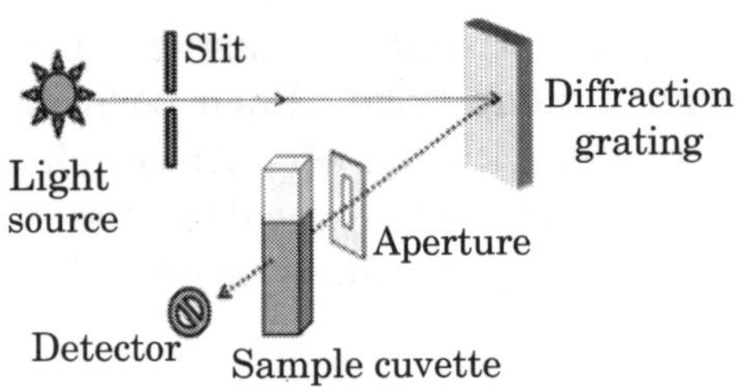

Figure 8.12: Schematic diagram of a Single Beam Spectrophotometers.

The radiation from the source is passed through a filter or a suitable monochromator to get a band or a monochromatic radiation. It is then passed through the sample (or the reference) and the transmitted radiation is detected by the photo detector. The signal so obtained is sent as a read out or is recorded. Typically, two operations have to be performed

- Firsly, the cuvette is filled with the reference solution and the absorbance reading at a given wavelength or the spectrum over the desired range is recorded.
- Secondly, the cuvette is taken out and rinsed and filled with sample solution and the process is repeated. The spectrum of the sample is obtained by subtracting the spectrum of the reference from that of the sample solution.

B. Double Beam Spectrophotometers

In a double beam spectrometer, the radiation coming from the monochromator is split into two beams with the help of a beam splitter. These are passed simultaneously through the reference and the sample cell. The transmitted radiations are detected by the detectors and the difference in the signal at all the wavelengths is suitably amplified and sent for the output.

The general arrangement of a double beam spectrometer is shown in Figure 8.13.

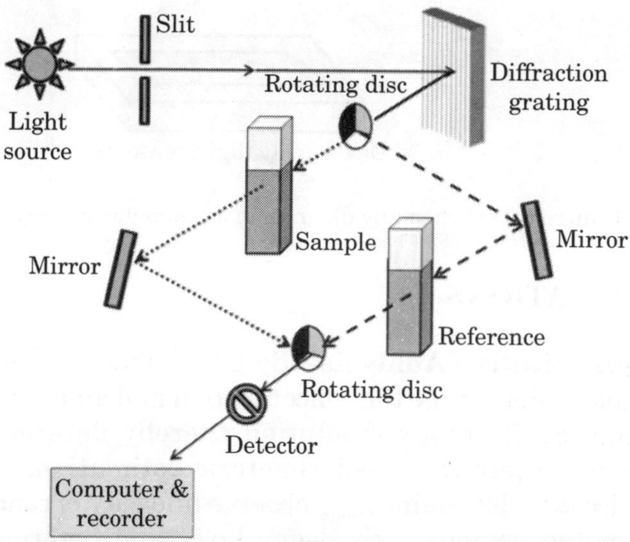

Figure 8.13: Schematic diagram of a double beam spectrophotometers.

C. Photodiode Array Spectrometer

In a photodiode array instrument, also called a multi-channel instrument, the radiation output from the source is focused directly on the sample. This allows the radiations of all the wavelengths to simultaneously fall on the sample. The radiation coming out of the sample after absorption (if any) is then made to fall on a reflection grating. The schematic arrangement of a diode array spectrometer is given in Figure 8.14. The grating disperses all the wavelengths simultaneously. These then fall on the array of the photodiodes arranged side by side. In this way the intensities of all the radiations in the range of the spectrum are measured in one go. The advantage of such instruments is that a scan of the whole range can be accomplished in a short time.

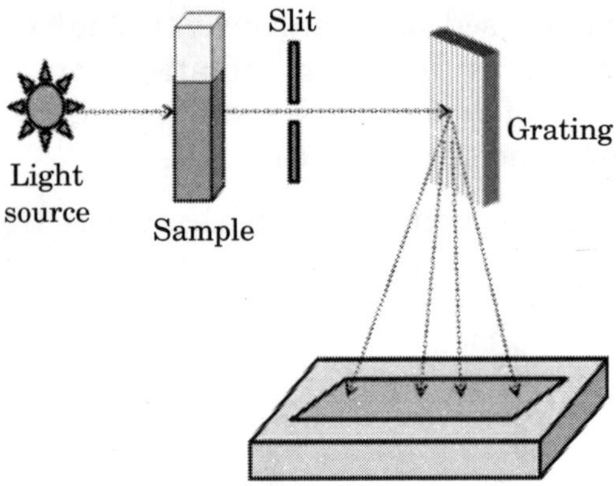

Figure 8.14: Schematic diagram of a photo diode array.

8.4 APPLICATIONS

- **In Quantitative Analysis**—By UV-visible spectroscopy it is possible to determine the concentration and amount of analyte by making its analyte solution thereby determination of percentage purity. In colorimeteric estimation, a common practice is to determine $\lambda_{max,}$ observe linearity of concentration ranges (agreement with Beer's law), and determination of $E_1^{1\%}$ cm.

- **Estimation of Functional Groups or Ions**—By visible spectroscopy it is possible to estimate various low concentration bearing functional groups or ions, using specific or non-specific chromogenic agents. Table 8.9 lists various chromogenic agents for specific compounds and functional groups.

Table 8.9: Chromogenic agents for specific organic compounds.

Chromogenic Agent	*Substance*	*Wavelength (nm)*	*Concentration Range (ppm)*
Ferric chloride	Phenols	530	0.1–9
Nessler's reagent	Ammonia	580	2–25
Ninhydrin	Glycine	565	0–0.04
Periodate	Manganese	515	1–25
8-Quinolinol	Magnesium	400	1–5

- **In Medicinal and Organic Compounds Estimation**—By visible spectroscopy only coloured medicinal and organic compounds can be estimated, so for compounds that are colourless a specific chromogenic agent is added for colouring. The benefit of this technique is that intensity of colour developed in the organic substance is proportional to drug concentration. The selection of chromogen depends upon the types of functional groups present in the molecule. Tables 8.10 and 8.11, list various chromogenic agents for specific medicinal compounds and functional groups.

Table 8.10: Some chromogenic agents for specific medicinal compounds.

Chromogenic Agent	*Substance*	λ_{max} *in nm*	*Solvent*
Cobalt thiocyanate	Bethanidine tablets	652	water
p-DMAB	Ergometrine inj	545	water
Methanol	Doxycycline	490	water
Methanolic HCl	Clofazimine	490	Chloroform
Phenyl hydrazine	Betamethasone	450	Alcohol
Picric acid (Alcoholic)	Digitoxin	495	Alcohol

Table 8.11: Some chromogenic agents for specific functional groups.

Chromogenic Agent	***Functional Group***
Acidic dye	Basic group
Basic dye	Acidic group
2, 4-DNP Hydrazine	Ketones or Aldehydes
Ferric ammonium sulphate or $FeCl_3$	Phenolic (OH)
Naphthyl EDA	Aromatic amines
Ninhydrin	Amino acids

- **In pKa value Estimation**—Through UV-visible spectroscopy the dissociation constant of an indicator (acid-base *e.g.*, Methyl red or orange) can be calculated by determining concentration relative to absorbance on varying pH.

$$pKa = pH - \frac{\log(\text{Ionized})}{(\text{Unionized})}$$

- **In Spectrometric Titrations**—In these titrations the absorption of reactant or product is measured at a specific λ_{max} of titrant, reactant or product. Nextly, following Beer's law one can easily determine the component concentration and purity of a given analyte. The benefits of spectrometric titrations are their sharp end point, absorbing species non interference, also on reaction incompletion species does not affects the end point.
- **In Detection of Impurities**—Through visible spectroscopy one can detect the impurities present in a sample under investigation. The presence of impurities can be determined by characteristic presence of additional peaks and change in absorbance ratio of sample molecules. For example, the pure cyanocobalamin absorbs at 550 nm, 278 nm & 361 nm; and exhibits an absorbance ratio of 0.57 at 278/361 nm and 0.3 at 550/361 nm. But this absorbance ratio of cyanocobalamin changes on exposure to impurities.
- **In Chemical Kinetics Study**—UV spectroscopy determines kinetics of a reaction by measuring absorbance changes when a UV radiation is passed through the reaction cell.
- **In Keto-Enol Tautomerism Study**—UV spectroscopy determines keto and enol form in a mixture.

- **In Structural Elucidation**—UV-Visible spectroscopy assists in deducing the structure of organic compounds. Some key features to elucidate the structure of compounds are
  - Unsaturation in the compound indicates presence of hetero atoms such as O, S, N or halogen can be determined.
  - The position of peak in a UV-visible spectrum further supports to structure of compound.
- **In Determination of Metal Complexes Ratio**—UV-visible spectroscopy makes possible to determine metal complexes ratio by generating plots for different ratios and testing the linearity.

8.5 PROBLEMS BASED LEARNING

After reading the UV-visible spectroscopy chapter, the readers are suggested/advised to test their learning by solving some below mentioned problems.

A. SUBJECTIVE PROBLEMS

1. Explain Principle of UV-visible spectroscopy.
2. Discuss various energy transitions in a molecule after absorption of UV radiations with suitable example.
3. Explain how polarity of solvent affects absorption maxima in UV spectroscopy.
4. Describe Beer's law and deviation from Beer's law.
5. What are essential requirement for a solvent to be used in UV spectroscopy.
6. Define chromophore and auxochrome.
7. Describe various components of a spectrophotometer.
8. Describe various essential requirements for a solvent to be used in UV spectroscopy.

B. OBJECTIVE PROBLEMS

1. State True or False

(*i*) $n \rightarrow \sigma^*$ and $\pi \rightarrow \pi^*$ are high energy transitions.

(*ii*) $n \rightarrow \pi^*$ low energy transitions.

(*iii*) K-bands possess >10000 intensity and are exhibited by $\pi \rightarrow \pi^*$.

(*iv*) R-bands possess <100 intensity and are exhibited by $n \rightarrow \pi^*$.

(*v*) E_1 bands are more intense than E_2 bands.

Ans.: 1. (*i*) True (*ii*) True (*iii*) True (*iv*) True (*v*) True.

2. Fill up the blanks

(*i*) Shift in λ_{max} on a longer wavelength is .... shift.

(*ii*) Shift in λ_{max} on a shorter wavelength is .... shift.

(*iii*) Increase in E_{max} is .... shift.

(*iv*) Decrease in E_{max} is ....shift.

(*v*) .... transitions possess $E_{max} > 10^4$.

(*vi*) .... transitions possess $E_{max} < 100$.

Ans.: 2. (*i*) Bathochromic (*ii*) Hypso (*iii*) Hyperchromic (*iv*) Hypochromic (*v*) Allowed (*vi*) Forbidden.

3. Match the following

(*i*) Beer's law (*a*) $I = I_0 \cdot e^{-k.c}$

(*ii*) Lamberts law (*b*) $I = I_0 \cdot e^{-k.t}$

(*iii*) Beer-Lamberts law (*c*) $A = \varepsilon.c.t$

Ans.: 3. (*i*) – (*a*), (*ii*) – (*b*), (*iii*) – (*c*).

4. Match the following

(*i*) $\sigma \rightarrow \sigma^*$ (*a*) $CH_3–CH_3$

(*ii*) $\pi \rightarrow \pi^*$ (*b*) $CH_2=CH_2$

(*iii*) $n \rightarrow \sigma^*$ (*c*) $CH_3–Cl$

(*iv*) $n \rightarrow \pi^*$ (*d*) $CH_3–C(=O)H$

Ans.: 4. (*i*) – (*a*), (*ii*) – (*b*), (*iii*) – (*c*), (*iv*) – (*d*).

5. Match the following

(*i*) $\sigma \rightarrow \sigma^*$ (*a*) $CH_3–CH_3$

(*ii*) $\pi \rightarrow \pi^*$ (*b*) K-band

(*iii*) n $\rightarrow \sigma^*$ (*c*) CH_3–Cl

(*iv*) n $\rightarrow \pi^*$ (*d*) R-band

Ans.: 5. (*i*) – (*a*), (*ii*) – (*b*), (*iii*) – (*c*), (*iv*) – (*d*).

8.5 SUGGESTED READINGS

[1] Pavia D.L., Lampman G.M., Kriz G.S., Vyvyan J.R., *Introduction to Spectroscopy*, 4th Edition, Brooks/Cole, 2009.

[2] Pretsch E., Buhlmann P., Badertscher M., *Structure Determination of Organic Compounds: Tables of Spectral Data*, Springer-Verlag, berlin Heidelberg, 2009.

[3] Brown W.H., Iverson B.L., *Organic Chemistry*, Brooks and Cole Cenagage Learning, USA, 2009.

[4] Lee D.C., Webb M., *Pharmaceutical analysis*, Black Well Publishing, Wiley India, 2008.

[5] Field L.D., Sternhell S., Kalman J.R., *Organic Structures from Spectra*, 4th Edition, John Wiley and Sons, 2007.

[6] Eric Anslyn V., Dougherty D.A., *Modern Physical Organic Chemistry*, University Science Books, 2006.

[7] Yadav L.D.S., *Organic Spectroscopy*, Kluwer Academics Publisher, Dordrecht, Netherland, 2005.

[8] Kalsi P.S., *Spectroscopy of Organic Compounds*, New Age International Publisher, 6th Edition, 2004.

[9] Fox M.A., Whitesell J.K., *Organic Chemistry*, 3rd Edition, Jones and Barlett Publisher, Canada, 2003.

[10] B.J. Clark, T. Frost, M.A. Russell, *UV Spectroscopy: Techniques, Instrumentation and Data Handling*, 1st Edition, *Chapman* & Hall, London, 1993.

Chapter 9

X-ray Spectroscopy

Objective

To make readers learn

- Concept of X-rays Production and Interaction with Matter.
- Basic Concepts, Theory and Underlying Principles of XRA, XRD and XRF.
- Instrumentation and Application of XRA, XRD and XRF.

9.1 INTRODUCTION

X-ray spectroscopy is a form of optical spectroscopy, which involves investigation of absorption, emission, and diffraction of X-rays. X-rays are electromagnetic radiations which lies between ultraviolet light and gamma rays in the electromagnetic spectrum. X-rays are characterized by shorter wavelengths of 0.01-100 Å ($10^{-10} - 10^{-6}$ cm). X-rays are produced by deceleration of high energy electrons or by electronic transitions of electrons in the inner orbital of atoms. The conventional X-ray spectroscopy is confined to 0.1 Å to 25 Å region. Based on use of different properties of X-rays following are the three types of X-ray spectroscopic techniques such as:

1. **X-ray Absorption Spectroscopy**—X-ray absorption technique is similar to other absorption techniques of spectroscopy. X-ray absorption method involves passage of X-rays through the analyte. Thereby fraction of X-ray photons determines the concentration of absorbing analyte. These methods are commonly used for elemental analysis, thickness measurements and for detection of imperfections in the physical body internal structures. Depending upon the density of matter

X-rays (*possessing shorter wavelength and high energy*) can penetrate into matter with great ease. For example bones being denser, can not be penetrated by X-rays, but fractures (*reduced density*) are easily seen in X-ray pictures.

2. **X-ray Diffraction Spectroscopy**—X-ray diffraction method involves scattering (*redirection of light*) of X-rays by the analyte crystals. These methods are commonly used to determine the crystal structures of solid compounds.
3. **X-ray Fluorescence Spectroscopy**—X-ray fluorescence methods are non-destructive techniques that firstly, involves generation of X-rays within the sample; and secondly, involves measurement of wavelength and intensity of generated X-rays. These methods requires very small amount of analyte for qualitative and quantitative determination.

To use X-ray spectroscopy as an analytical tool, it is necessary to understand the theory and basic principles of X-ray spectroscopy.

9.2 THEORY AND BASIC PRINCIPLES

9.2.1 X-rays Production

X-rays are short-wavelength (hence, high frequency, and hence, relatively high energy) electromagnetic radiation that are associated with electronic transitions in atoms. For analytical purposes, X-rays can be generated in three ways

1. Bombarding the metal target with high-energy electron beam (*in X-ray tube*).
2. Treatment of target material with primary X-ray beam and generates a secondary beam as X-ray fluorescence.
3. By making use of radioactive materials whose decay process includes emission of X-rays.

Like other electromagnetic radiations, the energy of X-rays is also inversely proportional to their wavelength as per Einstein expression expressed as:

$$E = h.\eta = \frac{hc}{l} \qquad ...(9.1)$$

Where,

E = Energy

η = frequency

h = Planck's constant

l = Wavelength

c = Velocity of light

9.2.2 X-rays Spectra

When target anode of the X-ray tube (*X-ray source*) is bombarded with electrons (*accelerated from the cathode filament*), it produces either continuous or characteristic spectra.

A. Continuous Spectra

Spectra that depends upon target material & voltage across the X-ray tube, and possess a range of wavelengths of X-rays with minimum wavelength and intensity (*counts per second*) are known as continuous spectra.

These spectra refer to radiation arising from the deceleration of particles. Increase in voltage of X-ray tube causes decrease in wavelength and increase in intensity. Figure 9.1 shows a continuous spectrum.

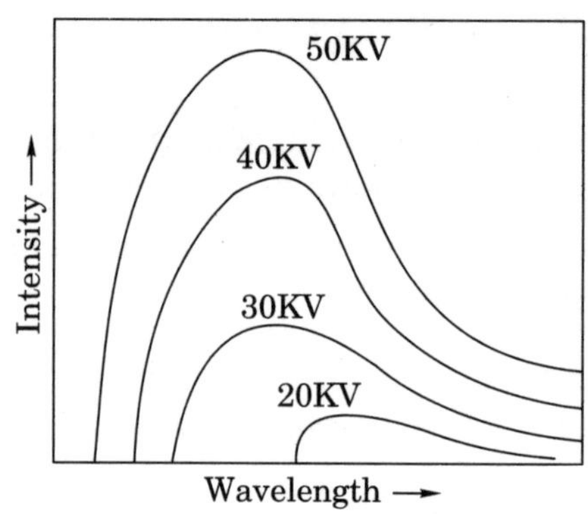

Figure 9.1: Continuous spectrum.

B. Characteristic Spectra

Spectra that are produced at high voltage as a result of specific electronic transitions that take place within individual atoms of the target material are called as characteristic spectra. The intensities of X-rays in characteristic spectrum are usually higher than continuous spectrum. A characteristic spectrum can be seen in Figure 9.2, which represents K_α X-rays having higher intensity than K_β X-rays.

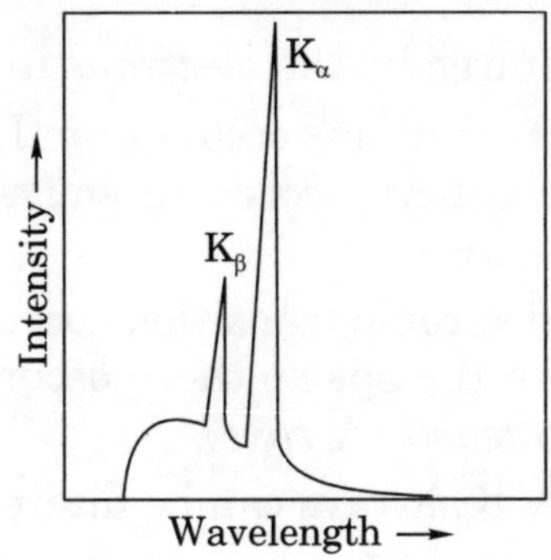

Figure 9.2: Characteristic spectrum.

The X-rays emission phenomenon can be easily understood using simple Bohr's model of an atom given in Figure 9.3.

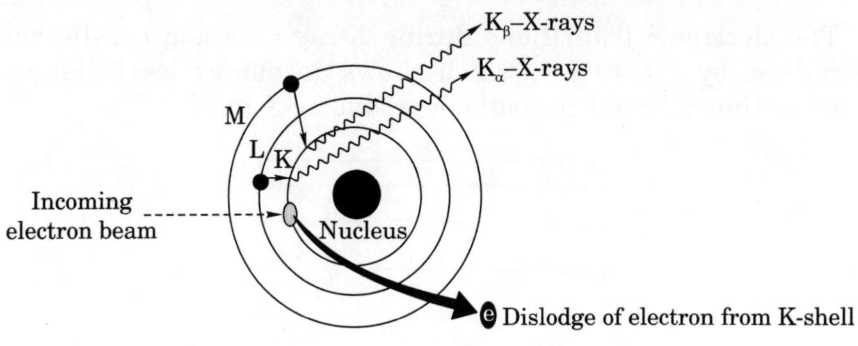

Figure 9.3: Bohr's model representing X-ray emission and electronic transition.

Key Concepts of Bohr's Model

- According to this model, atomic nuclei possess protons and neutrons.

- The nucleus is further surrounded by various shells of electrons.
- The innermost shell which surrounds nucleus is called as K-shell, which is further surrounded by the L and M-shells.
- When the energy of impinging electrons is accelerated to cause sufficiently high impact on target, it dislodges the electrons of innermost K-shell.
- As a result of electronic dislodging, there occurs space into the K-shell.
- This space is acquired by the electrons from L and M-shells.
- When electronic transitions occurs from L shell into K-shell to acquire the space, there occurs an emission of K_α radiations (*high intensity X-rays*).
- Whereas when electronic transitions occurs from M-shell into K-shell to acquire the space, there occurs an emission of K_β radiations (*low intensity X-rays*).
- The low intensity K_β X-rays can be filtered through a filter.
- The electronic transitions or production of X-ray wavelength depends on the exact structure of the atom being bombarded.
- So, wavelength of produced X-rays is characteristic and different for each atom.

The electronic transitions during X-ray emission can be best visualised by Figure 9.4, which shows an energy level diagram representing different transitions producing X-rays.

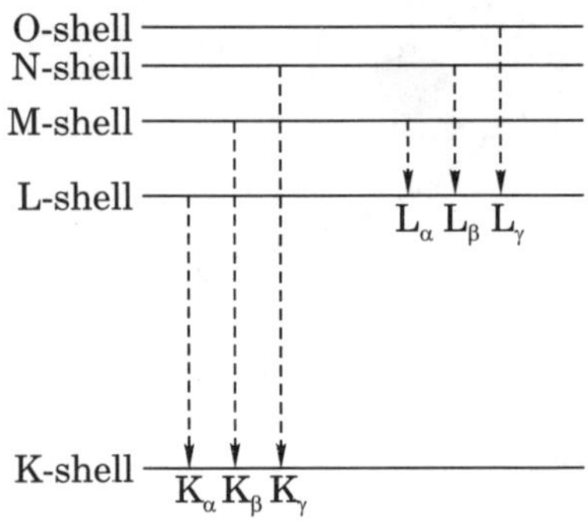

Figure 9.4: Energy level diagram representing different transitions producing X-rays.

9.2.3 Interaction of X-rays with Matter

Commonly, X-rays interacts with matter by three ways *i.e.*, absorption, scattering and diffraction.

A. Absorption

When X-rays are allowed to pass through a sample, they gets partly scattered and partly absorbed. *In case of absorption, the X-rays partially or totally transfers their energy to matter, and thereby causes excitation of an atom or increment in temperature.*

The energy of incident X-rays beam can induce electronic transitions and eject electrons from atoms (*photoelectric effect*). *The absorption of X-rays signifies absorption of energy by electrons of matter to get excited. Thereby these excited electrons emits secondary radiations which are characteristic for atoms.*

The X-ray absorption is followed by Beer's law such as:

$$I = I_0 . e^{-\mu . l . \rho} \qquad ...(9.2)$$

Where,

I_0 = Incident X-ray intensity
I = Transmitted X-ray intensity
l = Thickness of material in cm
μ = Mass absorption coefficient
ρ = Density of absorbing sample

As mass absorption coefficient (μ) of a given atom is usually independent of physical state but it decreases on decrease in the X-rays wavelength. This is explained as per the following expression.

$$\mu = \frac{CN}{A} . Z^4 . \lambda^3 \qquad ...(9.3)$$

Where,

C = Proportionality constant
N = Avogadro's number
A = Atomic weight of absorbing element
λ = Wavelength of X-ray radiation

When a plot of mass absorption coefficient (μ) is drawn against X-ray Wavelength (λ) it produces a curve which possesses a series of absorption edges or discontinuities. These absorptions are characteristic and are different for each element.

For example iodine present in the compounds like NaI or alkyl iodide, in all the three states such as solid, liquid or gas, always exhibits the same amount of absorption.

B. Diffraction

As X-rays beam possess a bundle of separate waves and these waves can undergo interaction with one another to cause interference. There can be two conditions

- *When all bundle waves are in phase*, then in their spectrum the crests and troughs will occur exactly at the same position. Thereby the waves will interfere with one another and their amplitudes will undergo summation. Ultimately a high amplitude wave will be produced (Figure 9.5).

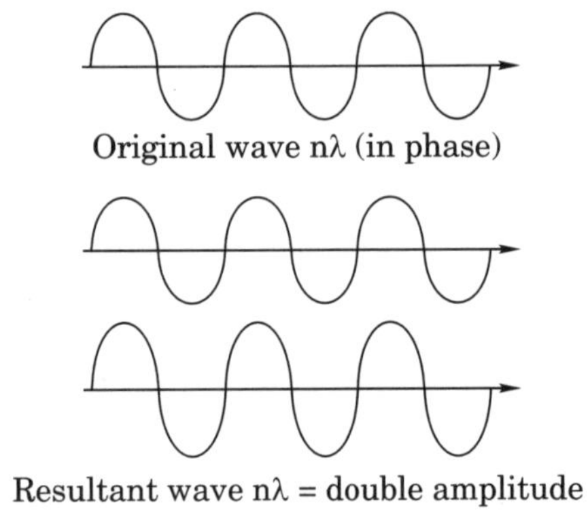

Figure 9.5: Wave diagram for in phase waves.

- *When waves are out of phase*, then destructive interference may occur with reduced amplitude of the waves. The resultant waves are completely destroyed and do not possess any amplitude (Figure 9.6).

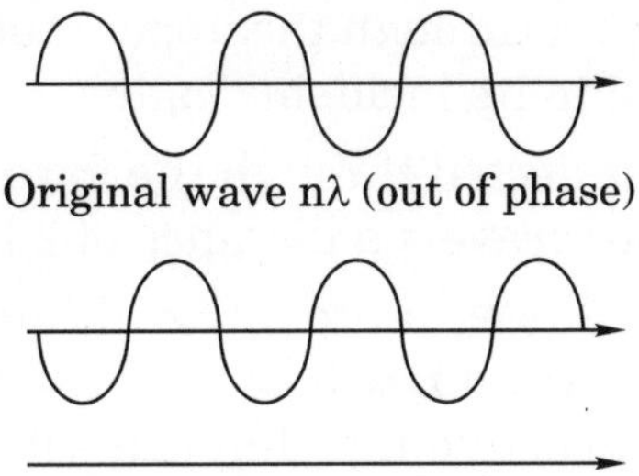

Figure 9.6: Wave diagram for X-rays.

Bragg's Law

According to Bragg's law, interaction of X-ray and atoms in crystals occurs in such a manner that it results in interference. This interaction is attributed to reflection of waves through atomic plane of a crystal structure.

This can be best understood from the Figure 9.7, according to which a beam of X-rays is imagined to enter into a crystal through one of the plane of atoms that are oriented at an angle θ to the incident X-ray beam.

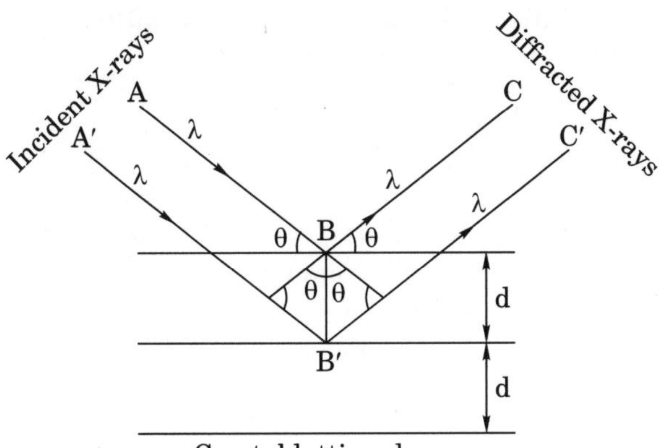

Figure 9.7: Bragg's Law reflection.

- Two such monochromatic X-ray beams A and A′ are shown, where the spacing between the atomic planes occurs over the distance, d.

- Wave A, is reflected through the upper atomic plane at an angle θ, which is equal to its incident angle.
- Wave B, is also reflected through the lower atomic plane at the same angle θ, but travels a distance of 2d farther than Ray A.
- When distance 2d is equal to integral number of wavelengths ($n\lambda$), then waves A and B will be in phase on their exit from the crystal and constructive interference will occur.
- When distance 2d is not an integral number of wavelengths, then destructive interference will occur and the waves will not be as strong as when they entered the crystal.
- Occurrence of constructive interference can be expressed as follows:

$$n\lambda = 2d \qquad \text{...(9.4)}$$

Thus, applying trigonometry, for distance 2d between the atomic planes gives,

$$n\lambda = 2d \sin \theta \qquad \text{...(9.5)}$$

Such an expression is called as Bragg's expression or Braggs law for X-ray diffraction.

Thus this diffraction occurs only if the rays are in phase when they emerge, and this will occur at only the appropriate value of n (1, 2, 3, etc.) and θ.

C. Fluorescence

Some time analyte irradiation with X-rays, causes analyte to emit radiations other than X-rays. Such a phenomenon is called as X-ray fluorescence. Fluorescence radiations wavelength is used to determine the type of element present in the analyte. On the other hand the intensity of X-rays is used for determination of amount of element present in the sample.

9.3 INSTRUMENTATION

Commonly the X-ray spectrometer are comprised of following components.

- Radiation Sources

- Collimator
- Monochromators
- Sample Handling
- Detector
- X-ray Absorbance Device
- X-ray Diffraction Device
- X-ray Fluorescence Device

9.3.1 Radiation Sources

Commonly X-rays can be produced by three methods such as X-ray tubes, Radioisotopes and secondary fluorescent sources.

A. X-Ray Tube (XRT)

X-rays are generally produced in a device called X-ray tube. The diagram of such a tube is illustrated in Figure 9.8.

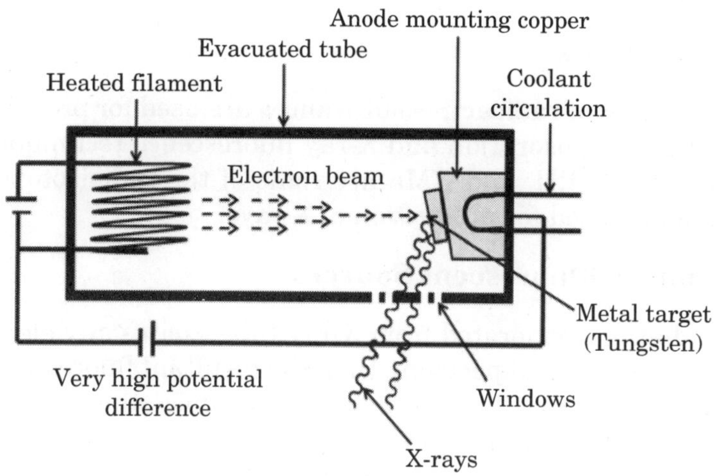

Figure 9.8: X-ray tube diagram.

Features and Functioning of XRT

- X-ray tube is composed of an evacuated chamber which possess a tungsten filament (*cathode*) at one end of the tube, and a metal target (*anode*) on the other end.

- When a current is passed through the tube, it causes tungsten filament to glow and emit electrons.
- Among the two electrodes a large voltage difference (*measured in kilovolts*) is applied.
- This causes the electrons to move at high velocity from the filament and strike on to anode.
- Due to high velocity impact of electrons onto the target, inner shell electrons of metal gets dislodge, which causes the outer shell electrons to jump to a lower energy shell to replace the dislodged electrons.
- These electronic transitions results in the generation of X-rays.
- The produced X-rays are allowed to move through a window of X-ray tube.
- And can be used to provide information on internal arrangement of atoms in crystals or the structure of internal body parts.

B. Radioisotopes

A large number of radioactive substances are used for production of X-rays in X-ray absorption and X-ray fluorescence techniques. For example ^{55}Fe, ^{101}RH, and ^{55}Mn are some of the radioisotopes that are commonly used for production of X-rays.

C. Secondary Fluorescent Sources

When radiations generated from X-ray tube excites any element to generate fluorescence spectrum, then the resultant fluorescence acts as a source in X-ray fluorescence spectroscopic studies. Although fluorescence spectra do not provides continuous spectra like primary sources.

9.3.2 Collimator

A collimator is a device that is used for narrowing (*i.e.*, collimated or parallel) of a beam of particles or waves. The diagram of such a device is illustrated in Figure 9.9.

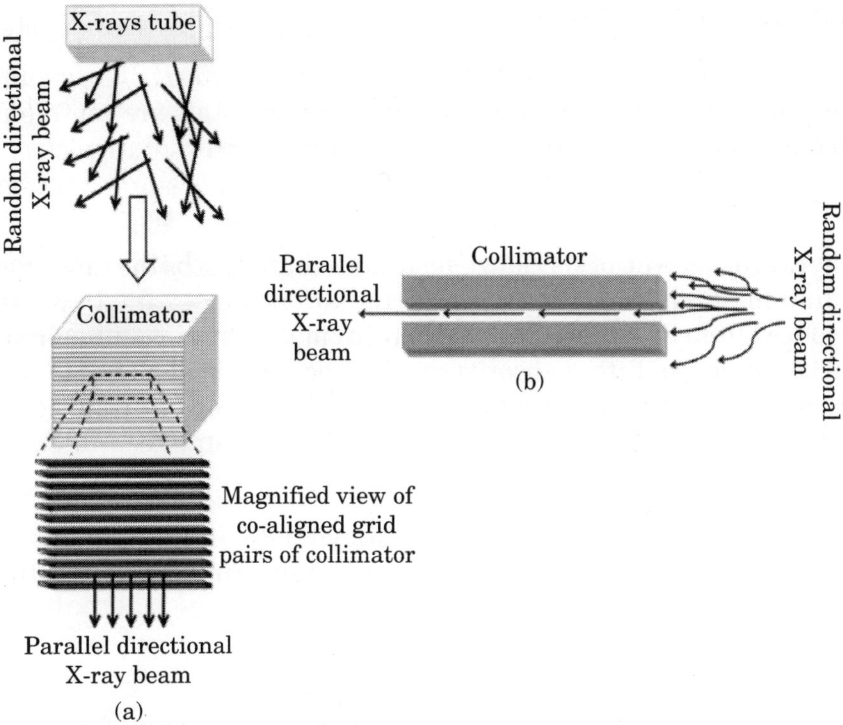

Figure 9.9: (a) Schematic diagram of collimator and (b) Mode of operation inside collimator.

Key Concepts

- Despite of the fact of resolution improvement, these devices reduces the intensity of radiations by blocking incoming radiation.
- Collimator makes, random directional X-rays (*from XRT*) to be narrow and parallel.
- Collimators are usually made up of tungsten, stainless steel and ceramics having grid pairs co-aligned so that they modulates the X-rays produced from source.
- These collimators range in pitch from 400 microns to 6 microns. Figure 9.9 (a) and (b), represents a schematic diagram and mode of operation of collimator.

9.3.3 Monochromators

A monochromator is a device that is used for removal of unwanted radiations. In X-ray spectroscopy filters and crystals are commonly used as monochromators for removal of unwanted radiations.

A. Filters

Filters are made up of specialized material that absorbs the unwanted radiations and passes the desired radiation. For example X-rays produced from molybdenum are monochromatize by zirconium filters. These zirconium filters absorbs the molybdenum and allows to pass the K_β lines. Similarly, for target elements like copper, iron, cobalt and nickel the monochromatizing filters like Ni, Mn, Fe, and Co are commonly used.

In old times X-ray filters were commercially used. Shortfalls of filters such as high background radiation and unsatisfactory monochromatization limit use of filters.

B. Crystal Monochromators

Crystal monochromators are the device, that partially polarize a nonpolarized X-ray beam on the basis of Bragg's diffraction. The most common materials for laboratory X-ray monochromators are pyrolytic graphite for broad band use and either silicon, germanium, or quartz for narrow band use. The crystal monochromators used in X-ray spectroscopy are of two types *i.e.*, flat crystal and curved crystal mono-chromators. Figure 9.10, shows the image of crystal monochromators.

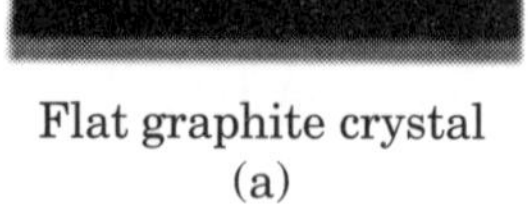

Flat graphite crystal
(a)

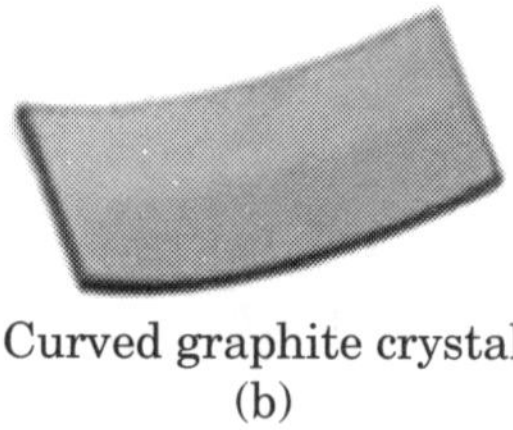

Curved graphite crystal
(b)

Figure 9.10: Images for crystal monochromators (a) flat graphite crystal and (b) curved graphite crystal.

9.3.4 Detectors

The intensities of the X-rays can be detected using two types of detectors i.e. photographic detectors and counter detectors.

A. Photographic Detectors

Photographic detectors uses a plane or cylindrical film to determine the position and intensity of X-rays. In these detectors cylindrical films are developed by exposing the detectors to X-rays.

The extent of blackening of developed film is expressed as density. Density (D) is a direct measurement of X-ray energy which causes blackening of photographic film. This relation is expressed as follows:

$$D = \frac{\log . I_0}{I} \qquad ...(9.6)$$

Where,

I_0 = Incident X-rays intensity
I = Transmitted X-rays intensity

This method is commonly used for X-ray diffraction studies.

B. Counter Detectors

(*a*) **Geiger-Muller Tube Counter**—Principle and mechanics of Geiger-Muller tube counter can be best understood from the schematic diagram given in Figure 9.11 (a) and (b).

Features

- The Geiger-Muller tube counter apparatus is composed of a glass tube having a diameter of about 19 mm.
- The tube is comprised of a half metal cylinder of about 4 inches length, made up of copper.
- Along the axis of cylinder a thin metal wire of tungsten is tied.
- The cylinder and wire are connected to an electrical voltage source.
- The tube is filled with a gas, usually argon, at a low pressure.

- A voltage is set up between the cathode (*cylinder*) and anode (*wire*) which is just a little less than needed to create an electrical discharge between two electrodes.

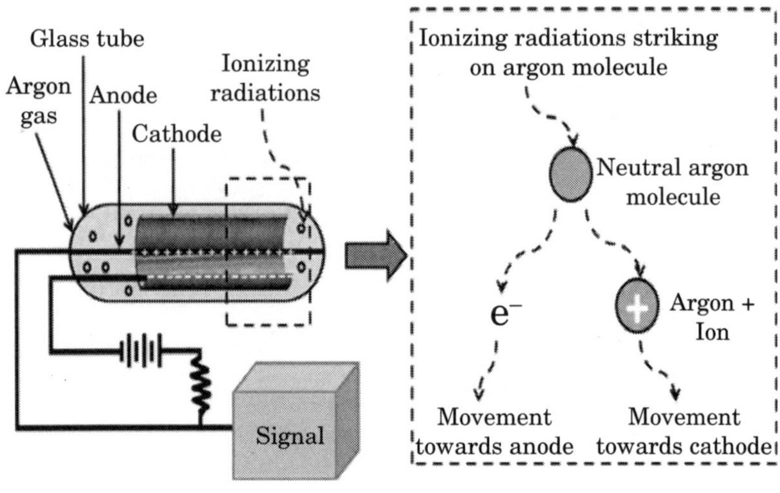

Figure 9.11: (a) Schematic diagram for Geiger Muller Tube and (b) Mechanics of ionization.

Working

- When X-rays enters the Geiger tube, a collision occurs between the gas molecules and X-rays.
- Thereby electrons are ejected out of the atoms of neural molecule of argon gas.
- This causes production of positive molecular ions and free electrons.
- These electrons, being negatively charged, moves towards anode (*wire*) and the positively charged argon ions moves towards cathode *i.e.*, half metal cylinder.
- *A potential gradient is applied to accelerate electrons. This causes electrons to pick much energy to eject more electrons out of atoms. This in turn picks up further energy and liberates even more electrons. Such a progressive process is called as electron avalanche.*

- On the other hand at the same time due to higher potential gradient, the positive ions hit the cathodic half cylinder with enough energy to eject further more electrons.
- Therefore avalanche of electrons inclines on wire which is detected as a pulse of electric current.
- The electric pulse so generated indicates passing of a charged particle through the tube.
- This pulse can be read or measured through a meter or circuit.

Merits

- Sufficient signals are obtained for a given X-ray intensity.
- Economical.
- Requires less maintenance.

Demerits

- Used for measuring low rate X-rays.
- Low efficiency below 1 A°.
- Unable to measure energy of ionizing radiations.

(b) Proportional Counter—The proportional counter apparatus is similar in construction to Geiger-Muller tube. But in this instrument instead of argon gas a heavy gas like P-10 (*mixture of argon 90% and methane 10%*) or xenon or krypton is used. The advantage of using heavy gas is that these heavy gases are easily ionized and the apparatus can be operated at low voltage.

- Proportional counter is a combination of two ionisation regions/ mechanisms in one chamber and those are:
  - *Ion drift region*—Region that exists in the outer volume of the chamber. This region assists in creation of number of ion pairs proportional to incident radiation energy.
  - *Avalanche region*—Region that exists in the immediate vicinity of the anode. This region assists in charge amplification of ion pair currents, while maintaining localised avalanches.
- A specialized circuit is connected to tube so that X-rays of particular energy can be measured. Therefore the output of proportional counter apparatus depends upon incident X-ray

intensity. The speciality of proportional counter is that it can distinguish between two α and β radiations.

Merits

- Count high rates without error.
- Good sensitivity and efficiency.

Demerits

- Expensive.
- Complex circuit.

(*c*) Scintillator Counter—*Scintillator is defined as a material, which exhibits the property of luminescence on excitement by ionizing radiations.* A scintillation detector is a combination of a scintillator with an electronic light sensor such as photomultiplier tube (PMT) or a photodiode.

- PMT absorbs the light emitted by the scintillator, and further re-emits in the form of electrons (*called photo-electrons*) via the photoelectric effect.
- The subsequent multiplication of those electrons results in an electrical pulse which on analysis provides information about incident radiation particles.
- There are different scintillation materials available such as sodium iodide, anthracene and napthalene etc. A scintillator counter possesses many advantages such as resolution of gamma rays for ^{60}Co & ^{137}Cs, linearity, density, speed, transparency and also manufacturing cost.

(*d*) Semiconductor Detector—The semiconductor types of detectors assists in promotion of X-rays through generated electrons into conduction bands. Therefore the current so generated is a direct measurement of X-rays intensity.

- These semiconductor detectors commonly makes use of Indium antimonide (IA) or lead tin telluride (LTT) or mercury cadmium telluride (MCT). In these detectors the thin layer of p-type of semiconductor is kept over the n-type surface to make a diffused p-n junction. The p-type surface is exposed to radiation which generates hole and electron pairs. These hole and electron pairs are separated by internal field existing on p-n junction. As a result voltage is generated. This assembly of semiconductor is arranged inside a Dewar type vacuum bottle cooling unit. The cooling is done through liquid nitrogen or Joule Thomson coolers of compressed nitrogen gas or liquid helium.

9.3.5 X-ray Diffraction Methods

Interaction of electromagnetic radiation with analyte (*solid substance*), whose repeated distance is almost similar to the wavelength of radiation it may lead to diffraction.

X-ray diffraction is a non-destructive method, which provides detailed information about the chemical composition and crystallographic structure of natural and manufactured materials.

For example, a visible light can be diffracted when directed onto a grating that contains several number of grooved lines, that are spaced apart with a few thousand of angstroms (Å = 10^{-10} m).

X-rays possesses a wavelengths order in angstroms (Å = 10^{-10}m), which is within the range of typical inter atomic distances in crystalline solids.

Therefore, X-rays can be diffracted from the repeating patterns of atoms that are characteristic of crystalline materials.

9.3.5.1 *Instrumentation of XRD*

A. Laue's Method

Reflection or transmission of white radiation through a fixed crystal is the basic principles of Laue's method. Laue's method determines the orientation of large single crystal. The Laue method can also be used to assess crystal perfection from the size and shape of the spots. In case, the crystal has been bent or twisted in anyway, the spots become distorted and smeared out.

Laue's method is categorized into two types *i.e.*, Back Reflection Laue and Transmission Laue.

(*a*) **Back Reflection Laue**—Principle and mechanics of Back reflection Laue's can be best visualized from the schematic diagram given in Figure 9.12.

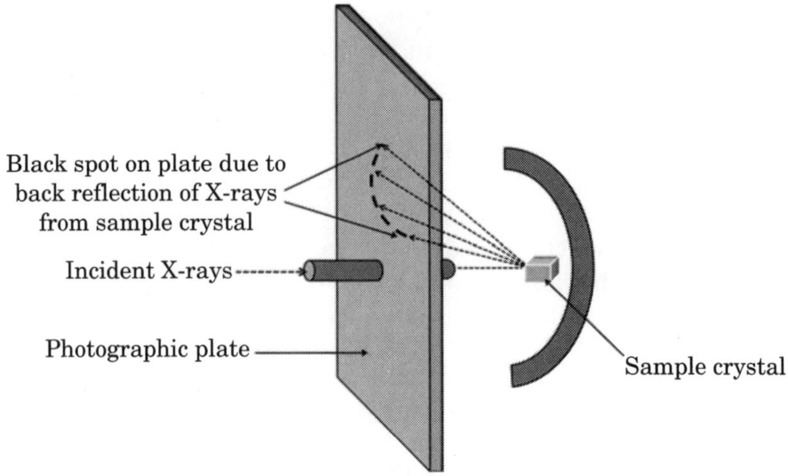

Figure 9.12: Schematic diagram for back reflection Laue.

Features

- In the back-reflection method, a film is placed between the X-ray source and the crystal to record the beams of X-rays that are diffracted in a backward direction.
- When X-ray falls on the crystal, one part of the X-ray beam is transmitted and other part is reflected back onto the surface of photographic film or plate, as black spots.
- With the pattern of position of back reflected black spots (*which is characteristic for different compounds*) one can easily determine the crystal orientation.
- Each spot is attributed to a particular plane, using special charts.
- For determination of back-reflection patterns commonly Greninger chart is used.
- The transmitted beam determines one side of the cone of Laue reflections. The film intersects the cone, with the diffraction spots lying on an hyperbola.

(*b*) Transmission Laue—Principle and Mechanics of Transmission Laue's can be best visualized from the schematic diagram given in Figure 9.13.

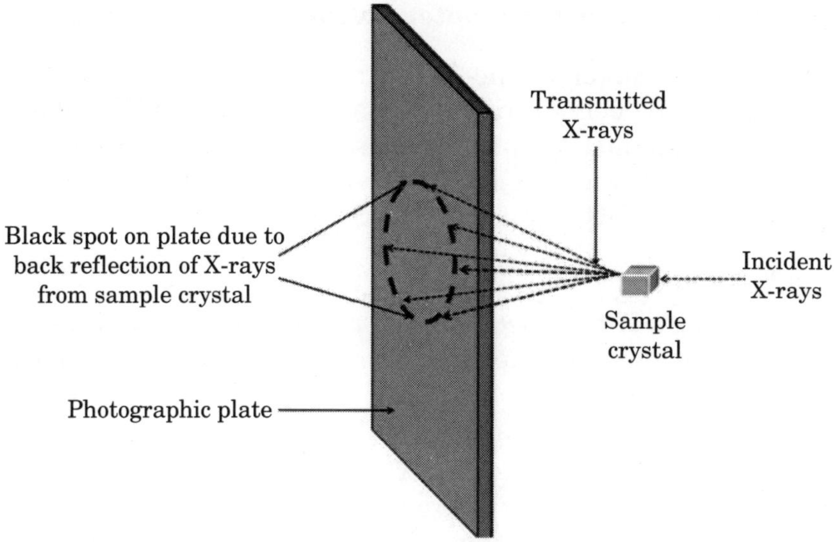

Figure 9.13: Schematic diagram for Transmission Laue.

Key Concepts

- In transmission Laue method, the photographic film is placed behind the crystal to record the X-ray beams that are transmitted through the crystal.
- In this method the transmitted beam defines one side of the cone of Laue reflections. The film intersects the cone, with the diffraction spots generally lying on an ellipse.
- The diffracted beam of X-ray creates an array of spots on curves on the film. For each set of crystal planes, the Bragg's angle is fixed. Also each set of planes picks out and diffracts a particular wavelength from white radiation in satisfaction to Bragg's law for d and θ.
- Therefore each curve is attributed to different wavelength and also spots lying on any one curve are attributed to reflections from planes belonging to one zone.
- Laue reflections from planes of the same zone, all lies on the surface of an imaginary cone whose axis is the zone axis.
- For study of transmission patterns commonly a Leonhard chart is used.

B. Bragg's X-Ray Spectrometer Method

Bragg's spectrometer is one of the most important methods for studying crystals using X-rays by measuring the intensity of X-ray beam. In this method, the photographs developed by Laue's method, are further investigated for analysis of crystals structures of compounds like of KCl, NaCl, and ZnS.

Key Concepts

- In Bragg's X-ray spectrometer method, the crystal analysis involves bombardment of crystal with X-rays of a fixed wavelength.
- On X-rays bombardment, at certain incident angles, some time a highly intense reflected X-rays are produced. This high intensity effect is attributed to constructive interference of wavelength of the scattered X-rays.
- To have such constructive interference, the differences in the travel path should be equal to integer multiples of the wavelength.
- During constructive interference, angle of diffracted beam of X-rays leaving the crystal analyte and incident beam are equal.
- Such, relationship between the wavelength of the incident X-rays, angle of incidence and spacing between the crystal lattice planes of atoms is known as Bragg's Law, which is expressed as

$$n\lambda = 2d \sin \theta$$

Where, n = Order of reflection

λ = Wavelength of the incident X-rays

d = Inter planar spacing of the crystal

θ = Angle of incidence.

Material of construction of Bragg's X-ray Spectrometer (Figure 9.14).

The Bragg's spectrometer is composed of

- An X-ray tube, which throws a narrow beam of X-rays via collimator on the crystal analyte.

- The crystal analyte is mounted on a rotating table, which is connected to a scale and Vernier for measurement of angle of incidence (θ).
- An arm (ionisation chamber) containing gas, that rotates in the same axis of crystal table.

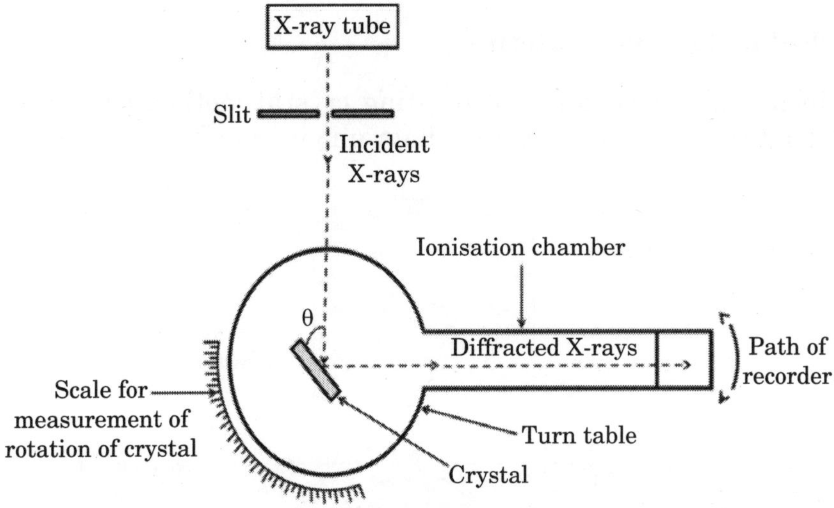

Figure 9.14: Bragg's ionization spectrometer.

Working of Bragg's X-ray Spectrometer

- The X-rays after reflection from the crystal enters the ionization chamber and thereby ionise the gas present inside.
- As a result of ionisation a current is produced which is measured by electrometer. This ionisation current is a direct measurement of intensity of reflected beam of X-rays from the crystal.
- This ionisation current is measured for different angles of incidence.
- The obtained values are plotted in the form of graph.
- For example, a crystal sodium chloride, the maximum reflection or peaks for 100 plane occurred at θ = 5.9°, 11.85° and 18.15°. The sines of these angles are 0.103, 0.205 and 0.312, which are in the ratio 1 : 2 : 3. These peaks represent the first, second and third order reflections.

- The ratio confirms the correctness of Bragg's equation.
- For example, a crystal of sodium chloride, exhibits maximum reflection for 100 planes, when $\theta = 5.9°$, $11.85°$ and $18.15°$. The sines of these angles are exhibited in a ratio of 1 : 2 : 3 at 0.103, 0.205 and 0.312. The obtained peaks represent first, second and third order reflections or diffraction.

C. Rotating Crystal Method

Principle and mechanics of rotating crystal method can be best understood from the schematic diagram given in Figure 9.15.

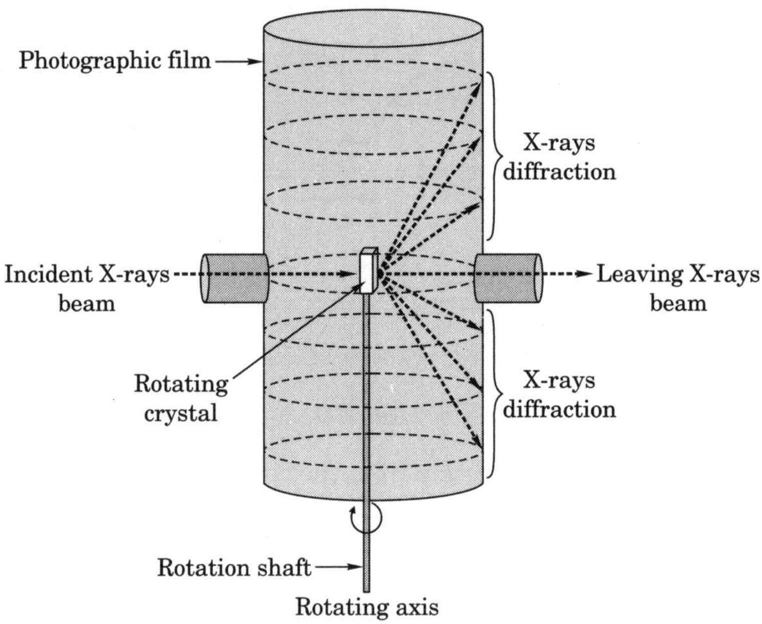

Figure 9.15: Schematic diagram of rotating crystal method.

Rotating crystal method was developed by Schiebold in 1919. This method underlines the basic principle of rotation of crystals, so as to determine the size of unit cell. The general features of this method are as follows:

- Generated X-rays from the source are allowed to pass through collimator via monochromator (filter).

- After the collimator X-rays are made to fall on a crystal, which rests over a shaft rotating with a uniform angular rate with the help of a motor.
- The shaft's slow and uniform rotation makes the crystal to rotate on its axis. Rotation of crystal at a particular point causes a set of lattice planes to make correct Bragg's angle for incident beam and formed diffracted beam.
- As a result several planes are generated in successive reflecting positions. Such that the θ value supports Bragg's law.
- When X-rays strikes on to the plane of slowly and uniformly rotating crystal, it causes each planes to generate a spot on photographic plate and produce a diffraction pattern. This diffraction pattern over photo plate is perpendicular to the X-ray beam.

The photography or rotation of crystal can be done by two ways

1. *Complete Rotation Method*—This method comprises complete revolutions of crystals resting over shaft. Due to which each plane of crystal diffracts four times during each rotation. Also the generated four beams distributes into a rectangle pattern about central point of photograph. For this A cylindrical film is placed around it and the crystal is rotated about the chosen axis.
2. *Oscillation Method*—This method comprises oscillation of crystals by 15° or 20° over a shaft. For this case the photo plate is moved back and forth with a similar time period of rotation of crystal. So, position of spot on the plate determines the orientation of crystal.

The most important use of the rotating crystal method is to determine unknown crystal structures.

D. Powder Crystal Method

X-ray powder crystal method is another useful method for determination of crystalline substances in a sample. The powder crystal method is used to determine value of the lattice parameters with accuracy. Lattice parameters are the magnitudes of the unit vectors a, b and c which defines the unit cell for the crystal.

As reorientation of the crystal, measurement of angle θ, and determination of d-spacing for all atomic planes are time consuming processes. So, to make the process fast a powder method is the best key.

Sequential Steps in Powder Crystal Analysis method

- The crystal analyte is grounded up to a fine powder.
- The powder is kept on a sample stage or strucked on a hair using gum and suspended vertically into the axis of cylindrical camera.
- Powder possesses thousands of grains with random orientations. Among random orientations most of the different atomic planes lies parallel to the surface in some of the grains.
- To determine all angles undergoing diffraction, the angle θ of incident X-ray beams is scanned from 0 to 90°. Each diffracting angles would be associated with a different atomic plane/spacing in terms of miler index (hkl).

The mechanics of powder X-ray diffractor can be best understood by Figure 9.16.

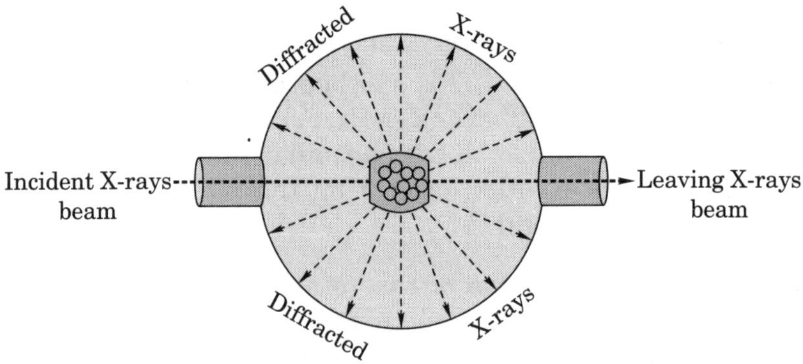

Figure 9.16: Schematic diagram for powder crystal method.

Features of Powder Crystal Method—A powder crystal method operated diffracting instrument is comprised of

- *An X-ray Tube*—that can be rotated to produce angles from 0 to 90°.

- *A Sample Stage*—above which powdered analyte is placed for irradiation by the X-rays.
- *An Electronic Detector*—to detect the diffracted X-rays of angles from 0 to 90°. The detector is also allowed to rotate to measure different diffracted angles. Detector records rate of diffracted X-rays coming out of other side of the sample (*other side of X-ray tube*) and passes such information to computer interface.
- *A Goniometer*—a device that is used to rotate both the X-ray tube and the detector is called as goniometer. Goniometer acts as an observer for all angle θ, and transfers such information to computer interface, while the detector records the rate of X-rays coming out the other side of the sample (in units of counts/sec) and sends this information to the computer.
- *A Computer Interface*—a scan of the sample is progressed and resultant X-ray intensity is plotted against the angle θ (or 2θ *depending upon older instrument construction ways*) to generate a chart. The angle 2θ for each diffraction peak can then be converted to d-spacing, using the Bragg's equation.
- *Structural Determination*—association of diffraction peaks and different atomic planes in form of Miller Index for plane (hkl) is done to deduce crystal structure.

9.3.5.2 *Applications of X-Ray Diffraction*

XRD possesses a wide variety of applications such as in the field of medical science, pharmaceutical science, engineering, physics, chemistry, biology, geology, forensic science, environmental science, and material science etc.

The most common uses of x-ray diffraction are in the structural analysis of chemicals, catalysts, ceramics, pharmaceutical corrosion products, clays, cement, composite materials, forensic materials, environmental studies, fly ash, industrial wastes, metals and alloys, mineral studies, mining, nano materials, oil drilling, patent process, plastic materials, soils, textiles, semiconductors, raw and fibres.

9.3.6 X-ray Fluorescence Methods

X-ray fluorescence (XRF) is a quantitative and qualitative analytical tool for elemental analysis of materials. XRF is used for determination of concentration of elements in solids and solutions, film thickness and composition, also identification of specific and trace elements in complex sample matrices.

XRF analysis is used in many industries such as pharmaceuticals, cosmetics, agriculture, plastics, rubbers, textiles, fuels, chemicals, and environmental analysis. Advantages of XRF method are its accuracy, non-destructive property, fast speed, and minimum sample preparation.

Theory—Mechanism of X-ray fluorescence can be well understood from Figure 9.17.

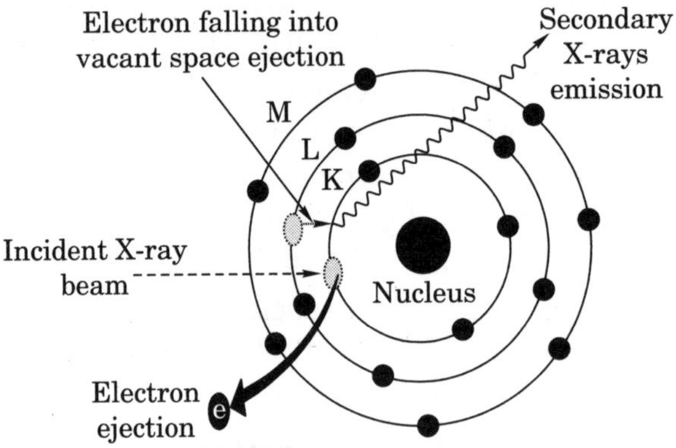

Figure 9.17: Schematic diagram for representing principle of X-ray fluorescence.

- Exposure of high intensity X-rays onto elements in sample analyte causes emission of fluorescent X-rays from the sample at energy levels unique to those elements.
- When the radiation energy possesses a sufficient energy to dislodge an electron from inner shell of atom, it causes dislodging of an electron from the inner shell.
- Thereby space is created in inner shell of atom.
- This makes the atom to become unstable and in turn an electron from outer shell of atom replaces the missing inner electron.
- During such process there occurs a release of energy due to decreased binding energy of inner electron orbital compared with an outer one.

- These emitted radiations are of lesser energy than incident X-rays and called as fluorescent radiations.
- As energy of emitted photon is characteristic of a transition between specific electron orbitals of any particular element, the fluorescent X-rays are used to detect concentration of elements in a given sample.

9.3.6.1 *Instrumentation of XRF*

Instrumentation of X-ray fluorescence can be well understood from block diagram given in Figure 9.18.

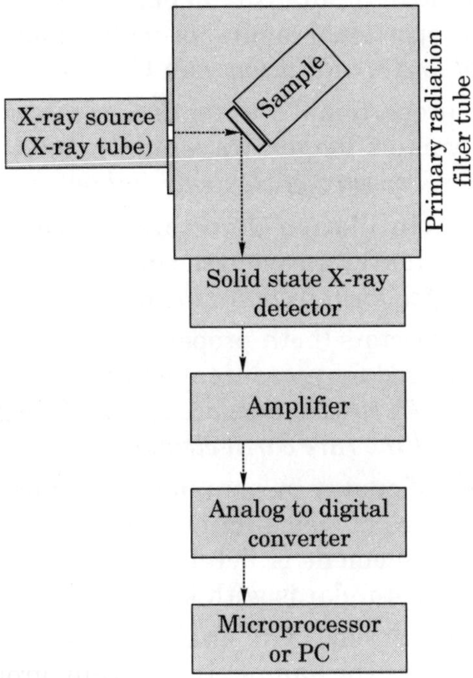

Figure 9.18: Block diagram representing X-ray fluorescence instrumentation.

Commonly, all XRF spectrometers comprises a source and a detection system. In XRF instruments in common practice an X-ray source (*XRT* or *radioactive source* or *synchrotron*) irradiates the sample and detector measures the fluorescence radiation emitted from the sample.

Features of XRF Experiment

- The primary X-ray beam illuminates sample to cause excitation.
- The sample excitation in turn emits X-rays along a spectrum of wavelengths which is characteristic and depends upon the types of atoms present in the sample.
- For a sample that possess many elements (*example in case most minerals and rocks*), a wavelength dispersive spectrometer is used.
- This allows separation of complex emitted X-ray spectrum into characteristic wavelengths for each element present.
- For measurement of intensity of emitted beam detectors such as gas flow proportional counter or scintillation or tandem (*both proportional and scintillation*) can be used.
    - Gas flow proportional counter detectors are commonly used for analysis of longer wavelength in X-rays spectrum (*Example: K spectra of elements lighter than zinc*).
    - Whereas scintillation detectors are commonly used for analysis of shorter wavelengths in X-ray spectrum (*Example: K spectra of element from Nb to I and L spectra of Th and U*).
    - Tandem detectors (both proportional and scintillation) are commonly used for measurement of X-rays of intermediate wavelength *(K spectra produced from Zn to Zr and L spectra from Ba and the rare earth elements*).
- The intensity of energy measured by detectors is proportional to the abundance of elements in sample. The proportionality value for each element is determined by comparing it with mineral or rock standards with known composition.
- In most of the cases like XRF analysis of rocks, ores, sediments and minerals etc. the sample is generally grounded to a fine powder and can be directly subjected for trace element analyses.
- But due to very wide range in abundances of different elements and sizes of grains in a powdered sample it becomes very difficult to compare proportionality to the standards.
- Hence, commonly the powdered sample is mixed with a chemical flux and melted. This melting generates a homogenous glass for analysis and calculation of abundance of elements.

The XRF instruments are of two types such as:

1. *Energy Dispersive XRF* and
2. *Wavelength Dispersive XRF*.

Both EDXRF and WDXRF X-ray make use of X-ray optics to enhance instrumental performance of XRF. In conventional XRF instruments, typical focal spot size diameter at the sample surface ranges from several 100 μm to several mm.

There are two ways to focus optics onto sample

- *Polycapillary focusing optics*—collects X-rays from source and directs as a small focused beam on sample surface with diameter less than tens of micrometers. The resultant effect is that increased intensity is delivered to sample in a small focal spot, which makes enhanced spatial resolution for small feature analysis and performance enhanced elemental analysis (for micro XRF applications).
- *Doubly curved crystal optics*—directs an intense micron-sized X-ray beam on to the sample surface for enhanced elemental analysis.

9.3.6.2 *Applications of XRF*

X-ray fluorescence possess a wide range of applications such as

- In Environmental studies.
- In ceramic and glass manufacturing.
- In petroleum industry for determination of sulfur content.
- In environmental studies for analyses of particulate matter.
- In cement industry.
- In sedimentary research.
- In igneous research.
- In metamorphic petrology research.
- In surveys of soils.
- In mining for measuring grade of ore.
- In metallurgy for quality control.
- In geological studies.

9.3.7 X-ray Absorbance Method

X-ray absorption spectroscopy (XAS) is a widely-used technique for determining the local geometric and electronic structure of matter in solids, solutions, or gaseous phase.

9.3.7.1 *Instrumentation*

X-ray absorption spectrometers are usually compact, inexpensive and give performance comparable to crystal monochromator instruments. But these undergo poor resolution of closely spaced lines.

The principle and mechanics of non-dispersive X-ray absorption spectrometer can be well understood from Figure 9.19.

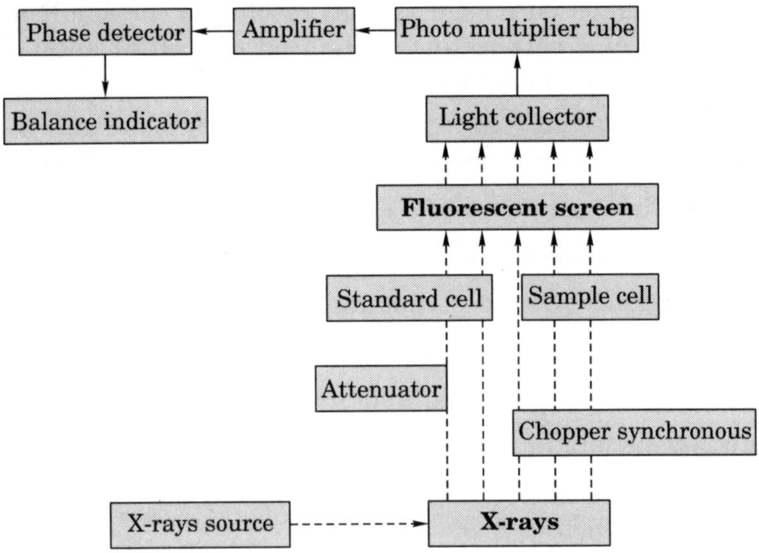

Figure 9.19: Block diagram representing Non-dispersive X-ray Absorbance Spectrometer.

An XAS is comprised of

- A tungsten X-ray tube.
- A synchronous chopper (*motor driven*) which is kept between cell and attenuator which is used to interrupt 1/2 of X-ray beam.

- A variable thickness attenuator between chopper and reference cell.
- Sample and reference cell in duplicate up to 65 cm length.
- Both halves of X-ray beams are made to fall on a phosphor-coated photomultiplier tube that is protected from visible light using thin filter.

Working of XAS

- Keep standard and samples in reference and sample cells respectively.
- Adjust attenuator until it balances the absorbance in two X-ray beams.
- Requirement for change in thickness of aluminium for different samples is a function of composition difference.
- For sample analysis a calibration curve is plotted, thereby the unknown sample concentration can be determined.

9.4 PROBLEMS BASED LEARNING

After reading the X-ray spectroscopy chapter, the readers are suggested/advised to test their learning by solving some below mentioned problems.

A. SUBJECTIVE PROBLEMS

1. Define X-ray diffraction? Explain Bragg's law.
2. Discuss instrumentation and application of Powder X-ray diffraction.
3. Explain instrumentation and application of X-ray Fluorescence.
4. Describe the non-dispersive X-ray absorption method.
5. Explain various methods for production of X-rays.
6. Discuss Bohr's model for X-rays.

B. OBJECTIVE PROBLEMS

1. State True or False

(*i*) Continuous spectra depends upon target material & voltage across X-ray tube.

(*ii*) Characteristic Spectra are produced at high voltage as a result of specific electronic transitions.

(*iii*) X-ray absorption is followed by beers law.

(*iv*) Geiger-Muller tube is filled with argon gas.

(*v*) Scintillator exhibits luminescence on excitement of ionizing radiations.

(*vi*) In back reflection Laue, film is placed before crystal.

(*vii*) In transmission Laue, film is placed behind crystal.

(*viii*) In Bragg's spectrometer rays reflected from crystal enters ionization chamber.

Ans.: 1. (*i*) True (*ii*) True (*iii*) True (*iv*) True
(*v*) True (*vi*) True (*vii*) True (*viii*) True

2. Fill up the blanks

(*i*) .... is a device used for rendering the beam parallel.

(*ii*) .... are made up of specialized material to absorb unwanted radiations.

(*iii*) .... tube undergoes electron avalanche.

(*iv*) .... counter is filled with xenon gas.

Ans.: 2. (*i*) Collimator (*ii*) Filters (*iii*) Geiger-Muller
(*iv*) Proportion.

3. Match the following

(*i*) XRD (*a*) Crystal structure

(*ii*) XRF (*b*) Qualitative and quantitative (Non-destructive)

(*iii*) XRA (*c*) Elemental analysis

Ans.: 3. (*i*) – (*a*), (*ii*) – (*b*), (*iii*) – (*c*).

9.5 SUGGESTED READINGS

[1] Van Der Heide P., *X-ray Photoelectron Spectroscopy: An introduction to Principles and Practices*, John Wiley and Sons, New Jersey, 2012.

[2] Als-Nielsen J., McMorrow D., *Elements of Modern X-ray Physics*, 1st Edition, John Wiley and Sons, Chichester, UK, 2011.

[3] Beckhoff B., Kanngieber B., Langhoff N., Wedell R., Wolff H., *Handbook of Practical X-ray Fluorescence Analysis*, Springer-Verlag Berlin Heidelberg, 2006.

[4] Kasai N., Kakudo M., *X-ray Diffraction by Macromolecules,* Springer, New York 2005.

[5] Singh A.K., *Advanced X-ray Techniques in Research and Industry*, IOS Press, Netherlands, 2005.

[6] Zachariasen W.H., *Theory of X-ray Diffraction in Crystals*, 1st Edition, Dover Publications, Mineola, NY, 2004.

[7] Günter G., Vo-Dinh T., *Handbook of Spectroscopy*, Volume 1, Wiley-VCH, Verlag Weinheim, 2003.

[8] Wilson K., Walker J., *Principles and Techniques of Practical Biochemistry*, 5th Edition, Cambridge University Press, 2000.

[9] Jenkins R., *X-ray Fluorescence Spectrometry,* 2nd Edition, John Wily and Sons, New York, USA, 1999.

[10] Suryanarayana C., Grant Norton M., *X-ray Diffraction: A Practical Approach*, Plenum Press, New York 1998.

[11] Woolfson M.M., *An Introduction to X-ray Crystallography*, 2nd edition, Cambridge University Press, UK, 1997.

[12] Lippard S.J., Berg J.M., *Principles of Bioinorganic Chemistry*, University Science Books, CA, 1994.

[13] Guinier A., *X-ray Diffraction: In Crystals, Imperfect Crystals, and Amorphous Bodies*, General Publishing Company, USA, 1994.

[14] Williams D., *Methods of Experimental Physics: Spectroscopy*, Part A, Vol. 13, Academic Press, New York, 1976.

Subject Index

C

D

E

F

G

H

I

J

K

L

M

N

O

P

Q

R

S

T